CYANOTYPE

The Blueprint in Contemporary Practice

Cyanotype: The Blueprint in Contemporary Practice is a two-part book on the much admired blue print process. **Part One** is a comprehensive how-to on the cyanotype process for both beginner and advanced practitioners, with lots of photographs and clear, step-by-step directions and formulas. **Part Two** highlights contemporary artists who are using cyanotype, making work that ranges from the photographic to the abstract, from the traditional to the conceptual, with tips on their personal cyanotype methods alongside their work. These artists illustrate cyanotype's widespread use in contemporary photography today, probably the most of any alternative process.

Book features include:

- A brief discussion of the practice of the process with some key historical points
- How to set up the cyanotype "dimroom"
- The most extensive discussion of suitable papers to date, with data from 100+ papers
- Step-by-step digital negative methods for monochrome and duotone negatives
- Chapters on classic, new, and other cyanotype formulas
- Toning to create colors from yellow to brown to violet
- Printing cyanotype over palladium, for those who want to temper cyanotype's blue nature
- Printing cyanotype on alternate surfaces such as fabric, glass, and wood
- More creative practice ideas for cyanotype such as handcoloring and gold leafing
- Troubleshooting cyanotype, photographically illustrated
- Finishing, framing, and storing cyanotype
- Contemporary artists' advice, techniques, and works

Cyanotype is backed with research from 120 books, journals, and magazine articles from 1843 to the present day. It is richly illustrated with 400 photographs from close to 80 artists from 14 countries. It is a guide for the practitioner, from novice to expert, providing inspiration and proof of cyanotype's original and increasing place in historical and contemporary photography.

Christina Z. Anderson's work focuses on the family snapshot and contemporary *vanitas* printed in a variety of 19th century photographic processes. Anderson's work has shown internationally in 110+ shows and 40+ publications. She has authored books which have sold in 40 countries—*The Experimental Photography Workbook*, *Gum Printing and Other Amazing Contact Printing Processes*, *Gum Printing, A Step-by-Step Manual Highlighting Artists and Their Creative Practice*, and *Salted Paper Printing, A Step-by-Step Manual Highlighting Contemporary Artists*, along with *Handcrafted: The Art and Practice of the Handmade Print* (in Chinese) coauthored with Samuel Wang, S. Carl King, and Zhong Jianming. *Cyanotype* is her sixth book. Anderson is Series Editor for Focal Press/Routledge's Contemporary Practices in Alternative Process Photography series and Professor of Photography at Montana State University. To see her work, visit **christinaZanderson.com**.

Contemporary Practices in Alternative Process Photography Series

Series Editor: Christina Z. Anderson

The Contemporary Practices in Alternative Process Photography series focuses the lens on a variety of alternative, historical processes from the medium's 180-year history. Each book outlines a step-by-step approach to a particular medium, and features contemporary artists who use that process regularly in their practice. The richly illustrated books in this series serve as guidebooks for those new to alternative processes, refresher courses for professionals already familiar with each medium, and a source of inspiration for all.

CYANOTYPE

The Blueprint in Contemporary Practice

CHRISTINA Z. ANDERSON

Routledge
Taylor & Francis Group
NEW YORK AND LONDON

First published 2019
by Routledge
52 Vanderbilt Ave., New York, NY 10017

and by Routledge
2 Park Square, Milton Park, Abingdon, Oxon OX14 4RN

Routledge is an imprint of the Taylor & Francis Group, an informa business

Library of Congress Cataloging-in-Publication Data

Names: Anderson, Christina Z., author.
Title: *Cyanotype : the blueprint in contemporary practice* / Christina Z. Anderson.
Description: London; New York : Routledge, 2019. | Series: Contemporary practices in alternative process photography series | Includes bibliographical references and index.
Identifiers: LCCN 2018049120| ISBN 9781138338845 (hbk : alk. paper) | ISBN 9781138338838 (pbk : alk. paper) | ISBN 9780429441417 (ebk)
Subjects: LCSH: Blueprinting. | Photography, Artistic.
Classification: LCC TR415 .A53 2019 | DDC 770--dc23
LC record available at https://lccn.loc.gov/2018049120

Typeset in Adobe Caslon Pro and Century Gothic
by Christina Z. Anderson

Cover image: *Spring Breeze* © Sam Wang 2018

Publisher's Note:
This book has been prepared from camera-ready copy provided by the author.

Also by Christina Z. Anderson:

The Experimental Photography Workbook (2012)
Gum Printing and Other Amazing Contact Printing Processes (2013)
Handcrafted: The Art and Practice of the Handmade Print (2014, 2015)
1st edition Chinese only, 2nd with an English insert, coauthors Jianming, King, Wang
Gum Printing: A Step-by-Step Manual Highlighting Artists and Their Creative Practice (2017)
Salted Paper Printing: A Step-by-Step Manual Highlighting Contemporary Artists (2018)

Exercise caution in the handling of all photographic chemicals. Use of any such chemicals constitutes some risk, and some are poisonous. The publisher and author accept no responsibility for injury or loss arising from the procedures or materials described in this book whether used properly or improperly. The workplace should be well ventilated. Chemicals should be mixed only in the manner described. Avoid contact between the chemicals and eyes, skin, clothing, and furniture. Do not eat or drink while using chemicals. Keep them away from pets and children. Wear protective eye wear and gloves if necessary. In short, be mindful of all safety procedures for yourself and others.

Dedicated to
Sam Wang, John Jackson,
and Sandy King,
my partners
in cyanotype research

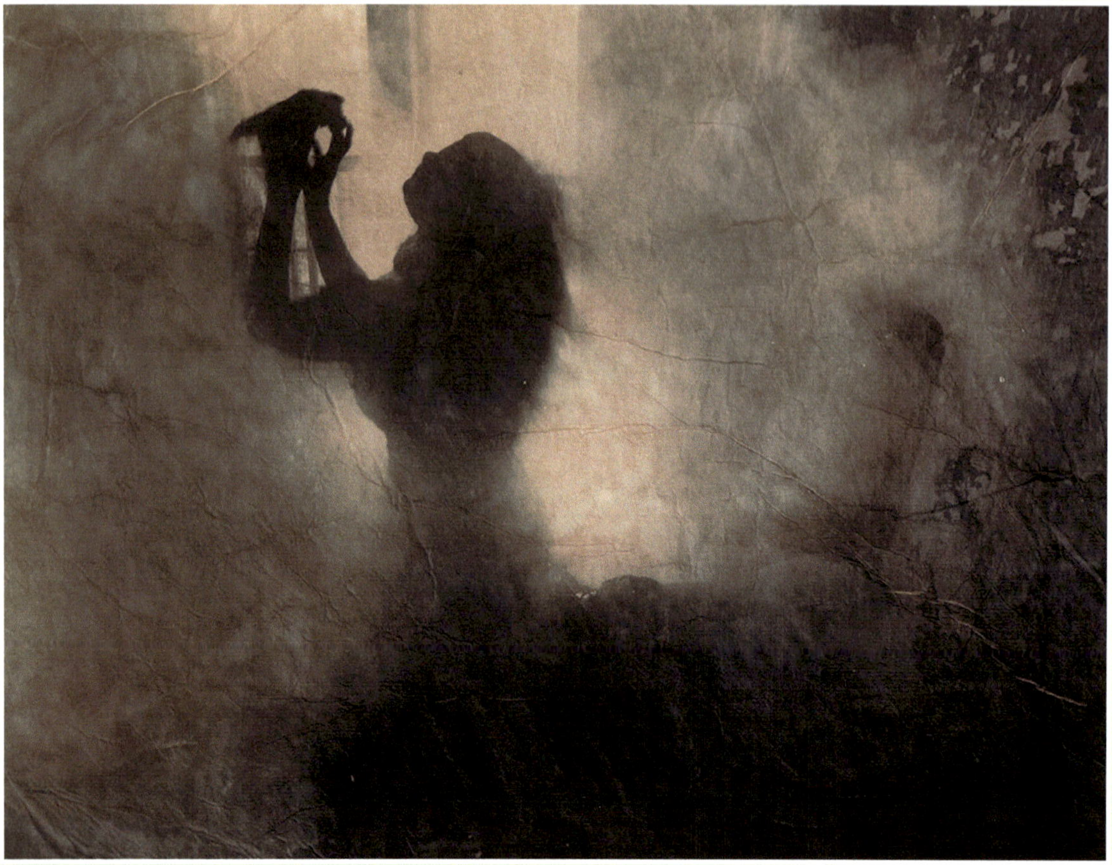

Figure P.1. *Veneration*, from the *Untamed* series, traditional cyanotype stained with Lipton tea on Japanese Kitakata paper, 16″ x 20″ © Jaime Aelavanthara 2014. "Poet Mary Oliver stated, 'I think when we lose the connection with the natural world, we tend to forget that we're animals, that we need the Earth. And that can be devastating.' *Untamed* articulates humankind's capacity to decay as a marker of our identity. Set in the natural places where one confronts life and death, growth and decay, these self portraits reflect on our transient, temporal lives that are each connected through the shared human condition. I reimagine the feral woman archetype, a strong woman who collects the bones, branches, and flora and treads with the animals, both dead and living. The otherworldly narrative is emphasized through the use of the cyanotype process, which shifts focus from potentially colorful landscapes and figures to patterns, textures, and the relationships of forms within the images. Tea-staining the prints dulls the blue and adds warmth, mimicking the earth tones encountered in the wilderness. Printing on Japanese Kitakata paper, which is prone to ripping, tearing, and wrinkling during processing, reflects the deterioration of nature and adds a feeling of fragility. Displayed unframed, the tea-stained prints are reminiscent of animal hides, and the prints are fastened delicately to the walls like tanned skins. *Untamed* ultimately reflects upon the forms, the impermanence, and the interconnectedness of natural life." Jaime Aelavanthara is an Assistant Professor at the University of Tampa in Photography & Foundations. She received her BFA from the University of Mississippi in Imaging Arts and her MFA in Photography from Louisiana Tech University. To see more of her work, visit JAelavanthara.com.

Table of Contents

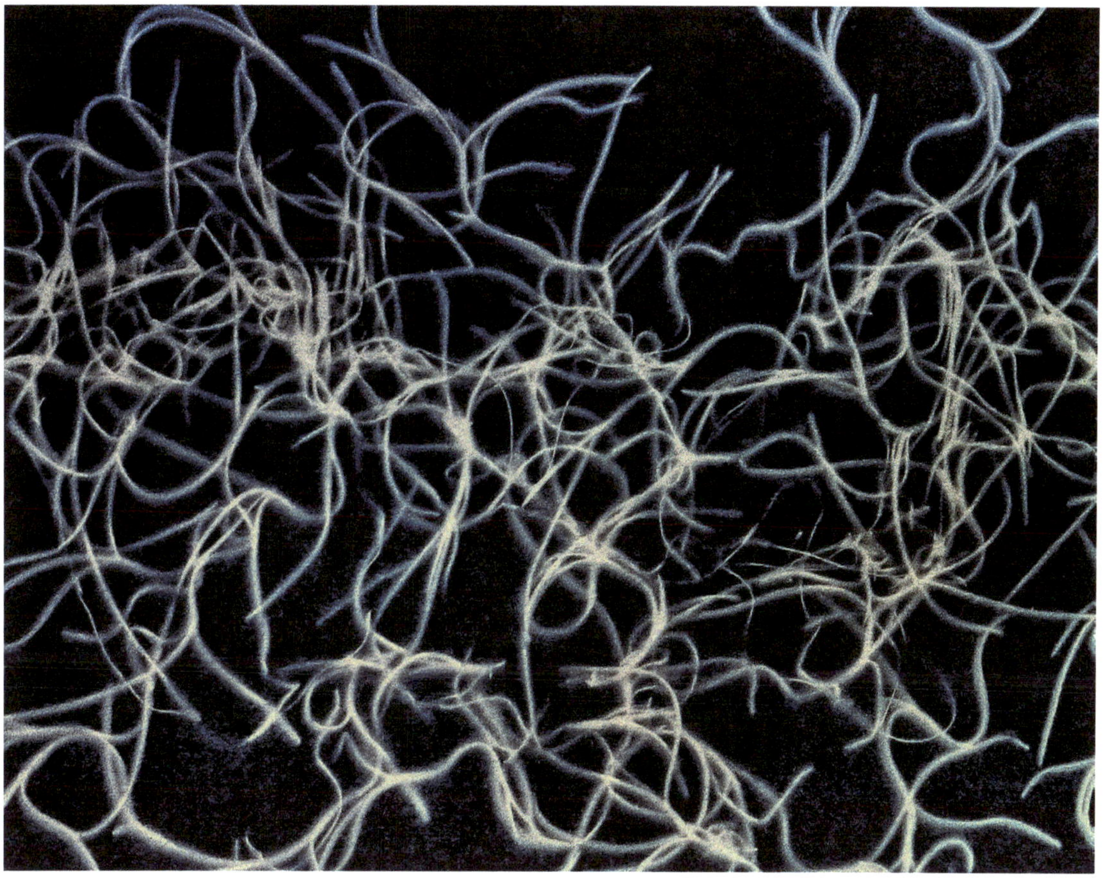

Figure P.2. *Spanish Moss*, sun-exposed cyanotype photogram on Revere Platinum, 16″ x 20″ © Diana Hooper Bloomfield 2018

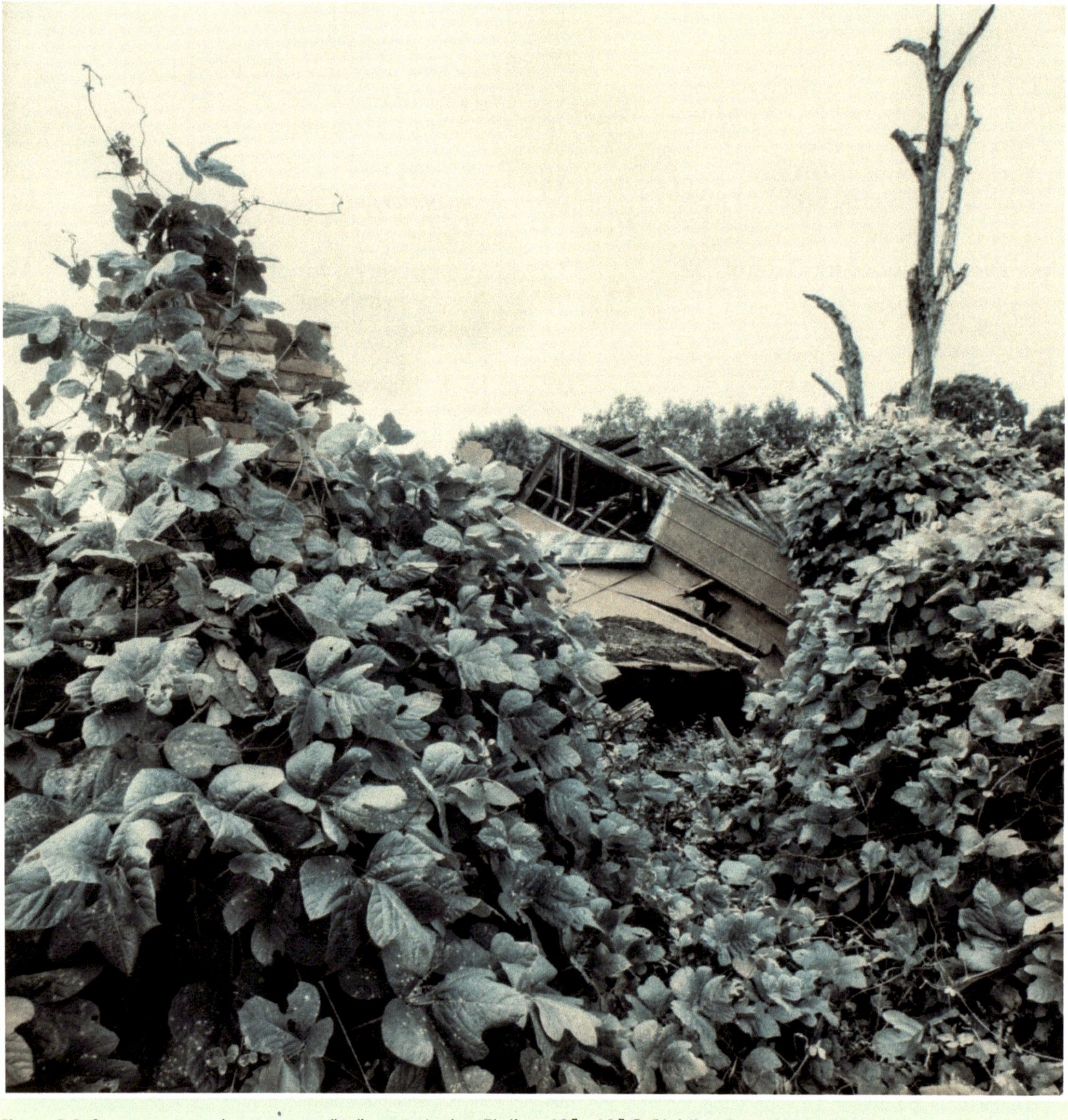

Figure P.3. *Seneca*, cyanotype over palladium on Arches Platine, 12″ x 12″ © Christina Z. Anderson 2015

Preface

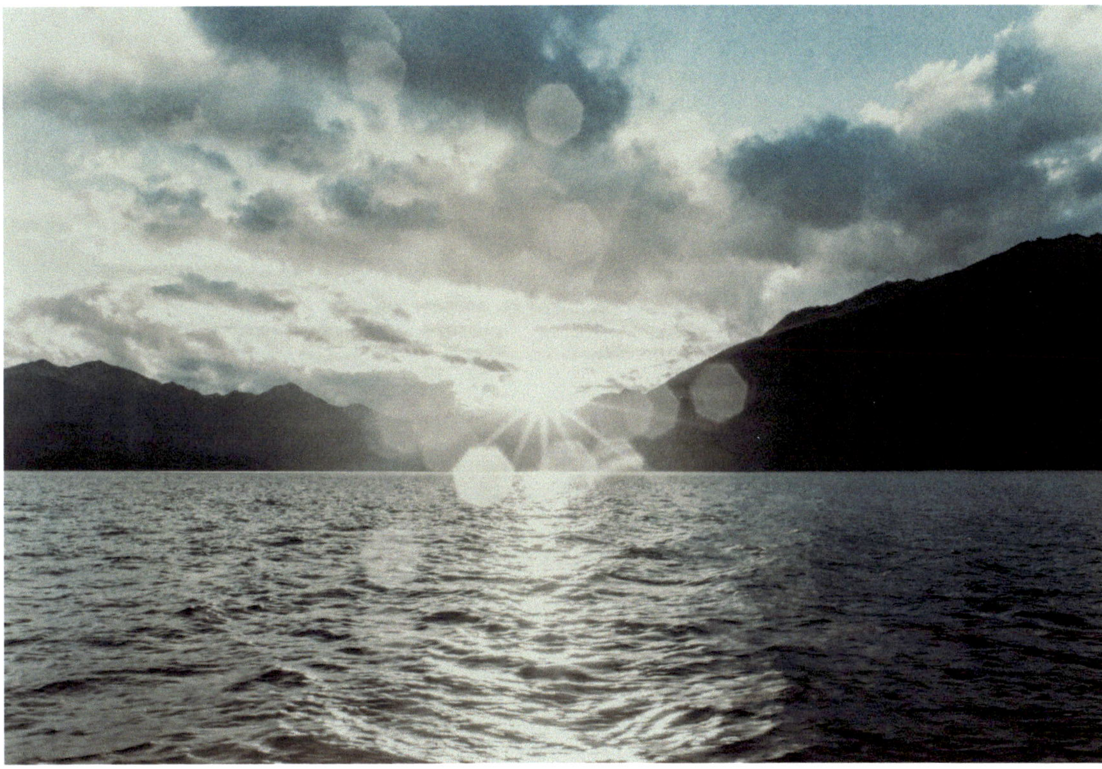

Figure P.4. *Te Anau*, cyanotype over palladium on Arches Platine, 17¾″ x 11¾″ © Christina Z. Anderson 2015

Over the last two decades I have been deeply involved in alternative process photography, a natural progression from a background in painting. In the beginning I was the proverbial kid in a candy shop, dabbling with abandon in this process or that process. In time, my processes of choice narrowed down to a select few: gum/casein bichromate, salted paper, mordançage, chemigram/chromo, and cyanotype, the latter most often in combination with platinum or gum. I have written extensively about these processes in five books to date. It was time to delve more deeply into the cyanotype process and give it as much attention as I have given to gum bichromate and salted paper.

Cyanotype is deceptively simple. Two key ingredients can be mixed in many different proportions and still produce an acceptable print. Perhaps because of its simplicity it has earned a less precious reputation than other more complex alternative processes. Historically it is often described as a process for the amateur photographer. The purpose of this book is to showcase cyanotype's complexity. It is one thing to make an *acceptable* cyanotype print. It is another to make an *exquisite* cyanotype print.

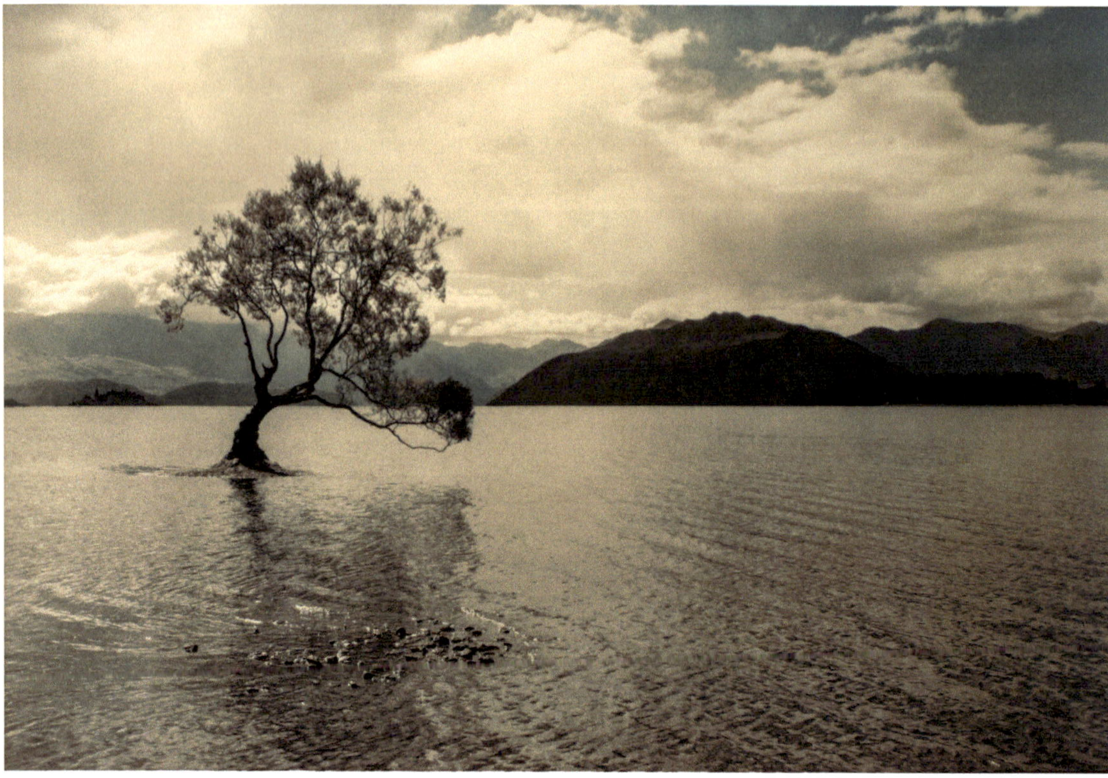

Figure P.5. *Wanaka*, cyanotype over palladium on Arches Platine, 18″ x 12″© Christina Z. Anderson 2015

The biggest challenge to writing about cyanotype is having confidence that I may be able to add to the sum knowledge about cyanotype that already exists in print. It wasn't until I recorded data from testing 136 papers and printing 1088 step wedges that I realized I actually had something to bring to the table.

I am not a scientist, nor am I a chemist. I am an artist first and foremost, and a professor who has taught for almost twenty years in a four-year college photography program. This has allowed me a unique perspective: how to share these processes in a user-friendly, unintimidating way which ultimately results in advanced practice made evident in technically and conceptually strong student final projects. A side benefit to teaching at the college level is having a class of sixteen prove you right—or wrong! It's equivalent to having a lab with sixteen personal assistants.

When I began research for this book, I corresponded with Sam Wang (South Carolina) and John Jackson (Tennessee) about our findings. Sam was my graduate school professor at Clemson University, alt process printer for decades longer than most of us, and now friend and fellow kudzu-hunter. John Jackson is an avid photographer and alt process printer who does a lot of densitometry work to quantify process information for the purpose of fitting a process to analog film negatives.

Soon Sandy King (South Carolina) joined the fray. Sandy King also taught at Clemson, and has expertise in multiple processes, carbon transfer being his first love. For over a year the four of us independently tested variables that were in the cyanotype "canon." Each one of us brought certain strengths (and equipment) to the testing, at the very least our vastly different climates from the cold, dry Northwest to the warm, humid Southeast. If all

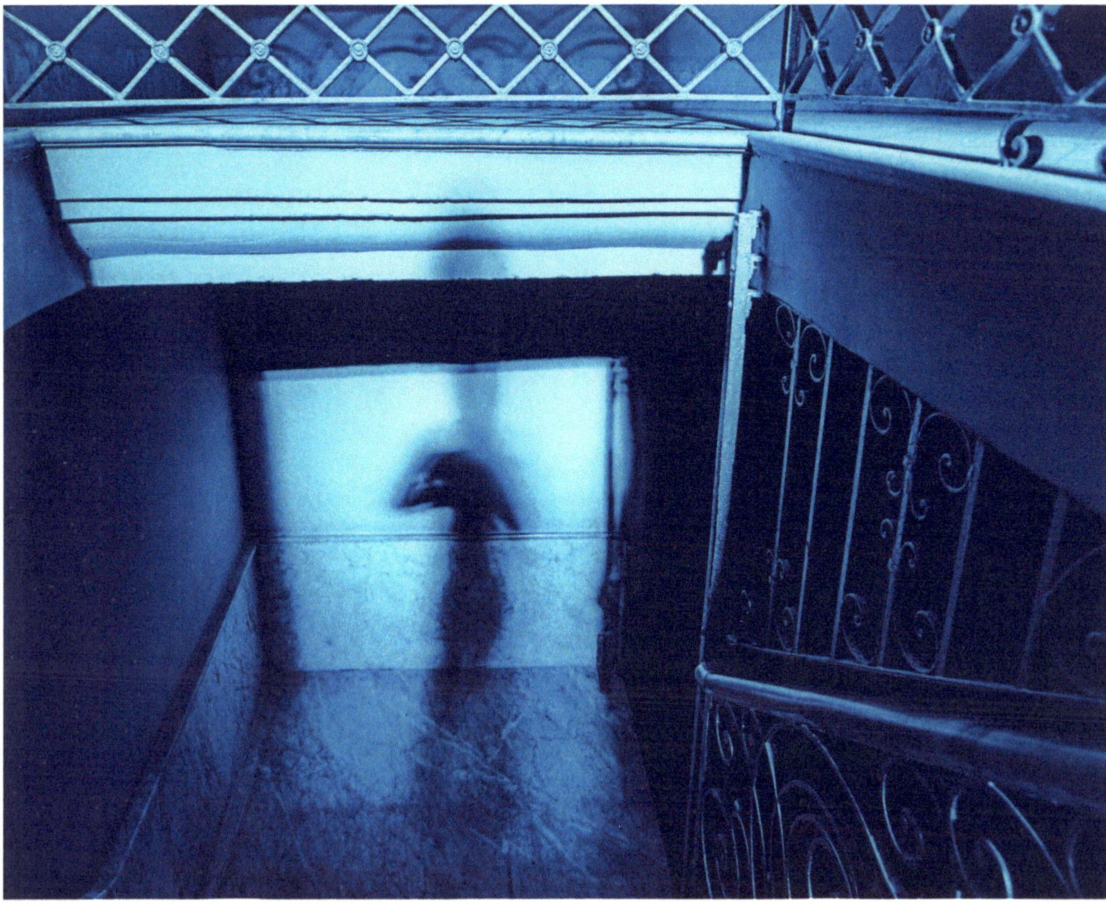

Figure P.6. *Self, Havana, Cuba*, new cyanotype on Arches Platine, 10″ x 7.75″ © Christina Z. Anderson 2018

four of us agreed on certain results, it was easy to write with confidence. If there were discrepancies in our findings, then the adage your results may vary (YRMV) applied. This part of writing was sometimes frustrating, but always rewarding. I thank these men for their insights and traveling this journey with me.

Two chemists who have been my "rocks" when I needed consultation are Dr. Mike Ware and Dr. Alberto Novo. Both have researched photographic chemistry extensively as well as being practicing artists in their own right. When I am stumped, I ask them for help which they have given freely. We in the alt pro movement are indebted to their research.

I thank all the artists who took the time to send work and answer questions for the mere gift of seeing their work in print. Without them this book means nothing.

I thank the editor Sheni Kruger and the assistant editor Sarah Pickles. These two women are nothing short of amazing, professionals and experts in their field. I thank Carolyn Jenkins, also, personal editor and friend, who has now patiently read and reread five of my books as my very first line of defense before Routledge even sees my draft. Sophie Cox, the final proofreader, was beyond awesome to spot every misplaced jot and tittle.

Finally I thank my husband Tom Anderson who has been through this book writing business now six times with me, along with my children and grandchildren, who have learned to be quiet when I sit in my Stressless™ chair, writing for hours and days on end.

Christina Z. Anderson
christinaZanderson.com

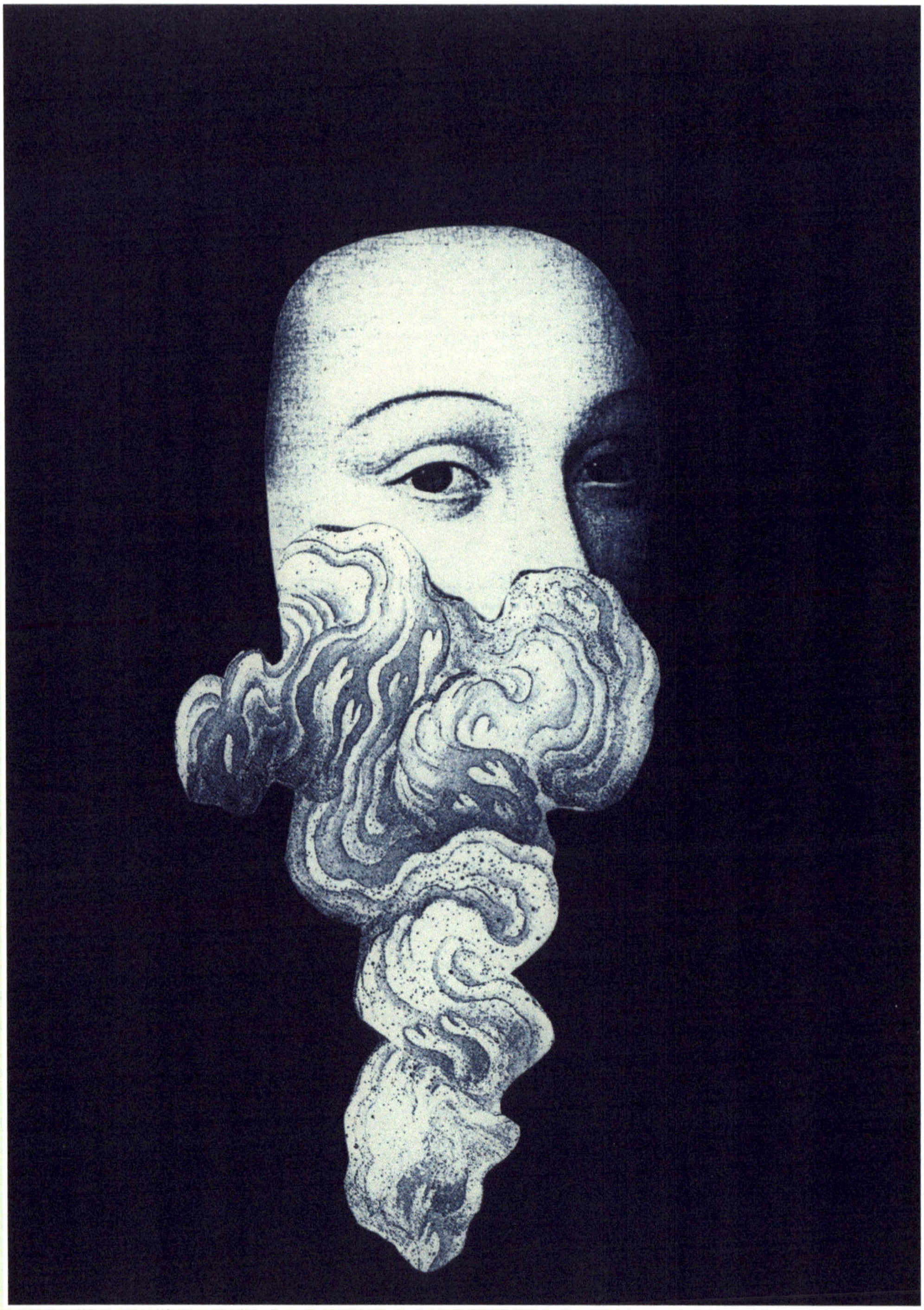

Figure 1.1. *Little Cyan Face*, cyanotype on paper, 9″ x 12″ © Brenton Hamilton 2017. "Working with cameraless techniques I find and recreate pictures from culture, refashioning the cut and collaged image works into something new. I am especially drawn to cyanotype and all of its hues. My storytelling and yarn spinning with images takes on and implies a radical new possibility. Paint, gold leaf, silver and the iconic deep blue-black are the palette of my cyanotype works." For over 20 years Brenton Hamilton has created a sustained body of work, largely concentrated within historic processes. Hamilton is a well-known teacher on the campus of Maine Media Workshops, Rockport, Maine. To see more of his work visit brentonhamilton.com.

Chapter 1

Cyanotype–History and Practice

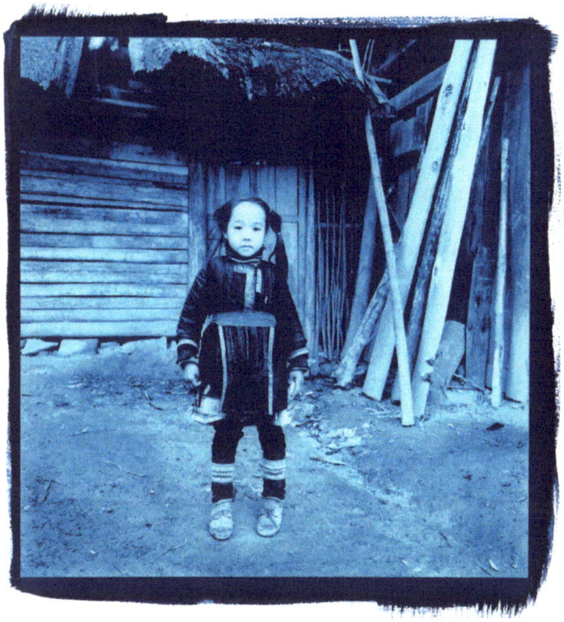

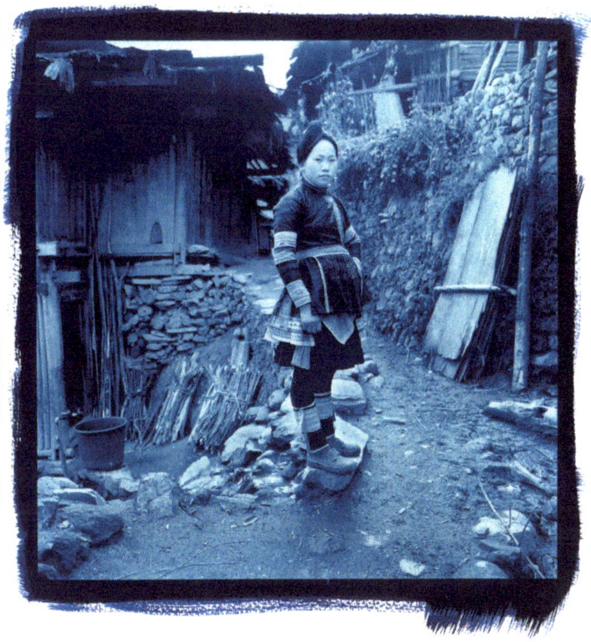

Figures 1.2 and 1.3. Left, *Miao Child*; right, *Miao Maiden*, new cyanotype © Hua Cheng 2018. "The photographic print as an end product and the appreciation for the fine craft associated with its production was foreign to those of us in China until at least the 1980s. The image as existed in newspapers or magazines was all we saw, besides some gelatin-silver prints made in the darkroom for mostly political and news coverage. So the print for an aesthetic end was a more recent idea. When Sam Wang and Sandy King showed us handcrafted prints, they opened the floodgate and we became very interested to learn to make them. The traditional cyanotype was the first process we tried, partly for its seeming simplicity and low cost. Our first attempts were far from good because of the difficulty of finding right ingredients locally. Instead of the green ferric ammonium citrate we found only the brown kind of unknown purity and the prints wouldn't clear in the wash. When we came across new cyanotype, we tried it and it worked! It overcame the problems Dr. Ware mentioned as reasons for developing the new formula and we were able to find the right chemicals domestically. These images came from scanned medium format negatives. The paper I found worked best was an inexpensive Canson sold by the local art stores. For exposure, I use an LED UV exposure unit made by a factory in Hangzhou, whose owner was very interested in helping us tune the UV to the exact wavelength that we needed. For developing, I use a 5% citric acid bath, and begin the wash cycle when the image looks right." After a very successful professional photographic career, Hua Cheng joined the photography faculty at the Nanjing Arts Institute, now called The Nanjing University of the Arts, about two hours from Shanghai, China. Now retired from teaching, he enjoys traveling, continues to photograph, and has been making mostly carbon transfer prints.

In 1900, almost sixty years after the invention of cyanotype, one author penned, "The indifference of the photographic world to the 'blue print' is one of the seven wonders of that little world."[1] He would therefore be surprised to see that today, cyanotype is probably the most widely practiced alternative process there is.

The process has its beginnings in the year 1842, merely three years after the invention of photography itself. Sir John Herschel discovered that ferric ammonium citrate in combination with potassium ferricyanide would become a photosensitive emulsion that yielded a beautiful blue color. Since that year, what is now known as the blueprint or cyanotype process has been taught all over the world from kindergarten classrooms to university labs. Yet cyanotype has hardly achieved the same cachet as other processes.

Cyanotype's benefits

It is easy to understand why cyanotype is so widely practiced. It is the perfect first-foray into alt for the following reasons:

- It doesn't require a darkroom
- It's inexpensive
- It's easy to mix up the two required chemicals
- The chemicals are for the most part non-toxic
- The cyanotype formulas vary widely but they all still work, making cyanotype somewhat foolproof; you could almost throw a little of this and a little of that into some water and it will still work
- Although exposure times are slower than most alt processes, it is a printing out process (POP), which can easily be inspected during exposure to monitor completion
- It only requires a water wash for development
- It can be used on many different surfaces aside from paper and is especially suited to fabric
- It is not so humidity-picky as other alternative processes such as platinum, meaning it works well in all climates[2]
- It is very archival.

Cyanotype's detriments

Cyanotype has a few drawbacks. It is slow to expose. It has a reputation for being a short exposure scale *aka* high contrast process,[3] which may have been problematic back in analog days when film negatives were the norm. The process was always blamed, as one author noted:

> Blue Prints…are extremely beautiful if properly made, and possess great artistic merit, the bad repute into which they at one time fell being due to improper coating of the paper and careless manipulation on the part of the operator. Apparently the easiest to handle of all photographic papers, every tyro in the art of printing felt himself to be above instruction, and licenses to the most careless and slovenly manipulation. Feeling himself a past master in the art of blue printing, the paper was invariably blamed for every defect, and any suggestion that the negative or the operator was at fault met with the universal comment, "Do you suppose I don't know how to make a blue print?" Well, usually he didn't! This paper renders faithfully every gradation of the negative, and is remarkably rich in soft and delicate definition, the color of the print being a rich Prussian blue with beautiful china-white highlights.[4]

Nowadays with digital negatives, contrast is changed with the press of a button. Cyanotype does not have to be high contrast with blown out highlights and blocked up shadows. When practiced properly with a well-executed digital negative and suitable paper choices, it has delicate highlights and detailed shadows rivaling platinum.

Singing the blues

A complaint often repeated is about cyanotype's blue color. When blue is the most preferred color of all colors, with 40% of people naming it as their number one color of choice,[5] why would a blue print not be preferred? When did black and white, or brown for that matter, become photographic norms, and not blue, especially since cyanotype was at the very beginning of photographic practice? Adelaide Skeel, an astute practitioner of the cyanotype process in the late 1800s, and well aware of the prejudice against cyanotype, theorized that brown, black, and white, but not blue, became photographic norms because nature was translated to those colors in printed books and journals.[6]

Cyanotype's blue has been accused of being insistently blue,[7] disagreeably blue, not for the production of real photographs,[8] cloying,[9] ugly in tone,[10] forbidding,[11] a color not suitable for the average subject,[12] a color "severed from any neutral relation to observed reality,"[13] and the list goes on. This is odd, given the popularity of Delft and Wedgwood china, with exactly the same color palette of a cyanotype print.

Cyanotype was pigeonholed into strictly defined "appropriate" subject matter. P. H. Emerson, an influential writer in the 1800s, famously stated, "No one but a vandal would print a landscape in red, or in cyanotype."[14] Acceptable subjects were ice and glacier scenes,[15] seascapes, cloudscapes, river, lake, and water scenery,[16] interiors, winter scenes,[17] moonlight "fakes,"[18] and cool, serene imagery.[19] One author elaborated:

> The negative should determine when to use "Blue Print Paper." Moreover, the type of picture should also be considered. It does not seem to me that groups of people or pictures of animals look well in this brand, and I have always restricted its use, not entirely of course, but nearly, to pictures where water and foliage entered in greater or less proportions. A

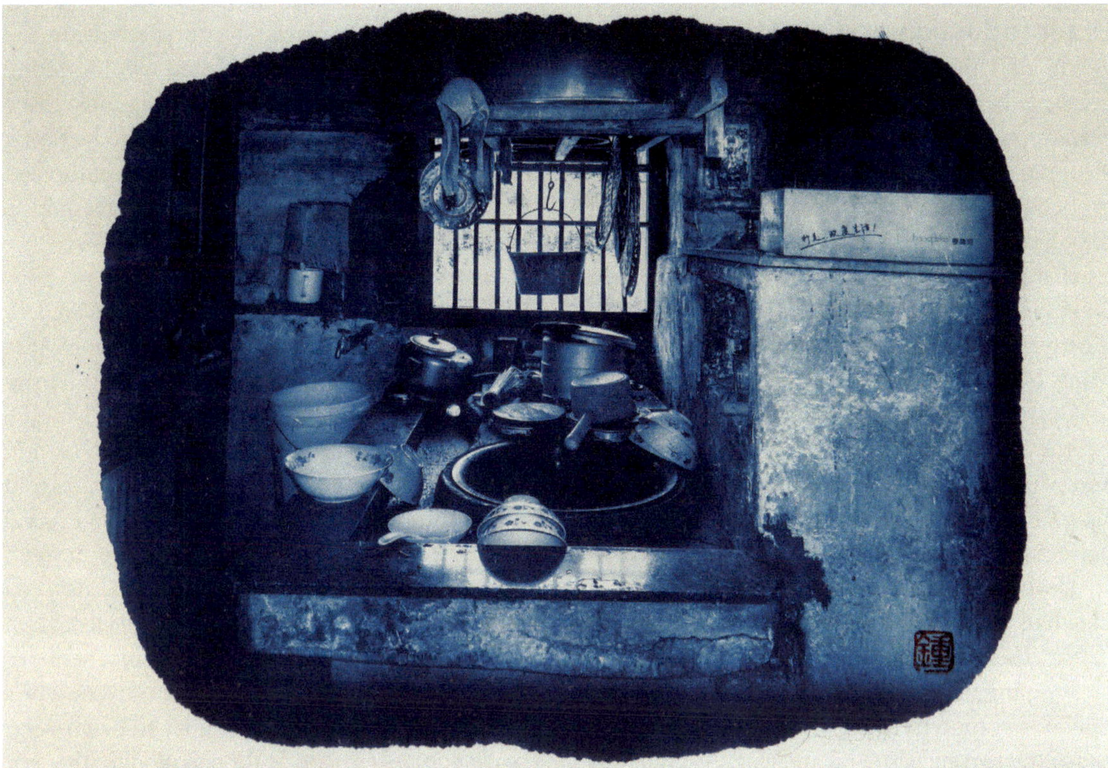

Figure 1.4. *Old House #2*, 31 x 24 cm, © Jianming Zhong 2017. "These were done to document the fast disappearing old buildings in China, most of which have now been abandoned, but still retain evidence of lives lived. With the rapid rate of change experienced everywhere in China, I felt the great need to make images before these places disappear. The prints were all made with traditional cyanotype on rice paper, the paper of choice in Chinese painting. The paper is unsized and allows the liquid to flow, a sense of spontaneity I felt important to retain." Jianming Zhong began his photography career at age 20. Twenty-five years later, after a successful professional photography practice, he joined the faculty at the Nanjing Arts Institute, now the Nanjing University of the Arts, and founded its photography program in 2001. He was introduced to alternative processes a few years later by visiting professor Sam Wang, but did not begin seriously making alternative process photographs until 2015. A prolific writer, Zhong has authored 5 books. One was *Handcrafted: Art and Practice of the Handmade Print*, coauthored with Christina Z. Anderson, Sandy King, and Sam Wang. His work has been exhibited in Australia, South Korea, the US, and China. He has curated more than 10 exhibitions, and organized numerous conferences on photography in higher education in addition to workshops on photographic printmaking practices.

> water scene with a somewhat distant background is lovely in this print, while for a pure black and white tone the same picture might not look so well. Rocks, fences, cedar trees, and buildings all show up nicely in the blue print, because in nature there is enough of this tint about those things to have a correspondence to the view. The horizon seems blue, cloud effects are bluish and white, and so on; while human beings in groups would not look well or even natural in blue tints. If the presence of people in a scene is only incidental and does not play any great importance in it, then it is legitimate to use the blue print.[20]

Though this prejudice of pigeonholing cyanotype to particular subject matter still exists today, this book will show cyanotype can "do it all."

The psychology of blue

Our emotional responses to color in general are culturally and individually conditioned.[21] Blue is the most popular color, which carries both positive (serenity, peace, spirituality) and negative connotations as do all colors. It is linked to depression in the term "feeling blue." It is linked to woeful music, the Blues. It is linked to pornography since at least the 1960s in the term "blue movie" although that term is not as well known today. It is linked to death at least since Shakespeare's time when "to burn blue" was an omen of death, evil spirits, or the devil himself.[22] Today it's probably most linked with tranquility tinged with a bit of melancholy.

Of reproduction and amateurs

With all this prejudice, and other printing processes being invented left and right in the 1800s, cyanotype was only occasionally practiced 1840–1880, and didn't become popular until the late 1880s to the 1920s.[23] There are, however, five well-known photographers who worked in cyanotype: Edward Curtis, Frederick Holland Day, Arthur Wesley Dow, Henry Bosse,[24] and Frederick Coulson. It was mostly considered a process for quick proofing of prints but not for finished works.

Cyanotype was also a popular postcard medium, but the cyanotype postcard was replaced with black and white postcards when Kodak came out with the 3A Folding Pocket Camera with negatives that perfectly fit the postcard format.[25]

Despite such a lengthy presence, Beaumont Newhall didn't even mention cyanotype in his classic *The History of Photography*.[26]

Proofing negatives and making postcards—not only was cyanotype cheap and easy, it was a means to a better end and also a bit amateurish. Cyanotype was described as "cheap, reliable, and within the reach of all."[27] It was said, "The paper may be prepared by the amateur himself, or he may use with success some standard brand found on the market ready prepared."[28] From the 1870s through World War I it was most popular with amateurs,[29] akin to photo scrapbooking today.

Certainly, the most common use of the cyanotype process was for blueprints in architecture. In my research it was more often referred to as the blueprint process or the ferroprussiate process instead of cyanotype. By at least 1880 it was said, "The chief use of the process is for copying plans and drawings such as are used by engineers and architects…"[30] This iteration of the cyanotype process would not earn cachet in photographic history. Herschel himself used cyanotype to copy his notes, like a handmade Xerox machine. This seed of reproduction has been there since cyanotype's infancy. Emerson said in 1889, "…blue prints are only for plans, not for pictures."[31]

Nancy Kathryn Burns' hypothesis in *Cyanotypes: Photography's Blue Period*, is that the use of cyanotype by amateurs and in industry in the form of blueprints placed cyanotype outside the margins of what was considered photographic, and placed the process in an ambiguous space between reproduction and photo reproduction.[32] That contemporary photography has embraced cyanotype and its "otherness" stands to reason.

Where to next?

The lack of cyanotype's cachet can be related to a number of factors outlined here. Cyanotype's drawbacks of assumed short scale and long exposure, its overtly blue color, the historical lineage of its industrial use, its pigeonholing into certain subject matter, its adoption by amateurs, etc., can all be contributory factors, but given the current level of practice and press, its first museum show devoted to cyanotype in 2016, Facebook groups devoted to the process, and its use by certain well-known artists—John Dugdale, Robert Rauschenberg, Christian Marclay, Marco Breuer—cyanotype's acceptance has been and continues to be growing. The cyanotype is a good fit for postmodern thought where its vivid blue calls attention to objectness,[33] to fiction not fact.

A few key events in cyanotype's history

Following is a timeline that touches upon a few interesting events and people who figured largely in the process from 1842 to present day. For a more comprehensive history see Larry Schaaf's and Mike Ware's books cited in the **Bibliography**.

- Cyanotype's history starts with Prussian blue. In 1706 Prussian blue was discovered by a Swiss color maker living in Berlin, Johann Jacob Diesbach.[34] Blue was costly before this time, so the discovery of an inexpensive blue was revolutionary. Diesbach shared his discovery with Berlin chemist Johann Frisch who started selling it in 1708. Until 1724 the method of making Prussian blue was kept a secret, but in 1724 John Woodward disclosed it to the public.[35]
- In the 1820s–1830s Prussian blue pigment made its way through the trade routes. The Dutch imported Prussian blue to Japan and China.[36] Hokusai's *Thirty-Six Views of Mt. Fuji* were in Prussian blue on white,[37] and Hokusai's

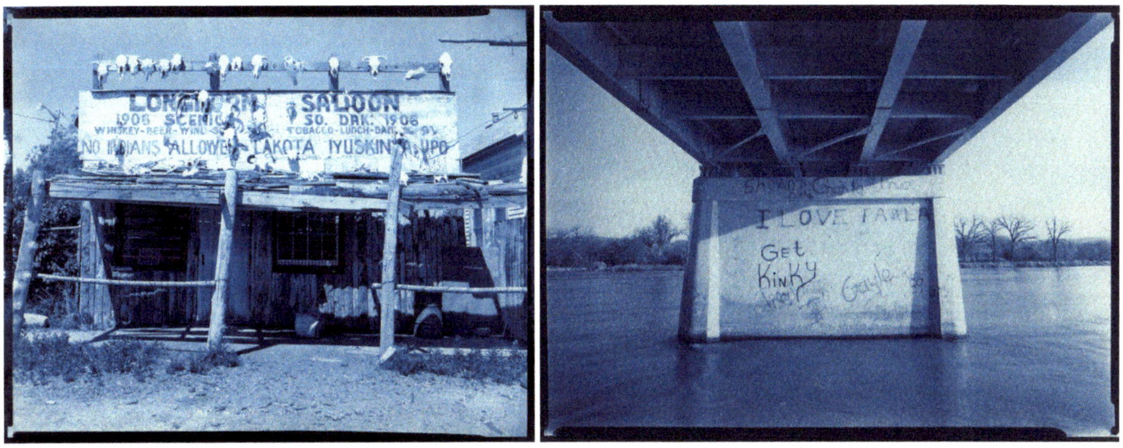

Figures 1.5 and 1.6. Left, *Scenic South Dakota* (n.d), 10″x 8″; right, *I Love Paula* (n.d.), 7″ x 5″, traditional cyanotype on Arches Cover © Michael Flecky 2018. Michael Flecky teaches fine art photography at Creighton University. His photographs have appeared in over 30 one-person exhibitions in the US. His work has taken him to the UK and Ireland, Scandinavia, the Dominican Republic, and North Africa. He has received artist residency grants from Ucross Foundation; Dorland Mountain Arts Colony; Hambidge Center; and the New York, North Carolina and Nebraska Arts Councils. Flecky's subjects include natural landscapes, figure studies, studio and found objects. In addition to cyanotype, he practices traditional photographic processes, platinum, palladium, and Van Dyke brown. His *Hopkins in Ireland: Pictures and Words* was published by Creighton University Press in 2008.

Great Wave Off Kanagawa is probably as well recognized as Andy Warhol's Marilyn Monroe silkscreen prints.

- The 1840s was a perfect storm for cyanotype. Alfred Smee presented his paper on how to make potassium ferricyanide from potassium ferrocyanide to the Royal Society June 18, 1840. Not too long thereafter, 1842, Smee sent Sir John Frederick William Herschel potassium ferricyanide. On April 23, 1842 Herschel made note of Prussian blue being formed on paper, and therefore the photographic properties of potassium ferricyanide.[38] It was Smee who told Herschel about ferric ammonium citrate and ferric ammonium tartrate which could increase the sensitivity of potassium ferricyanide.[39] Herschel took Smee's information and ran with it. His paper "On the Action of the Rays of the Solar Spectrum on Vegetable Colours, and on Some New Photographic Processes" was read before the Society on June 16, 1842, and printed in September of that same year in *Philosophical Transactions*.[40] He used the name "cyanotype" in his Memoranda August 16 of that year, and on August 29 he used "cyanotype" to refer to all photographic processes that resulted in Prussian blue.[41]
- The very first photo book was created by a woman, Anna Children Atkins (1799–1871), in 1843, the year before Talbot's *Pencil of Nature*.[42] Atkins was the daughter of zoologist John George Children.[43] Children was a member of the Fellowship of the Royal Society from 1807 on, so he would have been aware of Herschel's discovery in 1842.[44] Atkins was a botanist and a member of the Botanical Society of London since 1839.[45] Atkins combined her passion in botany with cyanotype and created organic photograms which resulted in 3 volumes of *British Algae: Cyanotype Impressions* 1843–1853.[46] She issued the work in fascicles, the first being released in October of 1843.[47] Ten years later, Atkins had completed 389 pages of captioned plates and 14 pages of titles and text. More than a dozen copies were produced and distributed, and one to Talbot.[48]
- 1864–1933 Frederick Holland Day used cyanotype to make prints.[49]
- During the 1870s and through World War I cyanotype was a popular process with amateurs.[50]

Commercial cyanotype papers were available for purchase from the 1870s to about 1930.[51]

- Marion's Papier Ferro Prussiate was released in 1872.[52]
- Henri Le Secq (1818–1882) used cyanotype to print his photographs because the silver photographs he had made did not last.[53] Close to the same time, Eadweard Muybridge (1830–1904) printed his locomotion prints in cyanotype.[54] Le Secq used cyanotype to make finished prints, whereas Muybridge used the process to proof prints, though they are now part of his photographic legacy.
- The main uses of the cyanotype process since 1842 were for botanical purposes (still popular today in the ubiquitous organic cyanotype photogram), in architecture for the blueprint, for proofing negatives, and for making postcards. Even in the latter two decades of the 19th century it was criticized as acceptable for pictorial work.[55]
- Henry P. Bosse (1844–1903) used the cyanotype process extensively to chronicle the Mississippi River between Minneapolis and St. Louis. Between 1883 and 1893 he made close to 900 cyanotype prints. The prints are quite beautiful, oval, somewhat large (10.25″ × 13″), on off white paper. Bosse's work is considered a landmark in the use of photography to redefine cartography. There were 345 different scenes in five albums, although each album does not contain all scenes.[56] Bosse's work chronicled the first systematic effort to recast the upper Mississippi from a natural river to a modern commercial highway.[57]
- Washington Teasdale (1831–1903) did a lot of cyanotype during the 1880s.[58]
- In 1891 it was described as a "popular process" but more popular in America than in England.[59]
- Frederick Coulson (b. 1869) did a minimum of 340 cyanotypes between 1890 and 1908. The cyanotypes date from his years as a draftsman when he was working in downtown Worcester in the Walker Building, home to the offices of architects and commercial photographers. He himself was an architect. In 1884 the Worcester Camera Club was formed and its studio was in Coulson's building. The club held meetings in the Worcester Natural History Society of which Coulson was a member. Photography was becoming recognized as an art form during this time, all a perfect storm for his photographs.[60]
- Utah State University has 2600+ cyanotypes of its history 1896–1916, many done by alumnus George Melvin Turpin.[61]
- Arthur Wesley Dow (1857–1922) made forty-one cyanotypes of his hometown Ipswich in New England, around 1899, and dedicated his personal album to his friend the poet Everett Stanley Hubbard.[62]
- In 1900 cyanotype was used during the Boer War, when the defenders of Mafeking ran short of money and stamps. Lieutenant Colonel Baden-Powell had amateur photographer, E. J. Ross issue these in cyanotype. Five negatives were made from Baden-Powell's design of a pound note, and 100 prints copied from each. The "Mint" was a dugout, and Ross succeeded in turning out twenty pound notes a day, each with two signatures. For stamps two different designs were used, for the values of 1d and 3d, the latter with a portrait of Powell. Both bear the inscription "Siege of Mafeking."[63]
- In the 1900s to 1920s pre-coated cyanotype postcards were commercially available in the USA.[64]
- Paul Burty Haviland (1880–1950) was a lesser-known member of the Photo-secessionists. He did a lot of cyanotypes including portraits and nudes. However, in the catalog of the works it is noted they were most likely proofs of his negatives.[65] Paul Burty Haviland's and Henri Le Secq's cyanotypes are in French museums today.[66]
- In 1903 Kodak came out with the 3A Folding Pocket Camera, which featured rolled film 3¼″ × 5½″ negatives specifically for postcards. Postcards went from cyanotype to gelatin silver, as did commercially prepared printing out papers like cyanotype and platinum. By the end of World War I, cyanotype largely disappeared except in blue prints for architecture and engineering.[67]
- In 1916 Kodak quit manufacturing their commercial cyanotype paper.[68]
- From around 1910 or so Edward S. Curtis made cyanotypes, which can be seen in the Canadian

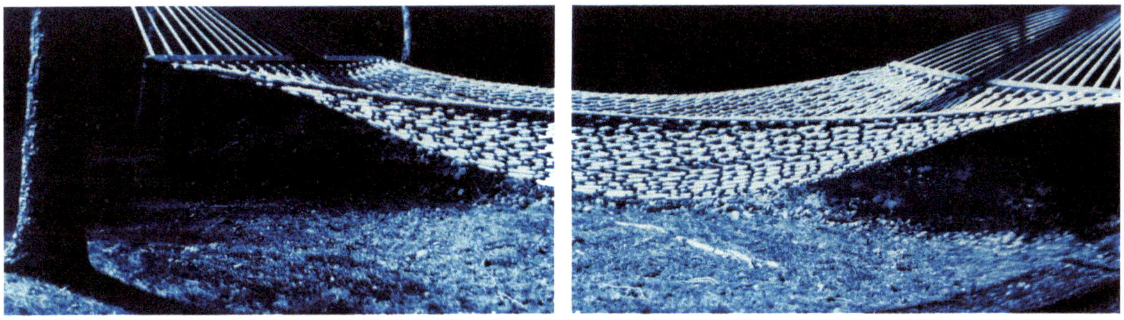

Figure 1.7. *Hammock at #91*, new cyanotype printed on Arches HP watercolor paper, hatbox pinhole camera and ortho film, 30″ x 14″ © Dianna Rust 2014. "Whatever means I use to capture my images, whether it be from a homemade hatbox pinhole camera to a modern scanner, or, as with photograms, no camera at all, I continue to be fascinated by what is mirrored back is no longer a literal interpretation of what was in front of me. In a time when we are so surrounded and bombarded by the ills and sadness of this earth, I hope that the work I do can bring some tranquility, peace, and nourishment for the soul. My vision is often guided by a quote from Charles Burchfield (painter, 1893-1967), 'The only divine reality is the unspeakable beauty of the world.'" To see more of Rust's work, visit **www.diannarust.com**.

National Gallery in Ottawa. He used cyanotype to proof negatives in the field. Curtis made about 40,000 negatives, and a lot of his cyanotypes still exist.[69]

- During World War I, a journal remarked that ferric ammonium citrate was particularly costly as a consequence of the war.[70]
- Cyanotype is surprisingly absent from the literature from the 1920s to the 1950s, although it was in wide use by engineers in 1931. Blue print paper was obtainable in three grades, fast, medium, and slow, and in most cut sizes and rolls of various lengths.[71] Its role as copy machine was vast, but its role as fine art, not so much. Gelatin silver was the norm at this time—think Ansel Adams. During the 1950s the blue print was also replaced by diazo and other processes.[72]
- In the 1960s a few contemporary photographers broke away from the limitations of straight photography.[73] The artists had a common goal to expand, test, and change the definition of photography through the use of nontraditional photographic materials.[74] This was a generation that rebelled against big box stores like Kodak and returned to hand coated processes and making one's own paper. Nineteenth century processes in general and cyanotype in particular finally became part of art photography.[75]
- In 1994 Mike Ware invented a new form of cyanotype using ferric ammonium oxalate instead of ferric ammonium citrate which he published in 1995. This solved the slow exposure problem inherent in classic cyanotype, among other things. Dr. Ware was the first person to delve deeply into the history and practice of cyanotype. His book *Cyanotype, The history, science, and art of photographic printing in Prussian blue* was published in 1999 and has been the definitive cyanotype text since then. Ware has continued to update this book, now out of print, and offers it as a free download on his website (see the **Bibliography**). It is an excellent text from a scientific and conservator perspective.
- In 2016 the Worcester art museum mounted the exhibition *Cyanotypes: Photography's Blue Period* from January 16 to April 24, the first museum exhibition devoted to the cyanotype process.[76]
- In 2018, in honor of the 175th anniversary of Anna Atkin's *British Algae*, the New York Public Library mounted two exhibitions devoted to Atkins and the cyanotype, one called *Anna Atkins Refracted: Contemporary Works*, an exhibition of nineteen contemporary artists working in cyanotype. It is a fitting year for the publication of this book, as tribute to the woman who created the first photographically illustrated book, and who initiated the organic cyanotype photogram still widely practiced today.

Endnotes

1. Tennant, John. A., ed. "The 'Blue Print' and its Variations" in *The Photo-Miniature*, Vol. 1, No. 10, January 1900, p. 481.
2. Hawkes, Lieut. H. P. *Photography in a Nutshell, by "The Kernel."* London: Iliffe, Sons & Sturmey Ltd, 1898, p. 91.
3. "Printing Process" in *The Photographic News, The Journal for Amateur Photographers*, Vol. XLI. London: Photographic News, April 2 1897, p. 222.
4. Worthington, J. C., and J. C. Millen. *The Photographic Primer, A Manual of Practice*. Riverton: The Riverton Press, 1897, pp. 91–92.
5. Museum of Fine Arts, Boston. *Blue: Cobalt to Cerulean in Art and Culture*. San Francisco: Chronicle Books, 2015, p. 17.
6. Skeel, Adelaide. "Something More About the Blues" in *The Photographic Times*, Vol. XXI. New York: Scovill Manufacturing Company, January 30 1891, pp. 54–55.
7. Burns, Nancy Kathryn and Kristina Wilson. *Cyanotypes: Photography's Blue Period*. Worcester, Massachusetts: Worcester Art Museum, 2016, p. 13.
8. Burns and Wilson, p. 14.
9. Rexer, Lyle. *Photography's Antiquarian Avant-Garde*. New York: Harry N. Abrams, 2002, p. 106.
10. Hamblin, R. A. "Hints to Beginners on Choosing a Printing Process" in *The Photographic News, The Journal for Amateur Photographers*, Vol. XLI. London: Photographic News, July 16 1897, p. 456.
11. MacLean, Hector. *Popular Photographic Printing Processes*. London: L. Upcott Gill, 1898, p. 89.
12. "The Workroom-Cyanotype" in *The Photographic Journal of America*, Vol. LV. Philadelphia: Edward L. Wilson Company, Inc., February 1918, pp. 93–94.
13. Rexer, p. 106.
14. Emerson, P. H. *Naturalistic Photography for Students of the Art*. London: Dawbarn & Ward, Limited, 1899, p. 139.
15. Wall, E. J. "The Iron Salts I" in *American Photography*, Vol. XVI, November 1922, p. 682.
16. Brown, George E. *Ferric and Heliographic Processes: A Handbook for Photographers, Draughtsmen, and Sun Printers*. New York: Tennant & Ward, 1900 [First edition 1900, reprinted 1905, this edition n.d.], p. 7.
17. *The Modern Way in Picture Making: Published as an Aid to the Amateur Photographer*. Rochester: Eastman Kodak Co., 1905, p. 120.
18. Hamblin, p. 456.
19. House, Suda. *Artistic Photographic Processes*. New York: American Photographic Book Publishing, 1981, p. 35.
20. Wike, B. H. "Choice of Papers" in *Popular Photography*, Vol. II. Boston, Massachusetts: F. R. Fraprie, July 1914, pp. 415–416.
21. Burns and Wilson, p. 28.
22. Wood, John. *The Photographic Arts*. Iowa City: University of Iowa Press, 1997, p. 42.
23. Reilly, James. *Care and Identification of 19th Century Photographic Prints*. Rochester: Eastman Kodak, 1986, p. 70.
24. Burns and Wilson, p. 13.
25. Ibid.
26. https://photographmag.com/reviews/cyanotypes-photographys-blue-period-at-the-worcester-art-museum/
27. Burns and Wilson, p. 17.
28. Worthington and Millen, p. 127.
29. Burns and Wilson, p. 13.
30. Pitt, Marion. *Practical Guide to Photography*. London: Marion and Co., 1880, pp. 210–212.
31. Burns and Wilson, p. 18.
32. Ibid., pp. 13–19.
33. http://worcestermag.com/2016/01/07/worcester-art-museum-presents-cyanotypes-photographys-blue-period/39218
34. Ware, Mike. *Cyanomicon II, History, Science and Art of Cyanotype: photographic printing in Prussian blue*. Buxton: self-published, 2016, pp. 26–27. http://www.mikeware.co.uk/mikeware/downloads.html.
35. Ibid., pp. 33–34.
36. Museum of Fine Arts, Boston, p. 132.
37. Wehrenberg, Charles. *Mississippi Blue, Henry P. Bosse and his Views on the Mississippi River Between Minneapolis and St. Louis, 1883–1891*. Santa Fe, New Mexico: Twin Palms Publishers, 2002, p. 13.
38. Ware, pp. 44–46.
39. Ibid., p. 48.
40. Ibid., p. 51.
41. Ibid., pp. 53–54.
42. Barnier, John. *Coming Into Focus, a Step-by-Step Guide to Alternative Photographic Printing Processes*. San Francisco: Chronicle Books, 2000, p. 38.
43. Coe, Brian and Mark Haworth-Booth. *A Guide to Early Photographic Processes*. London: Victoria & Albert Museum, 1983, p. 78.
44. Ware, p. 136.
45. Schaaf, Larry J. *Out of the Shadows. Herschel, Talbot, & the Invention of Photography*. New Haven: Yale University Press, 1992, p. 130.
46. Coe and Haworth-Booth, p. 19.
47. Ware, p. 141
48. Schaaf, p. 13.
49. Burns and Wilson, p. 13.
50. Ibid.
51. Barnier, p. 38.
52. Ware, p. 64.
53. Ibid., p. 150.
54. Ibid., p. 155.
55. Ibid., pp. 11–14.
56. Wehrenberg, pp. 8–12.
57. Ibid., pp. 155–159.
58. Ware, p. 13.
59. Harrison, Jerome. "The Chemistry of Silver Printing, Printing with Salts of Iron—Cyanotype and Kallitype" in *Photographic Times*, Vol. XXI, September 4 1891, pp. 440–442.
60. Welu, James A. *Frederick Coulson, Blueprints of a Golden Age*. Worcester, Massachusetts: Worcester Art Museum, 2016, pp. 11–13.
61. Ware, p. 154.
62. Ibid., p. 152.
63. Gernsheim, Helmut, p. 170, from the *British Journal of Photography* August 24, 1900 and Professor Eric Stenger, *Photo Magazin*, June 1952, p. 50.
64. Ware, p. 154.
65. Ibid., p. 151.
66. Ibid., p. 14.
67. Burns and Wilson, p. 13.
68. Ibid., p. 21.
69. Ware, pp. 154–155.
70. Fraprie, Frank Roy, ed. *American Photography*, Vol. XVI, 1922. Boston: American Photographic Publishing Company, 1922, p. 119.
71. Neblette, C. B. *Photography, Its Principles and Practice*, 2nd ed. London: Chapman & Hall, Ltd, 1931, pp. 460–461.
72. Ware, p. 17.
73. John Michael Kohler Arts Center. *The Alternative Image*. Sheboygan, Wisconsin: John Michael Kohler Arts Center, 1983, pp. vii–viii.
74. Kohler, p. viii.
75. Burns and Wilson, p.13.
76. http://www.worcesterart.org/exhibitions/cyanotypes-photographys-blue-period/

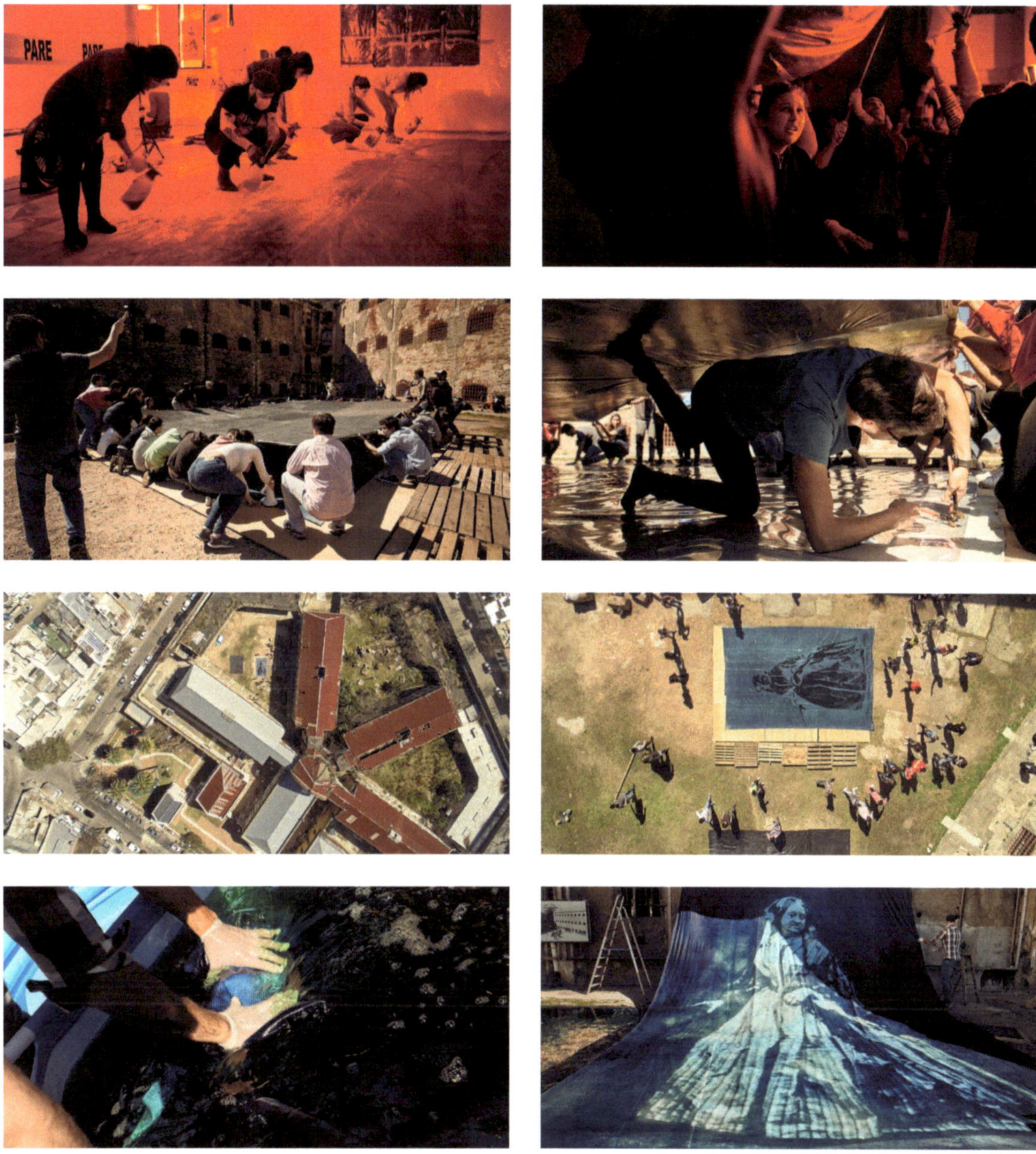

Figures 1.8–1.15. © Dario Invernizzi, Jessica Stebniki, Martín Tarallo, Alexis Magnone, Mauricio Castro, Luis E. Sosa and the Producción Gráfica 2D Team, Uruguay, South America, photos courtesy of Tali Kimelman. "This project was born in a class at the Public University of Uruguay, South America. 160 university students who work regularly with graphic arts and alternative processes embarked on a large cyanotype photograph, officially the largest made by contact printing. The photograph is one of the most recognized portraits of Anna Atkins, considered the first woman photographer in history, precursor of the photographic print on paper with her first books on *British Algae* (1843). This is a tribute to women photographers of all times and places. The original image was freely sourced from Wikipedia. The negative was digitally printed on PVC film in modules of 3 x 1 meters and glued with transparent adhesive tape to form a single negative that covers the entire surface. The cotton fabric has a surface of 5 x 7.5 meters (sewn as a single piece) and was sensitized with garden sprinklers. Once sensitized and dried, the negative was placed on the fabric and covered with black plastic held by the students. Then it was exposed for 20 minutes to winter sun at noon. The development was done in two kiddie pools with running water and a mild peroxide bath, and then dried in open shade. This great photograph will be hung in the Hall of the Faculty of Architecture, Design and Urbanism in Montevideo, Uruguay, and then travel to different places of cultural interest to honor Atkins' legacy." Dario Invernizzi is an instructor in the Visual Communication Design Degree of the University of the Republic (Montevideo). Check out the instagram account here: **producciongrafica2d**.

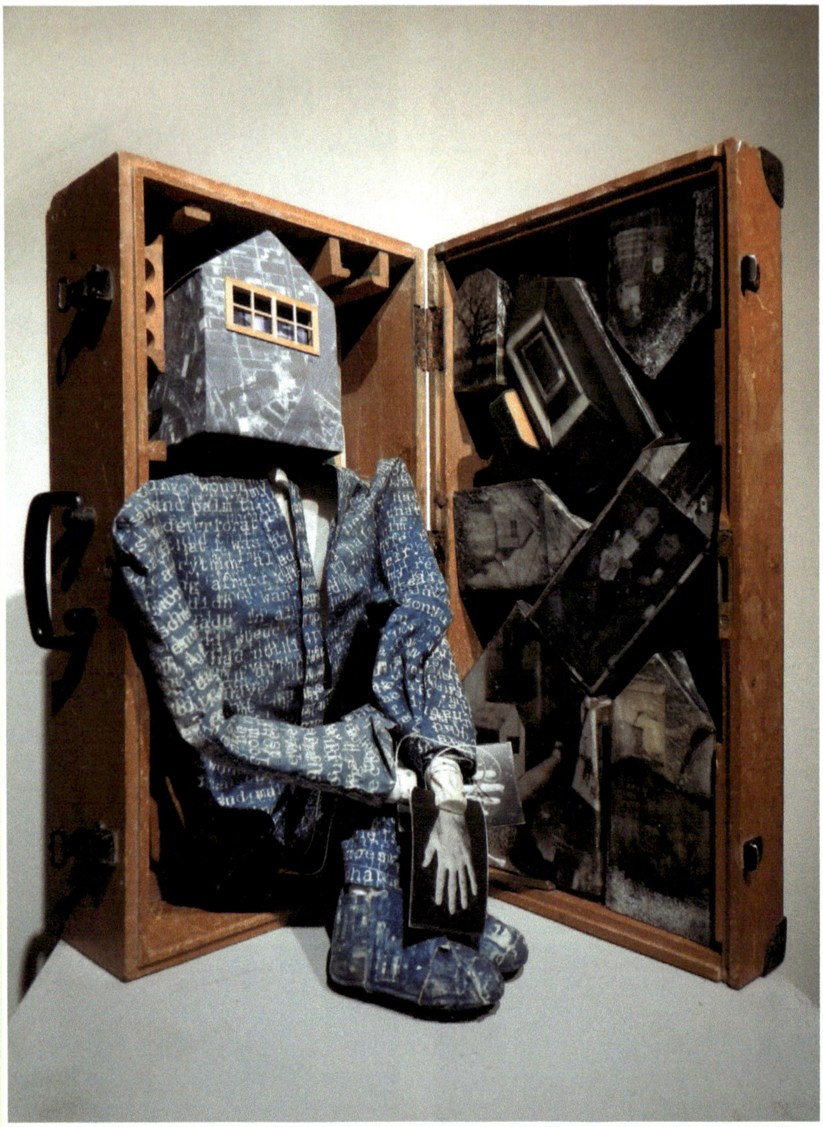

Figure SI. *Househead Puppet*, classic cyanotype on cotton, 3´ x 1´ x 5˝ © Melanie Walker 2018. "I frequently print on fabric, either cotton or silk. I have been working with these materials since graduate school in the 1970s and used to coat my own fabrics which was a huge mess. Lately I have been purchasing pre-sensitized fabrics from Blueprints on Fabric. I have never been disappointed with their materials. I started working with cyanotype and other heliographic processes when I was very young and learned them from my father, Todd Walker, who was one of the pioneers of historical photographic processes in the 1960s. My father was pretty much self-taught and had an incredible grasp on the chemistry behind photography. He is rarely given credit for the resurgence of interest in the historical photographic processes; however, in my mind, he was one of the pioneers. I continue to work with these processes to honor my father and history." Melanie Walker has been a practicing artist for over 50 years. Her expertise is in the area of alternative photographic processes and mixed media as well as large scale immersive photographic installations and public art. She received her BA from San Francisco State University and an MFA from Florida State University. She has received a number of awards including an NEA Visual Arts Fellowship, Colorado Council on the Arts Fellowship and an Aaron Siskind Award. She currently teaches at the University of Colorado Boulder and has work in many collections. To see her work visit https://www.melaniewalkerartist.com.

PART ONE
Cyanotype Step-by-Step

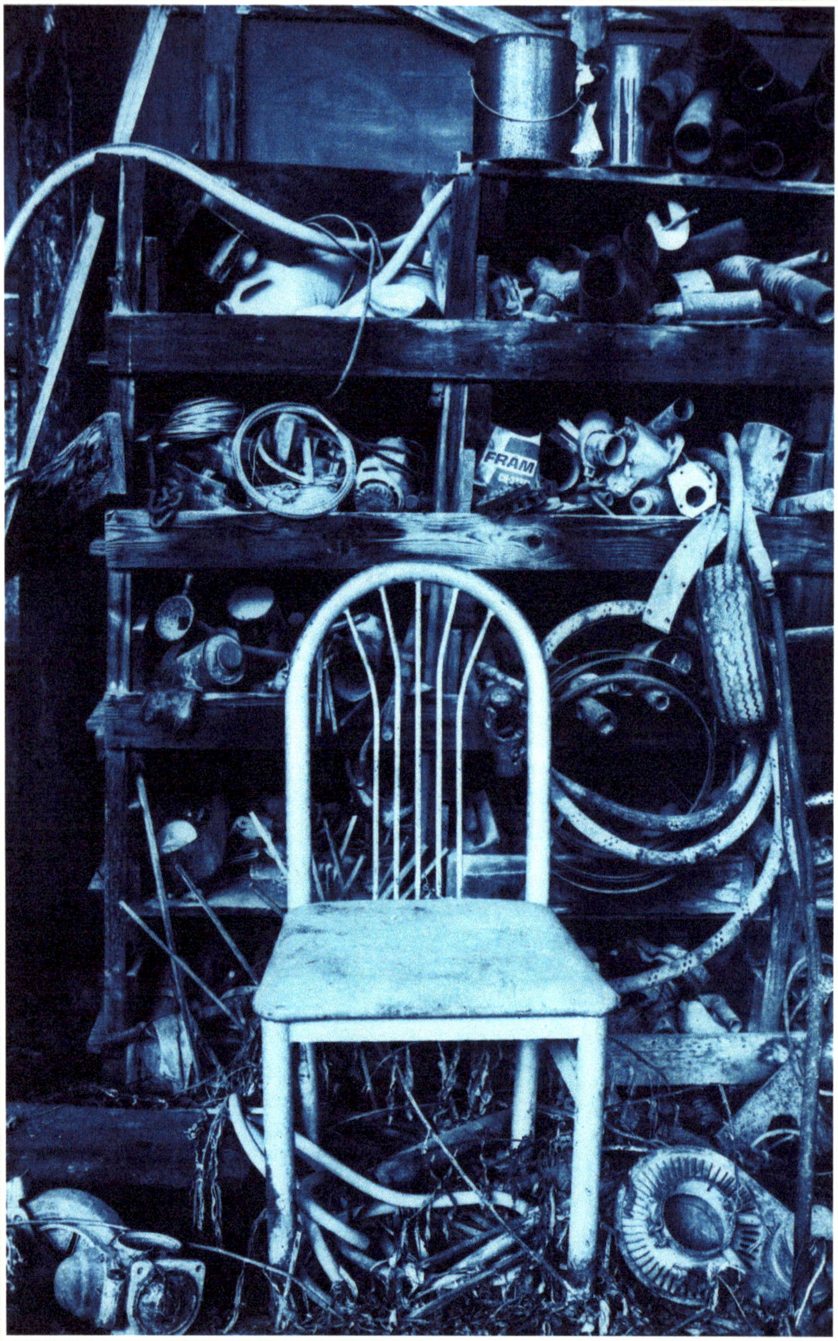

Figure 2.1. *Snowline Estate*, classic cyanotype on Indigo handmade watercolor paper, 45-minute UVBL exposure, 2-minute citric acid wash, 10-minute water wash, 6″ × 10″ © Megan Crawford 2018. Megan Crawford is an alternative process photographer based in northwestern Montana. She received a Bachelor of Arts in Film & Photography from Montana State University in 2017. Her work has exhibited nationally and is featured in various publications including *Salted Paper Printing*. Crawford's background in history and her interest in handmade photographs led her to pursue processes such as gum bichromate and salted paper. Her work often focuses on history and landscape: how they interact with modern and historical societies, how they interact with each other. To see more of her work, visit **meganlcrawford.com**.

Chapter 2

Setting Up the Cyanotype Dimroom

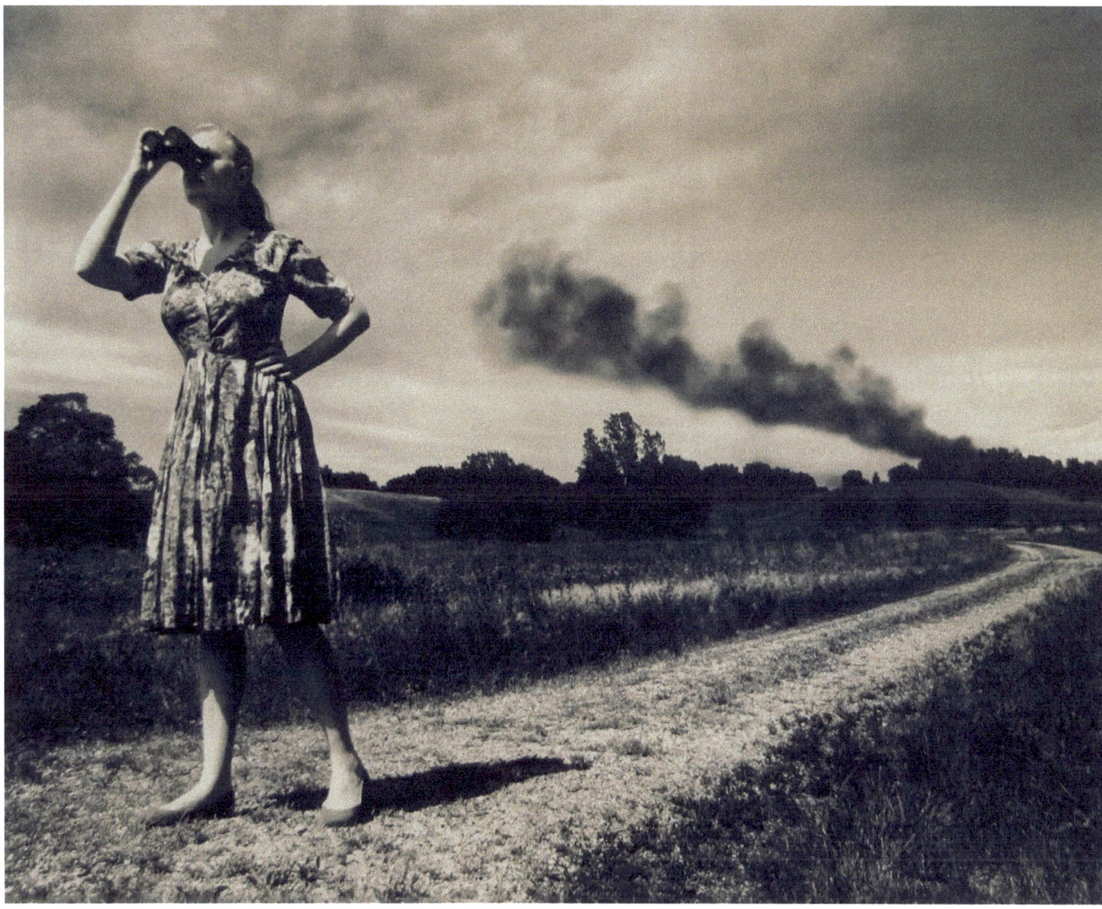

Figure 2.2. *Where There Is Smoke*, cyanotype toned with tea and wine tannin and a bleaching agent (usually highly diluted silver gelatin paper developer or sodium carbonate), 11″ x 14″ © Emma Powell 2012

The cyanotype "dimroom" doesn't have to be light-tight. It doesn't require a "darkroom" like the traditional black and white or color darkroom that has to have a safelight or complete darkness. As long as there is no sun shining through a window, normal room light such as that provided by a ceiling light fixture is no problem, because the cyanotype formula is slow to expose and not as sensitive to light. Tip: to test how much room light is safe, coat a small piece of paper with solution. Leave it out for a total of one hour, covering parts of it in 15-minute increments. Rinse the piece of paper for 15 minutes in a tray of water. Note where the paper remains paper white (which may be the entire sheet of paper!) and that will indicate a safe amount of room light time.

Merely two chemicals, one tray, water, and sun can start you on your journey to making excellent cyanotype prints. Thereafter, you'll want other chemistry and equipment which are outlined in the following pages.

Safety first!

All chemicals should be treated with utmost respect. Go to www.msds.com and download the MSDS (Material Safety Data Sheets) for every chemical used in the dimroom and store them in a readily available notebook. Have the local poison control number handy and prominent! Label all chemistry. Immediately clean up any spills that occur. Always keep hands clean. Don't look directly into UV bulbs—they are equivalent to bulbs in a tanning bed. Dispose of hazardous materials according to state regulations. Minimize absorption of chemicals through breathing, ingestion, or skin contact. Always keep chemistry locked up and away from children and pets. Have an eyewash kit in the dimroom in case splashes into the eye occur.

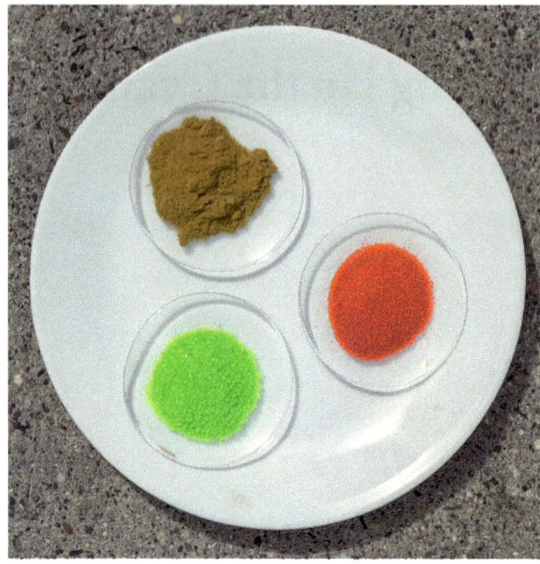

Figure 2.3. The three chemicals used in these cyanotype processes are (clockwise, from top left) ferric ammonium citrate, potassium ferricyanide, and ferric ammonium oxalate. Ferric ammonium citrate is actually quite chartreuse, but in comparison with the brilliant green of ferric ammonium oxalate it looks brown! If left out in sunlight, it will turn brown quite quickly.

Classic cyanotype chemistry

- Ferric ammonium citrate (FAC)
- Potassium ferricyanide (PF)
- Oxalic acid (optional)
- Distilled water
- White vinegar or citric acid for development
- Hydrogen peroxide (optional)

New cyanotype chemistry

- Ferric ammonium oxalate (FAO)
- Potassium ferricyanide (PF)
- Distilled water
- Tween 20™ surfactant
- Ammonium dichromate (optional)
- Citric acid for the formula and development
- Sulfamic acid for development

Toning chemistry

- Alkalies such as ammonia, sodium bicarbonate (baking soda), sodium carbonate (washing soda), or trisodium phosphate (TSP)
- Tannic acid
- Gallic acid
- Pyrogallic acid
- Catechu powder

Palladium chemistry

- Na2 Platinum/Palladium Kit for Digital Negatives from Bostick & Sullivan (**bostick-sullivan.com**). It comes with 10 ml 5% Na2, 25 ml palladium, 25 ml ferric oxalate, a developer and a clear, enough for 25 8″ × 10″ prints.
- Di-EDTA, citric acid, Heico Permawash for clearing, if desired.

Equipment and supplies

- Computer with Photoshop
- Epson Ultrachrome inkjet printer
- Contact printing frame no smaller than 11″ × 14″. Buy one with a split hinge back so the progress of the print can be checked
- UVBL exposure unit: Edwards makes a really nice one (**eepjon.com**) and so does Freestyle Photo (**freestylephoto.biz**)
- Digital scale accurate to 0.1 g. These are not terribly expensive. There are some weight-to-

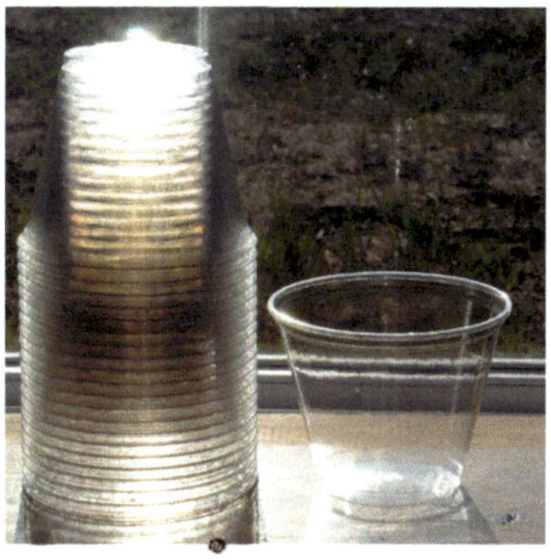

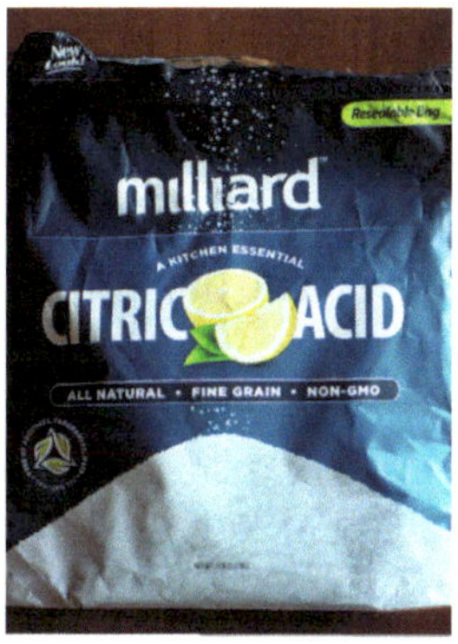

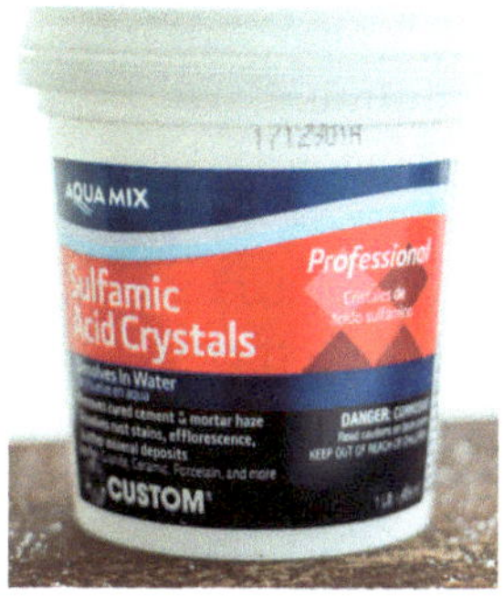

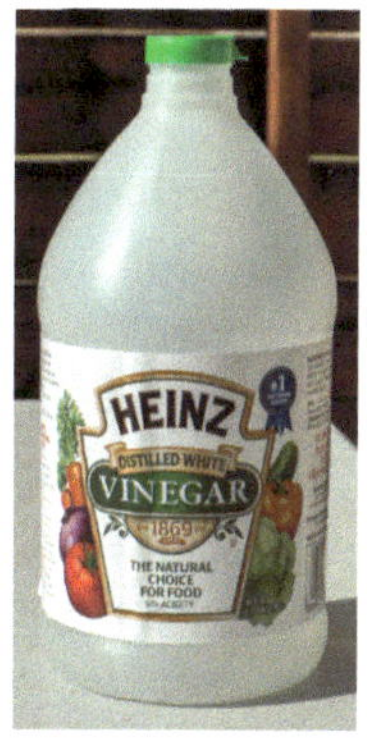

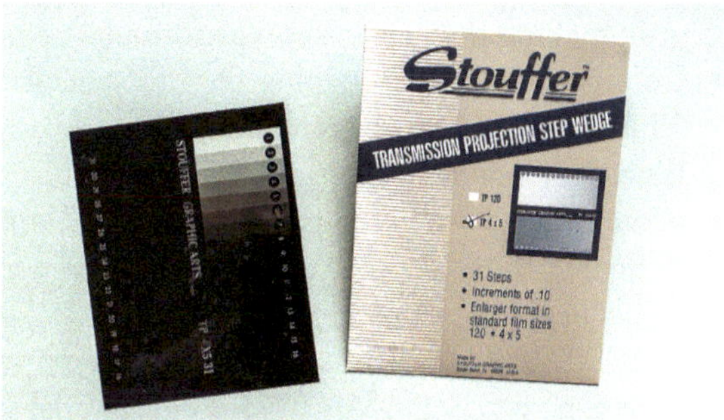

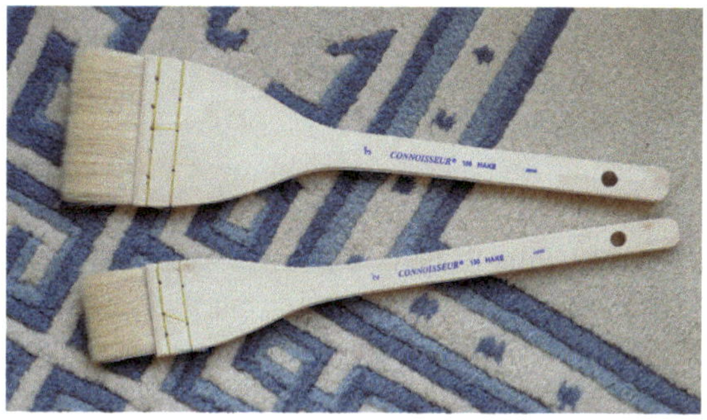

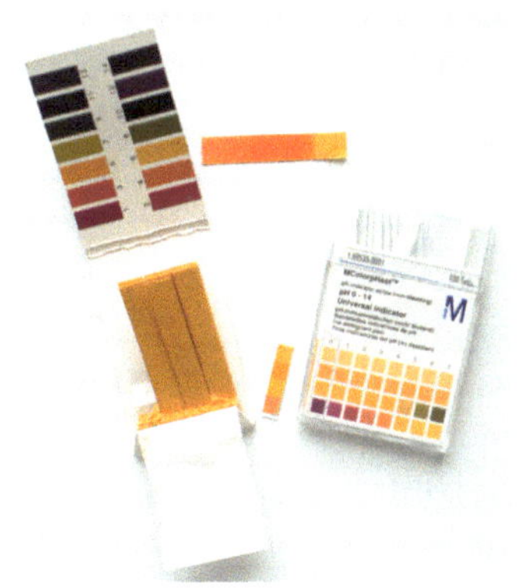

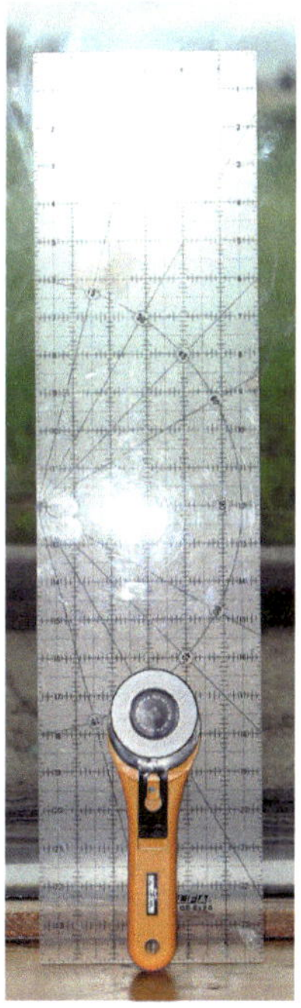

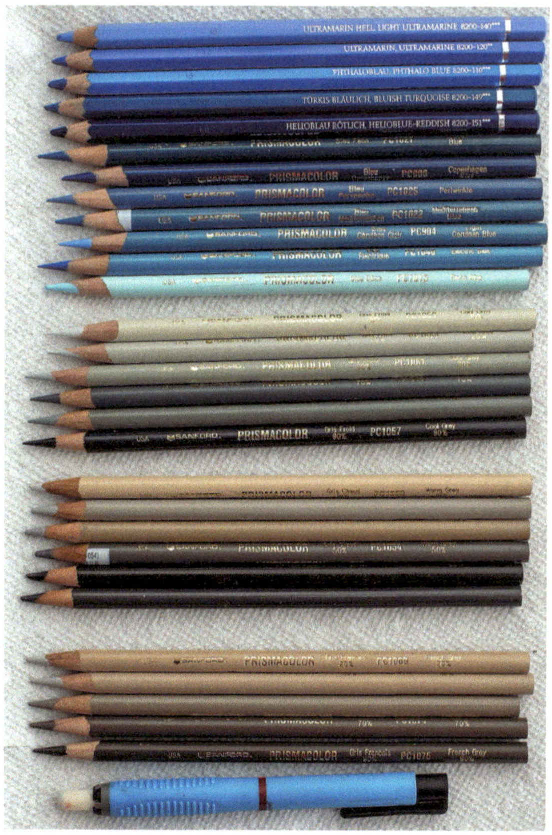

Figures 2.4–2.12. From left to right, top to bottom, aids to the cyanotype dimroom: my favorite squatty plastic "wine" cups available at any grocery store for mixing, soaking, weighing, and measuring; citric acid, sulfamic acid, and vinegar for development; the Stouffer 31-step film step wedge which I use to calculate all exposure times; my *favorite* Connoisseur 150 Series hake brushes; litmus strips to keep tabs on pH when cyanotype printing; my Olfa rotary cutter and clear gridded plastic ruler for cutting and measuring paper; a variety of colored pencils useful for spotting a print: green-blues to red-blues, percentage sets of cool grays, warm grays, and French grays to add density if necessary underneath the color desired.

volume measurements throughout this book that are accurate enough, but a scale is best

- Plastic trays, minimum size 11″ × 14″, for developing and toning
- Tongs
- Hot plate and thermometer for new cyanotype
- Brown plastic storage bottles, 500 ml and 1000 ml, for storing bulk chemistry and toners
- Beakers: 50 ml, 100 ml, 250 ml, 500 ml, 1000 ml
- Graduates: 10 ml, 25 ml, 50 ml
- Brown glass eyedropper bottles, 100 ml
- Funnels, various sizes
- Glass stirring rods, one or two
- Hair dryer
- Hand towels, washcloths, apron
- Vinyl-coated metal baby hangers and plastic clothespins to hang prints to dry. The hangers are about a foot wide. The plastic retains no chemistry, and the print hangs perpendicular to the shower curtain rod (my usual "clothesline"), so many prints can be hung that way, only several inches apart. The hangers can also be hung on each other in pyramid fashion for more space
- Scissors
- Set of plastic teaspoon measures to include ¼″, ½″, 1 teaspoon, and 1 tablespoon
- Stop watch/timer on watch or cell phone
- Stouffer step wedge calibrated in ⅓ stop increments
- Connoisseur 150 Series 2″ stitched ferrule hake brushes. They have stitched ferrules instead of metal—metal tends to rust when left soaking in water for a long time. After an initial period of use, shedding is minimal or nonexistent. They get better with age. After *every* use, immediately put the brush in water with the bristles down. Tip: if excess shedding occurs, try gently scratching the hairs of the brush on a piece of sandpaper, to take away the loose hairs. Also, take a piece of paper and paste around the ferrule, a little bit higher than the ferrule and fill this paper "well" with super glue and let dry.
- **Pictorico Premium** (not Ultra), **Fixxons**, Arista, Inkpress, or other suitable OHP film
- 8.5″ × 11″ notebook sleeves to store negatives
- Litmus paper to test acidity/alkalinity
- 9 oz clear plastic wide-mouth "wine" cups, the small, squatty ones for mixing, weighing chemicals on the gram scale, and soaking brushes. Indispensable, infinitely reusable, they fit a 2″ or 3″ brush perfectly, only weigh about 9 g, and are rinse 'n' go. You can also use cupcake papers that come in two sizes, normal cupcake size and one much smaller. There are also weigh boats or "watch glasses" from a scientific supply house
- Coffee filters for filtering out mold in Solution A or in making the new cyanotype formula
- Color pencils for retouching (e.g. Prismacolor China Blue, Copenhagen Blue, Indanthrone Blue, etc., Derwent Prussian Blue)
- Scotch Brand 810 green plaid box "magic" tape or Scotch Brand 811 blue box removable tape to tape a negative in place
- Rubylith tape, used for taping off borders so they don't expose, in ¼″–2″ wide rolls available here: **http://www.uline.com/BL_6413/3M-616-Lithographers-Tape?keywords=rubylith**. It is $34 for 72 yards, enough for 72 8″ × 10″ prints. Tip: add a very slight black border to each negative (less than ¼″) and then further mask the edge with the rubylith; this makes it easier to get the edge straight. If you put the tape on the contact frame, the edge will be softened. Also, it is *very* sticky, so get some of that adhesive dulled by sticking it to your pants a few times first. You can also use Kleenedge or Loparex low tack tape for masking borders during coating, if desiring a no-border look (Kleenedge doesn't hold back enough light when exposing)
- 2H pencils for recording all information on the edge of the print and signing
- Black permanent sharpies for marking negatives
- Plastic, latex, or nitrile gloves

Useful dimroom aids

- Olfa rotary cutter and self-sealing cutting mat, largest size available, and a 6″ × 24″ see-thru gridded plastic quilting ruler; of all the tools in the darkroom, these are tops. The blade of the cutter is sharp and accurate. The self-sealing mat is impervious to the cuts of the blade. The plastic 6″ × 24″ gridded ruler is perfect for trimming paper accurately (if displaying edges, use a tear bar). These three tools are available in any sewing store. Buy a circular Olfa blade sharpener while you're at it

- Portatrace light box, or an LED light pad, useful for registration if doing cyanotype over platinum. Other methods of registration are pin registration with a two-hole registration punch, and the ¼″/o pin height registration pins. A guerrilla registration method is just pins and a piece of cardboard, pinning all negatives in registration at once to create holes
- Skoy cloths or paper towels. You will want to have some quick way of mopping up a drop here and there, and Skoys are incredibly absorbent and infinitely reusable, not to mention much kinder to the environment than paper towels

Solid measurements

1 gram = 15.4323 grains
1 grain = 0.0648 gram
1 ounce = 28.4 grams
1 pound = 454 grams
1 gram = 0.0353 ounces
1000 grams = 2.2 pounds

Liquid measurements

An important measurement is 1 teaspoon = 5 ml. The next is British volume is 28.413 ml to the ounce; US volume is 29.574 ml to the ounce. Throughout the book formulas will be given in milliliters (ml) and occasionally teaspoons. See **unitconversion.org**.

1 teaspoon =	⅓ tablespoon
	5 ml
	75–100 drops
3 teaspoons =	1 tablespoon
	½ ounce
	15 ml
2 tablespoons =	1 ounce
	30 ml
8 ounces =	48 teaspoons
	16 tablespoons
	1 cup
	237 ml
2 cups =	1 pint
	473 ml
2 pints =	32 ounces
	1 quart
	1 (scant) liter/946 ml
4 quarts =	128 ounces
	1 gallon
	3785 ml

Ounces to milliliters: × 29.57
Milliliters to ounces: × 0.0338
Liter × 2.10 = pints
Liter × 1.06 = quarts
Liter × 0.26 = gallons
1 liter = 33.81 ounces or 1.057 quarts
1 cc water = 1 g = 1 ml — all interchangeable

Sources for supplies

Artcraft Chemicals
800.682.1730
www.ArtcraftChemicals.com

B&H Photo
877.848.1244
www.BHphotovideo.com

Bostick and Sullivan
877.817.4320
www.Bostick-Sullivan.com

Fixxons OHP film
714-738-9009
www.fixxons.com

Freestyle Photographic Supplies
800.292.6137
www.FreestylePhoto.biz

Photographers' Formulary
800.922.5255
www.PhotoFormulary.com

Step wedges
Stouffer Industries
574.252.5772
http://www.stouffer.net

UVBL lightboxes and bulbs
Jon Edwards UVBL lightboxes:
http://eepjon.com/
Freestyle markets a wonderful lightbox:
www.FreestylePhoto.biz
For replacement UVBL bulbs:
Bulbdirect.com or **Topbulb.com**

Sources for paper

United States

A.I. Friedman (New York)
www.aifriedman.com
All Media Art Supply (Ohio)
www.allmediaartsupply.com
Art Center Memphis (Tennessee)
www.artcentermemphis.com
Art Supply Warehouse (California)
www.artsupplywarehouse.com
Artist & Craftsman (all over US)
www.artistcraftsman.com
Binders (Georgia, North Carolina)
www.bindersart.com
Bostick & Sullivan (New Mexico)
www.bostick-sullivan.com
Cheap Joe's Art Stuff (North Carolina)
www.cheapjoes.com
David Art Center (Louisiana)
www.davidartcenter.com
DaVinci Artist Supply (New York)
www.davinciartistsupply.com
Dick Blick (all over US)
www.dickblick.com
Dolphin Papers (Indiana)
www.dolphinpapers.com
Freestyle Photographic (California)
www.freestylephoto.biz
Graphic Chemical (Illinois)
www.graphicchemical.com
Hiromi Paper (California)
www.Hiromipaper.com
Hollander's (Michigan)
www.Hollanders.com
Hyatt's (New York)
www.hyatts.com
Jerry's Artarama (all over US)
www.jerrysartarama.com
McClain's Printmaking Supplies (Oregon)
www.imcclains.com
Michaels (all over US)
www.michael's.com
Pygmalion's Art Supplies (Indiana)
www.pygmalions.com
Renaissance Graphic Arts (Pennsylvania)
www.printmaking-materials.com
Takach Paper (New Mexico)
www.takachpaper.com
Talas (New York)
www.talasonline.com
Twinrocker (Indiana)
www.twinrockerhandmadepaper.com
Utrecht (All over US)
www.utrechtart.com

Canada

Above Ground Art Supplies
www.abovegroundartsupplies.com
Artists' Emporium
www.artistsemporium.net
Curry's
www.currys.com
Japanese Paper Place
www.japanesepaperplace.com
Opus Art Supplies
www.opusartsupplies.com
Studio Six
www.studio-six.com

Europe

De Middelste Molen (Netherlands)
www.demiddelstemolen.nl
Garza Papel (Spain)
www.garzapapel.com
John Purcell (England)
www.johnpurcell.net
Les Papiers de Lucas (France)
www.les-papiers-de-lucas.com
Moulin de Larroque (France)
www.moulindelarroque.com
Peter van Ginkel (Netherlands)
www.petervanginkel.nl
Polymetaal (Netherlands)
www.polymetaal.nl
Ruscombe Paper (France)
www.ruscombepaper.com/uk
Schut Papier (Netherlands)
www.schutpapier.nl
Van Beek Art Supplies (Netherlands)
www.vanbeekart.nl

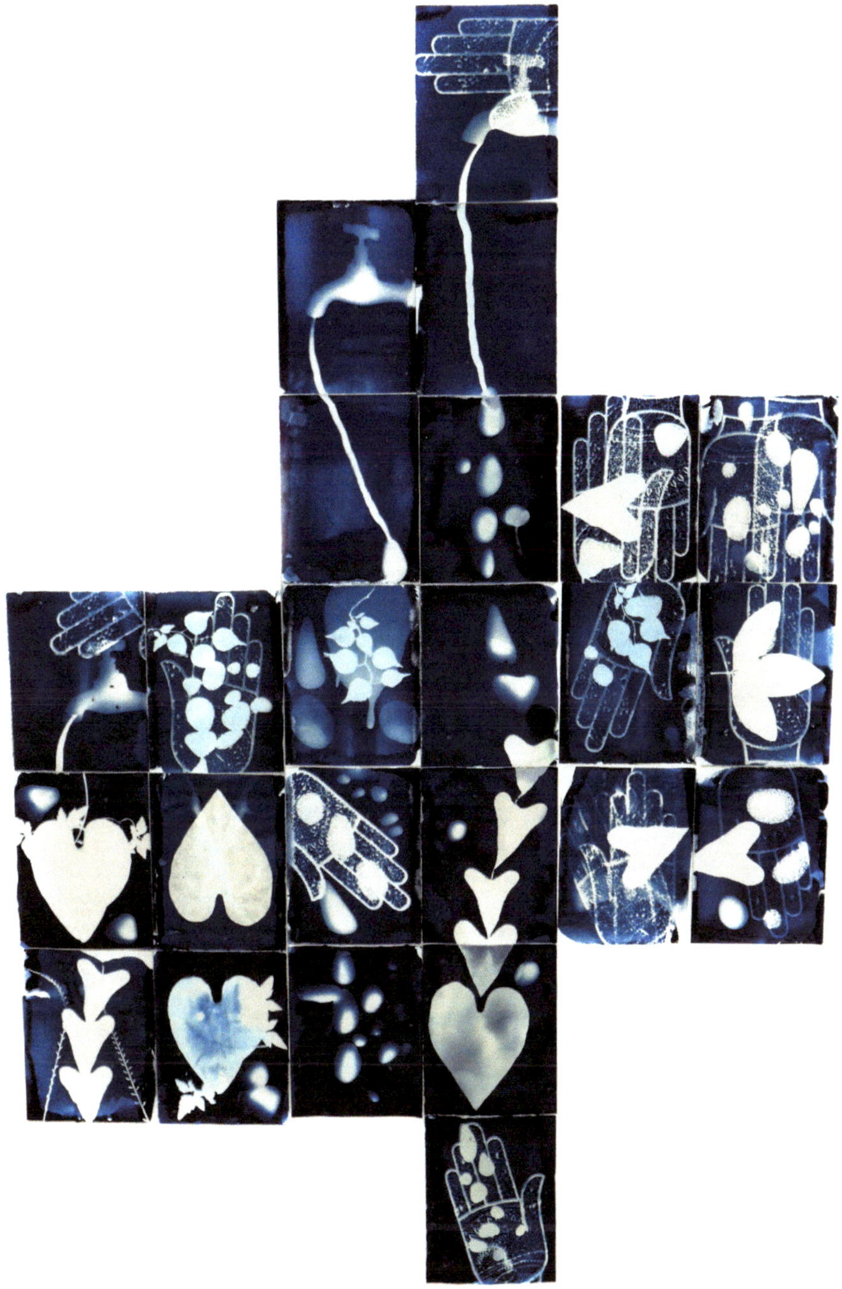

Figure 2.13. *Khetarra, Sharing Water from the Deep Well*, cyanotype and linen tape, 1120 x 740 mm © Susan Purdy 2018. "Khetarra is a folding, gridded, portable cyanotype work created in April 2018, under the Saharan sun on a Moroccan journey, appointing the colour blue as a motif and catalyst for travel and creative play. An adventurous, open-ended approach to investigating the colour blue inevitably drew my attention towards water. In discovering the unique system of wells, called *khetarra* in rural Morocco, and learning about the remarkable water management skills of nomadic Berber tribes in their dry land, I made a work that responds, in blueprint, to the intricate social cooperation required to share an essential resource, and my powerful apprehension of the significance in such action." Susan Purdy lives and works in the Strzelecki Ranges, in Gippsland, on the Southeastern corner of Australia. Her work is focused upon reinterpretation of early photographic practices. It is in a number of significant public collections including the National Gallery of Australia and the National Gallery of Victoria. To see more of her work, visit **www.susanpurdy.net** and **https://culturaltoursmorocco.com/2018-art-photography-with-susan-purdy/**.

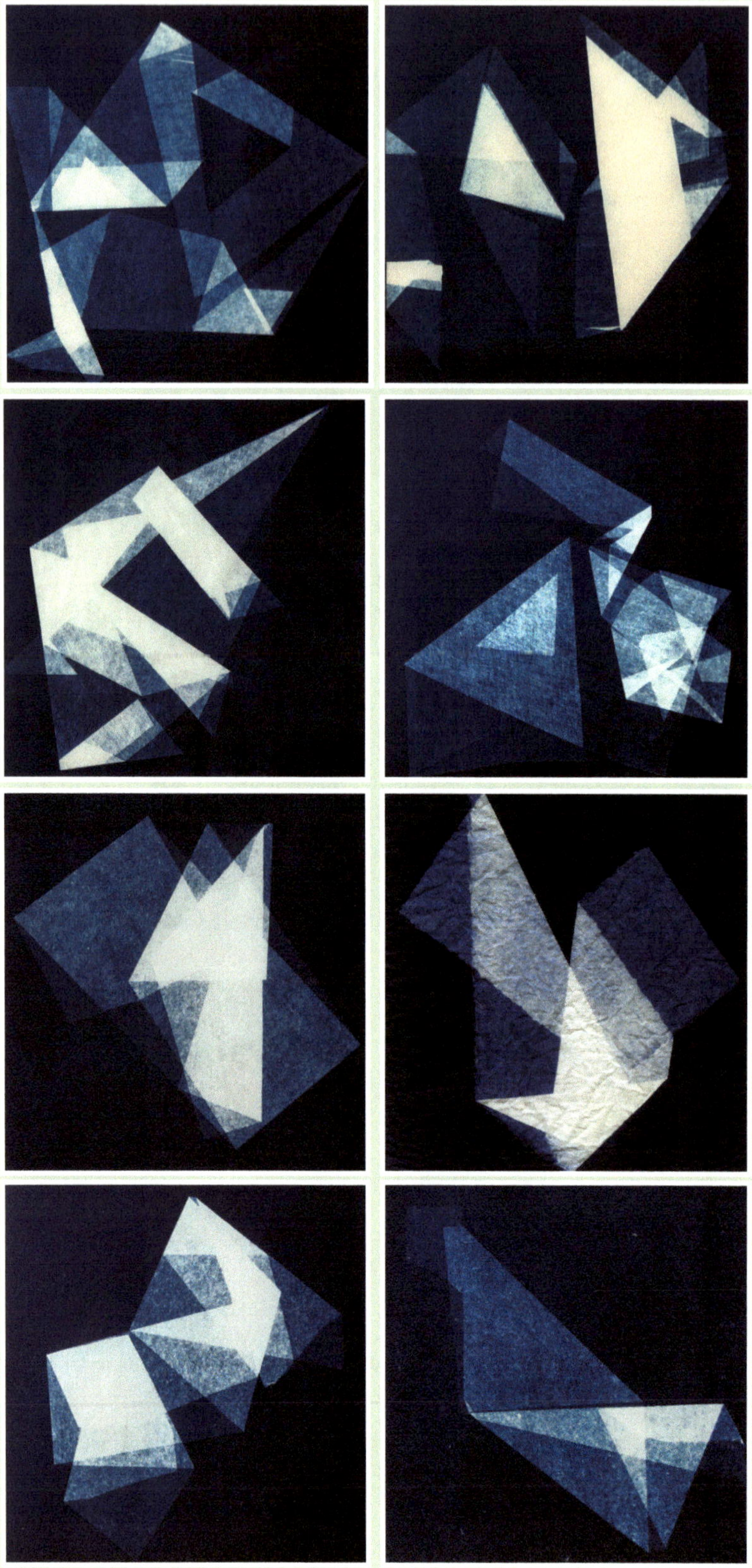

Figures 3.1–3.8. From left to right, top to bottom, *Folding Light #8*, cyanotype on Japanese paper, 33.3 x 33.3 cm; *Folding Light #6*, cyanotype on Japanese paper, 15 x 15 cm; *Folding Light #9*, cyanotype on Japanese paper, 33 x 33 cm; *Folding Light #7*, cyanotype on Japanese paper, 18.5 x 18.5 cm; *Folding Light #15*, cyanotype on Japanese paper, 16.5 x 17 cm; *Folding Light #1*, cyanotype on fabric, 21.1 x 16.8 cm; *Folding Light #13*, cyanotype on Japanese paper, 17 x 14 cm; and *Folding Light #12*, cyanotype on Japanese paper, 17 x 14 cm © Andrés Pantoja 2015.

"Folding light is an ongoing project that explores the intersections between light and paper through cyanotype. When paper is folded in different ways, new frontiers are created and demarcated by diagonals, horizontals and verticals; but it is when light hits the new folded planes, that new 'objects' that relate to scale, gravity, weight, movement, and rhythm appear on the sensitized surface."

Andrés Pantoja is a visual artist based in London, working primarily with photography. Initially trained as a graphic designer, he moved towards a career in the arts, graduating with honours from his degree in Visual Arts in Bogotá, 2010. Since then, Pantoja has lived and worked in London as a freelance visual artist, graphic designer and photographer. He is also codirector of Lux Darkroom, a space dedicated to teaching and researching analog and historical photographic processes. Pantoja's artistic practice is concerned with the relationship between digital, analog, and historic photographic processes, and the intersection between light and shadows.

Chapter 3

Papers for Cyanotype

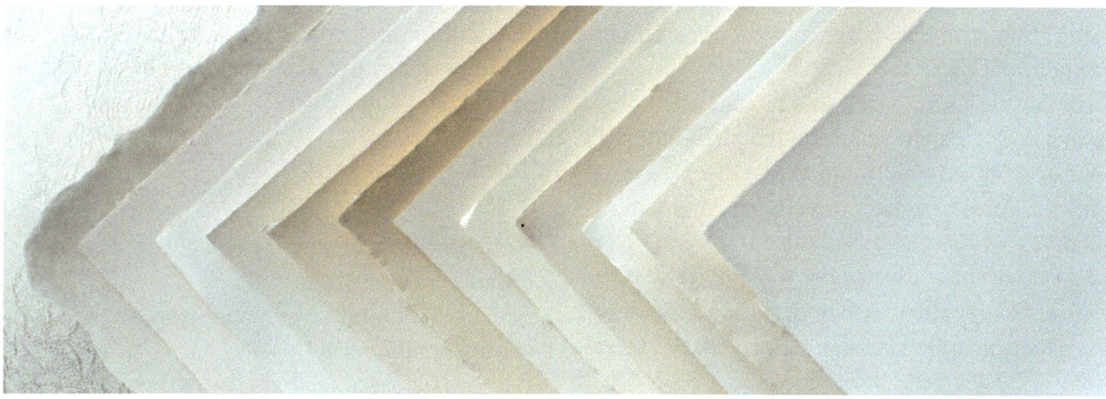

Figure 3.9. Twinrocker papers, pictured left to right: Bleached Abaca HT (abaca), May Linen LA (cotton), B-Buff T (cotton), Oyster HT (cotton), Yale T (cotton rag), Rotunda T (abaca, flax, and cotton rag), Da Vinci T (cotton rag), Cream LA (cotton), Biscuit HT (cotton rag), Bone T (cotton), Dickinson TT (cotton), White Cotton Rag TT. Twinrocker papers in text (T), heavy text (HT), and light art (LA) weights are my favorites. Most Twinrocker papers print classic cyanotype beautifully.

Cyanotype is probably the least expensive alternative process there is, less than a nickel's worth of chemistry for an 11″ × 15″ print. Coupled with water-only development, you can print with abandon, without ever having to think about the cost of chemistry, only the cost of the paper on which it is printed. Furthermore, because cyanotype is a one-coat process, it allows for broad experimentation with all sorts of papers from delicate to sturdy.

Nowadays most papers are buffered with an alkali, because an alkaline paper is more archival. Unfortunately, alkalinity is not good for cyanotype, both in the making of the print and in the print's long-term preservation. The lower the paper's pH the faster it will print and the greener the blue. The higher the pH the slower the paper will print and the grayer the blue; it results in a dull, washed out, grainy, gray-blue color, sometimes with brownish highlights. These kinds of prints give cyanotype a bad reputation. Plus, a cyanotype print will fade over time in contact with alkalinity, although internal paper buffering is unlikely to migrate once the paper is dry.[1]

Prompted by the recent demise of several excellent alternative process papers, I set out on a quest to test papers in an effort to broaden cyanotype's paper repertoire. This chapter is a compilation of results after having put 136 papers through the tests.[2]

I tested all papers with both classic cyanotype and new cyanotype. Classic cyanotype is a more paper-friendly formula. New cyanotype is "the canary in the coal mine" as Mark Nelson describes it, a process that will sniff out a buffered paper a mile away. This chapter is about papers suitable for classic cyanotype; the list of acceptable papers for classic cyanotype is much longer than those for new cyanotype. Papers suitable for new cyanotype will be listed in the **New Cyanotype** chapter as well as indicated in the paper chart at the end of this chapter.

Paper weight is expressed in *gsm*/grams per square meter and *lb*/pounds. Gsm measures are a little more "apples to apples" than pound measures so that is what is used here. Papers tested were mostly from 30 to 300 gsm; 300 gsm equates to a typical 140 lb watercolor paper, more than enough wet strength for the cyanotype process.

Paper is described as hot press (smooth), cold press (bumpy texture), or rough (*very* bumpy texture). Most prefer hot press for cyanotype. There are also other paper surfaces such as plate and vellum. Plate is ultra-smooth. Vellum is smooth but slightly modulated, like calfskin or parchment from days of old.

Sizing varies in paper and it directly affects how a paper prints. Paper can be unsized, sized internally only, or sized internally and externally. Sizing can refer to animal, vegetal, or synthetic ingredients. Sizing gives the paper body but more important for a cyanotypist, affects how absorbent the paper is. A lot of Japanese/Asian washi is unsized and the paper absorbs chemistry like a paper towel. Since washi is very thin, this is not a problem. If the paper is this absorbent and thick, it can become problematic, because the chemistry remains embedded in the paper fibers and won't clear well—Zerkall Copperplate, Hahnemühle 1587, and Arches 88 are three examples. Although all lovely papers, they are better suited to the printmaking processes for which they were designed. Sizing is a good thing as long as it is not too little and not too much.

Papers range in cost dramatically, from under a dime to six dollars per 11″ × 15″ print. Often the less expensive papers are made from less expensive ingredients. They are perfect for your first cyanotype prints, but as you progress you will want to use more expensive natural fiber papers such as cotton, cotton rag, gampi, kozo, linen, abaca, and mitsumata papers. Fibers are often combined with each other, or other ingredients such as high alpha cellulose, sulphite, esparto, and bamboo. There are also archival high alpha cellulose sulphite papers. For those who want *just the facts*, consult the **Quick Peek** highlights, or the paper chart at the back of the chapter where the information is summarized.[3]

Unbuffered papers made for alt

The following unbuffered papers were created specifically for alternative processes and work for both classic and new cyanotype. I applaud these manufacturers for supporting our niche market and making quality papers of the highest archival standards. These papers are perfect for cyanotype over platinum and all are sturdy enough to go through multiple wet processing steps that the combination of two processes entails. Six are 100% cotton and one is 100% linen. At the time of this writing, prices average about $1.70–$3.60 per 11″ × 15″.

Quick peek

Arches Platine 310 gsm (100% cotton, unbuffered)
Arches Platine Lightweight 145 gsm (100% cotton, unbuffered)
Bergger Cot 320, 320 gsm (100% cotton, unbuffered)
Bergger Cot 160, 160 gsm (100% cotton, unbuffered)
Hahnemühle Platinum Rag 300 gsm (100% cotton, unbuffered)
Legion Revere Platinum 320 gsm (100% cotton, unbuffered)
Ruscombe Mill Herschel Platinotype 200 gsm (100% linen, unbuffered)

Arches Platine

Platine has a very smooth front side and a slightly textured back side. It is surface-sized with starch on the front side and should be printed on the front. Platine is white, 100% cotton, moldmade, has a watermark, two natural and two torn deckle edges, is acid free, chlorine free, unbuffered and alkaline free. It comes in two weights, the normal 310 gsm weight and a 145 gsm thinner weight that is perfect for a one-step process such as cyanotype. Both print beautiful turquoise blues with a long tonal range.

Bergger Cot 320

Cot 320 is a 100% cotton paper that performs similarly to Arches Platine and even looks like Arches Platine, but it is a bit less absorbent when coating. It is slightly thicker at 320 gsm, bright

Figure 3.10. From bottom to top, Canson Bristol Recycled's white compared to the white of four of the unbuffered papers made for alt: Bergger Cot 320, Hahnemühle Platinum Rag, Arches Platine, and Legion Revere Platinum. All of these papers *look* bright white until they are placed next to an ultra-white paper such as Canson Bristol Recycled. Color is always relative.

white, no watermark, unbuffered, with a smooth front surface and a slight texture on the back. There is also a lighter weight 160 gsm version, excellent for cyanotype. Both print beautiful turquoise blues with a long tonal range.

Hahnemühle Platinum Rag

"HPR" is internally sized, acid free, no buffering, no optical brighteners, 300 gsm, pH-neutral, and 100% cotton. It is a smooth, sharp, crisp, bright white paper, smoother and ever so slightly whiter than Platine. It can be printed on the front or back which can be a cost saving if you make mistakes. HPR is a firmly sized paper, less absorbent than others. Prehumidify, use Tween with new cyanotype, brush carefully and thinly, and you'll be rewarded with a beautiful, sharp, dark, turquoise blue.

Legion Revere Platinum

Revere Platinum is a 100% cotton, 320 gsm, neutral pH, acid free, chlorine free paper with excellent wet strength, no buffering, four deckled edges, and is specially sized for platinum printing with vegetal gelatin surface sizing and Aquapel (AKD) internal sizing. It is quite modestly priced.

Ruscombe Mill Herschel Platinotype

If you are looking for a textured paper with a distinctive handmade look, Herschel fits the bill. This 200 gsm paper is made from 100% linen and has a pebbly, vellum-like surface that absorbs cyanotype chemistry in a textured manner. It is internally sized with AKD, has no added buffering, and no surface sizing such as gelatin. The paper has limited availability in the States but is readily available in Europe from the Ruscombe Mill in France.

Washi and delicate Asian papers

"Washi" refers specifically to the long tradition of handmade *Japanese* paper (wa = Japanese and shi = paper). There are also delicate, handmade papers from China, Thailand, Korea, India, and other countries, too. Washi is usually made from gampi, kozo, sulphite, abaca, cellulose, mitsumata, and other fibers. [Note: the term "rice paper" is not only a misnomer because this paper is not made from rice but it is actually a politically incorrect term to use when referring to these papers, cultures, and the culture of handmade paper in general.]

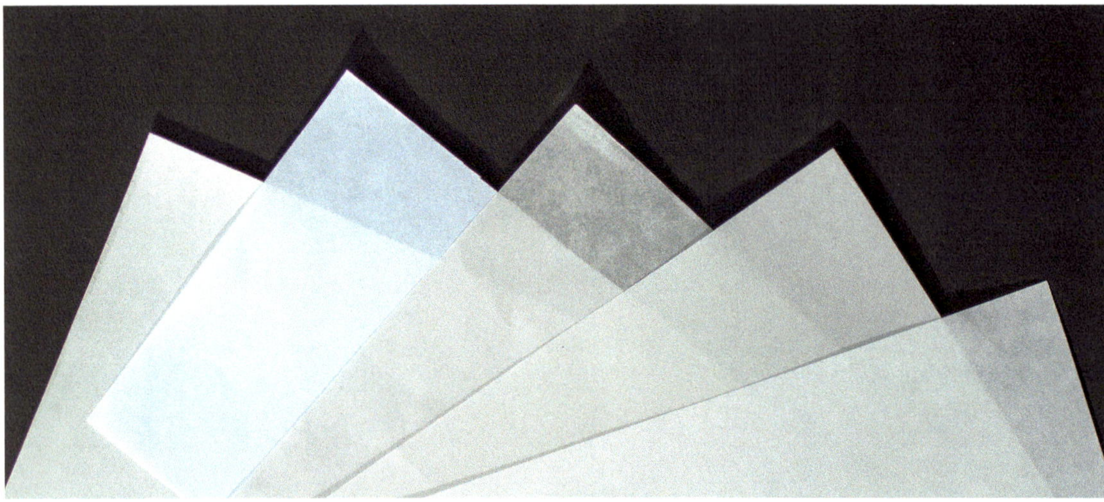

Figure 3.11. From left to right, Japanese Mulberry, Legion Masa, Legion Thai Kozo, Awagami Mitsumata, and Hahnemühle Sumi-e. Note how much thicker Japanese Mulberry is than the others, such as Thai Kozo which is paper-napkin thin. Legion Masa is by far the whitest of these papers; many are a more natural ivory.

Some washi I tested were eliminated because they disintegrated like toilet paper in water, or easily ripped off at the edges. Even though these types of papers can be handled from tray to tray supported with a piece of fiberglass screen or mylar, the following papers didn't require such coddling. Washi may require different coating procedures because of its absorbency; see **Making the Cyanotype Print** for techniques.

Quick peek

Awagami Platinum Gampi 30, 60 gsm (100% gampi, unbuffered)
Awagami Platinum Mitsumata 60 gsm (100% mitsumata, unbuffered)
Hahnemühle Sumi-e 80 gsm (100% cellulose, unbuffered)
Japanese Mulberry/Heavy Kozo 90 gsm (50% kozo/50% abaca, unbuffered)
Legion Goyu 50 gsm (50% kozo/50% sulphite, acid free)
Legion Masa 77 gsm (100% sulphite, neutral pH)
Legion Thai Kozo 35 gsm (100% kozo, unbuffered)

Awagami Platinum Gampi

This paper is actually an unbuffered alt paper. Awagami Platinum Gampi has a unique natural ecru sheen. An organic solution has been hand brushed onto each sheet adding to it a water-resistant quality that also increases the paper's overall wet-strength. It comes in two sizes, approximately 8″ × 11″ and 12″ × 19″, and two weights, 30 and 60 gsm. Buy the 60 gsm. The paper is strong, expensive, and worth it.

Awagami Platinum Mitsumata

This is another unbuffered paper made for alt. Made from mitsumata fibers, this paper has a soft dewy gloss and fine surface texture. An organic solution has been hand brushed onto each sheet adding to it a water-resistant quality that also increases the paper's overall wet-strength. Water resistant, warm white, made without optical brighteners, it comes in two sizes, approximately 8″ × 11″ and 12″ × 19″, and one 60 gsm weight. I prefer the Awagami Gampi.

Hahnemühle Sumi-e

This paper is an 80 gsm, 100% cellulose, natural white, unsized, and also unbuffered paper. It is extremely absorbent and strong; you can actually hang it by clothespins to dry without tearing.

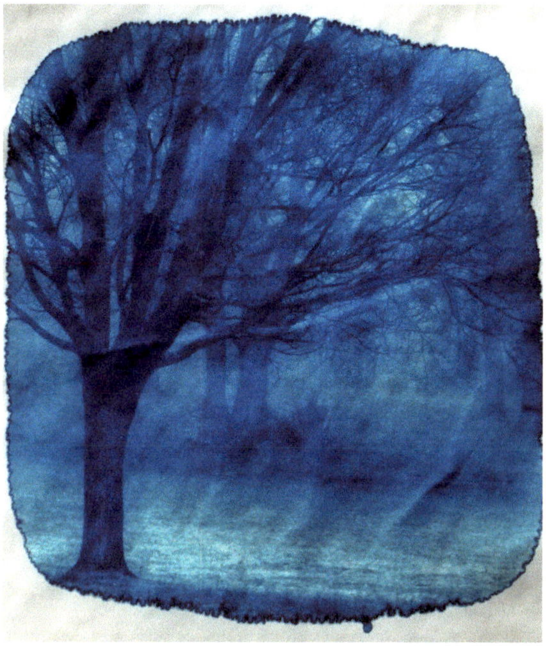

Figure 3.12. *Tree in Fog*, cyanotype on Hahnemühle Sumi-e paper, 8″ x 10″ © Diana Hooper Bloomfield 2018. Note the frilly edge effect that happens with sumi-e paper.

The back and front are hard to tell apart, but the back has a slightly more fibrous surface. It prints a smooth, deep, turquoise blue. Sumi-e comes in 19.68″ × 25.59″ sheets and—way handier—in 20-sheet pads. Each sheet in the pad is interleaved between a thin piece of white paper, ultra-protected and sturdy. The pad sizes are 9.44″ × 12.59″ and 11.81″ × 15.74″ and come at a student-friendly price.

Japanese Mulberry/Heavy Kozo

This pH neutral, unbuffered, natural white, smooth paper, available from Bostick and Sullivan, has a wove surface with a light satin or velvet-like sheen. At 90 gsm this paper is very strong when wet. The paper is made from 50% kozo and 50% abaca and comes in 12½″ × 19″ sheets with no deckles. It is one of my favorite papers.

Legion Goyu

This 50 gsm paper is a handmade, acid free, kozo/sulphite, neutral pH paper with four deckle edges. It has a natural white color, smooth surface, and is semi-translucent when held up to light. Subtle laid lines from the paper mold are visible but they don't print.

Legion Masa

This 77 gsm paper is a very affordable Japanese paper. It is a 100% sulphite, bright white, neutral pH paper with both internal and surface sizing. One side is smooth with a slight sheen. The other side is fibrous with tooth, two very different surfaces. Some object to the brightness of the paper but it looks great with cyanotype.

Legion Thai Kozo

This 35 gsm paper is machine made in Thailand of 100% kozo and features two natural deckles and two cut edges. Even though a mere 35 gsm, I found it quite strong in wet processing.

Vellum papers

There are papers that are referred to as vellum merely because their surface texture resembles the vellum from ancient times which was made from the skin of a calf. Then there are papers that are labeled vellum because of their transparency. Two of the papers listed here are translucent, not transparent—Clearprint and Bienfang Graphics 360. Three are transparent—Opalux, Vidalon, and Wyndstone. I've also included two imitation parchment papers for fun—Pergamenata, which is a milky-transparent speckled vellum, and Parchtone which is a mottled gold paper.

Some vellum papers take a bit of getting used to when working with them for the first time. They zing up into a scroll when coating unless taped down. The edges will cockle. Certain measures such as a) coating a smaller portion of the center of the paper and cutting off the cockled borders, b) using an iron or dry mount press on medium heat to flatten during the different stages, and c) drying the finished print between blotter paper weighted down with books may be in order. For a unique look, these papers are worth a try.

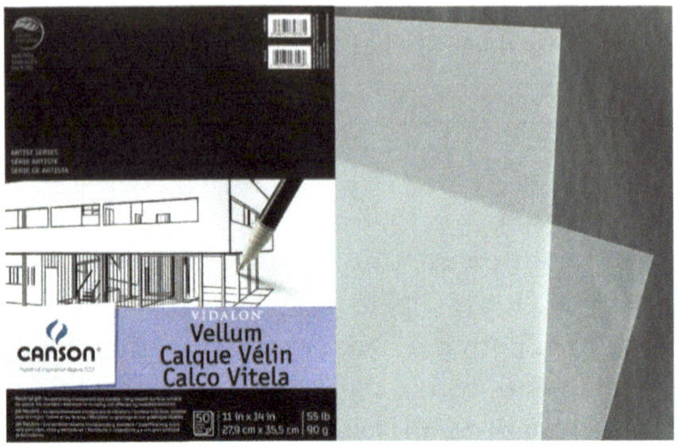

Figure 3.13. Canson has two vellum papers, a thinner Vidalon vellum in pads and a thicker Opalux vellum. Pictured to the right of the pad is Opalux on top of Vidalon on top of a Kodak gray card. Both of these papers print quite translucent.

Figure 3.14. Here is a comparison between Pergamenata, top, and Legion Parchtone, bottom, on top of a piece of soft white and black paper to illustrate transparency. Both are imitation parchment papers but they behave quite differently. Pergamenata is more like a vellum paper. Parchtone is more like a regular, smooth, very gold toned paper. Both are somewhat translucent.

Quick peek

Bienfang Graphics 360, 50 gsm (100% cotton rag, unspecified)
Canson Opalux Vellum 180 gsm (100% alpha cellulose, unspecified)
Canson Vidalon Vellum 90 gsm (100% alpha cellulose, pH neutral)
Clearprint 1020H Drafting Vellum 80 gsm (100% cotton, unspecified)
Legion Parchtone 90 gsm (100% sulphite, unspecified)
Pergamenata Parchment 110, 160, and 230 gsm (100% sulphite, pH neutral)
Wyndstone Vellum Tracing Paper 21, 29, and 40 lb (recycled chemical wood, acid free)

Bienfang Graphics 360

This 50 gsm 100% cotton rag translucent paper comes in convenient 9″ × 12″, 11″ × 14″, 14″ × 17″, and 19″ × 24″ pads. At 25¢ for a sheet of 11″ × 14″ cotton paper it is a steal.

Canson Opalux Vellum

This 100% alpha cellulose, translucent vellum has a satin finish, and is very strong and thick. It comes in a 180 gsm weight and a 19″ × 24″ size. It's a stubborn paper but worth printing on for its unique translucent look.

Canson Vidalon Vellum

90 gsm, 100% alpha cellulose, Vidalon vellum is half the thickness and half the stubbornness of Opalux but surprisingly sturdy for such a thin paper. It would be a good paper to buy when starting to work with vellum papers because it is inexpensive, comes in pads, and is easier to coat.

Clearprint 1020H Drafting Vellum

Clearprint comes in a number of weights and formats, from pads to rolls. The 1000H is 60 gsm/16 lb, the 1020H is 90 gsm/24 lb, and there is a heavier weight one of 180 gsm/48 lb. I prefer the 1020H. Pad sizes are 9″ × 12″, 11″ × 14″, 14″ × 17″, and 19″ × 24″. This paper is 100% cotton, archival, and prints beautifully.

Legion Parchtone

Parchtone is an imitation printable parchment-type paper, a semi-translucent golden yellow. The paper has a smooth surface with that slightly modulated parchment look. It is sized both internally and externally, 100% sulphite, and is archival. It is a fun paper to use when you want to experiment with a very gold color and parchment look.

Pergamenata Parchment

This Italian-made paper simulates aged parchment with the crisp snap, mottled look, and hard feel you would expect from an actual sheet of animal parchment. It is a translucent with an ever-so-slight speckle, 100% sulphite, neutral pH paper, available in white and natural and in 3 weights (110, 160, and 230 gsm).

Wyndstone Vellum Tracing Paper

This vellum comes in three weights: 2.5 point/21.25 lb, 3 point/29 lb/42 gsm, and 5 point/40 lb. It comes in two sizes, 19″ × 25″ and 25″ × 38″. It works similarly to Canson Opalux, but is not as readily available as the Canson vellums, unfortunately.

Bristol papers

Bristol papers are usually measured in plies, e.g. 2-ply, 3 ply, and 4-ply. 2, 3, or 4 sheets of paper are glued together to form the various thicknesses. 2-ply is 260 gsm and is my preference. 3-ply is 390 gsm, heftier than a typical watercolor paper, and 4-ply is 520 gsm, way overkill. Bristol can come in vellum or plate finishes and plate is the smoothest. These three Bristol papers print smooth, sharp, and dark. My favorite is Legion Rising Drawing Bristol. Canson Bristol Recycled is ultra-white and a great first foray paper, but I am not sure of its archival qualities nor its makeup. I do know it is not good for new cyanotype because some of the fibers turn the chemistry dark blue immediately and print very fibrous and speckly.

Quick peek

Canson Bristol Recycled 260 gsm (30% post-consumer materials, unspecified)
Legion Rising Drawing Bristol 2 ply 250 gsm (100% sulphite, unspecified)
Strathmore 500 Series Bristol 2 ply 260 gsm (100% cotton, unspecified)

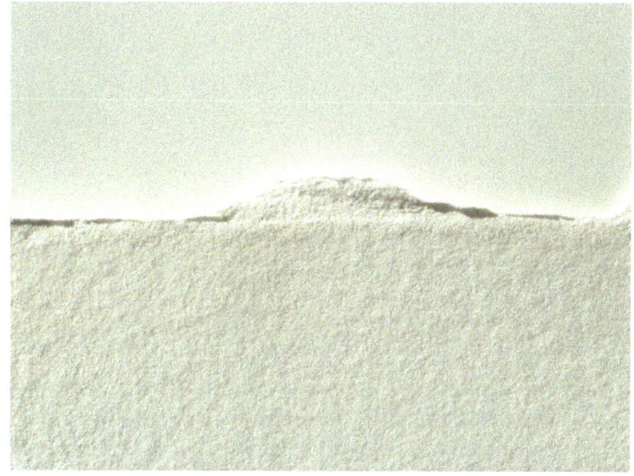

Figure 3.15. Top, Bristol board; bottom, cold pressed paper. Bristol board is ultra-smooth. Cold pressed paper is always textured, but cold press textures can range from very minute eggshell-like bumps to a large pebbly surface.

Canson Bristol Recycled

This 260 gsm paper is probably the whitest paper I tested, making all other papers look off white by comparison. It contains 30% post-consumer materials, making it the only recycled Bristol on the market. One surface is smooth and one is vellum textured. It is available in multiple sizes and pads.

Legion Rising Drawing Bristol

Rising Drawing Bristol's surface texture is either plate or vellum. Plate finish is as smooth as glass, vellum finish is a lightly textured surface. The paper is made from 100% sulphite, neutral pH, acid free, chlorine free, and heavily sized. Oddly, new cyanotype will show a slight striated texture that classic does not. I love this paper.

Strathmore 500 Series Bristol

This 100% cotton Bristol also comes in an ultra-smooth plate finish or a vellum surface with a more modulated, toothy finish. It comes in 2-, 3-, or 4-ply, in 23″ × 29″ and 30″ × 40″ sizes, as well as convenient 11″ × 14″ and 14″ × 17″ pads. It is the only cotton Bristol of these three.

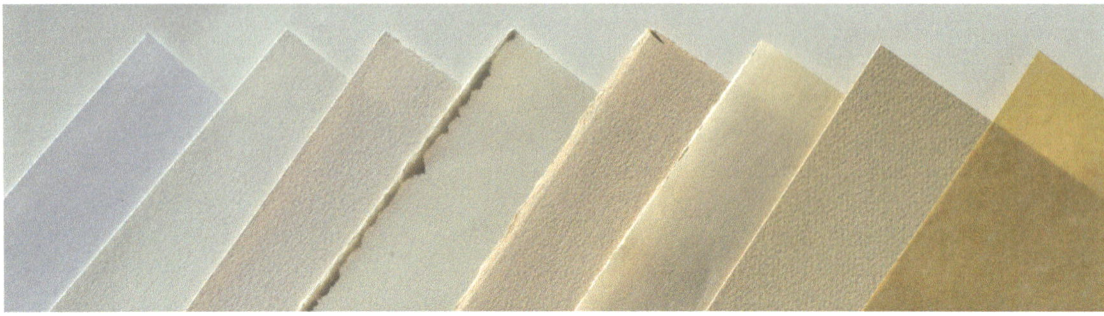

Figure 3.16. Cream or natural papers complement cyanotype's blue. The color can run the gamut from almost white to tan. The above papers are placed on top of a piece of Hahnemühle Platinum Rag for a comparison to a natural white paper. From left to right is Canson Bristol Recycled (to compare papers to an ultra-white), Coventry Rag, Arches Cover white, Rives BFK 175 cream, Arches Cover cream, Zerkall Frankfurt, Bockingford oatmeal, and Legion Parchtone to compare to a gold paper.

Papers for your first foray

The following papers represent a range of weights from very thin to watercolor weight, and at the time of this writing, all are under 40¢ an 11″ × 14″. They are great entry-level papers to hone your cyanotype practice because they are easy to coat, come in convenient pads, and are readily available in big box stores like Michaels. Content and buffering are unspecified—never a good thing—but for practice prints, they work wonderfully.

Quick peek

Canson Mix Media 160 gsm (unspecified)
Canson Pro Layout Marker 70 gsm (unspecified)
Canson Universal Sketch 96 gsm (unspecified)
Canson XL Watercolor 300 gsm (unspecified)

Canson Mix Media

This very white, smooth, 160 gsm paper comes in spiral bound pads of 5½″ × 8½″ up to 18″ × 24″ and when printed "long and strong" with an extended exposure of 48 minutes it is deep, dark, and contrasty. It is thicker and more substantial than Pro Marker and U-Sketch, and clears quickly, so if time is not of the essence or you are exposing out in the sun, this paper is a good choice.

Canson Pro Layout Marker

This 70 gsm quite white paper is available in pads 9″ × 12″, 11″ × 14″, 14″ × 17″, 19″ × 24″, and is an acid free, very smooth paper, whiter than Canson Universal Sketch and quite a bit thinner. It wrinkles and curls up when coating and can coat unevenly, so use a loaded brush to coat; don't pour it on. I prefer U-Sketch.

Canson Universal Sketch

This 96 gsm paper coats easily, prints smooth and fast, and develops quickly. It comes in a spiral bound pad, is slightly off-white and has the feel of a sturdy typing paper. It is my preference of the thinner papers in this group.

Canson XL Watercolor

300 gsm Canson XL, the typical weight of a watercolor paper, has a unique cold press surface with a sort of striated texture. It is acid free and prints a deep navy blue and is quite sharp. You can't beat its price for a watercolor weight paper, albeit with an unspecified content.

Papers with color options

A number of the papers in this category have multiple color choices aside from the typical white/ivory/cream colors, which is a great option.

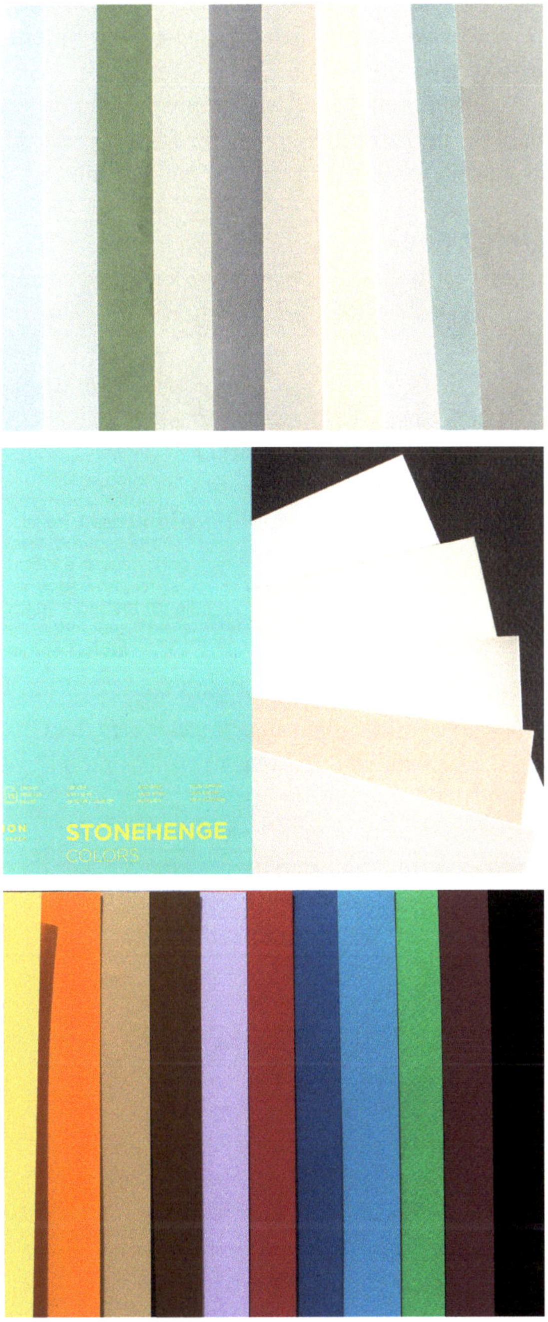

Figures 3.17–3.19. Top is Canson Mi-Teintes which comes in 42 shades and in color packages grouped into bright, fine art, pastel, muted, and portrait colors. The pastel-themed color pack, shown here, is the lightest and most suitable for cyanotype. Middle is Stonehenge Colors, which provides multiple colors in one pad. Bottom is a Fabriano Tiziano "pochette" of colors.

Quick peek

Canson Mi-Teintes 160 gsm (50% cotton, acid free)
Fabriano Tiziano 160 gsm (40% cotton/60% sulphite, pH neutral)
Legion Stonehenge Colors 250/320 gsm (100% cotton, unspecified)
St. Armand Colors 320 gsm (100% cotton, unbuffered)
St. Armand Old Masters 90 gsm (100% cotton, unbuffered)
Zerkall German Ingres 90 gsm (100% cellulose, pH neutral)

Canson Mi-Teintes

This paper has a vellum texture on the top side and a flat surface on the reverse, sort of like waffle-weave versus eggshell (best). It is 160 gsm, 50% cotton, and comes in 42 shades colored in the pulp to be highly light resistant and acid free. Mi-Teintes papers are available in individual 19″ × 25″ sheets and in 25-sheet packages of 8½″ × 11″ sheets, also color packages grouped into bright, fine art, pastel, muted, and portrait colors. The pastel-themed color pack is the most suitable for cyanotype.

Fabriano Tiziano

Tiziano has a natural woven herringbone-like texture with more texture on one side of the paper, giving a choice of surfaces. It is made from 40% cotton, weighs 160 gsm, has a neutral pH, and comes in 30 lightfast colors. It is similar to Canson Mi-Teintes in regards to weight, color offerings, and quality, but Mi-Teintes is smoother on one side.

Legion Stonehenge Colors

This paper comes in two weights, 250 gsm and 320 gsm. It is a 100% cotton deckled paper made to rival the more expensive European moldmade papers. It is buffered with calcium carbonate, but for some reason prints classic cyanotype well with acid development, with a long, smooth, and deep blue exposure scale. It comes in black, cream, fawn, kraft, natural, pale blue, pearl gray, polar white, steel gray, warm white, and white. The lighter, warm colors juxtapose nicely with cyanotype's blue. It is internally sized and has a smooth vellum finish.

St. Armand Colors

The St. Armand colored papers are made from cotton and also linen, flax, straw, and jute. The pulp is internally sized. The sheets have four deckled edges and come in white and eighteen colors: white, red, tomato red, blue, black, gold, ivory, sand, brown, dark brown, slate gray, charcoal, caribou, elephant gray, prune, olive, spring green, mineral green, and pink. Papers are 22″ × 30″ and 320 gsm. This is a very textured paper, so if smooth is your preference, stick with St. Armand Old Masters.

St. Armand Old Masters

These papers are thin at 90 gsm and made from linen and cotton rag. The paper is internally sized with four deckled edges. This paper comes in white, ivory, tan, taupe, ochre, light blue, black, denim blue, and jade green. Sheet size is 18″ × 24″. I love Frobisher which is a white paper with a beautiful eggshell texture.

Zerkall German Ingres

This 10% cotton/90% high alpha cellulose, acid free paper has an antique look and comes in fourteen colors. The paper has a lightly textured laid surface and four deckles.

And the list goes on…

The following category contains papers that can be used "as is," without pre-acidification discussed further along in this chapter, or they are too delicate to go through pre-acidification and further processing. The papers may or may not contain buffering agents; manufacturers don't always specify if, how, how much, or with what. Where a manufacturer specifies, it is noted. These listed have printed well in classic cyanotype even though they may contain buffering; I'm not sure why some buffered papers print well, where others don't, but new cyanotype is always the acid test (no pun intended) for the presence of buffering. Note on the paper chart which of these are not suitable for new cyanotype.

I am quite enamored with thinner weight papers because they absorb better, make for less wash off, and print, for the most part, smoother tones.

Quick Peek

Papers in italics sit on the border between not needing and needing pre-acidification.

Arches Aquarelle 300 gsm (100% cotton, alkaline reserve)
Arches Lavis Fidelis/En Tout Cas 220/280 gsm (25% cotton, 75% sulphite, pH neutral)
Arches Text Wove 120 gsm (100% cotton, pH neutral)
Canson Edition Etching 250 gsm (100% cotton, unspecified)
Canson Montval (300 gsm, acid free)
Fabriano Unica 250 gsm (50% cotton, acid free)
Garzapapel Drawing 180 gsm (100% cotton, unbuffered)
Garzapapel Ink-Pen 140 gsm (100% cotton, unbuffered)
Hahnemühle Cézanne 300 gsm (100% cotton rag, calcium carbonate)
Hahnemühle Turner 300 gsm (100% cotton rag, calcium carbonate)
Indigo 200/300/450/650 gsm (100% cotton, buffered)
Indigo 200/300/450/650 gsm (75% cotton/25% flax, buffered)
Khadi Cotton and Plant Papers 210 gsm (100% cotton rag and plant fibers, unbuffered)
Khadi Cotton Rag Paper 150/210/320/640 gsm (100% cotton rag, unbuffered)
Khadi Lightweight Cotton Rag 100 gsm (100% cotton rag, unbuffered)
Legion Bamboo 265 gsm (90% bamboo/10% cotton, calcium carbonate)
Legion Coventry Rag 290 gsm (100% cotton, calcium carbonate)
Legion Domestic Etching 175 gsm (50% cotton/50% sulphite, calcium carbonate)
Legion Folio 250 gsm (100% cotton, acid free)

Legion Lenox 100/250 gsm
(100% cotton, unspecified)
Magnani Acquerello 300 gsm
(100% cotton, pH neutral)
Magnani Pescia 300 gsm
(100% cotton, pH neutral)
Magnani Revere Book 120/175 gsm
(100% cotton, pH neutral)
Rives Heavyweight 175 gsm
(100% cotton, unspecified)
Rives Lightweight 115 gsm
(100% cotton, pH neutral)
Ruscombe Mill Chateau Vellum 110/180 gsm
(100% linen, pH neutral)
Schut Noblesse 300 gsm (100% cotton, acid free)
St. Armand White Cotton Rag 320 gsm
(100% cotton, unbuffered)
Twinrocker Papers 80/115/260/400/460 gsm
(100% cotton, cotton rag, etc., unbuffered)
Velke Losiny Moldau 280/360 gsm
(60% cotton, 40% linen, pH neutral)
Velke Losiny Moravia 280/360 gsm
(85% cotton, 15% linen, pH neutral)
Velke Losiny Prague 120/170 gsm
(85% cotton, 15% linen, pH neutral)
Zerkall Book 100/145 gsm (10% cotton/90% cellulose, calcium carbonate buffered)
Zerkall Frankfurt 120 gsm (cotton/high alpha cellulose, calcium carbonate)
Zerkall Nideggen 120 gsm
(cotton/high alpha cellulose, pH neutral)

Arches Aquarelle/Watercolor

This 100% cotton watercolor paper is gelatin sized through and through. It has a neutral pH, is acid free, chlorine free, with no optical brighteners. Sheets have four deckles. The paper comes in two colors, natural and bright white, and three surfaces, cold, hot, and rough press.

Arches Lavis Fidelis/En Tout Cas

This unique paper is moldmade in France of 25% cotton and 75% sulphite, neutral pH and has one side cold pressed and the other side hot pressed. White sheets are 19″ × 25½″.

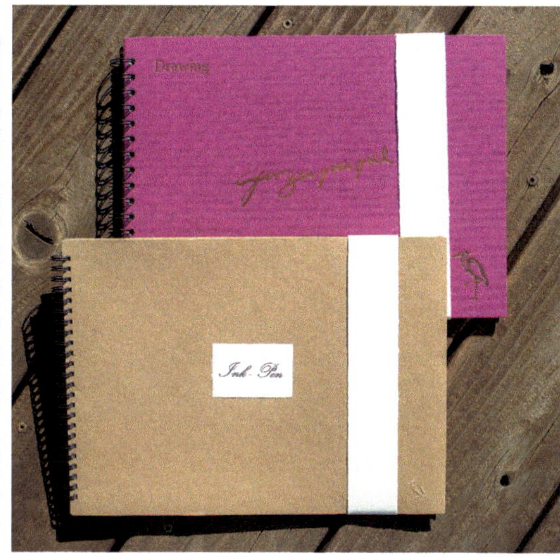

Figure 3.20. Garzapapel does the most beautiful packaging of their handmade paper. The paper is spiral bound but each sheet has a tear perforation that results is a fourth soft deckle. Both are suitable for new cyanotype.

Arches Text Wove

This is a thin, crisp, softly textured, eggshell-like, fine-grained, cold pressed surface in a natural white, 120 gsm 100% cotton paper. It is nicely absorbent without leaking through to the back. It is acid free and has four deckles. Sheet sizes are 19½″ × 25½″ and 25½″ × 40″. It prints a beautiful deep blue with vinegar or citric acid development.

Canson Edition Etching

Edition Etching comes in four colors—bright white, antique white, cream, and wheat—and is a 250 gsm 100% cotton moldmade paper from France. It has two distinct surfaces, one smooth, one textured.

Canson Montval

Canson Montval watercolor paper is a French-made student watercolor paper with cut edges on all sides. It comes in convenient tape-bound and wire-bound pads, watercolor blocks, and 22″ × 30″ watercolor sheets. It only comes in cold press or rough but the cold press is surprisingly sharp. It prints deep blues and is very reasonably priced for a watercolor paper.

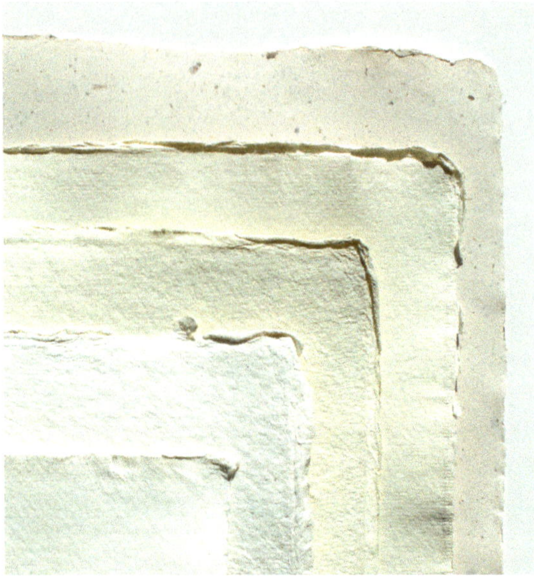

Figure 3.21. The handmade papers here all have distinctive deckled edges and textured surfaces. From top to bottom: Khadi banana paper, Khadi Cotton Rag, and three Indigo papers.

Fabriano Unica

Unica is a 250 gsm acid free, environmentally friendly paper made from 50% cotton. It is available in sheets or pads in ivory and white.

Garzapapel Drawing or Ink-Pen

Garzapapel is a very soft, elegant, natural white, velvety paper that comes in handcrafted pads or sheets. The paper is 100% recycled cotton with no added buffering as it is slightly alkaline naturally. Deckled on all four sides, it is pressed between wet wool felts on both sides to give an artisanal look. Sizes range from about 5″ × 8″ to 40″ × 60″ and from 180 to 600 gsm. I prefer the Drawing paper, a very smooth cold press, and the deckled edges are soft and fuzzy. The paper is made in Spain.

Hahnemühle Cézanne

This 300 gsm paper is 100% cotton rag, internally and surface-sized, mould-made, natural white, acid free, lightfast and extremely resistant to ageing. It is available in matte, rough, and hot pressed surfaces. Of all the Hahnemühle papers other than the best, Platinum Rag, I prefer Cézanne and Turner.

Hahnemühle Turner

Moldmade in Germany, 100% cotton, natural white, acid free and neutral pH, chlorine free, 300 gsm, Turner has a medium cold-pressed surface. It is internally sized with no surface sizing. Sheets are 21.6″ × 28.7″. I like its velvety quality, slight texture, and ability to print deep blues.

Indigo Papers

Indigo papers come in 100% cotton or 75% cotton/25% flax. They are cold press, acid free, internally and externally sized with "natural" gelatin, and have four natural deckle edges and a watermark. They are environmentally friendly papers; no chlorine, bleach, or harmful chemicals are used in their manufacture. The papers come in various weights such as 200, 300, 450, and 650 gsm—under 300 gsm is plenty—and in various sizes up to 22″ × 30″. The color is very yellow-cream, and the texture is heavily woven. They do contain buffering agents.

Khadi Cotton and Plant Papers

Khadi paper is made from cotton T-shirts. The paper is sized internally with an alkaline wax polymer and also externally tub sized with gelatin. These 210 gsm papers are acid free, rough or smooth surfaced, made with no chlorine or bleaching agents. Sheet size is 22″ × 30″ but paper dimensions are approximate because the paper has irregular deckle edges on all four sides. The cotton and plant papers have added plant-based material for natural color and texture. Gunny paper is made from jute sacking, Banana paper is made from chopped banana leaves, and Bagasse paper is made from sugar cane fiber. Gunny is rough, Banana is rough, and Bagasse is rough and smooth. I printed on the Banana paper and it is a rosy color with distinctive flecks and prominent laid lines for a unique look.

Khadi 100% Cotton Rag

This paper is made in India of 100% cotton. It is internally sized with an alkaline wax polymer and surface-sized with gelatin. It comes in various weights (150, 210, 320, 640 gsm) and sizes (6″ × 8″,

12″ × 16″, 22″ × 30″, 27″ × 39″, 8″ × 12″, and 12″ and 22″ diameters) and smooth and rough surfaces. It has a very handmade look and natural color that complements cyanotype well.

Khadi Lightweight Cotton Rag

This comes in a 100% cotton rag or cotton rag/jute mix for the jute colored paper. The sheet size is 17″ × 25″ and weighs 100 gsm. Surface choices are smooth or medium textured. It is a tan-cream color that complements cyanotype well. It performs like a washi or a thin, text weight, soft, handmade paper. I prefer this paper out of the three Khadi offerings listed here.

Legion Bamboo

This paper is an eco-friendly 90% bamboo/10% cotton paper with an acid free neutral pH. It comes in three weights; stick with the thin 105 gsm text weight or the sturdy 265 gsm cover weight because the 530 gsm is more like posterboard.

Legion Coventry Rag

This slightly cream-white paper is internally sized with a modest amount of surface sizing and is dimensionally stable. It has two surfaces, textured vellum and smooth. It is 100% cotton, neutral pH, acid free, and buffered with calcium carbonate.

Legion Domestic Etching

The paper is smooth on one side and has a slight vellum finish on the other. It comes in a larger size, 26″ × 40″, and is 175 gsm. It is internally and externally sized, has a neutral pH, is acid free, is a cream white color that complements cyanotype's blue, and is made from 50% cotton/50% sulphite.

Legion Folio

Folio is a versatile, American-made 100% cotton paper with a slight vellum finish. The bright white has two deckled edges and two cut edges. The antique white has four cut edges. It is neutral pH, acid free, and 250 gsm.

Figure 3.22. Sometimes papers with laid lines like this print well and sometimes they do not. The bottom paper, Twinrocker Yale, prints beautifully and the inch-wide parallel lines do not show up in the final print. Other times those inch-wide lines show up and do not work with an image as was the case with Canson Ingres, an otherwise beautiful paper.

Legion Lenox 100

This paper is a 100% cotton paper with no deckles, is off white, medium textured, 250 gsm, dimensionally stable, and with a neutral pH.

Magnani Acquerello

This watercolor paper from Italy is 100% cotton, and available in the two typical watercolor paper weights and surface textures, a cold press (Italia) which is a less-textured cold press, and a hot press (Portofino). It has a neutral pH, is acid free, chlorine free, has two natural and two tear deckles (sheets), and is internally and externally sized. Buy this in sheets because the watercolor block performs differently.

Magnani Pescia

This smooth, 300 gsm paper is moldmade in Italy from 100% cotton. It is chlorine free, with a neutral pH. Deckles can vary from no to two natural and two tear deckles. It comes gray, white, cream, soft white, and light blue in 22″ × 30″.

Magnani Revere Book

This is a 100% cotton, moldmade, neutral pH, chlorine free paper, internally sized, with two natural and two tear deckles. It is available in two weights, 120 and 175 gsm, and two colors, white and ivory. Sheet size is 26″ × 40″, and it is quite reasonably priced considering its large size which is 1½× the area of a 22″× 30″. The paper prints a *beautiful* blue both in classic cyanotype and new cyanotype; one of my favorites.

Rives Heavyweight

This is not a particularly heavy weight paper since it is only 175 gsm. It is a 100% cotton paper, moldmade in France, neutral pH, internally sized, with two natural deckles and two tear deckles, and comes in white and cream (buff) and sheet sizes of 19″ × 26″ or 26″ × 40″. The paper is actually sturdy enough to pre-acidify but I found it worked well enough un-acidified with acid development. Like all Rives papers, the feel is soft, absorbent, and velvety but it can tend to speckle and require extra solution when coating.

Rives Lightweight

Rives Lightweight is 115 gsm, 100% cotton, moldmade in France, acid free, neutral pH, internally sized, with two natural deckles and two tear deckles. It comes in white and cream (buff) and two sheet sizes of 19″ × 26″ and 26″ × 40″. Like all Rives papers, the feel is soft, absorbent, and velvety but it can tend to speckle and require extra solution when coating. Citric acid development worked well with this paper.

Ruscombe Mill Chateau Vellum

This paper is made from 100% linen, is pH neutral, and is wet and dry cold-pressed to produce a very smooth surface. It comes in bright white and 110 and 180 gsm. Sheet size is 22″ × 30″. This is an elegant paper, much the same as Herschel but unfortunately it does have buffering, so Herschel is the better choice.

Schut Noblesse

This is the premier watercolor paper produced by the Schut mill in the Netherlands. The surface of this 300 gsm paper has undergone a proprietary gelatin sizing process. Made from 100% cotton, acid free, off-white, it comes in sheets and blocks, cold press only.

Figure 3.23. Cream colored paper can range from a peach-cream to a yellow-cream, often not visible unless the papers are placed side by side.

St. Armand White Cotton Paper

This paper is made of 100% cotton rag in a large variety of sizes and shapes with double internal starch sizing and non-animal gelatin surface sizing. The paper varies in weight from 320 to 640 gsm and comes in extra rough, medium, and smooth. It is very textured so buy smooth. Watch clearing! St. Armand papers clear slowly.

Twinrocker Papers

These papers are handmade, sheet by sheet, in Brookston Indiana, and come in all sorts of sizes, weights, and colors. When first looking at the Twinrocker website, it can be a bit daunting because there are so many choices, but here's how to navigate it: the names such as B-Buff, Biscuit, Dickinson, refer to different colors, not different papers. That narrows down the choices considerably because for cyanotype you'll probably want to purchase from white to light tan colors (color offerings go all the way to charcoal).

Papers come in six weights: thin text (TT/80 gsm), text weight (T/115 gsm), heavy text weight (HT/200 gsm), light art weight (LA/260 gsm), art weight (A/400 gsm), and heavy art weight (HA/460 gsm). TT is very thin, like a washi, though sturdy. Art weight and heavy art weight are thicker than necessary. I prefer T, HT, and LA weights.

There are choices in paper composition. Cotton and cotton rag are my preferred choices because they print smoothly and evenly.

Papers come in rough, cold, or hot press. Hot press requires a nominal charge of 5–10% because the paper is run through a calendering press. You can choose to have only one side calendered for both a cold and hot pressed surface on one sheet of paper.

In summary my advice is to purchase the cotton or cotton rag papers, calendered on one side, in T, HT, or LA weights, in these colors: May Linen, Cream (these first two are my favorites), Da Vinci, Dickinson, Oyster, Bone, Yale, B-Buff, Biscuit, and White.

All Twinrocker papers are neutral pH without added buffering agents and internally sized with vegan-friendly size. The papers I tested are nicely absorbent; on even the thinnest papers, the solution beads up ever so slightly on the paper surface, and then absorbs into the paper without shooting through to the back. The internal sizing is nicely proportioned. Only the watercolor papers have an additional surface size, made from animal gelatin.

The papers print fast, beautiful turquoise blue, and with a long exposure scale. What intrigues me about Twinrocker papers aside from all that handmade goodness and cyanotype performance is the range of sizes and colors, all with beautiful deckled edges that point to their elegant handmade quality. These are papers you will want to frame showing the paper edges. Note: not all papers are available at all times. It costs $500 each time Twinrocker makes a fresh paper batch, even before one sellable sheet is created. The website is updated weekly with what is in stock. If you are printing a series of cyanotypes that need to match, I recommend buying a quantity of sheets at once.

Velke Losiny Moldau

Moldau 280/360 gsm white paper is handmade in the Czech Republic of 60% cotton and 40% linen, pH neutral, has four natural deckles, a wove surface and a watermark. It is very textured, so if texture is not your thing, use the less textured Moravia or Prague. The Velke Losiny papers clear slowly so watch this. Moldau is available in 11½″ × 16½″, 16½″ × 23″, and 22″ × 30″.

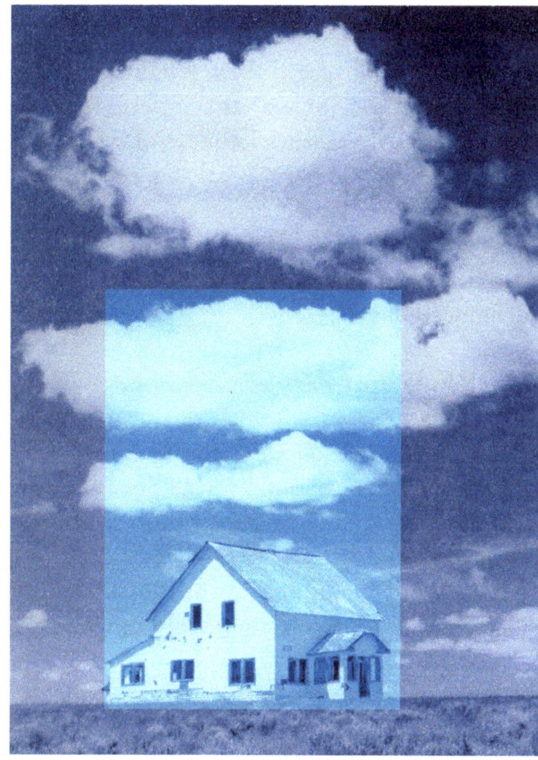

Figure 3.24. This illustrates the difference between a paper that prints more turquoise blue (in this case Hahnemühle Platinum Rag) versus a paper that prints more royal blue to blue-gray (any of the alkaline papers). The differences are usually not this radical, but generally a paper that is slow to print and produces a more gray-blue than green-blue color probably has some alkalinity involved.

Velke Losiny Moravia

Moravia is handmade in the Czech Republic of 85% cotton and 15% linen, pH neutral, four natural deckles and a watermark. It comes in 240 and 360 gsm weights, and a wove and vellum surface. Sizes are 8″ × 11½″, 11½″ × 16½″, 16½″ × 23″, and 22″ × 30″. The Velke Losiny papers clear slowly so watch this.

Velke Losiny Prague

Prague is 85% cotton/15% linen and handmade in the Czech Republic by Velke Losiny. Sheets have four deckles and come in white and cream, a laid surface, and 120 and 170 gsm weights. Sizes are 8″ × 11½″, 11½″ × 16½″, and 16½″ × 23″. Prague, for a thin paper, is still quite textured. This is my *favorite* of the Velke Losiny offerings here. The Velke Losiny papers clear slowly so watch this.

Zerkall Book

This paper is available in three weights (100 and 145 gsm wove, 125 gsm laid), two surfaces (smooth and vellum), three shades (white, soft white, and cream), and two sizes (20″ × 30″ and 24″ × 34″). It is moldmade in Germany of 10% cotton/ 90% high alpha cellulose. It has a neutral pH, is acid free, chlorine free, has four deckles, and is buffered with calcium carbonate.

Zerkall Frankfurt

This is a lightweight, cotton/high alpha cellulose, 120 gsm paper with four deckled edges that comes in two sizes, 22″ × 30″ and 25″ × 38″, two surfaces, vellum and smooth, and two shades, white and cream. It has a distinct wavy-lined laid pattern throughout the entire sheet, which makes for an intriguing surface with certain subject matter.

Zerkall Nideggen

Nideggen is a unique lightweight 120 gsm paper in a deep sand color (as deep as a brown paper bag) with the defining characteristic of irregular wavy laid lines. It is made from cotton and high alpha cellulose, has a neutral pH, is acid free, chlorine free, and has two natural and two tear deckles.

Pre-acidifying buffered papers

If the paper is buffered with calcium carbonate, it can be problematic for cyanotype. It will be slow to expose, result in a dullish gray-blue color, and over time the buffering can fade the cyanotype. Most papers are buffered nowadays, and some print so poorly that they're worthless for cyanotype unless a pre-acid bath of sulfamic acid is used to make the paper suitable. A sulfamic acid bath combines with the calcium carbonate in the paper to make water-soluble calcium sulfamate, which easily washes out of the paper and down the drain. There are plenty of papers already discussed that do not need this extra step, but there are a few excellent papers that are worth pre-acidifying: Fabriano Artistico and Saunders Waterford are two that come to mind.

Sulfamic acid has these benefits:[4]

- It is easily available at any hardware store
- It is a dry, odorless, non-volatile chemical
- It is moderately soluble in water
- In water it forms a very acid solution, as strong as mineral acids
- It is less corrosive than other acids (safer to use with stainless steel and uncoated iron materials)
- It is non-hygroscopic and can be stored in powder form indefinitely
- The reaction with the paper's alkaline buffer produces only co2 gas, and no hazardous gas is released
- The reaction's by-product is highly water-soluble calcium sulfamate so no water-insoluble by-product is produced
- It is cost effective and neutralizes the buffer completely with no loss of image quality
- The treated papers become suitable for every alkaline-sensitive alt process, with no loss of Dmax
- It does not damage the paper's fibers much like the other acids.

I know it is a pain to have to do one more step, but some papers are worth it. If it seems like too much work, stick with the papers already discussed.

Sulfamic acid pre-bath

Quadruple this to make a gallon if doing a lot of paper at once. Take caution when mixing and using sulfamic acid! Wear gloves and eye protection. It is a caustic cleaner. Note that this is a 10% solution, but others use as low as a 1–5% solution.

100 g sulfamic acid (about 6 tablespoons)
1000 ml water

1. Fill a graduate with 750 ml water.
2. Add the sulfamic acid to the water (AAATW *always add acid to water*, not the other way around!) and stir.
3. Add enough water to make 1000 ml.
4. Keep stirring until it is dissolved, which takes a bit. It is "moderately" soluble.
5. Once dissolved, pour into a tray.
6. Meanwhile, presoak paper in plain water until completely wet, 20–30 minutes.
7. Transfer the paper to the 10% sulfamic acid solution for 15–20 minutes, interleaving the sheets every several minutes. Some papers acidify more quickly.

8. Once the fizzing noise stops, transfer the paper to plain water and wash it for at least 20-30 minutes to wash out all the calcium sulfamate. In fact, wash the paper as long as possible because if all the acid is not washed out, the paper will fog blue stain in the highlights. Then hang to dry.

Quick Peek

Arches Cover 250 gsm (100% cotton, calcium carbonate)
Arches Johannot 240 gsm (75% cotton/25% esparto, unspecified)
Arnhem 1618, 245/320 gsm (100% cotton rag, acid free)
Bee Watercolor Paper 190 gsm (100% cotton, acid free)
Bockingford 300 gsm (100% alpha cellulose, calcium carbonate)
Canson Heritage 300 gsm (100% cotton, unspecified)
Fabriano Artistico 300 gsm (100% cotton, pH neutral)
Fabriano Tiepolo 290 gsm (100% cotton, acid free)
Fluid 100, 300 gsm (100% cotton, unspecified)
Hahnemühle Bamboo 125/265 gsm (90% bamboo/10% cotton, calcium carbonate)
Hahnemühle Lanaquarelle 300 gsm (100% cotton rag, calcium carbonate)
Legion Rising Barrier 250 gsm (100% cotton, calcium carbonate)
Magnani Vergata 130 gsm (100% cotton, pH neutral)
Rives BFK 280 gsm (100% cotton, calcium carbonate)
Rives de Lin 270 gsm (85% cotton/15% linen, buffered)
Saunders Waterford 190/300/425/640 gsm (100% cotton, calcium carbonate)
Somerset Satin White 250/300 gsm (100% cotton, calcium carbonate)
Winsor & Newton Professional Watercolor 300 gsm (100% cotton, acid free)

Arches Cover

Arches Cover is a moldmade, 100% cotton paper, internally sized and minimally surface sized. It has a lightly textured cold-press surface, with two natural and two tear deckle edges. Cover is dimensionally stable, acid free, chlorine free, very soft and absorbent, and thus it requires a lot of coating solution or it'll print grainy. It is buffered with calcium carbonate.

Arches Johannot

Johannot is a 240 gsm moldmade paper from France made of 75% cotton and 25% esparto. It is lightly textured, watermarked, buffered, and has two natural and two tear deckles.

Arnhem 1618

Arnhem is a 100% rag, acid free paper with a semi-smooth, vellum surface and strong, highly absorbent fibers. It is internally and externally sized. It comes in warm white and white, and two weights of 245 or 320 gsm, plus various sizes, including a 100 sheet box of the 245 gsm in 8.5″ × 11″ sheets. This size is under 50¢ a sheet, a great price for a cotton paper.

Bee Watercolor Paper

This is a 300 gsm 100% cotton, acid free paper that is internally and externally sized, cold pressed, has excellent wet strength, and comes in a variety of sizes.

Bockingford

This is a 100% alpha cellulose paper, internally and externally sized, and buffered with calcium carbonate. It is a bit rougher than a normal cold press. It comes in colors known as the Bockingford Tints: cream, oatmeal, blue, gray, and eggshell. The paper has a neutral pH, is acid free, chlorine free, with no optical brightening agents, and no deckles.

Canson Heritage

Heritage is a 100% cotton paper with a gelatin-free sizing that comes in cold, hot, or rough press, and sheets or hot press blocks sized 9″ × 12″, 10″ × 14″, 12″ × 16″, and 14″ × 20″.

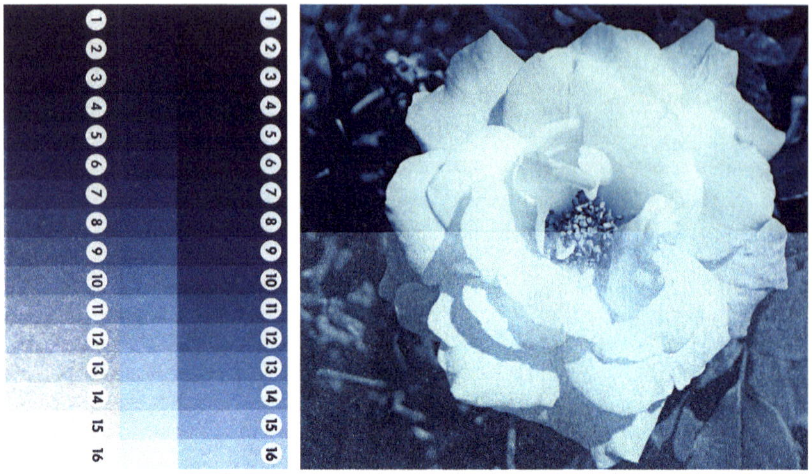

Figures 3.25 and 3.26. Some papers print very grainy. To the left are two paper comparisons, one that prints blue-gray and grainy (left Stouffer's step wedge and bottom half of the rose, Strathmore Aquarius) and one that prints deep blue and smooth (right side Stouffer step wedge and top half of the rose, Arches Platine). Graininess can result from paper fiber composition, paper sizing, or paper alkalinity. There is no need to settle for an unsuitable, grainy paper unless grain is a necessary visual or conceptual part of the image.

Fabriano Artistico

Artistico is a 100% cotton rag watercolor paper made from cotton balls instead of stretched and beaten cotton fibers which creates a more consistent and even water absorption over the surface and makes it strong and durable. Internal and external sizing without animal by-products produces a more consistent, non-yellowing, smell-free, and more resilient non-scuffing surface. Artistico has deckled edges with the "Fabriano+Artistico" watermark. There are no added whiteners, it is chlorine and acid free, and comes in cold, hot, rough, and soft presses, in traditional and extra white colors.

Fabriano Tiepolo

Featuring a traditional Italian soft white color, this 100% cotton 290 gsm, moldmade, acid free paper has four deckle edges and a beautiful watermark, with a particularly smooth surface.

Fluid 100

Fluid approximates more expensive moldmade papers at a good price. It is 100% cotton, acid free, natural white, internally and externally gelatin-sized and comes in 300 and 640 gsm weights and two surfaces, hot or cold press. It has a very smooth, subtle surface texture and two deckled edges.

Hahnemühle Bamboo

This 265 gsm paper is made from 90% bamboo and 10% cotton rag, with a unique surface and a natural white color. Available in sheets and blocks.

Hahnemühle Lanaquarelle

Lanaquarelle is 100% cotton, acid free, internally and vegan-friendly synthetic surface sized, and buffered with calcium carbonate. It is moldmade in France, pH neutral, with two natural and two tear deckles, and comes in cold press, hot press, and rough. It is a natural white color available in 300 and 640 gsm. Whiter than most watercolor papers, the sheet is popular because it is not heavily sized. It comes in sheets and convenient blocks.

Legion Rising Barrier Paper

This paper is a 250 gsm, single-ply paper with a slight textured surface and light surface sizing. It is 100% cotton, has a neutral pH, is acid free, lignin free, chlorine free, and buffered with calcium carbonate.

Magnani Vergata

This Italian 100% cotton paper is made with the cylinder mould machine slowed down considerably to add more textural laid and chain lines. It has a neutral pH, is acid free, chlorine free, and has four deckles.

Rives BFK

Rives BFK is moldmade in France, 100% cotton, neutral pH, acid free, with no optical brighteners. It is a very soft, velvety, absorbent paper, with moderate internal sizing and buffered with calcium carbonate. Each sheet is watermarked and deckles vary. It comes in multiple weights, colors and sheet sizes, but this one is the 22″ × 30″ 280 gsm.

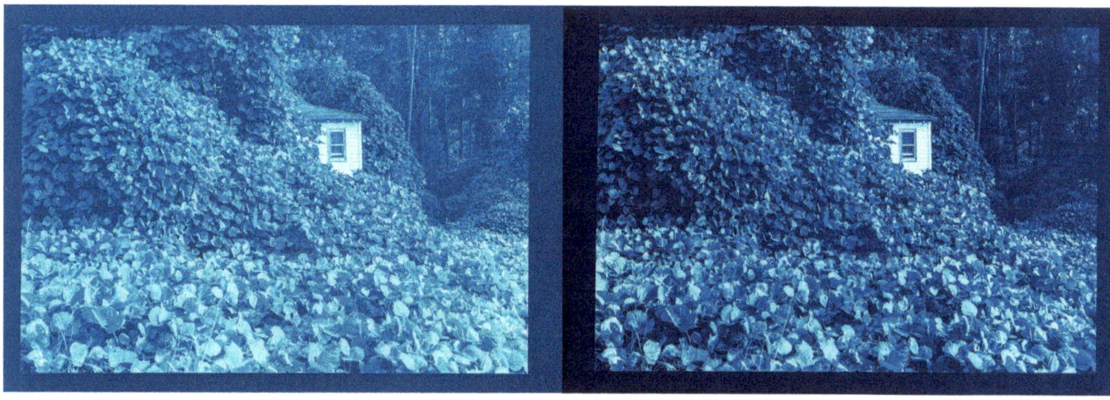

Figure 3.27. *Clemson House* © Christina Z. Anderson 2018. Both prints are made with the same negative, paper, and development (vinegar) but the left print was made with a 10/10 classic cyanotype solution (CC) and the right print with a combination classic/new cyanotype solution (SWC). The Dmax of the left print measures 1.36; the right print 1.50. Individually the prints look fully tonal and adequately dark. When compared side by side the difference of a 0.14 in Dmax is readily apparent.

Rives de Lin

Also known as Moulin de Gue, this white paper is a moldmade, acid free and buffered sheet of 85% cotton and 15% linen. It has the look and feel of a handmade paper with long fluffy deckle edges. It is lightly sized with gelatin. Rives de Lin is pH neutral and acid free. The 22″ × 30″ is 270 gsm.

Saunders Waterford

Waterford's attractive surface is created using natural woolen felts that give it a distinctive random texture. It is moldmade in England, 100% cotton, calcium carbonate buffered, two natural, two tear deckles, neutral pH, acid free, chlorine free, no optical brightening agents, internally starch sized and surface sized with gelatin, with a warm color and good wet strength. It is available in 22″ × 30″ sheets of various weights and rough, cold or hot press surfaces.

Somerset Satin

This is 100% cotton, has a neutral pH, is acid free, chlorine free, free of optical brighteners, and internally and surface sized, with two natural deckles. It is buffered with calcium carbonate. Somerset comes in radiant white, white, soft white, buff, antique, cream, and newsprint.

Winsor & Newton Professional Watercolor

This 300/640 gsm 100% cotton paper is moldmade and comes in sheets and blocks, cold, hot and rough press. It is acid free, with a bright, natural whiteness without optical brighteners. [Note: Winsor & Newton **Classic** is made from 100% cellulose and is only cold press, and a different paper than Winsor & Newton Professional.]

Dmax—the holy grail?

Dmax is a term that refers to *maximum density*, or how dark the deepest tone is in a print. A densitometer is used to take numeric measurements of Dmax and thus it enables it to be quantified and compared. Greater Dmax does not mean greater virtuosity in printmaking necessarily, because Dmax is affected by a number of factors outside a cyanotypist's control. And since color is never seen in a vacuum, some papers, though they may be lower in Dmax compared to others, will still seem beautifully dark, or as Dick Arentz of platinum fame termed it, "convincing black." We want prints that look convincingly deep and rich, fully tonal, with fully detailed deep blues and highest highlights. To that end, there are ways to maximize Dmax:

- *Choose the right paper*, and unbuffered papers or acidified buffered papers often deliver the greatest Dmax
- If you can't work in a humid environment, prehumidify the paper; if it is too dry the chemistry dries too quickly on the paper surface and wash off during development occurs

- Use more ferric ammonium citrate in the mix; 15–20% FAC will deliver slightly greater Dmax than 10% though other problems may occur such as grain and washoff.
- Use new cyanotype which often results in greater Dmax especially with sulfamic acid development
- Dilute the chemistry less so there is more chemical on the paper (i.e. new cyanotype diluted to the standard 100 ml instead of 150–200 ml)
- Use Tween 20™ with new cyanotype which will allow more chemistry to absorb into the paper
- Develop classic cyanotype in vinegar or citric acid and new cyanotype in citric or sulfamic acid, the latter producing the greatest Dmax with new cyanotype
- Develop fully and in an acidic bath but watch overall development time so it isn't longer than necessary; with too long development, cyanotype will wash out of the print
- Make sure your water is not alkaline; sometimes city water can be alkaline and will bleach the print. If this is the case, acidify all water washes to be slightly acid.

Dmax chart

Dmax was measured on all 136 papers. Around 30% of the papers were eliminated for various reasons such as unsuitability, unpleasing color, sizing issues, extreme grain and mottling, and even a few for cost compared to quality received.

Each paper was printed in classic cyanotype 10/10 and new cyanotype diluted to 200 ml. The particular strengths of formulas chosen was to avoid the inherent problems in each process: grain and washoff in classic cyanotype and splotching in new cyanotype. Thus if a 20/10 classic or a 100–150 ml new cyanotype formula is used, the resulting Dmax will most likely be higher.

Each formula was developed four different ways—no acid, vinegar, citric, and sulfamic—for a total of eight step wedges per paper. Whatever development produced the greatest Dmax is noted here, marked with an N, V, C, or S. For both cyanotype formulas it was often sulfamic acid development, but with classic cyanotype highlight staining has to be watched closely when using sulfamic acid, and vinegar and citric development may therefore be a better choice. *Never did non-acid development produce the greatest Dmax.* Vinegar development for new or SWC cyanotype is not only unsuitable but rarely produced the greatest Dmax. Although none of these measurements are cut in stone—you'd have to print many step wedges of each paper in a particular developer to reliably quantify—patterns emerge suggesting the suitability of at least vinegar and citric acid development for classic cyanotype and sulfamic acid development for new cyanotype.

Dmax was measured with a calibrated X-Rite 810 Densitometer on Visual (V) mode. The column on the far right is paper Dmin measured from the paper's "white." Dmax - Dmin = Density Range (DR), for those who want to calculate DR.

The higher the number, the darker the blue. A difference of 0.02 is hardly noticeable. A difference of 0.05 can be seen in a side-by-side comparison. 0.10 is noticeable. All prints looked maximum blue individually, but *comparatively*, Dmax above 1.4 is *very* dark. Note that a few of my favorite Twinrocker papers don't result in the highest Dmax. High Dmax is not always an indicator of excellent papers and, in fact, sometimes happens on unsuitable ones.

PAPER NAME	CC-Dmax	NC-Dmax	Paper Dmin
Arches Aquarelle	1.55s	1.53s	0.06
Arches Cover	1.48s	1.38c	0.05
Arches Johannot	1.35v	1.54s	0.05
Arches Lavis Fidelis	1.50c	1.46s	0.05
Arches Platine	1.52v	1.52v	0.04
Arches Platine Lightweight	1.59s	1.54s	0.06
Arches Text Wove	1.47c	1.39s	0.08
Arnhem 1618	1.49s	1.49s	0.05
Awagami Platinum Gampi	1.64s	1.66v	0.20
Bee Watercolor	1.41v	1.56s	0.04
Bergger Cot 160	1.42s	1.35s	0.04
Bergger Cot 320	1.42c	1.46c	0.06
Bienfang 360	1.35s	1.46s	0.09
Bockingford	1.44c	1.57c	0.03
Canson Bristol Recycled	1.49s	1.59s	0.06
Canson Edition Etching	1.36s	1.43c	0.05

PAPER NAME	CC-Dmax	NC-Dmax	Paper Dmin
Canson Heritage	1.42v	1.63s	0.04
Canson Mi-Teintes	1.60v	1.35s	0.09
Canson Mix Media	1.45s	1.49v	0.07
Canson Montval	1.47s	1.49s	0.04
Canson Opalux Vellum	1.48s	1.54v	0.38
Canson Pro Layout Marker	1.22v	1.43s	0.14
Canson Universal Sketch	1.25v	1.44s	0.09
Canson Vidalon Vellum	1.55s	1.52s	0.38
Canson XL Watercolor	1.53s	1.58s	0.04
Clearprint 1020HP Vellum	1.42s	1.45s	0.18
Fabriano Artistico	1.50v	1.48s	0.03
Fabriano Tiepolo	1.36s	1.41c	0.05
Fabriano Tiziano	1.46v	1.57s	0.09
Fabriano Unica	1.50s	1.46s	0.07
Fluid 100 HP	1.45s	1.51s	0.05
Garzapapel Drawing	1.43s	1.38s	0.06
Garzapapel Ink-Pen	1.42v	1.59s	0.07
Hahnemühle Bamboo	1.34c	1.49s	0.05
Hahnemühle Cézanne	1.49s	1.55s	0.04
Hahnemühle Lanaquarelle	1.48s	1.62s	0.04
Hahnemühle Platinum Rag	1.50c	1.63s	0.04
Hahnemühle Sumi-e	1.52c	1.59s	0.13
Hahnemühle Turner	1.60s	1.56s	0.05
Indigo 100% cotton	1.34c	1.26s	0.06
Japanese Mulberry/Heavy Kozo	1.50c	1.62s	0.11
Khadi Cotton and Plant Paper	1.43c	1.44s	0.13
Khadi Cotton Rag	1.48s	1.50s	0.09
Khadi Cotton Rag Lightweight	1.38s	1.58s	0.13
Legion Bamboo	1.53s	1.32c	0.04
Legion Coventry Rag	1.43s	1.49s	0.04
Legion Domestic Etching	1.48v	1.49s	0.07
Legion Folio	1.42v	1.74s	0.05
Legion Goyu	1.40c	1.58s	0.25
Legion Lenox 100	1.59s	1.49s	0.04
Legion Masa	1.55s	1.65s	0.15
Legion Parchtone	1.45c	1.57s	0.17
Legion Revere Platinum	1.37c	1.53s	0.04
Legion Rising Barrier	1.31s	1.38s	0.03
Legion Rising Drawing Bristol	1.45c	1.62s	0.05
Legion Stonehenge	1.40s	1.42s	0.05
Legion Stonehenge Colors	1.42c	1.43c	0.06
Legion Thai Kozo	1.45c	1.66s	0.24
Magnani Acquerello Portofino	1.44s	1.56s	0.04

PAPER NAME	CC-Dmax	NC-Dmax	Paper Dmin
Magnani Pescia	1.48s	1.43s	0.05
Magnani Revere Book 175 gsm	1.52v	1.54s	0.05
Magnani Vergata	1.59s	1.18v	0.11
Pergamenata Parchment	1.53s	1.55s	0.22
Rives BFK 280 gsm	1.33v	1.26c	0.04
Rives BFK Heavyweight 175 gsm	1.51v	1.63s	0.06
Rives de Lin	1.50s	1.49s	0.04
Rives Lightweight 115 gsm	1.32c	1.46s	0.05
Ruscombe Mill Chateau Vellum	1.47s	1.37s	0.08
Ruscombe Mill Herschel Pl	1.28s	1.46s	0.06
Saunders Waterford	1.36s	1.48s	0.07
Schut Noblesse	1.61s	1.42s	0.03
Somerset Satin	1.34v	1.40s	0.04
St. Armand Colors	1.44c	1.50s	0.04
St. Armand Old Masters	1.25c	1.37s	0.07
St. Armand White Rag	1.44c	1.39s	0.04
Strathmore 500 Bristol	1.42v	1.51s	0.05
Twinrocker B-Buff T	1.32c	1.04c	0.09
Twinrocker Biscuit HT	1.26s	1.28v	0.12
Twinrocker Bleached Abaca HT	1.46c	1.46c	0.11
Twinrocker Bone T	1.25s	1.27s	0.12
Twinrocker Cream LA	1.32c	1.30s	0.06
Twinrocker Da Vinci T	1.26s	1.28s	0.12
Twinrocker Dickinson TT	1.25c	1.24s	0.15
Twinrocker May Linen LA	1.47c	1.28s	0.07
Twinrocker Oyster HT	1.34c	1.44s	0.10
Twinrocker White Cotton Rag TT	1.16c	1.21c	0.10
Twinrocker Yale T	1.39s	1.37s	0.11
Velke Losiny Moldau	1.53c	1.41s	0.05
Velke Losiny Moravia	1.42v	1.48s	0.04
Velke Losiny Prague	1.51s	1.44s	0.09
Winsor & Newton Watercolor	1.53s	1.56s	0.04
Wyndstone Vellum	1.51c	1.60s	0.38
Zerkall Book	1.41s	1.47s	0.09
Zerkall Frankfurt	1.49c	1.44c	0.09
Zerkall German Ingres	1.43c	1.53s	0.10
Zerkall Nideggen	1.43c	1.53s	0.25

Quick peek paper chart

This is a summation of the various papers tested. As always, a caveat: take this chart as an overall guide but YRMV—your results may vary.

Although the chart is self-explanatory, there are a few things to note. One, sizing is not always specified and often proprietary, but a vegan customer can always ask if the paper is vegan friendly and get an answer (which means it is or is not sized with animal gelatin). Two, whether or not and how a paper is buffered is often not specified, or specified generically (pH neutral, acid free) but some manufacturers do say their paper is calcium carbonate buffered and it is best to pre-acidify these papers if possible. Three, under CC? note three answers: yes, acidify, or B. The B signifies the paper printed well enough unacidified with classic cyanotype, but was perhaps *b*orderline in need of acidification. Four, although all papers listed are good for classic cyanotype (unsuitable papers have been eliminated from this chapter), darker green highlighted papers are good for both classic and new cyanotype and lighter green are good for both but may require citric acid added to the new cyanotype formula if fogging is a problem.

PAPER NAME	CC?	NC?	MATERIAL	SIZING	BUFFERED?	SURFACE	COLOR
Arches Aquarelle 300 gsm	Yes	Citric	100% cotton	Gelatin	Alkaline reserve	CP/HP	Bright white, natural white
Arches Cover 250 gsm	Acidify	Acidify	100% cotton	Internal, external	Calcium carbonate	Textured	White, cream, black
Arches Johannot 240 gsm	Acidify	Acidify	75% cotton/ 25% esparto	Unspecified	Unspecified	Textured	White
Arches Lavis Fidelis 220/280 gsm	Yes	Yes	25% cotton/ 75% sulphite	Gelatin sized	pH neutral	Two surfaces	White
Arches Platine 145/310 gsm	Yes	Yes	100% cotton	External starch	No buffering	Smooth	White
Arches Text Wove 119 gsm	Yes	Citric	100% cotton	Surface gelatin sized	pH neutral	Textured	Natural white
Arnhem 1618 245/320 gsm	Acidify	Acidify	100% cotton rag	Internal/ external	Acid free	Semi-smooth vellum	White, warm white
Awagami Platinum Gampi 30/60 gsm	Yes	Acidify	100% gampi	Organic external	No buffering	Smooth	Ecru
Awagami Platinum Mitsumata 60 gsm	Yes	Acidify	100% mitsumata	Organic external	No buffering	Smooth	Natural
Bee Watercolor 190 gsm	Acidify	Acidify	100% cotton	Internal/ external	Acid free	CP	White
Bergger Cot 320 160/320 gsm	Yes	Yes	100% cotton	External gelatin	No buffering	HP	White
Bienfang Graphics 360 50 gsm	Yes	Yes	100% cotton rag	Unspecified	Unspecified	Smooth	White
Bockingford 300 gsm	Acidify	Acidify	100% alpha cellulose	Internal/ external	Calcium carbonate	Textured	Blue, cream, eggshell, gray, oatmeal, white
Canson Bristol Recycled 260 gsm	Yes	Acidify	30% post-consumer materials	Unspecified	Unspecified	Vellum/ Smooth	Very bright white
Canson Edition Etching 250 gsm	B	Acidify	100% cotton	Unspecified	Unspecified	Two surfaces	Antique white, bright white
Canson Heritage 300 gsm	Acidify	Acidify	100% cotton	Gelatin-free sizing	Unspecified	CP/HP/R	White
Canson Mi-Teintes 160 gsm	Yes	Yes	66% cotton	Unspecified	Acid free	Vellum	42 colors
Canson Mix Media 160 gsm	Yes	Acidify	Unspecified	Unspecified	Acid free	Smooth	White

PAPER NAME	CC?	NC?	MATERIAL	SIZING	BUFFERED?	SURFACE	COLOR
Canson Montval (300 gsm)	Yes	Yes	Unspecified	Unspecified	Acid free	CP/R	Natural white
Canson Opalux Vellum 180 gsm	Yes	Yes	100% alpha cellulose	Unspecified	Unspecified	Vellum	Transparent white
Canson Pro Layout Marker 70 gsm	Yes	Yes	Unspecified	Unspecified	Unspecified	Smooth	White
Canson Universal Sketch 96 gsm	Yes	Acidify	Unspecified	Unspecified	Unspecified	Smooth	White
Canson Vidalon Vellum 90 gsm	Yes	Yes	100% alpha cellulose	Unspecified	pH neutral	Vellum	Transparent white
Canson XL Watercolor 300 gsm	Yes	Acidify	Unspecified	Unspecified	Acid free	CP	White
Clearprint Vellum 1000H/1020H 60/80 gsm	Yes	Yes	100% cotton	Unspecified	Unspecified	Vellum	Transparent white
Fabriano Artistico 300 gsm	Acidify	Acidify	100% cotton	Internal/ external synthetic	pH neutral	CP/SP/HP	Traditional white, extra white
Fabriano Tiepolo 290 gsm	Acidify	Acidify	100% cotton	Unspecified	Acid free	Smooth	Soft white
Fabriano Tiziano 160 gsm	Yes	Yes	40% cotton/ 60% sulphite	Unspecified	pH neutral	Textured	30+ colors
Fabriano Unica 250 gsm	B	Acidify	50% cotton	Unspecified	Acid free	Smooth	White, ivory
Fluid 100 300 gsm	Acidify	Acidify	100% cotton	Internal/ external gelatin	Unspecified	HP/CP	White
Garzapapel Drawing 180 gsm	Yes	Yes	100% recycled cotton	Unspecified	Acid free, no added buf.	CP	Natural white
Garzapapel Ink-Pen 140 gsm	Yes	Yes	100% recycled cotton	Unspecified	Acid free, no added buf.	Medium grain	White
Hahnemühle Bamboo 265 gsm	Acidify	Acidify	90% bamboo 10% cotton rag	Material sized	Calcium carbonate	Cold pressed	Natural white
Hahnemühle Cézanne 300 gsm	Yes	Citric	100% cotton rag	Surface, material sized	Calcium carbonate	Matte/ R/HP	Natural white
Hahnemühle Lanaquarelle 300 gsm	Acidify	Acidify	100% cotton rag	Surface, material sized	Calcium carbonate	CP/HP/R	Bright white
Hahnemühle Platinum Rag 300 gsm	Yes	Yes	100% cotton rag	Material sized	Unbuffered	HP	White
Hahnemühle Sumi-e 80 gsm	Yes	Yes	100% cellulose	Unsized	Unbuffered	Very smooth	Off-white
Hahnemühle Turner	Yes	Citric	100% cotton rag	Internal	Calcium carbonate	CP	Natural white
Indigo	Yes	Acidify	100% cotton or 75% cotton/25% flax	Internal, external	Acid free	CP	Natural
Japanese Mulberry/ Heavyweight Kozo 90 gsm	Yes	Yes	50% kozo/ 50% abaca	Unspecified	Unbuffered	Smooth	Cream
Khadi Lightweight	Yes	Yes	100% cotton	Internal and external	Acid free	Smooth	Natural white
Khadi Plant paper	Yes	Citric	100% cotton etc.	Internal and external	Acid free	CP	Natural white
Khadi Watercolor 210, 320, 640 gsm	Yes	Citric	100% cotton	Internal neutral/ external gelatin	Unbuffered	Smooth/ Rough	Off-white/creamy
Legion Bamboo 105/265/530 gsm	Yes	Acidify	90% bamboo/ 10% cotton	Internal AKD/ external vegetal gelatin	Calcium carbonate	Textured	White
Legion Coventry Rag 290 gsm	B	Acidify	100% cotton	Internal/ external	Calcium carbonate	Textured	Natural white
Legion Domestic Etching 175 gsm	Yes	Citric	50% cotton/ 50% sulphite	Internal/ external	Calcium carbonate	Textured	Natural white
Legion Folio 250 gsm	B	Acidify	100% cotton	Unspecified	Acid free	Vellum	White
Legion Goyu 50 gsm	Yes	Yes	50% kozo/ 50% sulphite	Unspecified	Unspecified	Smooth	Natural white
Legion Lenox 100 250 gsm	B	Acidify	100% cotton	Unspecified	Unspecified	Textured	Natural white

PAPER NAME	CC?	NC?	MATERIAL	SIZING	BUFFERED?	SURFACE	COLOR
Legion Masa 77 gsm	Yes	Yes	100% sulphite	Unspecified	Unspecified	Smooth	Natural white
Legion Parchtone 90 gsm	Yes	Yes	100% sulphite	Internal/ external	Unspecified	Vellum	Parchment-colored (yellow cream)
Legion Revere Platinum 300 gsm	Yes	Yes	100% cotton	Internal AKD/ external vegetal gelatin	Unspecified	HP	White
Legion Rising Barrier 250 gsm	Acidify	Acidify	100% cotton	Internal AKD/ external vegetal gelatin	Unspecified	Textured	White
Legion Rising Drawing Bristol 2 ply 250 gsm	Yes	Yes	100% sulphite	Heavily sized, unspecified	Unspecified	Bristol	White
Legion Stonehenge 250/320 gsm	Yes	Acidify	100% cotton	Internal AKD, external vegetal gelatin	Unspecified	Smooth	Black, cream, fawn, pearl gray, kraft, natural, pale blue, polar white, steel gray, warm white, white
Legion Thai Kozo 35 gsm	Yes	Acidify	100% kozo	Unspecified	Unspecified	Smooth	Natural white
Magnani Acquerello 300 gsm	B	Acidify	100% cotton	Internal and external	pH neutral	CP/HP	White
Magnani Pescia 300 gsm	B	Citric	100% cotton	Unspecified	pH neutral	Smooth	gray, white, cream, soft white, light blue
Magnani Revere Book 120/175 gsm	Yes	Yes	100% cotton	Internal	pH neutral	Smooth	White, ivory
Magnani Vergata 130 gsm	Acidify	Acidify	100% cotton	Unspecified	pH neutral	Texture laid/ chain	Ivory
Pergamenata parchment 110/160/230 gsm	Yes	Yes	100% sulphite	Unspecified	pH neutral	Vellum	White, natural
Rives BFK 280 gsm	Acidify	Acidify	100% cotton	Internal/lightly sized	Calcium carbonate	Textured	gray, white, black, cream, tan, off white
Rives de Lin 270 gsm	Acidify	Acidify	85% cotton/ 15% linen	Lightly sized gelatin	Buffered	Textured	White
Rives Heavyweight 175 gsm	Yes	Citric	100% cotton	Internal	Unspecified	Textured	White, buff
Rives Lightweight 115 gsm	B	Citric	100% cotton	Internal	pH neutral	Textured	White, buff
Ruscombe Mill Chateau Vellum 110/180 gsm	Yes	Acidify	100% linen	Unspecified	pH neutral	Smooth	Bright white
Ruscombe Mill Herschel Platinotype 200 gsm	Yes	Yes	100% linen	Internal AKD, no external	No buffering	Textured	White
Saunders Waterford 90/140/200 lb	Acidify	Acidify	100% cotton	Internal starch, external gelatin	Calcium carbonate	CP/HP/R	White, high white
Schut Noblesse 300 gsm	B	Citric	100% cotton	External gelatin	Acid free	CP	Off white
Somerset Satin 250/300 gsm	Acidify	Acidify	100% cotton	Internal and external	Calcium carbonate	Satin, Textured, Velvet	White, cream, soft white, antique, radiant white, buff, newsprint gray
St. Armand Colors 320 gsm	Yes	Yes	100% cotton	Internally sized	No added buffering	CP	White, gold, ivory, 17 colors in all
St. Armand Old Masters 90 gsm	Yes	Yes	Linen and cotton rag	Unspecified	Unspecified	Laid	White, ivory, tan, taupe, ochre, light blue, black denim
St. Armand White Rag 320 gsm	Yes	Yes	100% cotton	Starch based double internal, external non-animal gelatin	No added buffering	CP	White
Strathmore 500 Series Bristol 2 ply 260 gsm	Yes	Acidify	100% cotton	Unspecified	Unspecified	Plate/ Vellum	White
Twinrocker B Buff T/HP	Yes	Citric	100% cotton	Internal gelatin size	Unbuffered	Smooth	Pale cream
Twinrocker Biscuit HT/HP	Yes	Citric	100% cotton rag	Internal gelatin size	Unbuffered	Smooth	Deeper cream
Twinrocker Bleached Abaca HT/HP	Yes	Acidify	100% abaca	Internal gelatin size	Unbuffered	Smooth	Pale cream

PAPER NAME	CC?	NC?	MATERIAL	SIZING	BUFFERED?	SURFACE	COLOR
Twinrocker Bone T/HP	Yes	Yes	100% cotton	Internal gelatin	Unbuffered	Smooth	Pale cream
Twinrocker Cream LA/HP	Yes	Citric	100% cotton	Internal gelatin	Unbuffered	Smooth	Pale cream
Twinrocker Da Vinci T/HP	Yes	Yes	100% cotton rag	Internal gelatin	Unbuffered	Smooth	Pale cream
Twinrocker Dickinson TT/HP	Yes	Citric	100% cotton	Internal gelatin	Unbuffered	Smooth	Deeper cream
Twinrocker May Linen LA/HP	Yes	Citric	100% cotton	Internal gelatin	Unbuffered	Smooth	Pale cream
Twinrocker Oyster HT/HP	Yes	Citric	100% cotton	Internal gelatin	Unbuffered	Smooth	Pale cream
Twinrocker Watercolor	Yes	Citric	100% cotton	Internal/ external gelatin	No added buffering	Textured	White
Twinrocker White Cotton Rag TT/HP	Yes	Citric	100% cotton rag	Internal gelatin	Unbuffered	Smooth	White
Twinrocker Yale T/HP	Yes	Citric	100% cotton rag	Internal gelatin	Unbuffered	Smooth	Pale cream
Velke Losiny Moldau 360 gsm	Yes	Yes	60% cotton/ 40% linen	Unspecified	pH neutral	Wove finish	White
Velke Losiny Moravia 360 gsm	Yes	Yes	85% cotton/ 15% linen	Unspecified	pH neutral	Vellum	White
Velke Losiny Prague 170 gsm	Yes	Yes	85% cotton/ 15% linen	Unspecified	pH neutral	Laid	White, cream
Winsor & Newton Professional	Acidify	Acidify	100% cotton	Unspecified	Acid free	CP/HP/R	White
Wyndstone Vellum Tracing 21, 29, 40 lb	Yes	Yes	Recycled chemical wood	Unspecified	Acid free	Vellum	Transparent white
Zerkall Book 100, 145 gsm	Yes	Citric	10% cotton/ 90% cellulose	Neutral sized	Calcium carbonate	Vellum, smooth	White, cream, soft white
Zerkall Frankfurt 120 gsm	Yes	Yes	Cotton/high alpha cellulose	Neutral sized	Calcium carbonate	Wavy laid	White, cream
Zerkall German Ingres 90 gsm	Yes	Yes	100% cellulose	Unspecified	pH neutral	Textured	White, gray, bluest, cream, sand, earth, tan, orange, scarlet, mint, light blue, light green, ivory
Zerkall Nideggen 120 gsm	Yes	Yes	Cotton/high alpha cellulose	Unspecified	pH neutral	Wavy laid	Sand

Endnotes

1. Ware, Mike. *Cyanomicon II, History, Science and Art of Cyanotype: photographic printing in Prussian blue*. Buxton: self-published, 2016, pp. 177, 276.

2. My tests consisted of Stouffer's step wedges printed at a baseline printing time (BPT) of 30 minutes on each paper for classic cyanotype,10 minutes for new cyanotype. From this initial exposure time I could then determine the standard printing time (SPT) for each paper, as papers have different exposure speeds, with some papers printing very slowly and some quite fast. I then developed the step wedges five different ways at first: in plain water, and in water acidified with vinegar, citric, hydrochloric, or sulfamic acid. Halfway through the testing I eliminated hydrochloric acid because of its outgassing nature and sulfamic acid's equal benefits. From these tests I could observe how the acid development affected paper speed, the color blue it printed, how dark the blue was (Dmax), the exposure scale of the paper whether short or long, if the paper printed grainy or smooth, and which acidified water development worked the best for that particular paper. Finally I used one particular image that has a tonal range from extremely bright highlights to deep shadows, and printed it on each paper, using that paper's SPT and the acid development that looked best.

3. Paper descriptions have been culled from various websites and catalogs, the paper manufacturers, and most often Legion Paper, Dick Blick, Jerry's Artarama, and Talas. Most of the time the information is consistent and appears to be paper manufacturers' official paper descriptions, but if there are mistakes, may the reader forgive me.

4. Bilici, Serdar, "Sulfamic acid for paper acidification": 2013 **http://www.apug.org/forums/forum42/118153-sulfamic-acid-paper-acidification.html**. Serdar Bilici wrote up an excellent summation of the sulfamic acid pre-bath and I owe him thanks for all his testing of this chemical. All of the sulfamic acid information is his. There are other chemicals that people use to pre-acidify paper—citric, acetic, hydrochloric, oxalic—and Bilici found sulfamic acid to be the best.

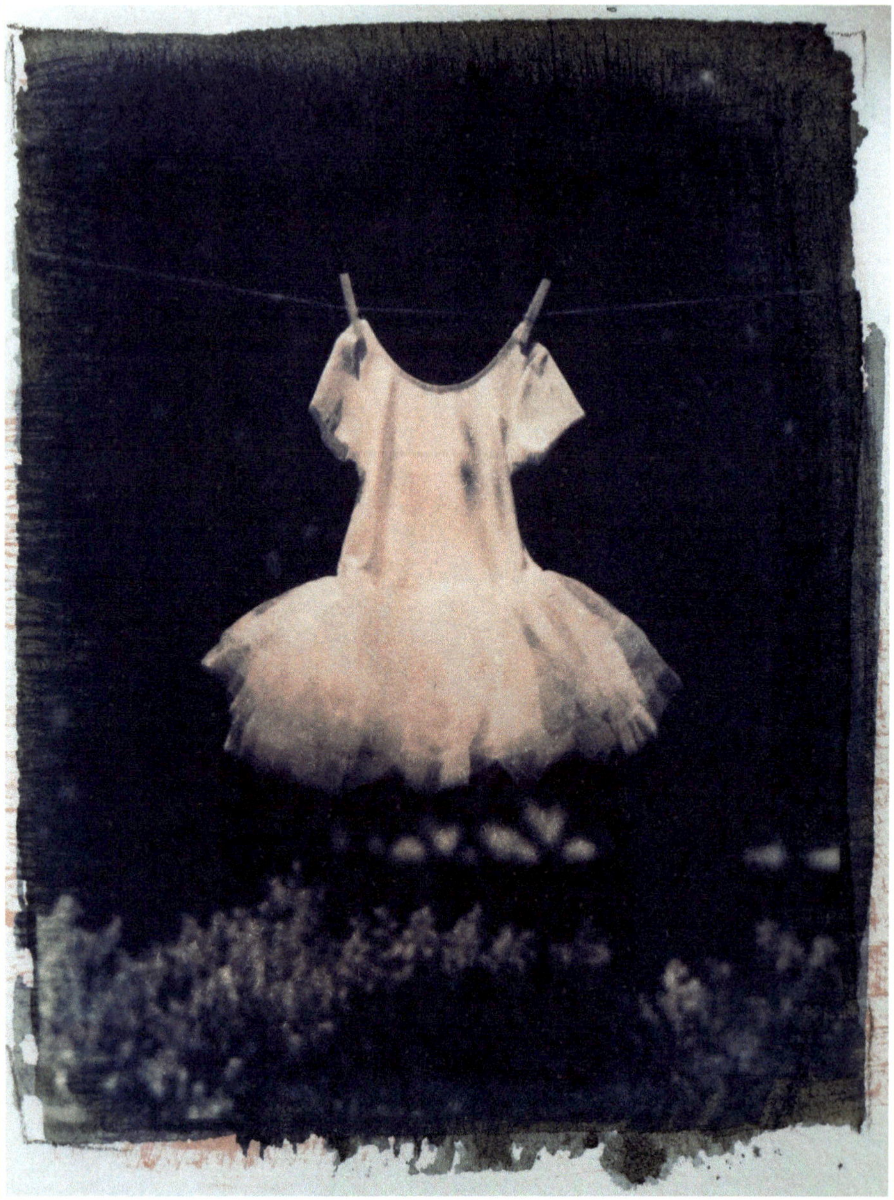

Figure 4.1. *Tutu*, gum wash over classic cyanotype, printed with a 4″ x 5″ film negative © Kim Sinclair 2018. Kim Sinclair is a fine art photographer who has been working with 19th century photographic processes for over twenty years. Born and raised in Toronto, Canada, Sinclair graduated with a diploma in Photomedia in 1996 from the Ontario College of Art & Design. At a time when digital photography was beginning its great revolution, Sinclair continued to work in traditional silver gelatin print methods and historic photographic processes, primarily cyanotype and Vandyke and now carbon, photogravure, and gum bichromate. Sinclair worked and exhibited for nearly ten years in Toronto before moving to Canberra, Australia in 2001, where she continues to work and teach classes in traditional darkroom and alternative photographic practices. Influence from an early drawing and painting background is evident in the loosely brushed, paint-like application Sinclair often utilizes. Multiple imagery, layering techniques and sequencing methods are also arrangements characteristically recognized throughout much of her work.

Chapter 4

Digital Negatives for Cyanotype

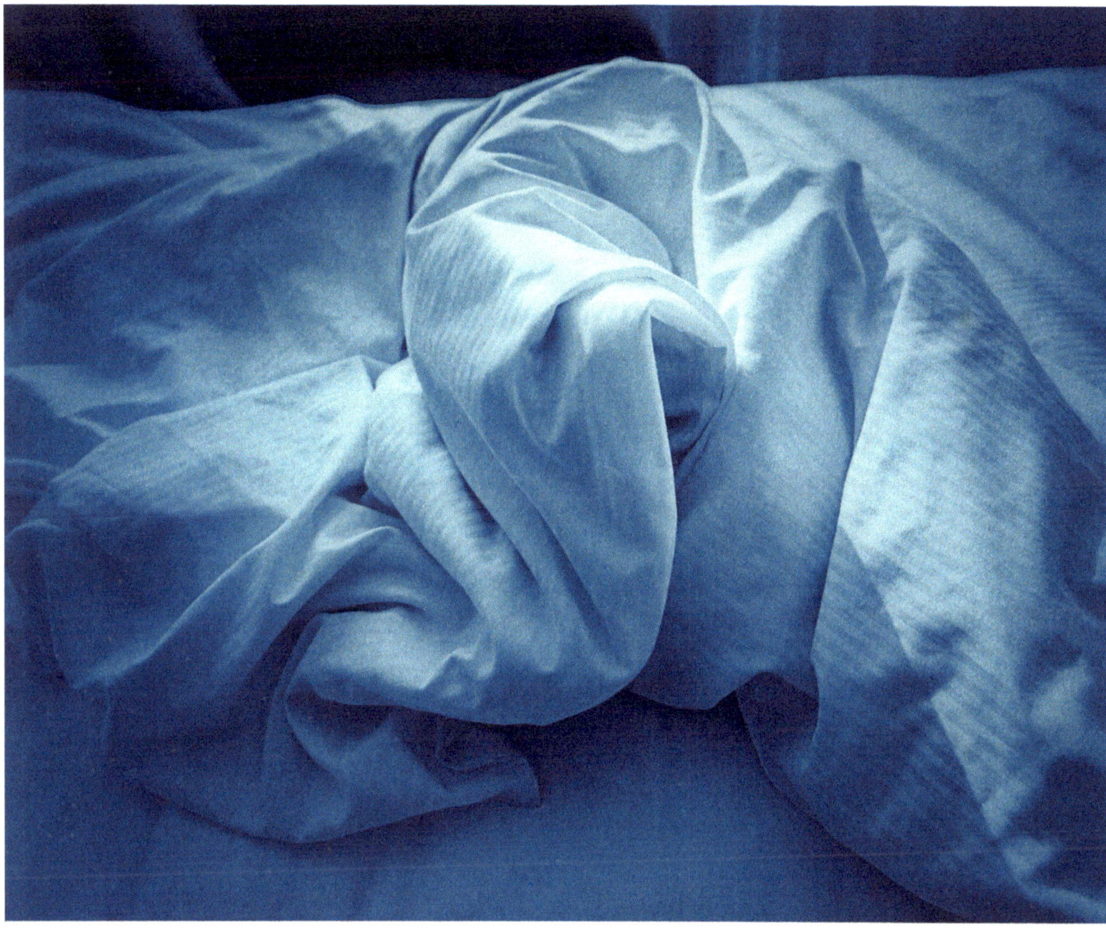

Figure 4.2. *Iceland*, SWC cyanotype curved with the SWC curve on Hahnemühle Platinum Rag © Christina Z. Anderson 2018

Cyanotype is often labeled a "short scale" process, one that doesn't have an extended, smooth tonal scale from deepest shadows to highest highlights, but blocked up shadows and blown out highlights. This is only the case when the process is improperly practiced with an incorrectly prepared negative and most likely printed on an unsuitable paper. This chapter is all about how to create a proper digital negative.

Since cyanotype is a *contact* printing process with the negative exposed in direct *contact* with the sensitive emulsion, the final image is the same size as the negative. Before digital photography, negatives were always from a large format camera. Today, most often the negative originates in an ink jet printer. This chapter will walk you through a step-by-step method of making ink jet negatives, starting with finding the correct exposure time to give the darkest blue without overexposure. This can be done two ways, one low tech and one high tech. Once that exposure time is determined—and it varies with each paper—then that exposure time

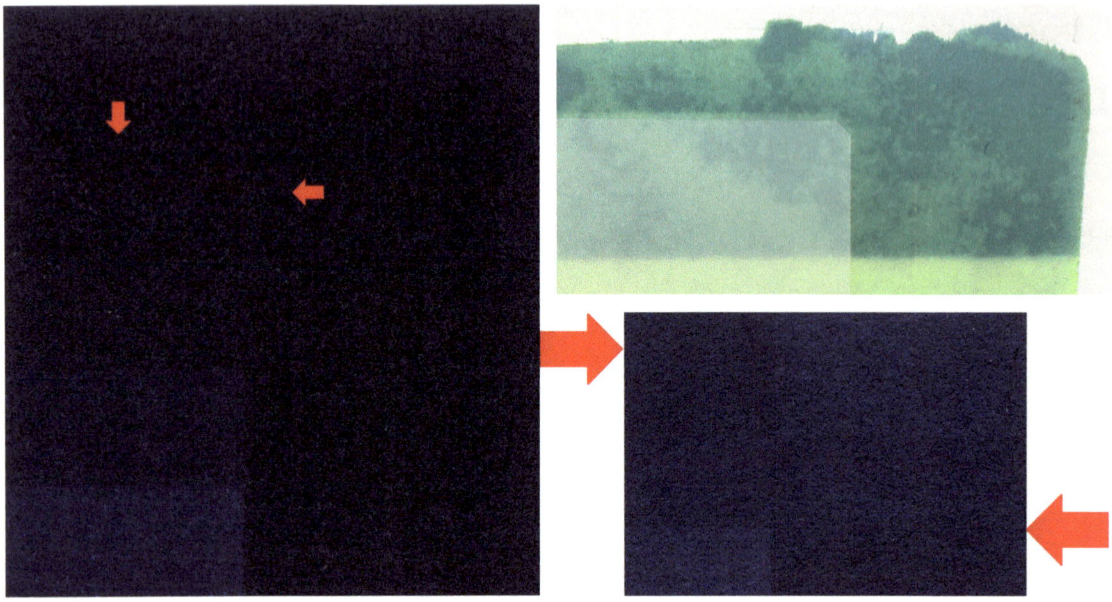

Figures 4.3–4.5. In the left image, the left half of the paper was covered with Pictorico and the right not (see unexposed image, top right). Note where two steps on the OHP side merge to the same maximum blue, and the side covered with OHP and the side not covered with OHP also merge to maximum blue, indicating the correct exposure time. The image on the right illustrates overexposure; cyanotype will lighten or "solarize" when overexposed. The step just below solarization is the correct exposure time—maximum blue without solarization.

is matched with a contrast adjustment curve which helps all tones in the image to be printed correctly, without blocking up on either the shadow or the highlight end. One final adjustment that is helpful with classic cyanotype is to the negative's ink density if desired (see p. 52).

Digital inks have a density range from black to white (full ink to no ink) greater than most alt processes can handle. Thus the curve is usually a contrast-*lowering* curve. With a correct exposure time and a correctly curved negative, perfect cyanotype prints will result. The good news: if one's practice remains consistent, these digital steps only have to be done once for a particular paper.

Transparency film

Transparency film (OHP or overhead projection film) is made by Pictorico, Arista, Fixxons, Inkpress, and others. My new favorite is Fixxons (fixxons.com) at a quarter the price of Pictorico. It tends not to curl. When OHP film curls ever so slightly, it will jam in the printer. My next favorite is Pictorico because it has an excellent ink-receiving layer on one side and also remains flat. Pictorico comes in Premium and Ultra which has a thicker ink-receiving layer. Use Premium.

Determining exposure the low tech way

The aim is to determine the *minimum* UV exposure needed to obtain *maximum* blue when printing through the OHP negative.

1. Coat paper with cyanotype solution and dry.
2. Place a sheet of OHP over half of the paper.
3. Place a sheet of opaque material so it covers both halves and leaves a strip exposed at the top.
4. Make a 5-minute exposure (1-minute for new cyanotype).
5. Move the opaque material down to uncover another strip and make another 5-minute exposure (1-minute for new cyanotype).
6. Continue this process for 45–60 minutes (15 minutes for new cyanotype), until there is a series of 5-minute (1-minute) exposures.
7. Develop the test sheet and dry it. Note the step where there is no difference between the OHP and non-OHP side. This is the minimum exposure that is capable of giving maximum blue. Tip: for classic cyanotype, it can be between 5 and 45 minutes under UVBL exposure but with 100+ papers I found it most often 19–30 minutes. For new cyanotype it can be between 2 and 15 minutes under UVBL exposure

but with 100+ papers I found it most often 5–8 minutes.

8. Once this standard printing time (SPT) is determined for that paper, it does not change.

Determining exposure the high tech way

The most accurate method to determine exposure is with a Stouffer film step wedge. The step wedge is used to find the amount of time it takes to get cyanotype as dark as it needs to be and no darker (overexposure; note cyanotype will get *lighter/* solarize at a certain point of overexposure). The Stouffer step wedge is calibrated in either ½ stop (21-step) or ⅓ stop (31-step) increments. The 31-step wedge is easier to fine-tune your time.

Each increment of the 31-step wedge corresponds to ⅓ stop and a mathematical number of 0.794 (.8 for ease) which is used to add time (by dividing the exposure time) or subtract time (by multiplying the exposure time) depending on how many maximum blue steps are showing on the print. This is found by first *overexposing* the step wedge to make sure two or more steps on the Stouffer's merge together with no differentiation. Once that is done the rest can be calculated mathematically. No standing there moving an opaque piece of material every 5 minutes while a test print is being made! The math is very easy. If math is not your thing, note the number of the first darkest step and consult Tables 4.1 (classic cyanotype) or 4.2 (new cyanotype).

1. For classic cyanotype use an initial overexposure time of 45 minutes which should be overkill enough to merge at least two steps. For new cyanotype use 15 minutes.
2. Note the step that is the first darkest step, e.g. most likely steps 2–8.
3. Find that step in Table 4.1 (classic cyanotype) or 4.2 (new cyanotype) and note your final exposure time. Once this exposure time is found it never changes for that paper.

The math: for every step moved back towards Step 1, multiply by .794. Example: if Steps 1–4 are all maximum black, Step 4 needs to move 3 steps back to Step 1. The initial 45-minute exposure time is multiplied 3x consecutively by .794. Then the film density is subtracted since you won't be printing through film (multiply by .89) and then add ½ stop to account for Pictorico OHP density by dividing the final amount by ½ stop or .707. Tip: if you forget these numbers, use easy to remember numbers .7, .8, and .9.

45-minute exposure CC (Baseline Printing Time BPT)	Final exposure (rounded up) (Standard Printing Time SPT)
If Step 1 is max black:	57 minutes
If Step 2 is max black:	45 minutes
If Step 3 is max black:	36 minutes
If Step 4 is max black:	29 minutes
If Step 5 is max black:	23 minutes
If Step 6 is max black:	18 minutes
If Step 7 is max black:	15 minutes
If Step 8 is max black:	12 minutes
If Step 9 is max black:	9 minutes
If Step 10 is max black:	7 minutes

15-minute exposure NC (Baseline Printing Time BPT)	Final exposure (rounded up) (Standard Printing Time SPT)
If Step 1 is max black:	19 minutes
If Step 2 is max black:	15 minutes
If Step 3 is max black:	12 minutes
If Step 4 is max black:	10 minutes
If Step 5 is max black:	8 minutes
If Step 6 is max black:	6 minutes
If Step 7 is max black:	5 minutes
If Step 8 is max black:	4 minutes
If Step 9 is max black:	3 minutes
If Step 10 is max black:	2 minutes

Tables 4.1 and 4.2. Use the top table 4.1 for classic cyanotype and the bottom table 4.2 for new cyanotype.

Curving the negative

These cyanotype curves are based on Epson printers that use Ultrachrome inks. The current models you may own are the 3880, the P600 or the P800. There are other printers on the market that may be excellent printers for other purposes, but for digital negatives Epson printers have the market cornered: the ink is dense enough, and the printer driver has the ability to lay down more or less ink when necessary, and with classic cyanotype (not new cyanotype!) it is often helpful to cut back on the ink density a bit even up to as much as -30% on a P800.

Table 4.3 shows the input/output curve points to build generic curves that will get you in the ballpark to making a great print. The curve creation process for all negatives is the same: you plug in a set of coordinates in the Photoshop curve dialog box. The difference is in the actual coordinate numbers.

There are five curve choices below. Two are for classic cyanotype (cc)—a generic curve for all papers and a slightly different one for washi like Hahnemühle Sumi-e and Japanese Mulberry. One is for new cyanotype (NC). One is for combining the two formulas (SWC). And one is for palladium. Surprisingly, the NC curve is very close to the palladium curve as to be almost interchangeable (which means a negative made for new cyanotype can be used for palladium, too).

CC	CC Washi	NC	SWC	Palladium
0/0	0/0	0/0	0/0	0/0
13/24	13/8	13/11	13/23	13/6
25/30	25/15	26/15	26/30	25/11
38/34	38/20	38/19	39/36	38/14
51/37	51/25	51/22	51/41	51/19
69/41	64/29	69/27	64/45	64/22
89/46	89/36	89/32	89/55	89/29
115/54	115/45	115/41	115/66	115/36
140/66	140/54	140/49	140/79	140/46
166/82	166/68	166/58	166/96	166/57
191/103	191/88	191/71	191/119	191/73
204/117	204/106	204/81	204/132	204/84
217/139	217/126	215/99	217/147	217/96
229/164	229/148	229/134	229/166	228/111
238/188	239/175	237/161	239/187	238/135
255/255	255/255	255/255	255/255	255/255

Table 4.3. Sets of coordinates for generic curves for classic cyanotype (CC), new cyanotype (NC), combination classic and new cyanotype (SWC), and palladium.

1. Open an image in Photoshop.
2. Add a curve layer (**Layer>New Adjustment Layer>Curves**) (Figure 4.6).

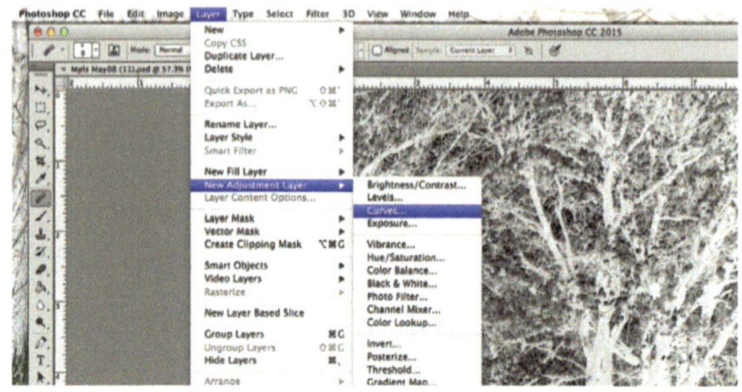

Figure 4.6.

3. Go to the dropdown menu in the Curves Panel located at the very top right corner and select **Curves Display Options** to make sure the Curves Panel is set to **Light** and **0–255** (Figure 4.7).

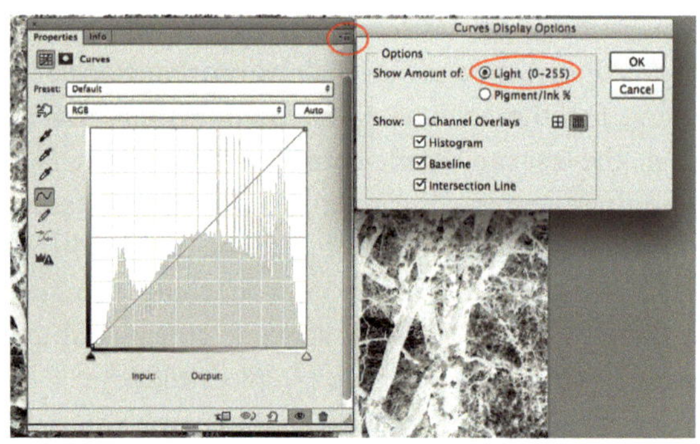

Figure 4.7.

4. Click on the diagonal line just above the bottom left to create a point on the curve. The **Input/Output** numbers will appear in the Curves panel at the bottom (Figure 4.8).
5. Enter the Input/Output numbers, starting with the 13/xx pair.
6. Continue to click a new point and add the next set of numbers. Don't enter 0/0 or 255/255 because the diagonal line already has those points anchored at the bottom and top. That means there are only 14 sets of numbers to enter.

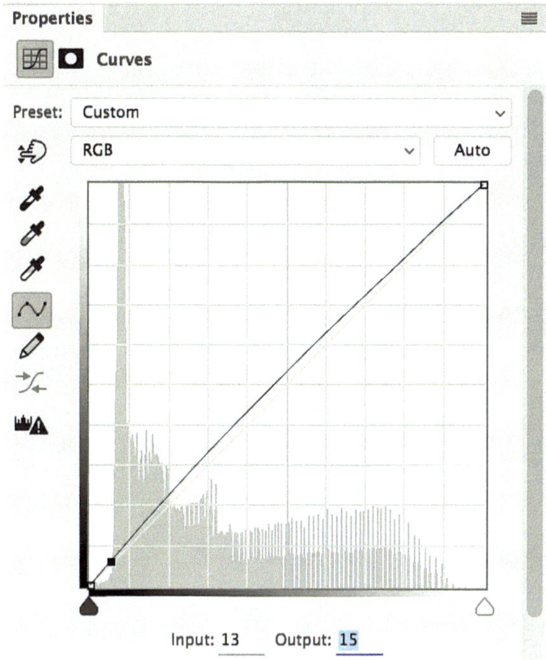

Figure 4.8.

7. The curve will look a bit funky until all the numbers are entered. When all are entered, it should be smooth. Select one of the points and use the +/- keys to toggle up and down the curve from point to point to check all input/output values.
8. In the dropdown menu in the **Curves Panel** click **Save Curves Preset** and name it with a name you will remember like *CyanotypeHPR19mn-20ink.acv* (Figure 4.9). Your curve work is done!

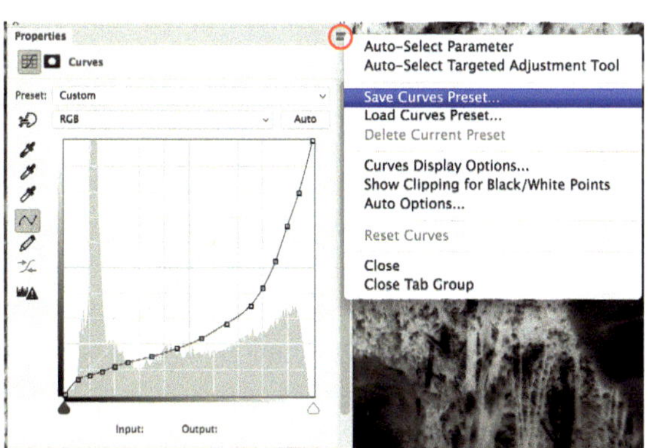

Figure 4.9.

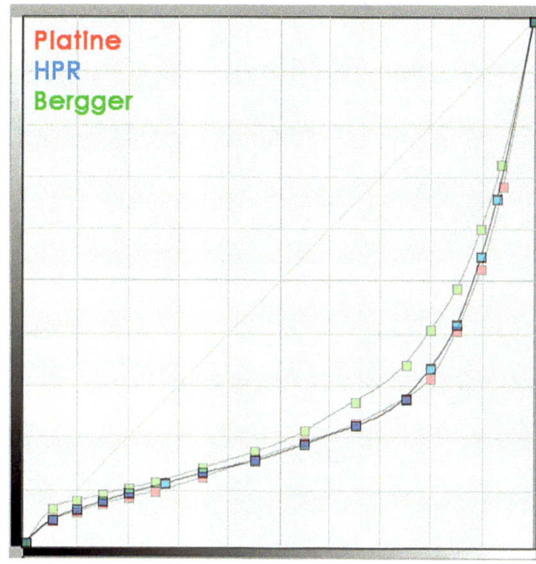

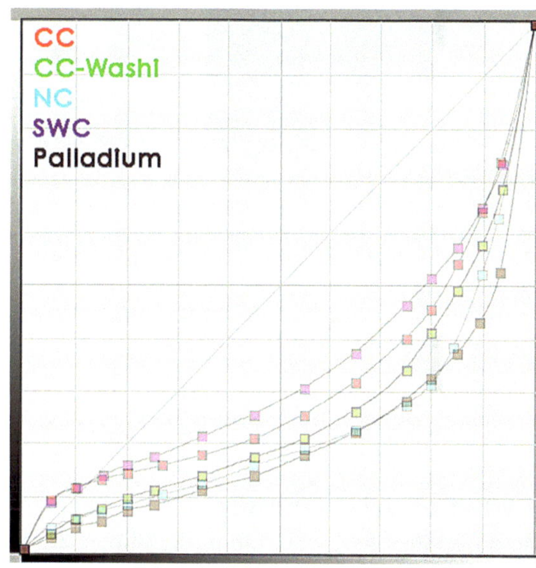

Figures 4.10 and 4.11. Note in the top screen shot that Arches Platine and Hahnemühle Platinum Rag respond almost the same to the cyanotype solution (NC). The Bergger Cot 320 curve is slightly different but still very close. The bottom screenshot is a visual of the five curves included in this chapter. Note that the palladium curve and the NC curve are almost interchangeable. It is important to remember that although generic curves come close, they are not the best practice in making alternative process prints. It is always best to calibrate each paper and process with more "high tech" curve systems like Precision Digital Negatives (PDN). Even though curves look so similar in shape, *small changes in points translate to larger changes in the print.*

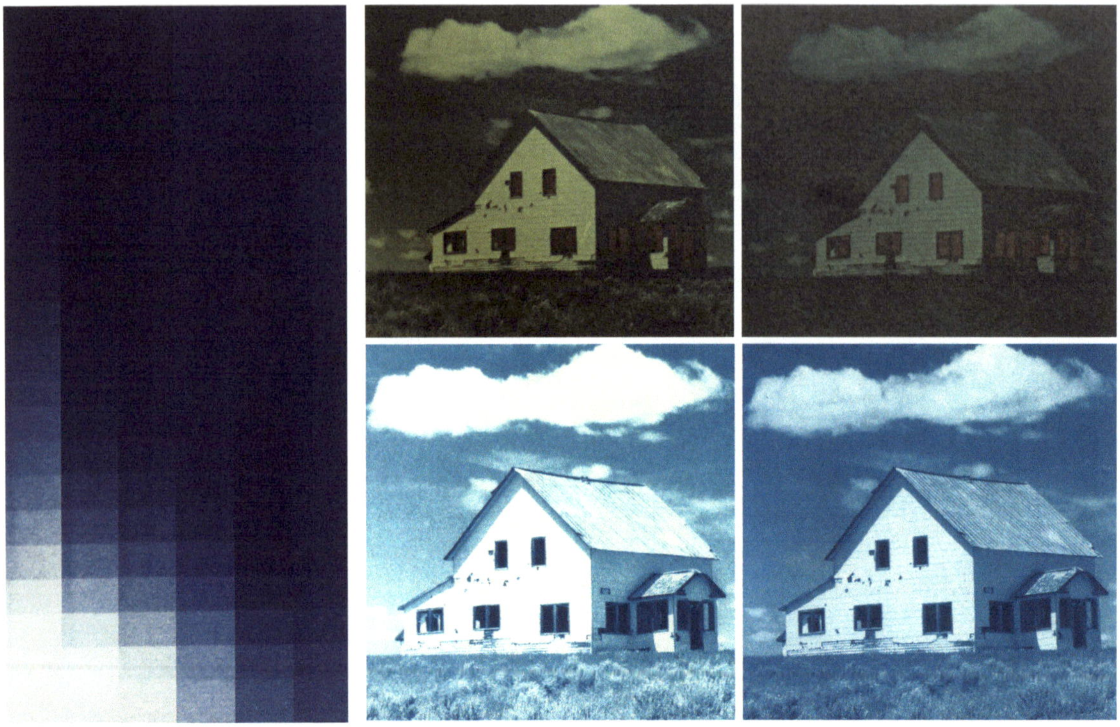

Figures 4.12 and 4.13. Left is a cut and paste of digital step wedge prints put side by side, made from 0% to -50% ink densities from left to right. Note how the highlights and shadows respond. The strip on the far left has blown out highlights. It's not until the -10% and -20% where the highlights are well printed (only one bottom square should be paper white). The far right strip has no paper white and thus is too thin. In the strip on the left the shadows are not maximum blue until the -10% and -20% strips. Then from -30% to -50% the shadows are overexposed and have begun to lighten/solarize. -15% would be the best choice for this paper. The house images on the right are printed exactly the same, but the left pair is a CC print with a 0% negative before and after development, the right pair a CC print with a -15% negative before and after development. I rarely use a 0% negative with classic cyanotype, unless the image is one with a limited tonal range. I always use a +/- 0% with new cyanotype. Note: this particular image is my test image because of the extreme tonal range from white to black. If the puffy white clouds and the minute black slats on the side of the house facing the sun both have detail, the negative density is correctly chosen.

A word about ink density

Epson printers have a feature in the printer driver, usually found under Advanced Media Control, where you can vary the amount of ink sprayed onto the paper from +50% to -50%. I hadn't given this feature much thought, or used it, until I bought my first 3880 printer. For the first time I had to use more ink to get a dense enough negative for alt. This need was greater when printing salted paper which needs a very dense negative, probably denser than all alt processes.

Cyanotype is on the opposite side of the spectrum from salt. One day I grossly overexposed a cyanotype print, but I noticed that the highest highlights—in this particular image it was in the puffy white clouds—were still paper white with little detail. I printed a series of step wedges: 0%, -10%, -20% -30%, -40%, and -50%, and was surprised to find that highlight detail improved, exposure time lessened, and the overall print looked remarkably better somewhere between 0% to -20% on an Epson 3880 and up to -30% on an Epson P800 which seems to print a denser black. Past -30% the negative was too thin and didn't hold back enough light to keep the highlights bright. I now often use a -10% to -20% negative for classic cyanotype but no ink density change for new cyanotype. I encourage you to test different negative densities, too.

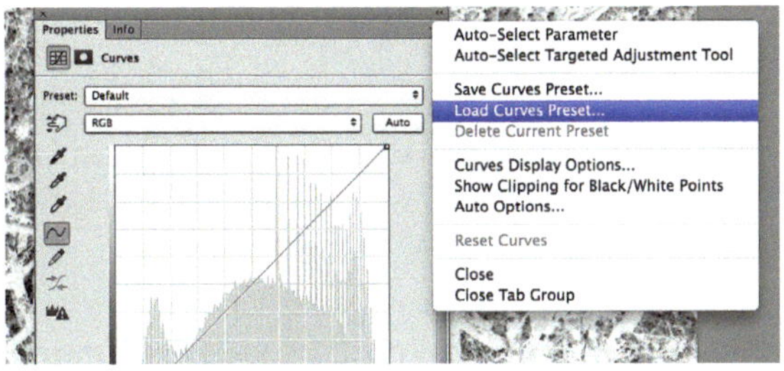

Figure 4.14.

Monochrome digital negatives

1. Open a digital image and convert it to 16 bit RGB (**Image>Mode>16 bit**) and Adobe 1998 (**Edit>Convert to Profile>Adobe 1998**).
2. Size it to 360 ppi no more than 8″ × 10″ to fit on 8.5″ × 11″ OHP film with enough borders, and "save as" (**File>Save As>imagename_negative.psd**) so as not to write over the original.
3. Do any kind of image adjustments needed and for the final step do one or two sharpening steps if desired: a) **Filter>Sharpen>Unsharp Mask Amount 50–150%, Radius 0.5, Threshold 0** and/or b) **Layer>Duplicate Layer** then with that layer selected, **Filter>Other>High Pass Radius 10**. Set that layer at blend mode of **Soft Light** on the layer's panel dropdown menu. View the image at 50%, and with the duplicate layer active, play with the opacity slider at the top right of the **Layers** panel to see when it looks sharp enough but not too sharp. Flatten the two layers (**Layer>Flatten Layers** or **Command+E**). Save (**Command+S**).
4. Make the image monochrome (**Image>Adjust>Black and White**). The image will still be in RGB mode though it will look monochrome.
5. Invert the image (**Command+I**).
6. Apply a curves adjustment layer to the image (**Layer>New Adjustment Layer>Curves**).
7. Click the dropdown menu in the curves panel, click **Load Preset** and load the cyanotype curve (Figure 4.14).
8. Save the image (**Command+S**). Do not flatten these two layers but save the negative with the curve always separate. That way the image can be printed with a different curve, if desired.
9. Go to **File/Print**. Document should be in Adobe 1998 (this lays down more ink).
10. In the print dialog panel under Color Handling select **Photoshop Manages Colors** and select **Epson Premium Glossy Photo Paper** or something similarly glossy in the Printer Profile menu.
11. Check **Send 16-bit data** if the image is in 16-bit.
12. Select **Normal** Printing.
13. Rendering Intent: **Perceptual**.
14. Black Point Compensation **unchecked**.
15. Dropdown menu click **Printer Settings**.
16. Under **Basic** click:
    - Media Type: **Premium Glossy**
    - Super Photo **2880**
    - High Speed **unchecked**
    - Mirror Image (or Flip Horizontal) **checked**
    - Finest Detail **unchecked**
    - 16-bit Output **checked**
17. Under **Advanced Color Settings** see that the Epson Driver Color Management is **off**.
18. For cyanotype, under **Advanced Media Control** choose between 0% to -20% ink density using the Advanced Media Control slider (even -30% on an Epson P800). For NC no need to change the slider from the normal +/- 0%.
19. Save and print. Be sure to print on the correct side of the OHP film. Tip: do the "lick and stick" test—lick a finger and see if it sticks to a corner of the OHP; if so, that is the side to print on. If the Pictorico cut-off corner is at the top right, that means the printable side is facing you.
20. Let a negative dry for at least an hour face up with nothing on top, or dry it with a blow dryer on warm. Then store it in a notebook sleeve.

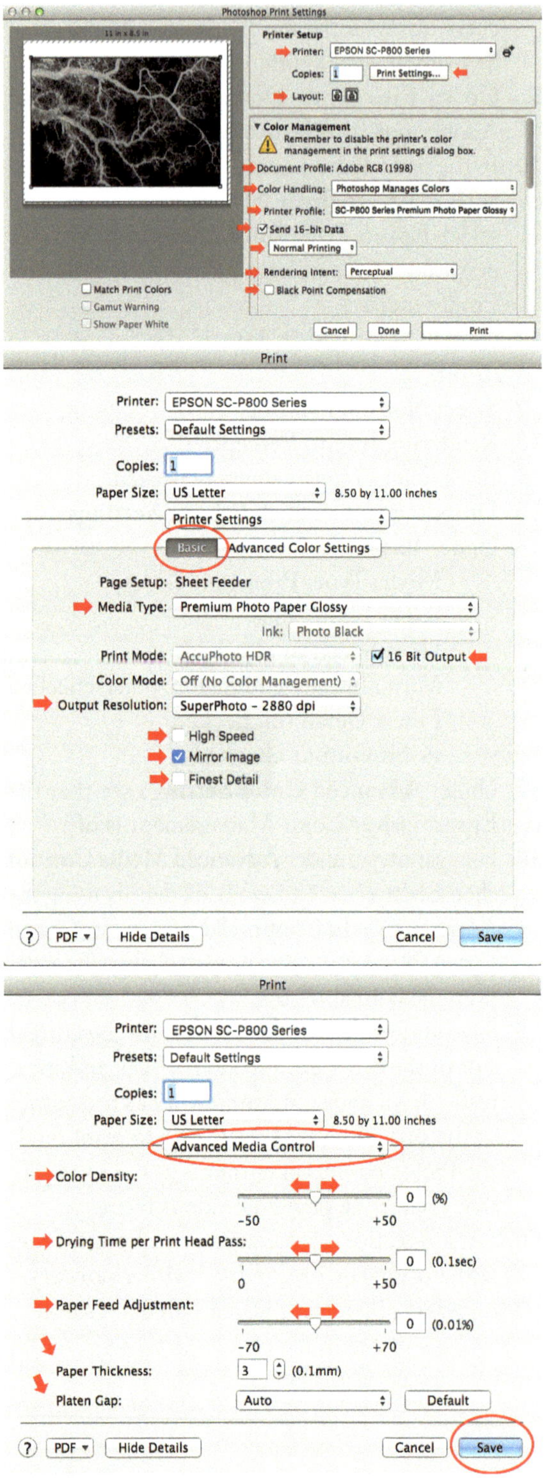

Figures 4.15–4.17. These show typical Epson printer settings from the first panel that pops up when selecting File/Print to the panels that appear in the dropdown menu under Print Settings.

Duotone negatives from RGB

I learned this ingenious method from Sam Wang. This digital negative workflow is for printing cyanotype over palladium. The method starts with a full color image which is then split into three color separation negatives (RGB). RGB negatives actually print their opposite colors of CMY—R (red) prints C (cyan or blue), G (green) prints M (magenta) and B (blue) prints Y (yellow). The two negatives that print the warm colors of the image are combined to yield a single negative that is used to print the palladium part of the equation, and the R which prints blue is used for cyanotype.

1. Open a color digital image and convert it to 16 bit RGB (**Image>Mode>16 bit**) and Adobe 1998 (**Edit>Convert to Profile>Adobe 1998**).
2. Size it to 360 ppi no more than 8″ × 10″ to fit on 8.5″ × 11″ OHP film with enough borders, and "save as" (**File>Save As>imagename_negative.psd**) so as not to write over the original.
3. Do any kind of image adjustments needed and for the final step do one or two sharpening steps if desired: a) **Filter>Sharpen>Unsharp Mask Amount 50–150%, Radius 0.5, Threshold 0** and/or b) **Layer>Duplicate Layer** then with that layer selected, **Filter>Other>High Pass Radius 10**. Set that layer at blend mode of **Soft Light** on the layer's panel dropdown menu. View the image at 50%, and with the duplicate layer active, play with the opacity slider at the top right of the **Layers** panel to see when it looks sharp enough but not too sharp. Flatten the two layers (**Layer>Flatten Layers** or **Command+E**). Save (**Command+S**).
4. Invert the image (**Command+I**) and save at this step because the next step you can't undo (**File>Save** or **Command+S**).
5. In the Layers Panel select the **Channels** tab; *image must be flattened* (**Layer>Flatten Image**) *or this will not work*!
6. In the dropdown menu on the Channels tab select Split Channels. You will suddenly have three duplicate *grayscale* images, named R, G, and B.
7. Convert each negative from grayscale back to RGB mode Adobe 1998 (**Image>Mode>RGB and Edit>Convert to Profile>Adobe (1998)**).

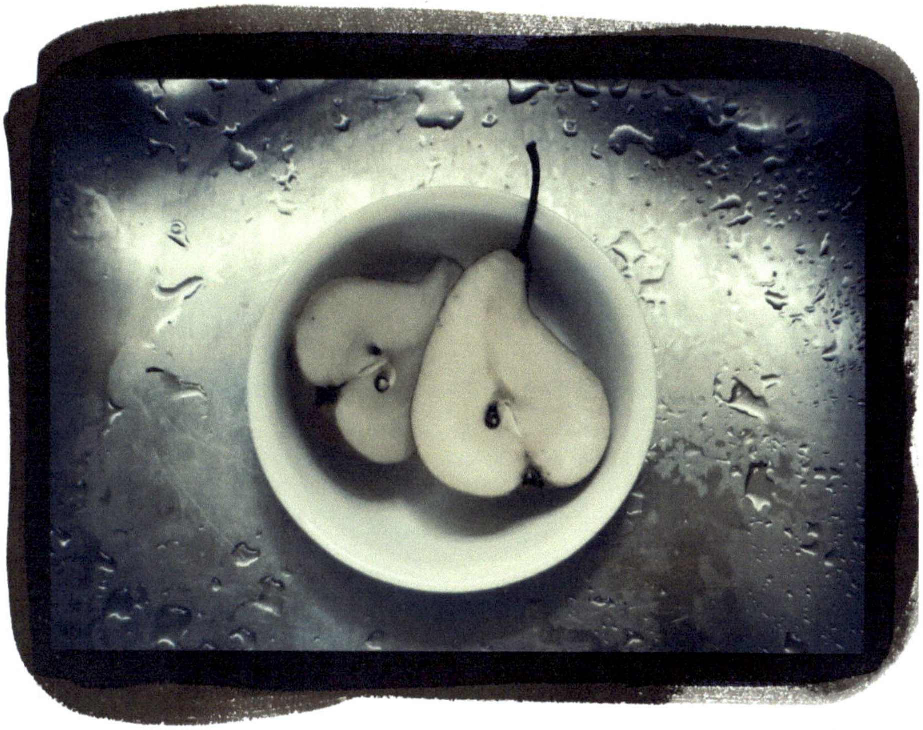

Figure 4.18. *Pear Halves*, cyanotype over palladium created with duotone negatives © Sam Wang 2018

8. Save the R negative immediately as *imagename_COOLneg.psd*. This negative prints cyanotype.
9. Apply a curves adjustment layer to the image (**Layer>New Adjustment Layer>Curves**).
10. Click the dropdown menu in the curves panel, click **Load Preset** and load the cyanotype curve. This negative is ready to print using 0% to -30% ink density as required.
11. Take the G image and with the shift key held down, the move tool selected in the Photoshop tools panel, and the mouse or track pad clicked, drag and drop the G image directly on top of the B image and release.
12. With the G layer selected, move the opacity slider in the **Layers** panel to 50% so that the G layer is at 50% opacity on top of the B layer at 100% opacity. This way both the G (which prints magenta) and the B (which prints yellow) are at 50/50, equally contributing. Flatten these two layers and save the G/B negative as *imagename_WARMneg.psd*—this negative prints palladium.
13. Apply a curves adjustment layer to the image (**Layer>New Adjustment Layer>Curves**).
14. Click the dropdown menu in the curves panel, click **Load Preset** and load the palladium curve. This negative is ready to print *with no adjustment to the ink density*.
15. Now you are ready to print both negatives. Select **File>Print**. In the first print dialog panel that comes up, makes these selections:
    Under **Color Handling** select **Photoshop Manages Colors**
    Select **Epson Premium Glossy Photo Paper** or something similarly glossy in the Printer Profile menu
    Check **Send 16-bit data** if the image is in 16-bit
    Select **Normal** Printing
    Rendering Intent: **Perceptual**
    Black Point Compensation **unchecked**
16. In the dropdown menu click **Printer Settings** and under **Basic** click:
    Media Type: **Premium Glossy**
    Super Photo **2880**
    High Speed **unchecked**
    Mirror Image (or Flip Horizontal) **checked**
    Finest Detail **unchecked**
    16-bit Output **checked**
17. Under **Advanced Color Settings** see that the Epson Driver Color Management is **off**.

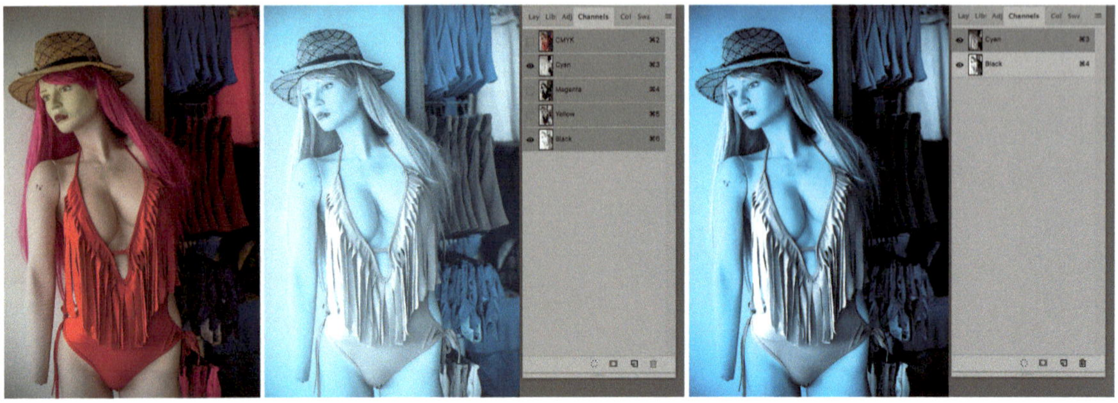

Figure 4.19. I chose as extreme a color example as I could find for this visual! Left is the original image. Middle, the image is converted to CMYK and in the Channels panel, the magenta and yellow layers are hidden to see what the image looks like. Note how the image is pale because of the missing layers. The image on the right has the magenta and yellow layers deleted, and a levels adjustment layer was used to darken and add contrast before the channels are split into the C and K negatives which will print cyanotype and palladium.

18. Under Advanced Media Control *only for the cyanotype negative* change the ink density slider to 0% to -30%; the platinum negative remains at 0%.
19. Save and print. Be sure to print on the correct side of the OHP film. Tip: do the "lick and stick" test—lick a finger and see if it sticks to a corner of the OHP; if so, that is the side to print on. If the Pictorico cut-off corner is at the top right, that means the printable side is facing you.
20. Let the negatives dry for at least an hour face up with nothing on top, or dry them with a blow dryer on warm. Then store them in notebook sleeves.

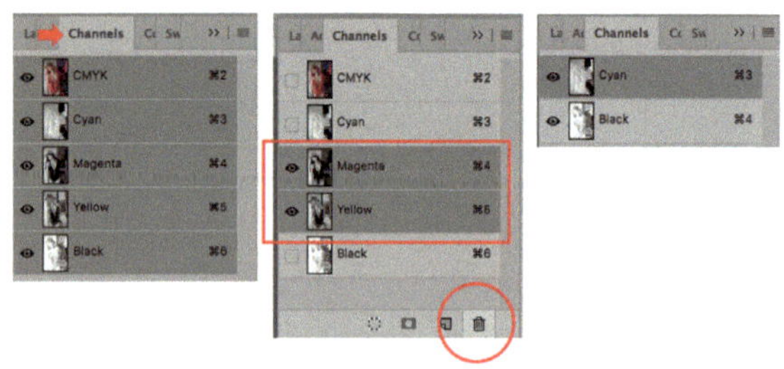

Figure 4.20.

Duotone negatives from CMYK

Note: when working with CMYK, each color prints itself and not its opposite as in RGB. In other words, C prints *cyan*/blue, M prints *magenta*, Y prints *yellow* and K prints blac*k*.

1. Open a color digital image and convert it to 16 bit (**Image>Mode>16 bit**) and CMYK (**Image>Mode>CMYK Color**).
2. Size it to 360 ppi no more than 8″ × 10″ to fit on 8.5″ × 11″ OHP film with enough borders, and "save as" (**File>Save As>imagename_negative.psd**) so as not to write over the original.
3. Do any kind of image adjustments needed and for the final step do one or two sharpening steps if desired: a) **Filter>Sharpen>Unsharp Mask Amount 50–150%, Radius 0.5, Threshold 0** and/or b) **Layer>Duplicate Layer** then with that layer selected, **Filter>Other>High Pass Radius 10**. Set that layer at blend mode of **Soft Light** on the layer's panel dropdown menu. View the image at 50%, and with the duplicate layer active, play with the opacity slider at the top right of the **Layers** panel to see when it looks sharp enough but not too sharp. Flatten the two layers (**Layer>Flatten Layers** or **Command+E**). Save (**Command + S**).
4. In the Layers Panel select the **Channels** tab; note that there are four channels, cyan, magenta, yellow, and black. With the shift key held down, click on the magenta and yellow channels to make them active, and click on the garbage can at the bottom right of the Channels panel and discard the two channels. You will be left with just the cyan and black channels (Figure 4.20).

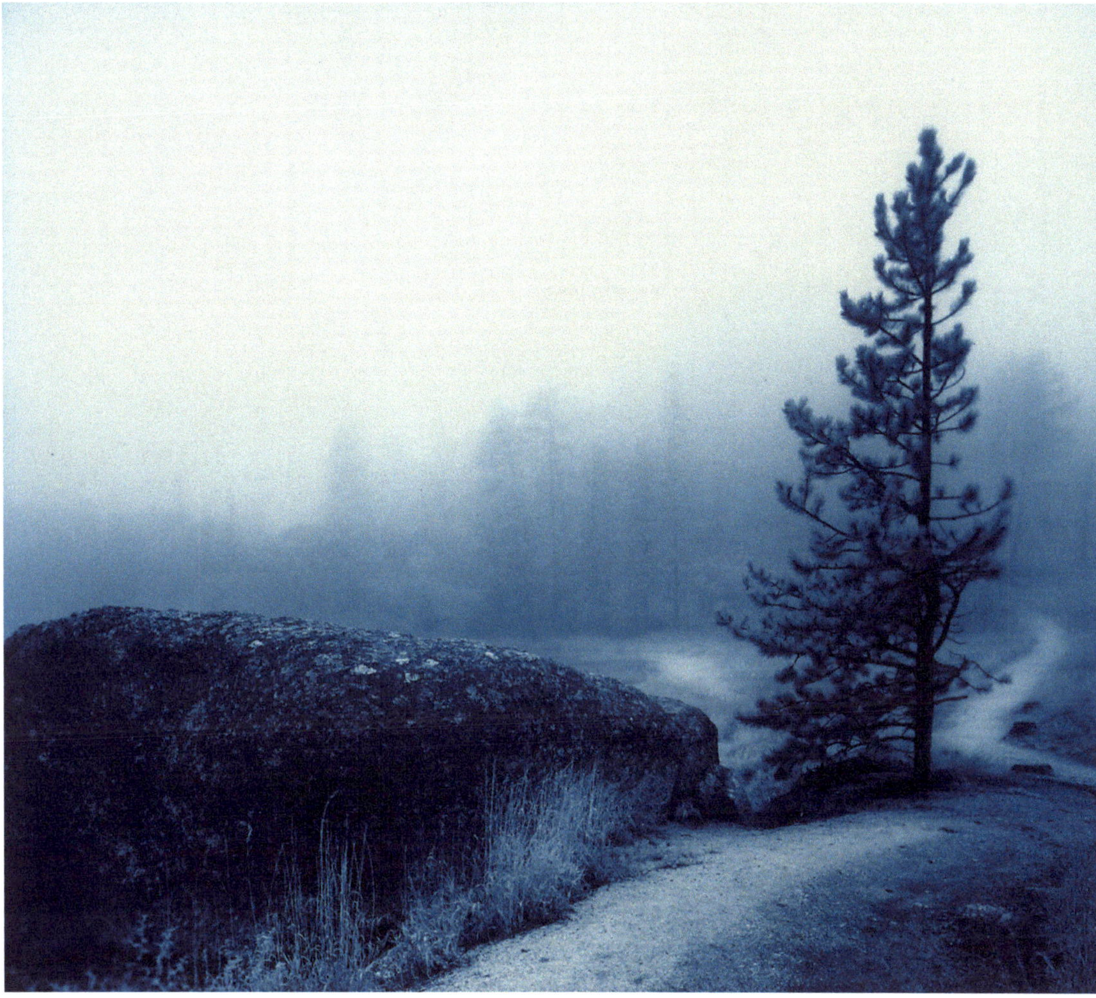

Figure 4.21. *Morning Fog* © Hua Cheng 2018. A perfectly calibrated digital negative will show all detail from highlights to shadows, including high key fog.

5. At this point, the image will look a bit pale and washed out because you have deleted the layers, so do a **Levels Adjustment** (**Image>Adjustments>Levels**) and play with the sliders until the image looks sufficiently dark and hit OK.
6. In the dropdown menu on the Channels tab select Split Channels. *Image must be flattened* (**Layer>Flatten Image**) *or this will not work*! You will have two *grayscale* images, named CYAN and BLACK.
7. Convert each image from grayscale back to RGB mode Adobe 1998 (**Image>Mode>RGB and Edit>Convert to Profile>Adobe (1998)**).
8. Invert each image (**Command+I**) and save each negative (**File>Save** or **Command +S**) as *imagename_CYANOneg.psd* and *imagename_PDneg.psd*.
9. Apply a curves adjustment layer to each negative (**Layer>New Adjustment Layer>Curves**); click the dropdown menu in the curves panel, click **Load Preset** and load the curves. The negatives are ready to print using 0% ink density for the palladium negative and 0–30% ink density for the cyanotype negative depending on whether using classic or new cyanotype.

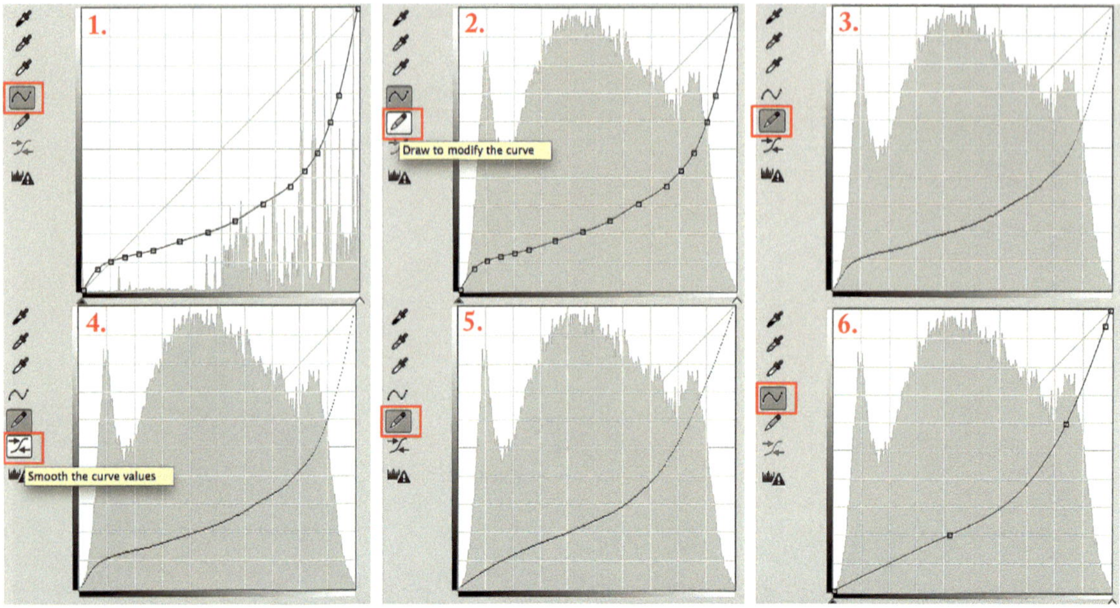

Figure 4.22. Guerrilla fix to adjust contrast. Click the Pencil icon to remove the points. Click the Smooth Icon from 1 to 7 times. Each click will smooth the curve closer to the diagonal line and raise contrast globally.

Troubleshooting the digital negative

- Posterization, where one tone in a continuous tone image plateaus and creates dull, flat areas in a print, has multiple causes. Photographing a high-contrast scene in jpeg mode is one. Over-editing an image, editing in 8-bit mode and not 16-bit mode, and using too drastic a curve are others. With any camera, use as lossless and uncompressed a capture system as you can (RAW for digital SLRs or at least TIFF, never JPEG) and always edit in 16-bit mode. However, if the *curves* produce posterized images, or ones that are slightly low in contrast, there is a "guerrilla" fix available in the Curves Panel that can be used in a pinch. This fix will move the adjustment curve towards the diagonal line in the curves panel in small increments. The guerrilla fix uses the **Pencil** icon to temporarily hide curve points and the **Smooth** icon to smooth the curve closer and closer to the diagonal line with each click. Contrast will rise globally across the negative. Sometimes a little drama with more contrast works.

1. Screen Shot 1 shows the original curve that needs a contrast tweak with the curve icon active.
2. Click on the **Pencil** icon shown in Screen Shot 2; when clicked and made active as it is in Screen Shot 3, it will remove all points and show the "naked" curve.
3. Just below the pencil icon is the **Smooth** icon as in Screen Shot 4. Click on this button to activate it.
4. Once activated, click from 1 to 7 times on this button to move the curve "bowl" towards the diagonal line. With each click towards that line, contrast increases. Be judicious; small point movements make big changes. Note in Screen Shot 5 that the adjustment is global, meaning it will affect the highlights as well as the shadows.
5. After the desired amount of clicks, click the curve icon and the points will appear, though there will be fewer of them (Screen Shot 6).

- Be sure to check your negatives for pizza wheel marks when there are large areas of highlights in the negative (e.g. clouds, sky), microscopic pinprick dots where the teeth on the track wheels that advance the paper in the printer mar the ink. They can only be seen by holding the negative obliquely against something dark and looking through a loupe. They end up printing microscopic black dots marching across

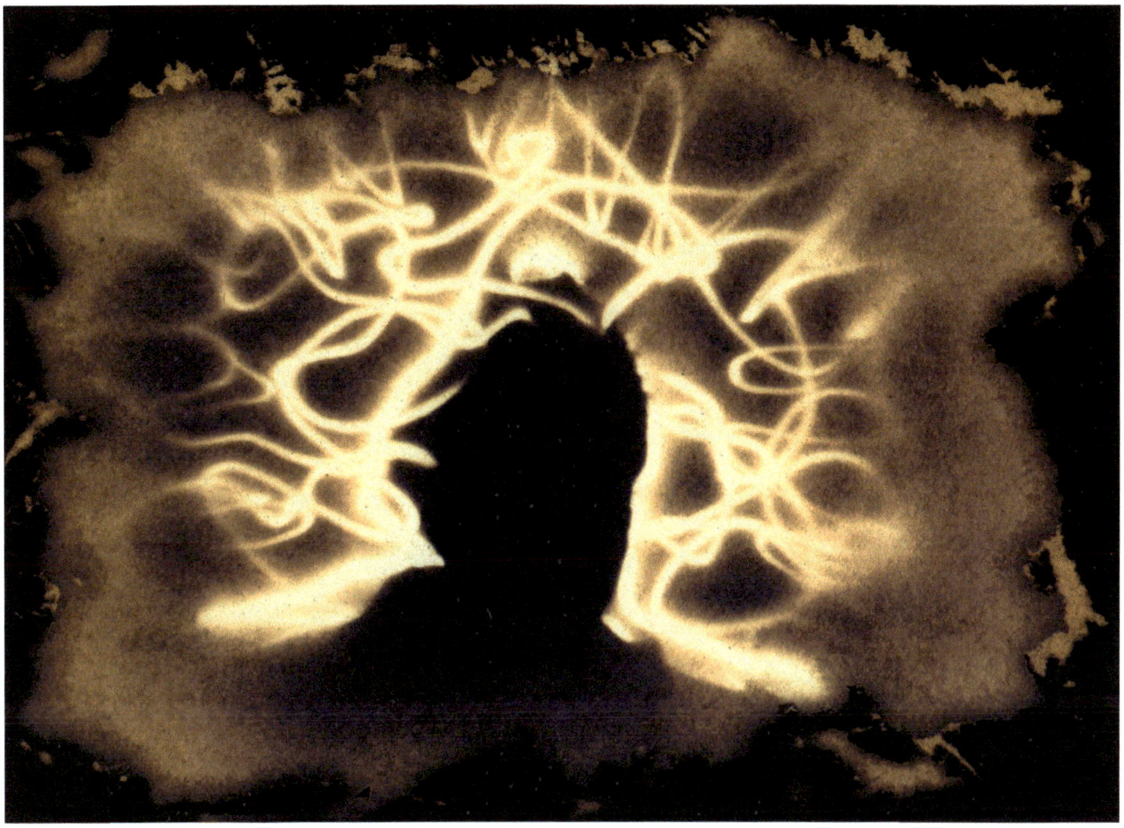

Figure 4.23. *Self Portrait with Sun*, digital print from a scanned in-camera cyanotype negative made from traditional cyanotype on Arches Hot Press, 14″ x 11″© John Beaver 2005. "I have used what I call *cyanonegative* photography—using cyanotype as 'film' in homemade cameras. The long time exposures combined with shallow depth of focus, necessary for cyanonegative photography, compresses time and erases movement. The result can be unsettling, but the process of it calms me." For nearly 20 years, John Beaver has used old processes to make new negatives, often in ways that can only be realized as a print with digital scanning and printing. This includes his development of the cyanonegative process, innovative work (in collaboration with Teresa Patrick) with instant film, and most recently his development of an accelerated, unfixed printing-out process he calls "Ephemeral-Process photography." He is Professor of Physics and Astronomy at University of Wisconsin—Fox Valley. His three-volume *The Physics and Art of Photography* is forthcoming from Morgan and Claypool publishers. To see more of his work, visit **www.JohnEBphotography.com**.

the image. To alleviate this somewhat, under **Advance Media Control** change the slider next to **Increase Drying Time per Pass** to +20–+50, the **Paper Thickness** to 15, and the **Platen Gap** to Wider. If that doesn't work, use the front load feeder of the printer. Some alt practitioners have devised a method of disengaging the pizza wheels entirely by tricking the printer into thinking the front load feeder is being used when it is not, and then using the top paper feeder with the pizza wheels disengaged; search the web for this method.

- As the curve increases in slope towards the diagonal line in the curve dialog box, contrast increases; as it moves away from the diagonal line contrast decreases.
- The bottom left of the curve box closest to 0/0 represents the blacks in the image which, since this is a curve on a negative, will affect the density and contrast of the blacks in the negative which correspond to the highlights. The top right represents the whites in the image which, since this is a negative, will affect the density and contrast of the clears in the negative which correspond to the shadows.

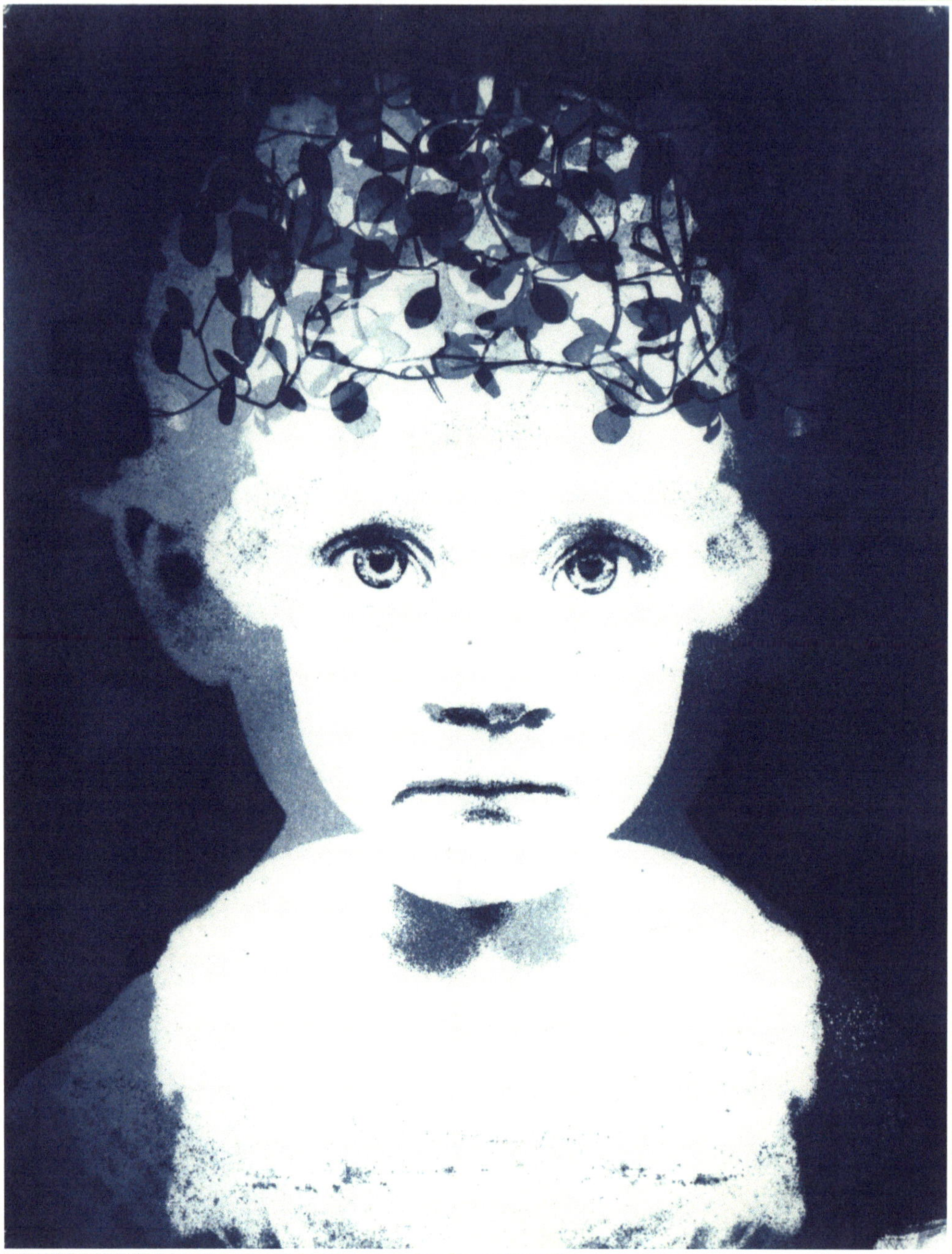

Figure 5.1. *Veronica Maidenhair*, from the *Home* series, traditional cyanotype toned in black tea, 6½″ x 8″ © Leslie Burns 2018. "I document, while simultaneously escaping, the world around me. Making alt prints in my backyard, with my kid and my dogs, isn't how I envisioned my studio "career." But it's a damn wonderful blessing, being bound by kiddie pools, clotheslines and water hoses. I make pictures of my friends, my kid, my plants, and the things I love. I cherish these beauties as they distract me from my inner darkness. Processes invigorate me. Variables enliven me. Final prints are proof I was here." Leslie Burns was born in 1975 in Florence, South Carolina. In 1999, she received her BA with concentrations in Photography and Sculpture from the College of Charleston. In 2011, she earned her MFA in Photography from Lamar Dodd School of Art at University of Georgia. In June 2012, Burns was named as one of the "100 Under 100: The New Superstars of Southern Art" by *Oxford American* magazine. In 2013, she married Scott Wallace, and gave birth to their daughter, Veronica. Currently, Mrs. Burns resides with her family in North Charleston, South Carolina, and works as an Adjunct Professor of Photography at the College of Charleston.

Chapter 5

Classic Cyanotype

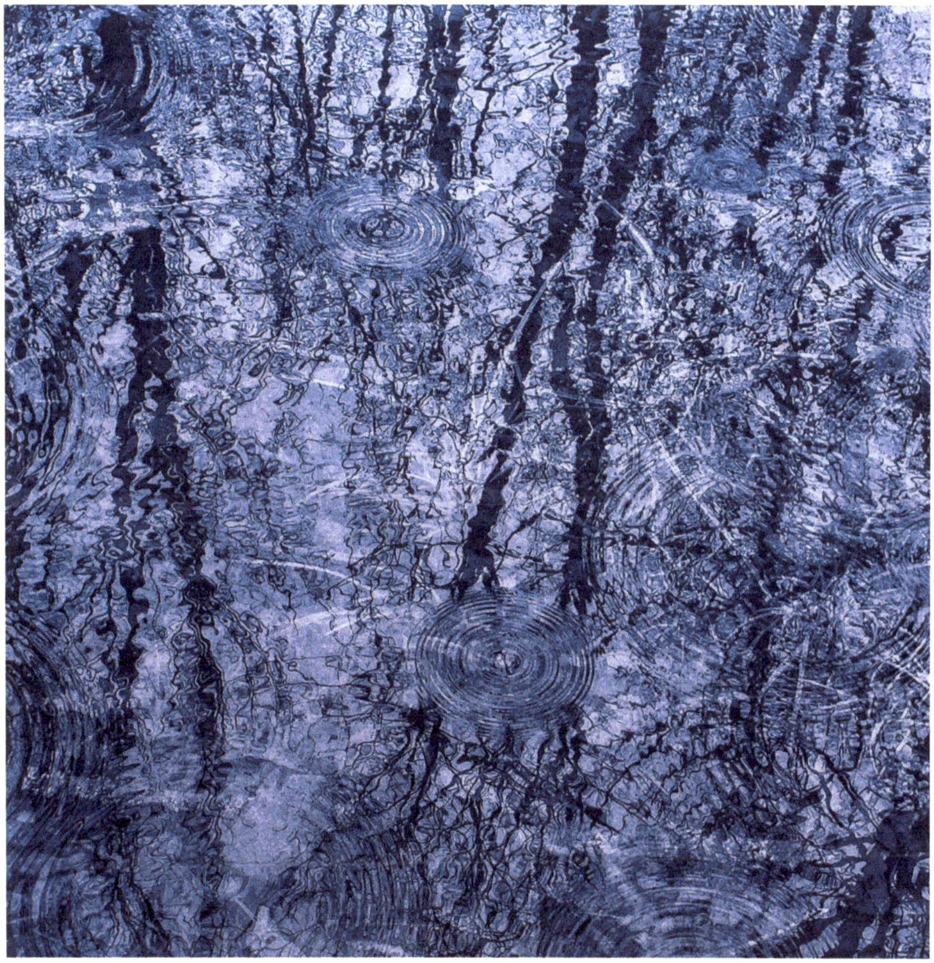

Figure 5.2. *Vernal Pool*, two coats of classic cyanotype, toned with gallic acid, 10″ x 10″ © Susan Davens 2018

Cyanotype formulas can be divided into essentially two types: the original formula from 1842 made with ferric ammonium citrate (FAC) and potassium ferricyanide (PF), and the new formula invented by Mike Ware in 1994 wherein FAC is replaced with ferric ammonium oxalate (FAO). Although both formulas result in a print composed of Prussian blue, they require somewhat different practices.

I chose to standardize nomenclature when writing about the formulas. FAC cyanotype versus FAO cyanotype is inelegant. Old cyanotype versus new cyanotype connotes something that has passed its usefulness and implies replacement of a better sort. To call the original formula "traditional" implies there was a tradition and considering the hundreds of ways to mix FAC and PF in the literature there is hardly a traditional

Download and read the MSDS sheets at ScienceLab.com before using all chemicals. Exercise caution in the handling of all photographic chemicals. Use of any such chemicals constitutes some risk, and some are poisonous. The workplace should be well ventilated. Chemicals should be mixed only in the manner described. Avoid contact between the chemicals and eyes, skin, clothing, and furniture. Do not eat or drink while using chemicals. Keep them away from pets and children. Wear protective eye wear and gloves if necessary. In short, be mindful of all safety procedures for yourself and others.

formula. I asked Mike Ware his preference, and he referred to his FAO formula as "new." Sam Wang cinched it for me by differentiating the two with the terms "classic" and "new," conveniently CC and NC, and that seemed right, because *classic* implies old in a *good* way.

It is surprising that there are so many classic cyanotype formulas, given there are only two ingredients, ferric ammonium citrate (FAC, the more light-sensitive part of the compound), and potassium ferricyanide (PF, also light-sensitive but not to the same degree as FAC[1]). It is precisely this immense flexibility that makes cyanotype so user-friendly. You could probably throw a handful of the two chemicals together in some water and get an acceptable print. For those who do not want to spend a lot of time reading about cyanotype and just want to make a print, the following classic cyanotype formula works well. Following this formula, we will discuss my favorite formula as well as others, and the finer details of chemistry, formula ratios, and added ingredients for those who want a more in-depth discussion of the process.

Classic cyanotype formula

For repeatable and exact results, it is best to measure everything out with a gram scale. However, the tablespoon and teaspoon measurements are close enough to the suggested gram amounts if you don't own a gram scale. As long as you are consistent with the way you measure, you can achieve consistent results.

Although the often-repeated formula in contemporary literature is a 20%/8% solution, this solution of 20%/10% is closer to the optimum ratio to be as fast as possible without the potential for the shadows bleeding into the highlights which problem will be discussed further on.[2]

Solution A 20% ferric ammonium citrate
20 g ferric ammonium citrate (5 tablespoons)
100 ml distilled water

Mixing Solution A:

1. Measure out 20 g of ferric ammonium citrate and add it to 75 ml distilled water and stir until dissolved.
2. Add water to make 100 ml total volume.
3. Store in a brown eyedropper bottle and label the bottle "Solution A."

Solution B 10% potassium ferricyanide
10 g potassium ferricyanide (2 teaspoons)
100 ml distilled water

Mixing Solution B:

1. Measure out 10 g of potassium ferricyanide and add it to 75 ml distilled water and stir until dissolved.
2. Add water to make 100 ml total volume.
3. Store in a brown eyedropper bottle and label the bottle "Solution B."

At time of use:

1. Shake both solutions before using to make sure the chemistry is evenly dispersed.
2. Mix equal proportions of Solution A and Solution B. 2 ml (about ½ teaspoon) of combined solution are enough for an 8″ × 10″, 4 ml (scant 1 teaspoon) for an 11″ × 14″, but mix a few ml extra to account for initial brush absorption. [Note: there are 5 ml in a teaspoon.]

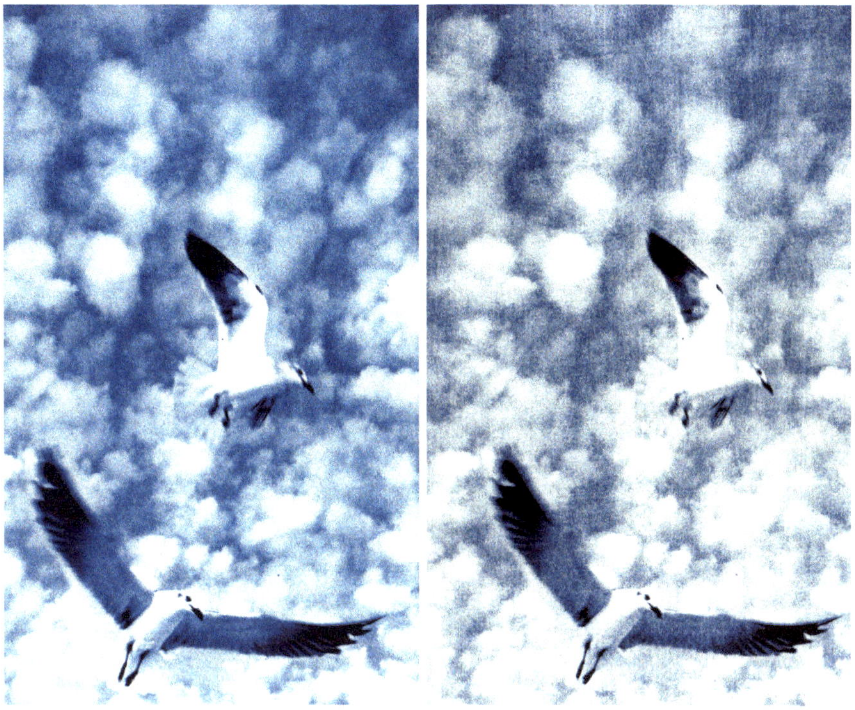

Figure 5.3. The detail of a seagull image on the left side is a 10/10 cyanotype print, and on the right side is a 20/10 cyanotype print, same paper. Note how that grainy, scratchy looking texture crops up with the increase in ferric ammonium citrate. This happens even more so in the winter months and in dry climates and with a hard-sized paper. If you are getting this kind of scratchiness, use less ferric ammonium citrate and more potassium ferricyanide as I now do. Note also the more turquoise blue of the 10/10, too, and as you can see in the deepest shadows, the Dmax of a 10/10 print is perfectly adequate.

My favorite classic cyanotype formula

With some papers—actually, more than a few—the typical 20/10 CC formula will produce gritty, grainy results wherein there is much wash off in the development baths and the final print has a scratchy appearance. This is greatly exacerbated if you live in a dry climate as I do, and in winter months when the humidity drops. My guess is those who live in humid climates rarely see this, and certainly if you are in the habit of humidifying your darkroom for other processes, you will probably not see this either.

When I first began testing papers for cyanotype I decided to level the playing field and use equal proportions of FAC and PF. I made my decision to do so based on the fact that there were so many varying proportions of these two ingredients in the literature. I decided to start with 10% solutions of both FAC and PF and then from there test increasing amounts of FAC.

I was surprised to find out that I *preferred* the more turquoise color of the 10/10, the washoff in the development baths was minimal, the exposure scale (ES) was long and smooth, rivaling the new cyanotype, and the exposure time was most often *faster* than that for 20/10! This is not what the literature says, including my previous books, I am embarrassed to say. It is thought the more Solution A, the faster the exposure, and, in fact, it is often suggested to double Solution A when coating, using a 2A:1B ratio to make exposure faster. I did not find this to be true with hundreds of step wedge prints.

What, therefore, is not to like about 10/10? The only drawback is that the Dmax, or how dark the darkest blue, is a modicum less, a difference not noticeable unless one were to put two prints side by side. For ease of use, I don't mind this trade-off.

I am not proposing to reinvent the cyanotype wheel, but I do encourage you to give the 10/10 formula a try and see for yourself if you prefer it, especially in the winter months and in areas of low humidity.

For repeatable and exact results, it is best to measure everything out with a gram scale. However, the tablespoon and teaspoon measurements are close enough to the suggested gram amounts if you don't own a gram scale. As long as you are consistent with the way you measure, you can achieve consistent results.

Solution A 10% ferric ammonium citrate
10 g ferric ammonium citrate (2 tablespoons)
100 ml distilled water
Mixing Solution A:

1. Measure out 10 g of ferric ammonium citrate and add it to 75 ml distilled water and stir until dissolved.
2. Add water to make 100 ml total volume.
3. Store in a brown eyedropper bottle and label the bottle "Solution A."

Solution B 10% potassium ferricyanide
10 g potassium ferricyanide (2 teaspoons)
100 ml distilled water
Mixing Solution B:

1. Measure out 10 g of potassium ferricyanide and add it to 75 ml distilled water and stir until dissolved.
2. Add water to make 100 ml total volume.
3. Store in a brown eyedropper bottle and label the bottle "Solution B."

At time of use:

1. Shake both solutions before using to make sure the chemistry is evenly dispersed.
2. Mix equal proportions of Solution A and Solution B. 2 ml (about ½ teaspoon) of combined solution are enough for an 8″ × 10″, 4 ml (scant 1 teaspoon) for an 11″ × 14″, but mix a few ml extra to account for initial brush absorption. [Note: there are 5 ml in a teaspoon.]

Bleeding

Bleeding is where the darkest blues or shadows of the image migrate into the highlight areas with a sort of blue, creeping stain. Alberto Novo, an Italian scientist, found in research through patents (specifically US Patent no. 2,350,991) that if the ratio of FAC:PF is no more than a 20–23% FAC to a 10% PF, bleeding is prevented, and the blue is deeper and the contrast better. This keeps the light sensitive iron in ferric ammonium citrate to no more than 2× the potassium ferricyanide (calculated with moles not grams).

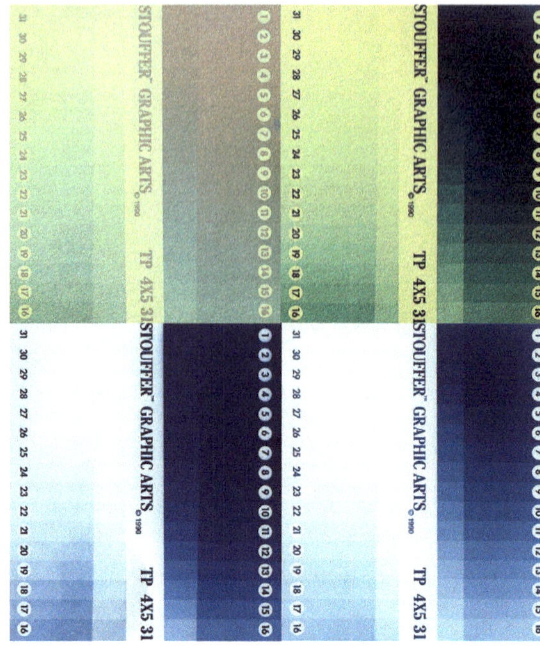

Figure 5.4. This is a composite of two step wedges before and after development. The left side is 4 FAC:1 PF and the right side is 1FAC: 4 PF. Note the bleeding that happens when too much FAC is used. Both are extremes that you would *never* use in practice (unless you are one who uses 2A:1B), but this visual illustrates how forgiving the process is if you don't mix it exactly to proportion, and also that using such an excess of PF, which is never done today, still produces a beautifully smooth tonal range with a more turquoise blue.

Novo's cyanotype formula

Solution A
18 g green ferric ammonium citrate (15.1 g brown)
1 drop of 50% thymol solution (5 g thymol in 10 ml isopropyl or ethyl alcohol)
100 ml distilled water
Mixing Solution A:

1. Add the ferric ammonium citrate to 75 ml water and stir.
2. Add water to make 100 ml total volume.
3. Store in a 100 ml brown eyedropper bottle with a drop of thymol added to prevent mold and label the bottle "Solution A."

Solution B
8 g potassium ferricyanide
100 ml distilled water
Mixing Solution B:

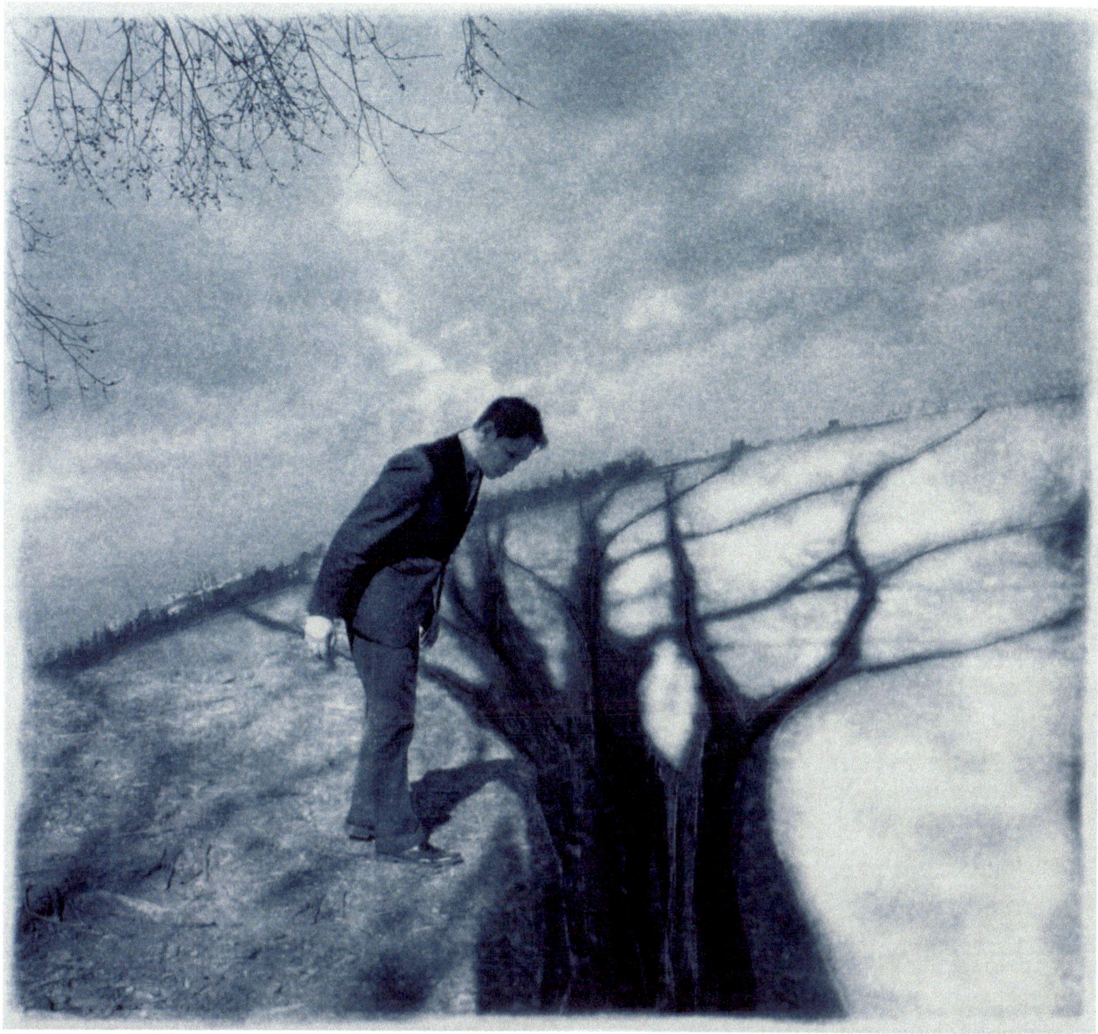

Figure 5.5. *The River Crossing*, collaged cyanotype print with digital drawing, 15.7″ x 17″ © Jonah Calinawan 2013

1. Add the potassium ferricyanide to 75 ml water and stir.
2. Add water to make 100 ml total volume.
3. Store in a 100 ml brown eyedropper bottle and label the bottle "Solution B."

At time of use:

1. Shake both solutions before using to make sure the chemistry is evenly dispersed.
2. Mix equal proportions of Solution A and Solution B. 2 ml (about ½ teaspoon) of combined solution are enough for an 8″ × 10″, 4 ml (scant 1 teaspoon) for an 11″ × 14″, but mix a few ml extra to account for initial brush absorption. [Note: there are 5 ml in a teaspoon.]

Brown vs. green FAC

There are two forms of ferric ammonium citrate, a brown (in historical literature sometimes referred to as "red"[3]) and a newer form that is chartreuse yellow-green. Before 1897 the commonly available kind was brown, which was slow to expose. Cyanotype is slow enough to expose as it is, so this was problematic. In 1897 green ferric ammonium citrate was introduced by Eduard Valenta. This new green form was said to make exposure eight times faster (a bit of a stretch, but the equivalent of three stops exposure)[4] and it of course became the preferred ferric ammonium citrate. Green ferric ammonium citrate is the most commonly used kind today, though you can still find the brown scales.

Valenta's cyanotype formula

This is the original Valenta formula from 1897 when he introduced the green form of ferric ammonium citrate for the first time, included here for historic interest.[5] Note that Valenta's ratio moves into a potential "bleeding range."

Solution A

25 g green ferric ammonium citrate

100 ml distilled water

Mixing Solution A:

1. Add the ferric ammonium citrate to 75 ml water and stir.
2. Add water to make 100 ml total volume.
3. Store in a 100 ml brown eyedropper bottle and label the bottle "Solution A."

Solution B

9 g potassium ferricyanide

100 ml distilled water

Mixing Solution B:

1. Add the potassium ferricyanide to 75 ml water and stir.
2. Add water to make 100 ml total volume.
3. Store in a 100 ml brown eyedropper bottle and label the bottle "Solution B."

At time of use:

1. Shake both solutions before using to make sure the chemistry is evenly dispersed.
2. Mix equal proportions of Solution A and Solution B. 2 ml (about ½ teaspoon) of combined solution are enough for an 8″ × 10″, 4 ml (scant 1 teaspoon) for an 11″ × 14″, but mix a few ml extra to account for initial brush absorption. [Note: there are 5 ml in a teaspoon.]

Ratio of FAC:PF

In over one hundred formulas from 1844 on, the ratio of ferric ammonium citrate to potassium ferricyanide varied from 9⁄10 as much FAC to PF to one outlier formula that had almost sixteen times as much FAC to PF (and perhaps was recorded mistakenly). How can such a simple process that involves only two ingredients become so diverse? It is primarily this ratio of the two chemicals that accounts for the wide variety in formulas over 175 years. One author seemed to think the reason for this variety is that Prussian blue (ferric ferrocyanide) and Turnbull's blue (ferrous ferricyanide) are different colors, and that "by varying the percentage of the two ferric salts forming the sensitizing solution, the color of the blue may be varied thereby. Hence the difference in the formulas given by different authors."[6]

I compiled a chart of the various formulas as best as I could and include it here at the end of this chapter for historical interest. Complicating factors are that it is often not stated whether the brown or the green form of ferric ammonium citrate is used; before 1897 it was probably all the brown form but after that date, who knows? It was often not stated whether the chemistry is added to the recommended volume of water or *up to* the recommended volume, which would result in slightly different percentage solutions. Also, it is often assumed we know what measuring system is in use, but measurement systems vary— US, British, metric, avoirdupois, apothecary, grains, grams, or who knows what else.

When Herschel originated the process, he did not state any specific ratio of the two chemicals. It was only in his Memoranda where ratios are stated, and in the beginning the ratios went from more potassium ferricyanide than ferric ammonium citrate to more ferric ammonium citrate than potassium ferricyanide.[7]

What was said in the late 1890s is probably still true today: "Success is not so much dependent upon the proportions of the ingredients, for the process allows considerable latitude in this respect, and the formulae given for general use vary somewhat; nor does it require chemically pure material, that sold by druggists generally being of the requisite purity; but it does require carefulness in the preparation and use."[8]

Suffice it to say that in over one hundred formulas from 1844 to present day, few have less ferric ammonium citrate than potassium ferricyanide, and since 1900 the amount of FAC in proportion to PF has increased, only partly because more green FAC is required to equal the brown variety. The give-and-take balance between the two chemicals can be summarized as follows:

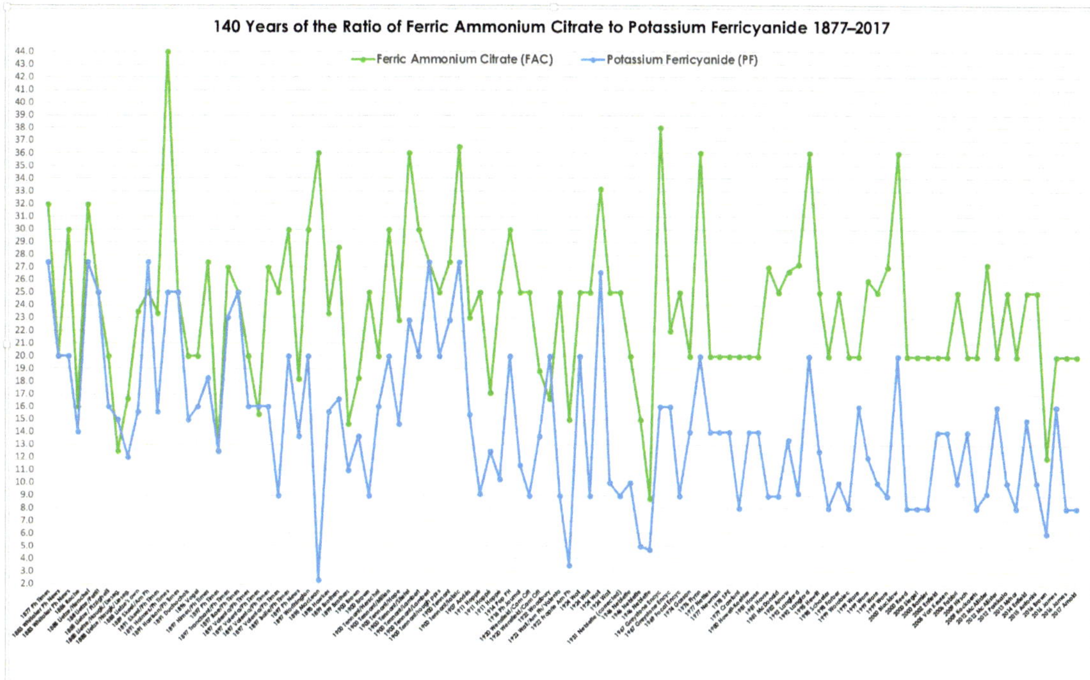

Figure 5.6. The takeaway from this chart is this: in 140 years of practice, there has been quite a variance in the gram amounts of ferric ammonium citrate (FAC) and potassium ferricyanide (PF) used and in the ratio between the two chemicals. In general, though, almost all formulas contain more grams of FAC than PF. And in general, a greater proportion of PF was used in the formulas before modern times. Today a 20 g/8 g solution is often repeated from one book to another as if it is cut in stone, when it is not, and when there are better ratios to use that produce smoother tones, less grain, and faster speeds.

More FAC

- Streaks more when coating
- Absorbs less readily into the paper (you can see minute specks on a freshly coated surface when viewed obliquely)
- Faster exposure speed (not in my practice!)
- More grain
- More bleeding
- More washoff
- More staining of the highlights
- More paper fogging
- More Dmax (darker darks)
- More reddish-blue

More PF

- Streaks less to none when coating
- Absorbs more readily into the paper
- Slower exposure speed (not in my practice!)
- Finer grain
- Less bleeding of darks into highlights
- Less washoff
- Less staining of the highlights
- Less paper fogging
- Less Dmax[9]
- More greenish blue

In 1883 Channing Whitaker tested cyanotype formulas from 2% ferric ammonium citrate: 20% potassium ferricyanide to 20% ferric ammonium citrate: 2% potassium ferricyanide. His conclusion was that having ferric ammonium citrate 1.5× the potassium ferricyanide was about the best, or 15 g ferric ammonium citrate and 10 g potassium ferricyanide in 100 ml water. Whitaker was using brown ferric ammonium citrate, not green, and in a single solution mix,[10] so his ideal formula would equate to 18 g green ferric ammonium citrate to 10 g potassium ferricyanide.

There is probably no one holy grail formula, but there is a sweet spot where the formula has enough ferric ammonium citrate to make it as dark as possible, but not too much so graininess, scratchiness, washoff, and bleeding occur.

Whitaker's cyanotype formula

This formula is the one that Whitaker settled upon as the best. If he was going to print coated paper right away, he would increase the ferric ammonium citrate, because he wasn't as worried that the increase in FAC would fog.[11] Note that this formula is fairly close to Novo's recommendation of 18/8, with slightly more potassium ferricyanide (which I find makes a less grainy print). Note also this is a combined solution stored in one bottle, but you can always mix separate solutions. One last point: this is double the strength of a normal formula; a normal two-solution formula is a combined total of 200 ml water whereas this is a combined total of 100 ml water, so more chemistry will be delivered to the paper surface at time of coating.

15 g brown ferric ammonium citrate (18 g green)
10 g potassium ferricyanide
100 ml distilled water

1. Measure out 75 ml water.
2. Add the ferric ammonium citrate and stir.
3. Add the potassium ferricyanide and stir.
4. Add water up to 100 ml.
5. Store as a single solution mix in a 100 ml brown eyedropper bottle.

At time of use:

2 ml (about ½ teaspoon) of combined solution are enough for an 8″ × 10″, 4 ml (scant 1 teaspoon) for an 11″ × 14″, but mix a few ml extra to account for initial brush absorption. [Note: there are 5 ml in a teaspoon.]

Combining solutions and shelf life

In the modern literature on cyanotype, ferric ammonium citrate and potassium ferricyanide are kept in separate solutions A and B and mixed at time of use. Often books will parrot the information that once Solution A and B are mixed, the combined solution only lasts a day until it goes bad, turns blue, gets contaminated, slows down, veils the highlights of the print, and doesn't work. This is part myth and part truth. Out of one hundred formulas spanning 175 years, close to one-third of the formulas called for the ferric ammonium citrate and the potassium ferricyanide to be combined in one solution. Two-thirds called for the two solutions to remain separate and be mixed together at time of use. The practice of mixing all the chemistry in one solution fell out of favor over the decades, but the fact that a combined solution existed in practice is proof that at the very least the combined solution does not expire in a day.

Separate or combined, the solutions are light sensitive and should be stored in brown or light-tight bottles. Best practice is to keep the solutions separate, but if combined, a solution will last, by some accounts, up to a year, even if it has turned blue.[12] I used a seven-month old formula and it was blue but worked fine. My practice is to store separate solutions, but whatever I combine, I use up until it is gone, even if it takes days.

Classic cyanotype with oxalic acid

Oxalic acid added to a cyanotype formula was said to print faster and result in brighter highlights.[13] The formula is often repeated in 250 ml amounts, probably because of the minute amounts of chemistry needing to be measured, but here it is translated to 100 ml amounts like the rest of the formulas in this book. It is merely the Valenta formula with added oxalic acid and ammonium dichromate.

Solution A

25 g ferric ammonium citrate

0.5 g oxalic acid

Distilled water to make 100 ml

Mixing Solution A:

1. Add the ferric ammonium citrate to 75 ml water and stir.
2. Add the oxalic acid and stir.
3. Add water to make 100 ml total volume.
4. Store in a 100 ml brown bottle and label the bottle "Solution A."

Solution B

9 g potassium ferricyanide

0.5 g oxalic acid

0.2 g ammonium dichromate

Distilled water to make 100 ml

Mixing Solution B:

1. Add the potassium ferricyanide to 75 ml water and stir.
2. Add the oxalic acid and stir.
3. Add the ammonium dichromate and stir.
4. Add water to make 100 ml total volume.
5. Store in a 100 ml brown bottle and label the bottle "Solution B."

At time of use:

1. Shake both solutions before using to make sure the chemistry is evenly dispersed.
2. Mix equal proportions of Solution A and Solution B. 2 ml (about ½ teaspoon) of combined solution are enough for an 8″ × 10″, 4 ml (scant 1 teaspoon) for an 11″ × 14″, but mix a few ml extra to account for initial brush absorption.

Double coating cyanotype

It is said that the more chemistry on the paper, the darker blue in the resulting print[14] and therefore some cyanotypists coat the paper twice to get deeper blues. I have never felt a need to double coat to get a deeper blue than is already provided by a single coat except on a few papers—almost always lightly sized or unsized printmaking papers—that are very absorbent and seem to require more solution to cover, in which case I use more solution when coating. In other words, although double coating is not necessary, it is a technique you may find fruitful to test.

Mold in Solution A

Over time Solution A will grow mold. It'll look like brown globs. If Solutions A and B are combined, mold doesn't happen. If Solution A is separate, which it usually is, and mold appears, merely filter the solution through a coffee filter and return it to the bottle. It tends to grow less mold after the first filtering. Mold is not something I have ever worried about, but the *smell* can be offensive to a sensitive nose. There are some quick fixes:

- Add rubbing alcohol to the tune of 5% of the formula (e.g. add 5 ml for every 100 ml of Solution A)[15]
- Add a few drops of 15% thymol dissolved in methylated alcohol[16]
- I always have thymol in my dimroom but if you do not, use Listerine. Your Solution A will smell minty fresh
- Add a few drops of formalin per 100 ml if you happen to have this on hand, which nowadays most people don't
- Add 1 gram of ammonium dichromate to the solution.[17] This affects exposure and contrast, though, whereas the other methods don't affect sensitizer performance.

Adding dichromate to the formula

Dichromate added to the classic cyanotype formula accomplishes three things: it prevents mold growth in Solution A as stated above, it helps sensitized paper last longer, and it increases print contrast. The drawbacks are that exposure slows, you need another chemical in the dimroom, and you have to be cautious with dichromate because it is toxic.

Dichromate is not necessary. If mold is a bother there are other methods to prevent it. For increasing contrast with dichromate, you can choose a higher

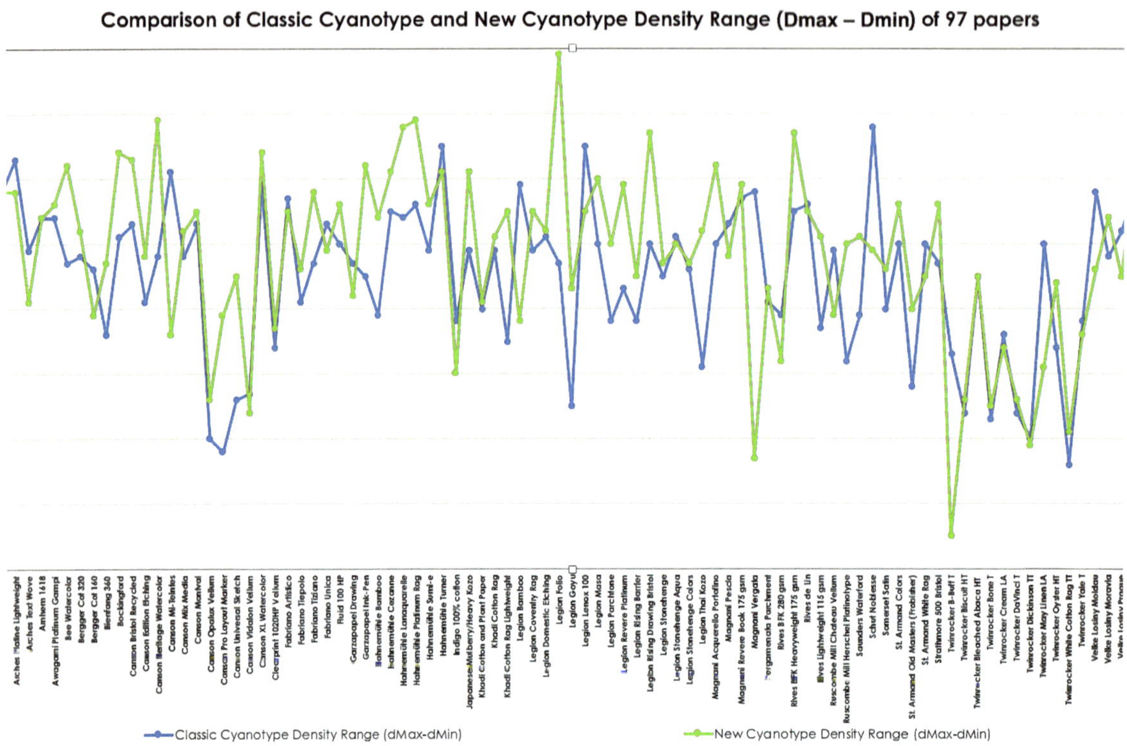

Figure 5.7. The fun thing about collecting data on 136 papers is being able to make visuals of your findings. Though charts drive some people crazy, I love them! Here is a visual to show the density range (abbreviated as DR) of classic cyanotype and new cyanotype measured on an X-Rite Densitometer in Visual Mode. A densitometer measures how much light is reflected off the surface of the paper in reflection mode. DR is the densitometer measurement of the deepest blue *minus* the paper white (Dmax–Dmin). This chart shows that the DR of new cyanotype is generally greater than the DR of classic cyanotype, but not on every paper. DR is both process and paper dependent, and on some papers the classic formula has a greater DR, as evidenced where the parallel points of the blue line rise higher than those on the green line. Is this information necessary to know how to make great cyanotype prints? Absolutely not, except to note that there are always exceptions to generalizations and that, again, paper choice is probably the most important variable at your disposal. Note that this comparison is between a 10/10 classic cyanotype and a 200 ml new cyanotype so if a 20/10 classic and 100 ml new are used, the graphed DR lines for both would most likely move slightly higher.

contrast curve. To keep coated paper good for a long time you can refrigerate it.

Dichromate does come in handy with low contrast *film* negatives. It also is good for preserving coated paper. You can add up to 1 g per liter of Solution A and coated paper will last 3–5 months.[18] To add it to the coating solution only at time of use, mix a 1% solution (1 g in 100 ml) and at time of coating, add 1–6 drops to every 2 ml sensitizer which will increase contrast as much as one stop.[19]

Tips

- Potassium ferricyanide is soluble up to 33%[20] whereas ferric ammonium citrate, which is hygroscopic, meaning it absorbs water from the atmosphere, is very soluble—1200 g per liter! Thus keeping the container tightly capped will prevent it turning into a lump of gooey green. It still works fine, but it's harder to measure it accurately.
- Potassium ferricyanide when exposed to light when dry or in solution is decomposed to

potassium ferrocyanide and can result in a blue precipitate when it is mixed with ferric ammonium citrate upon making your coating solution. This will result in stained highlights.[21] Back in the day they used to rinse the crystals and dry between blotting paper[22] or merely brush off the yellow crystals on top of the red crystals to get to the undecomposed potassium ferricyanide below.[23] If all of a sudden your cyanotype exposure is proceeding at a snail's pace when it used to proceed much faster, check your potassium ferricyanide.[24]

- Potassium ferricyanide should not be heated above 300°F or allowed to come in contact with any concentrated acid such as acetic acid in stop bath, because poisonous hydrogen cyanide gas can be produced![25]
- In my research I came across a single solution classic cyanotype formula with added gum Arabic. You can get the same effect at time of coating by just adding 2 drops of gum Arabic per ml just before coating the paper, if desired.
- An addition of 5 g of potassium oxalate per 100 ml of sensitizer (5%) was used back in the day to increase speed, but the paper was said to not keep as long.[26]

Endnotes

1. Ware, Mike. *Cyanomicon II, History, Science and Art of Cyanotype: photographic printing in Prussian blue*. Buxton: self-published, 2016, pp. 52–57, 96. http://www.mikeware.co.uk/mikeware downloads.html.
2. Ibid., pp. 91–92. Ware recommends mixing equal parts of a 20% brown ferric ammonium citrate with a 16% potassium ferricyanide, or a 26% green ferric ammonium citrate with a 12% potassium ferricyanide (w/v). The differences in formulas is because of the differing iron contents of the brown and green ferric ammonium citrate. His 26%/12% recommendation is equivalent to 20%/9.23%, which is slightly rounded up here to 10%.
3. Brown, George E. *Ferric and Heliographic Processes: A Handbook for Photographers, Draughtsmen, and Sun Printers*. New York: Tennant & Ward, 1900 [First edition 1900, reprinted 1905, this edition n.d.], p. 10.
4. Valenta, E. "The Use of Various Iron Salts in Printing Processes" in *The Photographic Times*, Vol. XXIX. New York: The Photographic Times Publishing Association, May 1897, pp. 237–238.
5. Valenta, pp. 237–238.
6. Duchochois, P. C. *Photographic Reproduction Processes*. New York: The Scovill & Adams Company, 1891, p. 34.
7. Ware, pp. 82–83.
8. Himes, Charles F. "Blue-Print Solution Will Keep" in *The Photographic Times*, Vol. XXIX. New York: The Photographic Times Publishing Association, May 1897, pp. 223–224.
9. James, Christopher. *The Book of Alternative Photographic Processes*. New York: Delmar Cengage, 2016, p. 173.
10. Whitaker, Channing. "Apparatus for Printing by the Blue Process" in *The Photographic News*, Vol. XXVII. London: Piper and Carter, December 7 1883, p. 359.
11. Ibid., p. 359.
12. Tennant, John A., ed. "The 'Blue Print' and its Variations" in *The Photo-Miniature*, Vol. 1, No. 10, January 1900, p. 489; Enfield, Jill. *Jill Enfield's Guide to Photographic Alternative Processes, Popular Historical and Contemporary Techniques*. Burlington, Massachusetts: Focal Press, 2014, p. 67.
13. Langford, Michael. *The Darkroom Handbook*. New York: Alfred A. Knopf, 1992, p. 328.
14. James, p. 172.
15. Bolle, Oscar. "Prints in Turnbull's Blue" in *The Photographic Times*, Vol. XXIX. New York: The Photographic Times Publishing Association, December 1897, pp. 579–580.
16. Seigel, Judy ed. *The World Journal of Post-Factory Photography*, Issue 5. New York: Post Factory Press, August 2000, p. 32.
17. Hirsch, Robert. *Photographic Possibilities*, 3rd ed. Boston: Focal Press, 2009, p. 199.
18. "Ferro-Prussiate Paper" in *The Photographic News, The Journal for Amateur Photographers*, Vol. XLI. London: Photographic News, January 15 1897, p. 48. The 3–5 month time frame is repeated in Hawkes, Lieut. H. P. *Photography in a Nutshell, by "The Kernel."* London: Iliffe, Sons & Sturmey Ltd, 1898, p. 91. In Brown, p. 18, it is less, no more than 1 g of dichromate per liter of combined solution.
19. Crawford, William. *Keepers of Light*. New York: Morgan and Morgan, 1979, p. 166.
20. Ware., p. 90.
21. Tennant, p. 491.
22. Brown, pp. 118–119.
23. Tennant, p. 491.
24. Blacklow, Laura. *New Dimensions in Photo Processes*, 3rd ed. Boston: Focal Press, 2000, p. 102.
25. Hirsch, p. 199; Farber, Richard. *Historic Photographic Processes, A Guide to Creating Handmade Photographic Images*. New York: Allworth Press, 1998, p. 61.
26. Wall, E. J. "The Iron Salts I" in *American Photography*, Vol. XVI, November 1922, pp. 677–688.

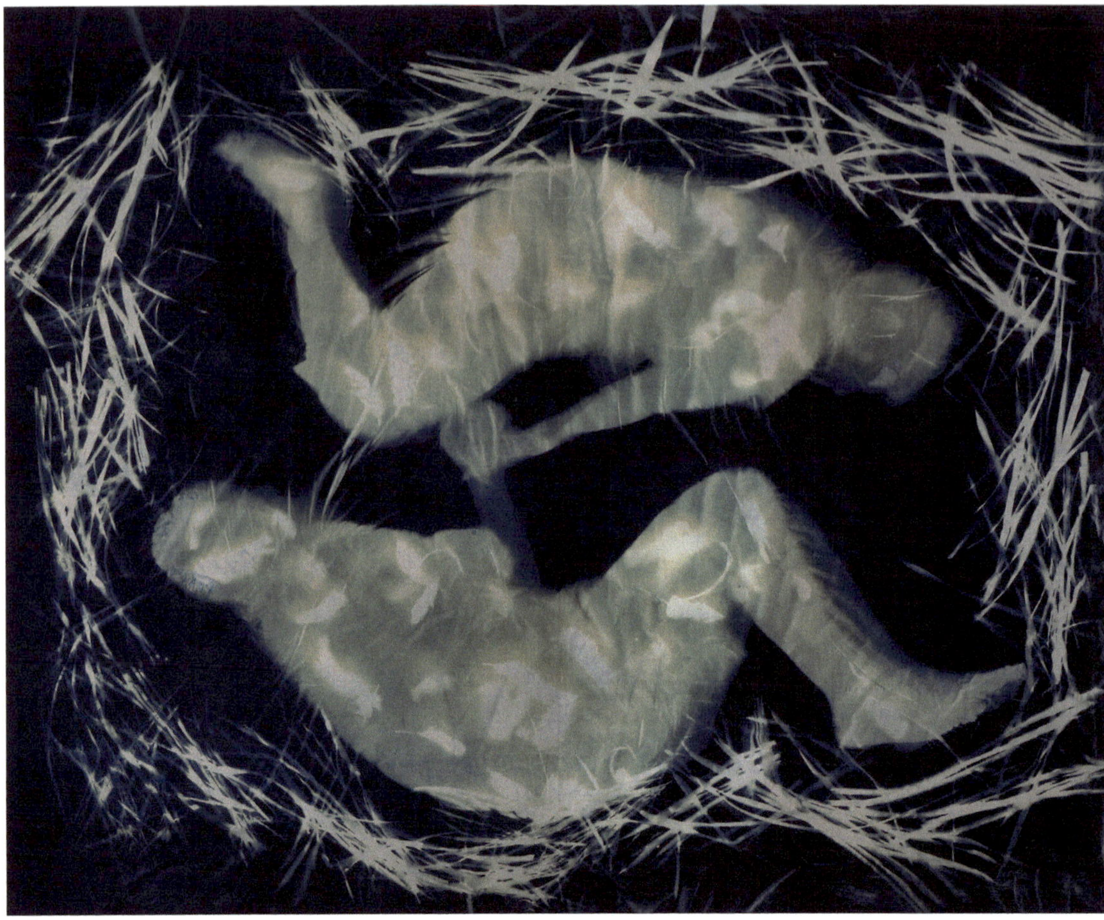

Figure 5.8. *The Nest*, classic cyanotype on fabric, 86″ x 72″ © Vicki Reed 2015. "When faced with my elderly parents fading away through dementia and memory loss I wanted to make a lasting record of them and all they were leaving behind as well as have a way to process my ongoing grief and feeling of loss. I realized it would be more meaningful if I could include my parents themselves. At their age (mid-eighties) I had to solve the mobility issue as they found it difficult to get up and down. I solved the problem by placing a queen size bed in my garage where I could place the fabric and then them on top of it in subdued light and then wheel them out in the sun for an exposure. They thought the entire experience was a lot of fun. We made it a party with Tony Bennett songs playing during the exposure time and then fruit smoothies and pastries on the back porch afterward.

"I use precoated fabric that I buy online from Blueprints On Fabric. When creating my fabric pieces, I use Inkaid transparencies for digital positives. I use the alcohol transfer process to add them to my cyanotype pieces. I usually expose my fabric pieces in the sun for 15–20 minutes depending on time of day, time of year and how sunny a day it is. If I am doing a double exposure, the second exposure is much shorter, about two minutes. I rinse the fabric in warm water until there is no more yellow or green color. Then I do a final soak/ rinse in water with hydrogen peroxide added. I hem the edges and create a pocket at the top to accommodate a curtain rod or pole to allow them to be displayed as wall hangings.

"I enjoy working with pre-coated cotton fabric to create life size portraits. To solve the problem of having enough time to design a piece I use a queen size bed in my garage. I can take my time in the subdued light to complete my design and then I wheel the bed out into the light for an exposure. I place the cyanotype fabric on the bed and then place the person and objects on top of the fabric. I use safety pins around the edges to pin the fabric to the bed to keep it from blowing in the breeze. With lightweight objects that can blow away I often place a plastic painter's tarp over the objects and fabric and use straight pins to secure the plastic and fabric together. The plastic keeps the objects intact and allows the sun to shine through and expose the fabric."

Vicki Reed is a former news photographer and magazine art editor, using plastic, vintage, pinhole and digital cameras to explore her personal and natural environment. She experiments with alternative processes including lith, encaustic, lumen, photogravure and cyanotype. Her work has been featured nationally and internationally in exhibits, books and magazines. To see more of her work, visit **vickireed.com**.

Historical formulas chart

Following is a chart of over one hundred formulas. To me, most interesting is how the ratio of ferric ammonium citrate rose in proportion to potassium ferricyanide through the years. It was closer to 1:1 to 1.3:1 in the early days and then quite obviously 2:1 today. Also of interest are the various additions to the formulas, noted in the right column of the chart. Third is noting how many of the formulas are used as a combined solution, denoted by "C."

Abbreviations probably don't need to be explained but are as follows: *brn* denotes brown FAC; *grn* denotes green FAC. FAO refers to ferric ammonium oxalate. G is grams, not to be confused with *gr* or *grs* which refer to *grains*. *Pt* refers to *part* or *parts*. S1:1 means separate solutions mixed in equal proportions at time of use. *Am di* and *pot di* are ammonium dichromate and potassium dichromate, am bromide and pot bromide likewise. *Ltr*, *ml*, and *oz* are liter, milliliter, and ounce as can be expected. *Opt* refers to an optional addition. To find the source for the formula, use the year and author and refer to the **Bibliography**.

YEAR	AUTHOR	Brn/ Grn	FAC	PF	WATER	Cmb/ Sep	ADDITIONS
1877	Ph Times	brn	140 grs	120 grs	2 oz total	C	
1883	Whitaker Ph News	brn	8 pt	8 pt	80 pt	C	1 pt gum
1883	Whitaker Ph News	brn	3 pt	2 pt	20 pt	C	
1886	Roche	brn	140 grs	120 grs	2 oz ea	C	
1888	Lietze/Herschel	brn	140 grs	120 grs	2 oz total	C	
1888	Lietze/Pettit	brn	1 oz	1 oz	4 oz ea solution	S1:1	
1888	Lietze/Pizzighelli	brn	10 pt	8 pt	100 pt total	C	
1888	Lietze/Haugk/Lieseg.	brn	5 pt	6 pt	80 pt total	C	
1888	Lietze/Haugk/Lieseg.	brn	1 pt	3 pt	6 pts FAC/25 pts PF	S1:1	
1888	Lietze's own	brn	1 oz	5 oz	4.25 oz FAC/32 oz PF	S1:1	
1889	Skeel/Am Ph	brn	110 grs	120 grs	1 oz ea	S1:1	8 grs pot. bromide
1891	Skeel/Ph Times	brn	1.875 oz	1.25 oz	8 oz ea	S1:1	
1891	Holmes/Ph Times	brn	1 oz	1 oz	2.25 oz FAC, 8 oz PF	S1:2	
1891	Harrison/Ph Times	brn	1 oz	1 oz	4 oz ea	S1:1	5 drops 10% am. bromide per oz
1891	Duchochois	brn	20 pt	15 pt	100 pt ea	S1:1	
1896	Vogel	brn	10 g	8 g	50 ml ea	S1:1	
1897	Himes/Ph Times	brn	60 grs	60 grs	4 drs FAC, 6 drs PF	C	
1897	Ph Times	brn	1 pt	1 pt	8 pt ea	S1:1	
1897	Tranchant/Ph Times	brn	27 pt	23 pt	100 pt ea	C	
1897	Rae/Ph Times	brn	1 oz	1 oz	4 oz ea	S1:1	
1897	Valenta/Ph Times	brn	10 pt	8 pt	40 or 50 pt ea	S1:1	
1897	Valenta/Ph Times	grn	7.7 pt	8 pt	40 or 50 pt ea	S1:1	
1897	Valenta/Ph Times	grn	13.5 pt	8 pt	100 pt total	C	
1897	Valenta/Ph Times	grn	12.5 g	4.5 g	50 ml ea	S1:1	
1897	Bolle/Ph Times	grn	15 g	8 g	50 ml FAC/40 PF	C	1 g pot bi per 900 ml
1897	Ph News		80 grs	60 grs	1 oz ea	C	2 grs pot bi per oz mixture
1897	Worthington		1.5 oz	1 oz	10 oz	C	

YEAR	AUTHOR	Brn/ Grn	FAC	PF	WATER	Cmb/ Sep	ADDITIONS
1898	MacLean	brn	475 grs	30 grs	6 oz	C	30 minims liquor ammonia
1898	Hawkes		1.875 oz	1.25 oz	8 oz ea	S1:1	I grain pot bi per 2 oz mixture
1899	Brothers		4 oz	2.5 oz	14 oz FAC, 15 oz PF	S1:1	
1899	Brothers		64 grs	48 grs	1 oz ea	S1:1	
1900	Brown	brn	183 g	137 g	1000 ml ea	C	opt. 1 g pot bi per ltr
1900	Brown	grn	250 g	90 g	1000 ml ea	C	
1900	Tennant/Herschel	brn	20 pt	16 pt	100 pt ea	S1:1	
1900	Tennant/Millen	brn	1.5 oz	1 oz	10 oz	C	
1900	Tennant/Nicol	brn	3 oz	2.25 oz	4 oz ea	S1:1	dilute 1:1 at time of use
1900	Tennant/Hope	brn	50 grs	256 grs	0.5 oz FAC, 4 oz PF	S1:1	1 gr pot bromide to 8 drams mix
1900	Tennant/Stiefel	brn	475 grs	300 grs	6 oz	C	30 minims stronger ammonia
1900	Tennant/Lambert	brn	150 pt	80 pt	500 pt FAC, 400 pt PF	S1:1	1 pt pot bi to PF
1900	Tennant/Lambert	brn	120 grs	120 grs	1 oz ea	S1:1	
1900	Tennant/rough pprs	brn	5 pt	4 pt	20 pt ea	S1:1	1 pt gum in FAC
1900	Tennant	grn	120 grs	100 grs	1 oz ea	S1:1	
1900	Tennant/fabric		160 grs	120 grs	1 oz ea	S1:1	
1907	Andés		15 pt	10 pt	65 pt ea	C	
1911	Hoppé	grn	1 oz	160 grs	4 oz ea	S1:1	
1911	Hoppé	brn	300 grs	1/2 oz	4 oz ea	S1:1	
1911	Hoppé		1 oz	180 grs	4 oz ea	S1:1	80 grs gum in FAC
1914	Pop PH	grn	1.5 oz	1 oz	10 oz	C	
1918	Ph Journal	grn	1 oz	200 grs	4 oz ea	S1:1	
1920	Wensfield/Cam Crft	grn	250 g	90 g	1000 ml ea	S1:1	
1920	Wensfield/Cam Crft	brn	188 g	137 g	1000 ml ea	S1:1	
1922	Woodbury		10 pt	20 pt	60 pt FAC/100 pt PF	S1:1	
1922	Wall/Am Ph/Valenta	grn	125 g	45 g	500 ml ea	S1:1	opt 0.5% oxalic acid, .05% pt bi
1922	Fraprie Am Ph	grn	150 g	35 g	1000 ml	C	
1924	Wall	brn	250 g	200 g	1000 ml ea	S1:1	
1924	Wall	grn	125 g	45 g	500 ml ea	C	opt 0.5% oxalic acid, .05% pt bi
1924	Wall	brn	166 g	133 g	1000 ml	C	33 g FAO
1924	Wall	FAO	250 g	250 g	1000 ml	C	250 g FAO
1924	Wall	grn	250 g	100 g	1000 ml ea	S1:1	for gelatinized plates
1931	Neblette (corrected)		250 g	90 g	1000 ml ea	S1:1	
1946	Neblette		800 g	400 g	8 ltr	C	150 g gum
1946	Neblette		600 g	200 g	8 ltr	C	150 g gum, 200 g FAO, 12.5 g ferric oxalate/1000 ml
1946	Neblette		350 g	187.5 g	8 ltr	C	187.5 g FAO, 300 g FAO, 50 g ferric oxalate, 150 g gum
1967	Greystone Encyc	grn	95 g	40 g	250 ml ea	S1:1	
1967	Greystone Encyc	brn	110 g	40 g	500 ml FAC/250 ml PF	S1:1	
1969	Focal Encyc	grn	250 g	90 g	1000 ml ea	S1:1	
1974	Gassan	grn	50 g	35 g	250 ml ea	S1:1	
1976	Flynn	grn	90 g	50 g	250 ml ea	S1:1	
1977	Nettles	grn	50 g	35 g	8 oz ea	S1:1	

YEAR	AUTHOR	Brn/ Grn	FAC	PF	WATER	Cmb/ Sep	ADDITIONS
1977	Newman	grn	50 g	35 g	250 ml ea	S1:1	
1978	SPE	grn	50 g	35 g	250 ml ea	S1:1	
1979	Crawford	grn	20 g	8 g	100 ml ea	S1:1	
1980	Howell-Koehler	grn	50 g	35 g	250 ml ea	S1:1	
1981	House	grn	50 g	35 g	250 ml ea	S1:1	
1981	House	grn	33.7 g	11.2 g	250 ml total	C	1.25 g oxalic acid, .25 g am di
1981	McDonald	grn	250 g	90 g	1000 ml ea	S1:1	
1982	Arnow	grn	31 g	15.5 g	116 ml ea	S1:1	
1992	Langford	grn	68 g	23 g	250 ml ea	S1:1	1.3 g oxalic FAC/1.3 g oxalic/0.5 g am di PF
1992	Langford	grn	90 g	50 g	250 ml ea	S1:1	
1995	Hewitt	grn	1 oz	1/2 oz	8 oz total	C	
1998	Schaefer	grn	10 g	4 g	50 ml ea	S1:1	1 drop gum per 12 mix; 1–2 drops 1% di per 1 ml mix
1998	Farber	grn	25 g	10 g	100 ml ea	S1:1	
1999	Worobiec	grn	20 g	8 g	100 ml ea	S1:1	
1999	Ware	brn	20 g	16 g	100 ml ea	S1:1	
1999	Ware	grn	26 g	12 g	100 ml ea	S1:1	
1999	Ware	grn	25 g	10 g	100 ml ea	S1:1	
1999	Ware	grn	27 g	9 g	100 ml ea	S1:1	0.5 g oxalic acid to FAC, 0.5 g oxalic acid/0.2 g pot di to PF
2000	Blacklow	grn	90 g	50 g	250 ml ea	S1:1	1.25 g oxalic acid
2000	Reed	grn	10 g	4 g	50 ml ea	S1:1	
2000	Siegel	grn	20 g	8 g	100 ml ea	S1:1	
2000	Barnier	grn	10 g	4 g	50 ml ea	S1:1	
2002	Enfield	grn	50 g	35 g	250 ml ea	S1:1	6 drops 1% pot bi per 2ml
2005	Van Keuren	grn	50 g	35 g	250 ml ea	S1:1	
2006	Fabbri	grn	25 g	10 g	100 ml ea	S1:1	
2009	Hirsch	grn	50 g	35 g	250 ml ea	S1:1	
2009	Kockaerts	grn	20 g	8 g	100 ml ea	S1:1	
2012	McAllister	grn	27.2 g	9.2 g	200 ml	C	1 g oxalic acid, 26.25 g gelatin
2012	McAllister	grn	20 g	16 g	200 ml	C	
2012	Porkkala	grn	25 g	10 g	100 ml ea	S1:1	
2013	Mrhar	grn	20 g	8 g	100 ml ea	S1:1	
2014	Enfield	grn	25 g	15 g	100 ml ea	S1:1	opt. 6 drops 1%/pot di/2ml
2015	Antonini	grn	25 g	10 g	100 ml ea	S1:1	
2016	Brown	grn	30 g	15 g	250 ml	C	
2016	James	brn	20 pt	16 pt	100 pt ea	S1:1	
2016	James	grn	100 g	40 g	500 ml ea	S1:1	opt 2 g oxalic to FAC and PF; 0.8 g am di to PF
2017	Arnold	grn	100 g	40 g	500 ml ea	S1:1	2 g oxalic added to FAC; 2 g oxalic and 0.8 g am di to PF

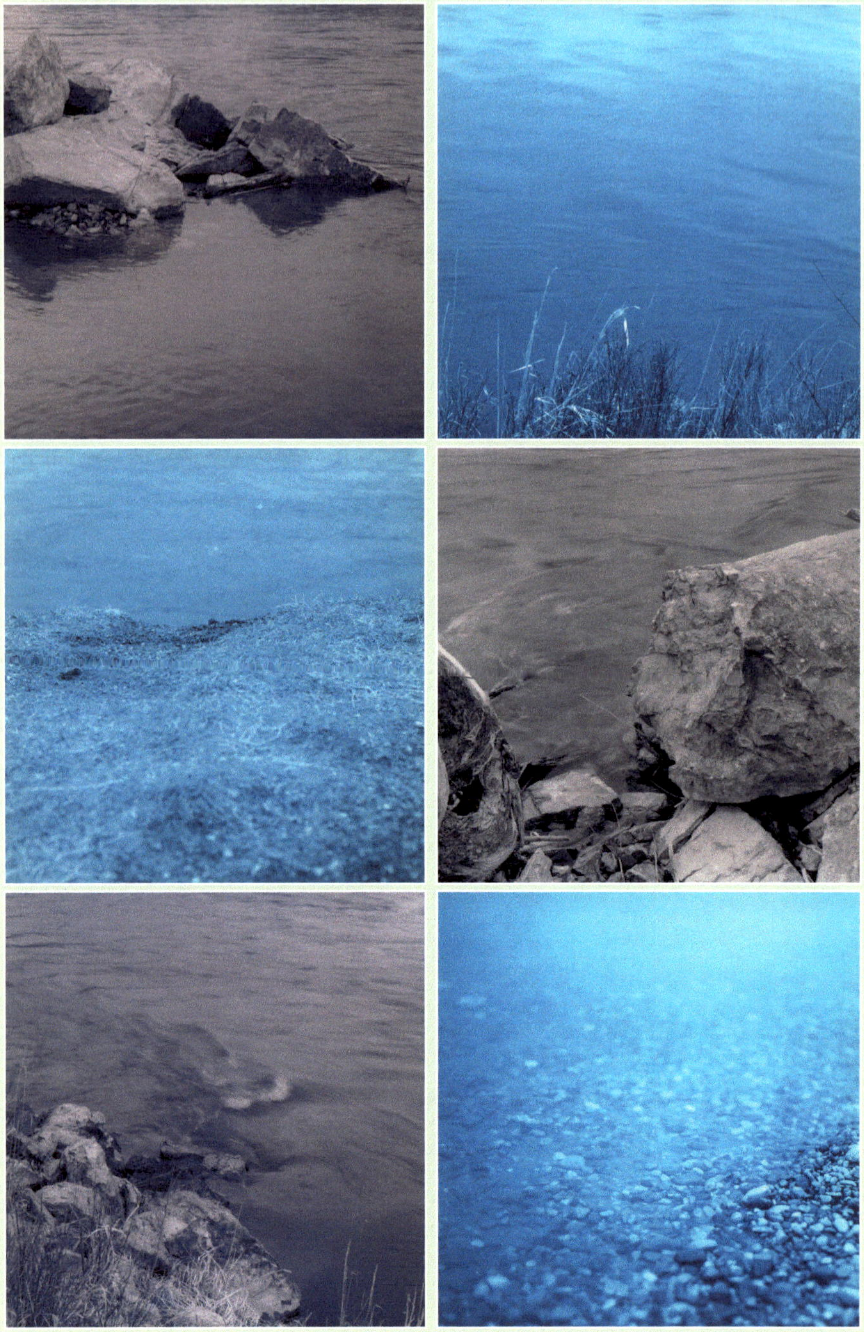

Figures 6.1–6.6. *Headwaters* series, 6″ x 6″ each © Emma Boardman 2018. "The Missouri Headwaters is a part of America's westward discovery and continues to play a role of sanctuary and historical evidence to this day. As you wander through the state park, you find yourself being guided by the three ever-bending and ever-flowing mighty rivers that create the Missouri. The Gallatin, the Madison, and the Jefferson entice all who pass by with either adventure or solace. These cyanotype prints accentuate the intimate details within this land and simplify the space into something that is felt deeply, as if for the first time. These prints are made with new cyanotype diluted to 200 ml on Arches Platine paper. Toning proceeded with a water soak, a 7–10 second ammonia bath, a water rinse, and a 2–3 minute tannic acid bath, followed by another water rinse of several minutes. This created a split-toned effect with deep dark blues, silver midtones, and pink highlights."

Emma Boardman (b. Washington) is a photographer and printmaker residing in Bozeman, Montana who prints a variety of19th century processes such as palladium, cyanotype, and gum bichromate. She received her Bachelor of Arts in Film and Photography at Montana State University, Bozeman (2018). To see more of her work, visit emmaprints.com.

Chapter 6

New Cyanotype

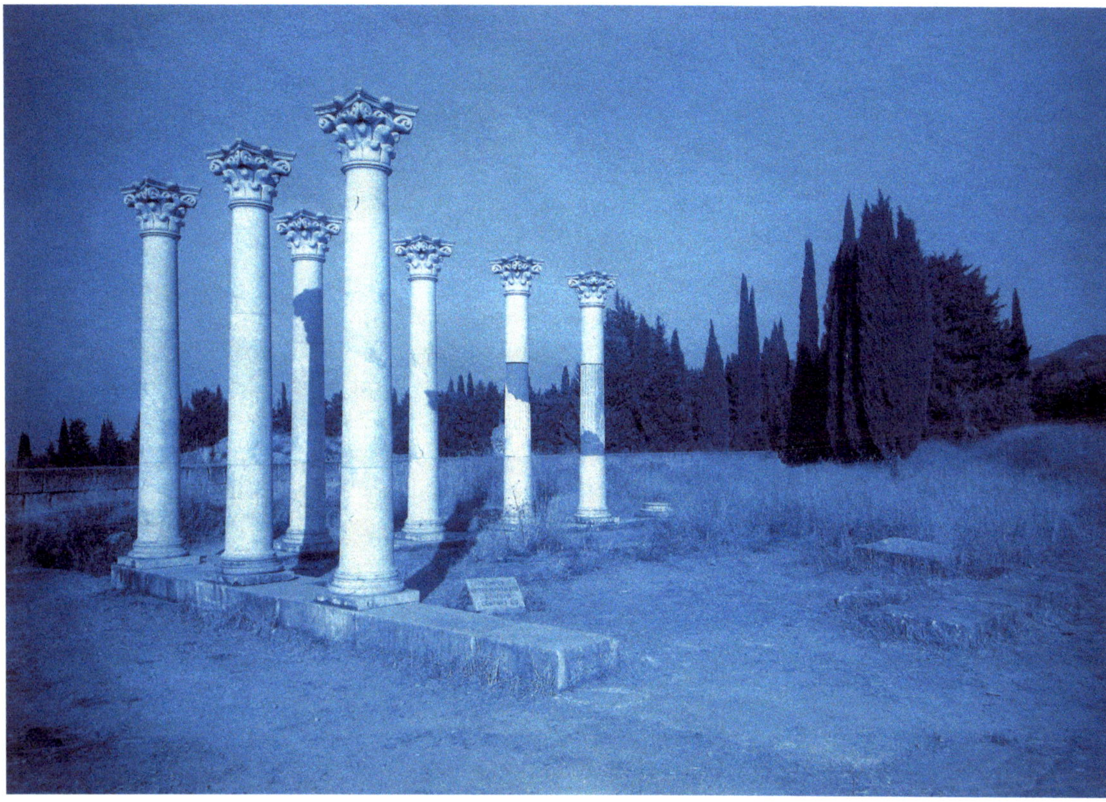

Figure 6.7. *Roman Temple, The Asclepion, Kos*, from an original 6 x 9 cm negative scanned and printed digitally with an Epson Photo R2880, 408 x 284 mm © Mike Ware 2008. "The blue of the Aegean sea and sky is the national colour of Greece. It's also the colour of sacredness, decorating Greek Orthodox churches. So the Prussian blue of my new cyanotype process seems fitting to render the images that I made in the Dodecanese islands of Tilos, Symi, Patmos, and especially Kos, the home of Hippocrates, the founder of medicine in Western culture. He gave us the Hippocratic Oath, the earliest expression of medical ethics, still sworn by qualifying physicians. The Asclepion of Kos is a sanctuary that he dedicated to the Greek and Roman god of medicine and healing, Asclepius. The marble Corinthian columns of the Roman temple make a counterpoint with the dark conifers of the surrounding sacred wood."

In 1994 a new cyanotype formula was invented by Dr. Mike Ware wherein he replaced the ferric ammonium citrate with ferric ammonium oxalate (FAO). This formula is referred to as *new cyanotype*, *cyanotype II*, or *Ware's cyanotype*. New cyanotype (NC) is a single solution formula and its original formula, to my knowledge, has remained exactly the same; one formula to learn, one solution to mix.[1]

There were precursors to new cyanotype's use of ferric ammonium oxalate,[2] though none of these formulas became mainstream. This is because when ammonium ferric oxalate and potassium ferricyanide are mixed, they create a not-very soluble salt, potassium ferric oxalate, which crystallizes out of solution.[3] Not only does it produce a crystallized effect in the print, but the overall Dmax is low and the color a pale washed out blue with a lot of wash off in the water bath. I know; I tried it. It wasn't until Dr. Ware figured out how to get ferric ammonium oxalate and potassium ferricyanide to happily work together that it became an excellent alternative to classic cyanotype for the reasons stated in this chapter. Ware solved the problem by mixing an initial excess of chemistry in a smaller amount of water, using heat during preparation, and forcing the potassium ferric oxalate to drop out of solution early on so it can be filtered out and tossed. The remaining chemistry is ammonium ferric oxalate and ammonium ferricyanide which together produce a deep blue and a *very* fast solution.

The benefits of new cyanotype

New cyanotype is an excellent improvement for the following reasons:

- FAO is a consistent chemical, eliminating the inconsistent character of FAC
- New cyanotype is faster to expose, by as much as 2–3 stops *which is ⅛ the exposure time*. This is *the* most incredible benefit to NC
- New cyanotype produces a deep navy blue, so if Dmax is your thing, this formula is for you
- New cyanotype doesn't mold
- New cyanotype is more resistant to alkalis
- New cyanotype prints a long and gradual exposure scale
- New cyanotype doesn't bleed
- New cyanotype is *ultra*-smooth and grainless
- New cyanotype absorbs into the paper fibers better so there is no pooling when coating and less wash off during development.

Drawbacks to new cyanotype?

There are a few reasons why classic cyanotype has not been completely replaced by new cyanotype. They are "practice adjustments" rather than drawbacks to my way of thinking.

- New cyanotype is more exacting to mix. You can't really throw together two chemicals and come out OK like you can with classic cyanotype. If mixing chemistry is not for you, Fotospeed sells new cyanotype in 50 ml bottles
- New cyanotype requires heat and a thermometer to mix (a stove or hot plate works great but buy a dedicated thermometer)
- New cyanotype involves more toxic chemistry
- New cyanotype is *expensive* to buy readymade; at the time of this writing it is $36 for 50 ml, enough for 30 prints. Compare this to mixing it from scratch, though! It is only slightly more expensive to make than classic cyanotype—$2.12 for 100 ml versus $1.71 for 100 ml of CC—and an extra 41¢ spread out among fifty prints is hardly breaking anybody's bank
- It is *definitely more paper-picky*, and when that happens it results in an unpleasant brown, gray, blue, or pink fog; however, pre-acidifying an unsuitable paper or adding citric acid to the formula at time of coating solves a multitude of paper-picky problems and besides, even if there are fewer good papers for NC compared to CC, following is a list of 39 papers that work "out of the box" and another 20 that probably just need a bit of citric acid in the formula
- New cyanotype prints detail so sharply, and is used so thinly, that it can show brush stroke marks so you may want to invest in a 2″ synthetic bristle brush with really soft, smooth bristles
- It works best with a stronger acid development bath than CC; using no acid or even vinegar both result in various shades of yellow to brown on many papers, even suitable ones, so stick with citric or sulfamic acid. The tonal range when using stronger acids is worth it.

Papers for new cyanotype

New cyanotype is a perfect litmus test for unsuitable, alkaline, buffered papers. Mark Nelson once referred to it as "the canary in the gold mine." Unsuitable papers show their true colors right after coating with solution. The paper while drying will turn various shades of blue, some quite dark, within an hour, which results in fogging of the print's highlights.

There are three ways to deal with an unsuitable paper: don't use it; pre-acidify it with a sulfamic acid bath; or add drops of citric acid to the coating

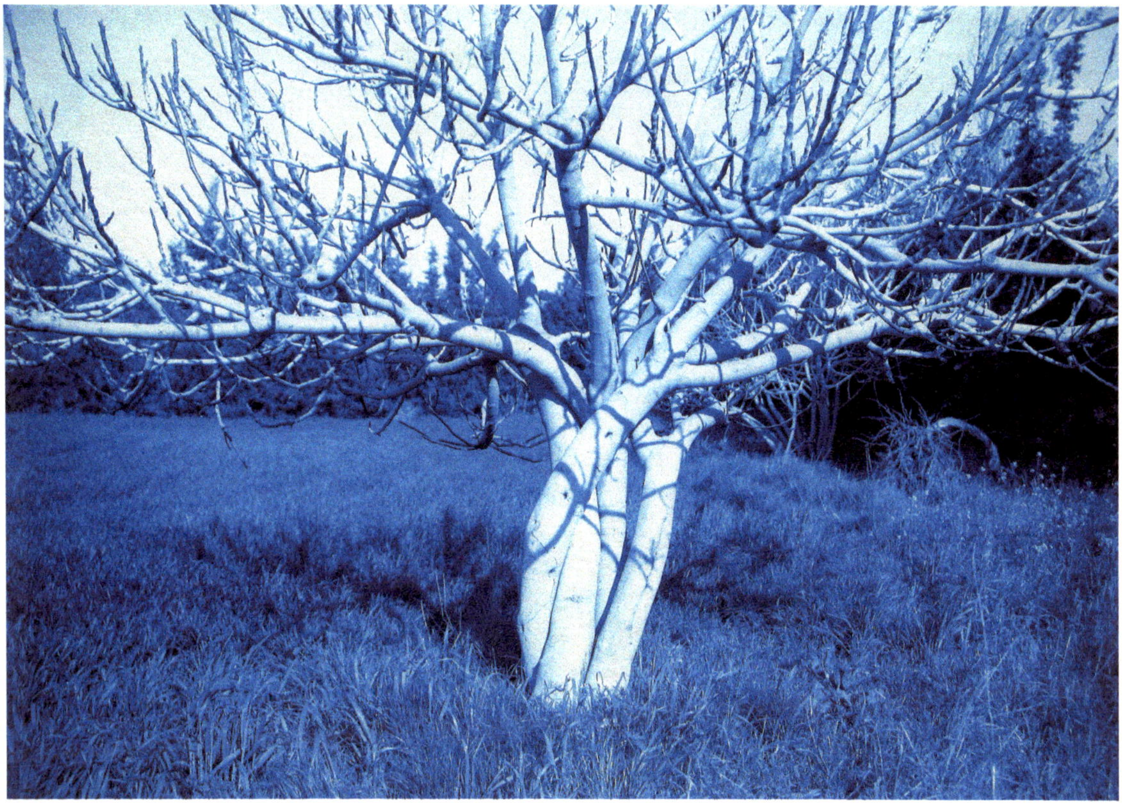

Figure 6.8. *Entwined Trees, Polis, Cyprus*, from an original 6 x 9 cm negative scanned and printed digitally with an Epson Photo R2880 408 x 284 cm © Mike Ware 2008

solution just before coating. The sulfamic acid bath is described in the **Papers for Cyanotype** chapter. The addition of citric acid to the coating solution is outlined below in the new cyanotype formula directions.

Following is a list of trouble-free papers for new cyanotype that do not require pre-acidification or citric acid added to the coating solution. Try one sheet at a time from the paper store and request that they label each sheet; you'll soon find your favorites. Refer to the **Papers for Cyanotype** chapter for paper descriptions. Although this list contains the usual suspects, unbuffered papers made specifically for alt, there are most likely a few papers that you have not yet tried.

- Arches Lavis Fidelis
- Arches Platine 145 gsm
- Arches Platine 300 gsm
- Bergger Cot 160
- Bergger Cot 320
- Bienfang 360
- Canson Mi-Teintes
- Canson Montval
- Canson Opalux Vellum
- Canson Pro Layout Marker
- Canson Vidalon Vellum
- Clearprint 1020H Vellum
- Fabriano Tiziano
- GarzaPapel Drawing
- GarzaPapel Ink-Pen
- Hahnemühle Platinum Rag
- Hahnemühle Sumi-e
- Japanese Mulberry/Heavyweight Kozo
- Khadi Cotton Rag Lightweight
- Legion Goyu
- Legion Masa
- Legion Parchtone
- Legion Revere Platinum
- Legion Rising Drawing Bristol
- Magnani Revere Book
- Pergamenata Parchment
- Ruscombe Mill Herschel Platinotype

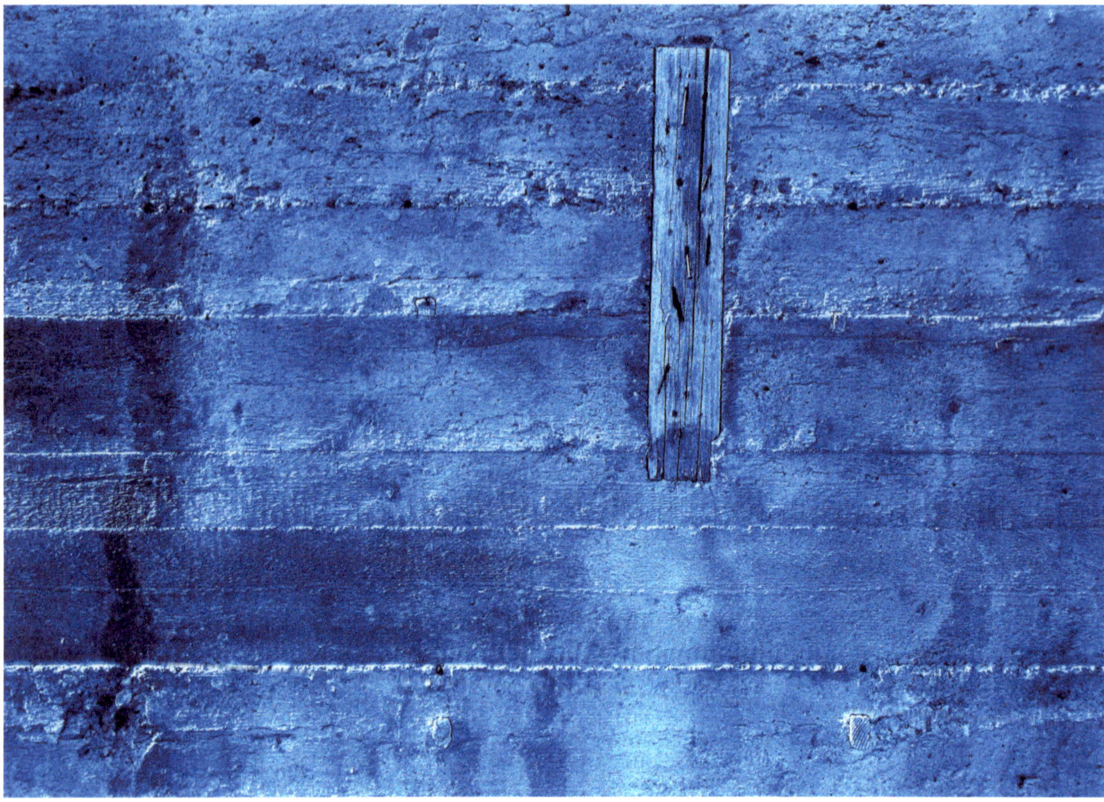

Figure 6.9. *Cemented Timber, Castellana Sicula, Sicily*, from an original 6 x 9 cm negative scanned and printed digitally with an Epson Photo R2880 408 x 284 cm © Mike Ware 2009

- St. Armand Colors
- St. Armand Old Masters (Frobisher white)
- St. Armand White Rag
- Twinrocker Bone (T, HT, LA)
- Twinrocker Da Vinci (T, HT, LA)
- Velke Losiny Moldau
- Velke Losiny Moravia
- Velke Losiny Prague
- Wyndstone Vellum
- Zerkall Frankfurt
- Zerkall German Ingres
- Zerkall Nideggen

Following is a list of more papers that are almost there, probably only needing a drop or two of citric acid added to the coating solution to prevent fogging. In general I have had excellent luck with all Twinrocker papers, so the ones mentioned below that slightly fogged may or may not need citric acid:

- Arches Aquarelle
- Arches Text Wove
- Hahnemühle Cézanne
- Hahnemühle Turner
- Khadi Cotton and Plant Paper
- Khadi Cotton Rag
- Legion Domestic Etching
- Magnani Pescia
- Rives BFK Heavyweight 175 gsm
- Rives Lightweight 115 gsm
- Schut Noblesse
- Twinrocker B-Buff (T, HT, LA)
- Twinrocker Biscuit (T, HT, LA)
- Twinrocker Cream (T, HT, LA)
- Twinrocker Dickenson (T, HT, LA)
- Twinrocker May Linen (T, HT, LA)
- Twinrocker Oyster (T, HT, LA)
- Twinrocker White Cotton Rag (T, HT, LA)
- Twinrocker Yale (T, HT, LA)
- Zerkall Book

Download and read the MSDS sheets at ScienceLab.com before using all chemicals. Exercise caution in the handling of all photographic chemicals. Use of any such chemicals constitutes some risk, and some are poisonous. The workplace should be well ventilated. Chemicals should be mixed only in the manner described. Avoid contact between the chemicals and eyes, skin, clothing, and furniture. Do not eat or drink while using chemicals. Keep them away from pets and children. Wear protective eye wear and gloves if necessary. In short, be mindful of all safety procedures for yourself and others.

New cyanotype formula[4]

10 g potassium ferricyanide (2 level *tea*spoons)
30 g ferric ammonium oxalate (2 level *table*spoons)
0.1 g ammonium dichromate (optional)
Distilled water to make 100 ml

1. Under low room light, place 10 g of potassium ferricyanide in a 100 ml beaker and add 20 ml water. Heat until about 160°F/70°C. Keep hot.
2. Dissolve 30 g ferric ammonium oxalate (FAO) in a separate 100 ml beaker in 30 ml distilled water and heat to 120°F/50°C until dissolved. Add the 0.1 g ammonium dichromate. [Note: dichromate may be omitted, but it helps with contrast and shelf life of the sensitizer. If dichromate is omitted, the solution lasts 6–12 months; if included, 2–3 years.[5] If there is no scale accurate enough, make a 20% solution by mixing 5 g ammonium dichromate in 20 ml distilled water, and then add ½ ml of this to the solution.]
3. While hot, add the potassium ferricyanide solution to the ferric ammonium oxalate solution and stir vigorously.
4. Let set in a dark place to cool to room temperature and crystallize, about 2 hours or overnight. I recommend overnight.
5. Filter the solution through a coffee filter. Weigh the crystals and observe the remaining volume of solution; there should be approximately 15 g crystals left in the filter, and about 62 ml formula remaining. If this is the case, you have done everything correctly. The crystals can be thrown.
6. Add water for a total volume of 100 ml. If desiring a less intense blue, or if you find that the solution splotches (see **Troubleshooting Cyanotype**), you can dilute the solution with an equal amount of distilled water.[6] I generally prefer it at 150 or 200 ml.
7. At time of use: Measure out a scant 2 ml of solution for an 8″ × 10″, a scant 4 ml for an 11″ × 14″, but use a few ml extra to account for the initial brush absorption. If necessary, add the following to the sensitizer as needed:
   - 1 drop of 40% citric acid per ml of sensitizer[7] to make coated paper last longer, and to help with paper incompatibility. It also makes clearing easier.[8] Citric acid cannot be added to the stock solution because it'll shorten the shelf life of the solution.
   - 1 drop of 5% Tween (5 ml Tween in a total volume of 100 ml) per ml of sensitizer to help the solution absorb into the paper better. Tween is a surfactant that enables the solution to penetrate the paper fibers. Do not mix this into the stock solution, but in the coating solution at time of use because it does not keep well in solution.[9] I often use Tween with new cyanotype because it helps more solution to penetrate the paper, less solution to wash off, and can result in a greater Dmax (how dark the darkest darks will get). This is especially important when using the 100 ml dilution on hard-sized papers.

Endnotes

1. Ware, Mike. *Cyanomicon II, History, Science and Art of Cyanotype: photographic printing in Prussian blue*. Buxton: self-published, 2016, pp. 204–205. **http://www.mikeware.co.uk/mikeware downloads.html**
2. Wall, E. J. *Photographic Facts and Formulas*. Boston: American Photographic Publishing Co., 1924, p. 259; Neblette, C. B. *Photography, Its Principles and Practice*, 4th ed. NY: Van Nostrand Co., Inc, 1946, p. 699.
3. Ware, p. 97.
4. Ware, Mike. *New Cyanotype, Siderotype Workshop Notes*. Buxton: self-published, 2009, p. 6. **http://www.mikeware.co.uk/mikeware/downloads.html**
5. Ware, *Cyanomicon II*, p. 210.
6. Barnier, John. *Coming into Focus, a Step-by-Step Guide to Alternative Photographic Printing Processes*. San Francisco: Chronicle Books, 2000, p. 205.
7. Ware, *Cyanomicon II*, p. 216.
8. Barnier, p. 207.
9. Ibid., p. 206.

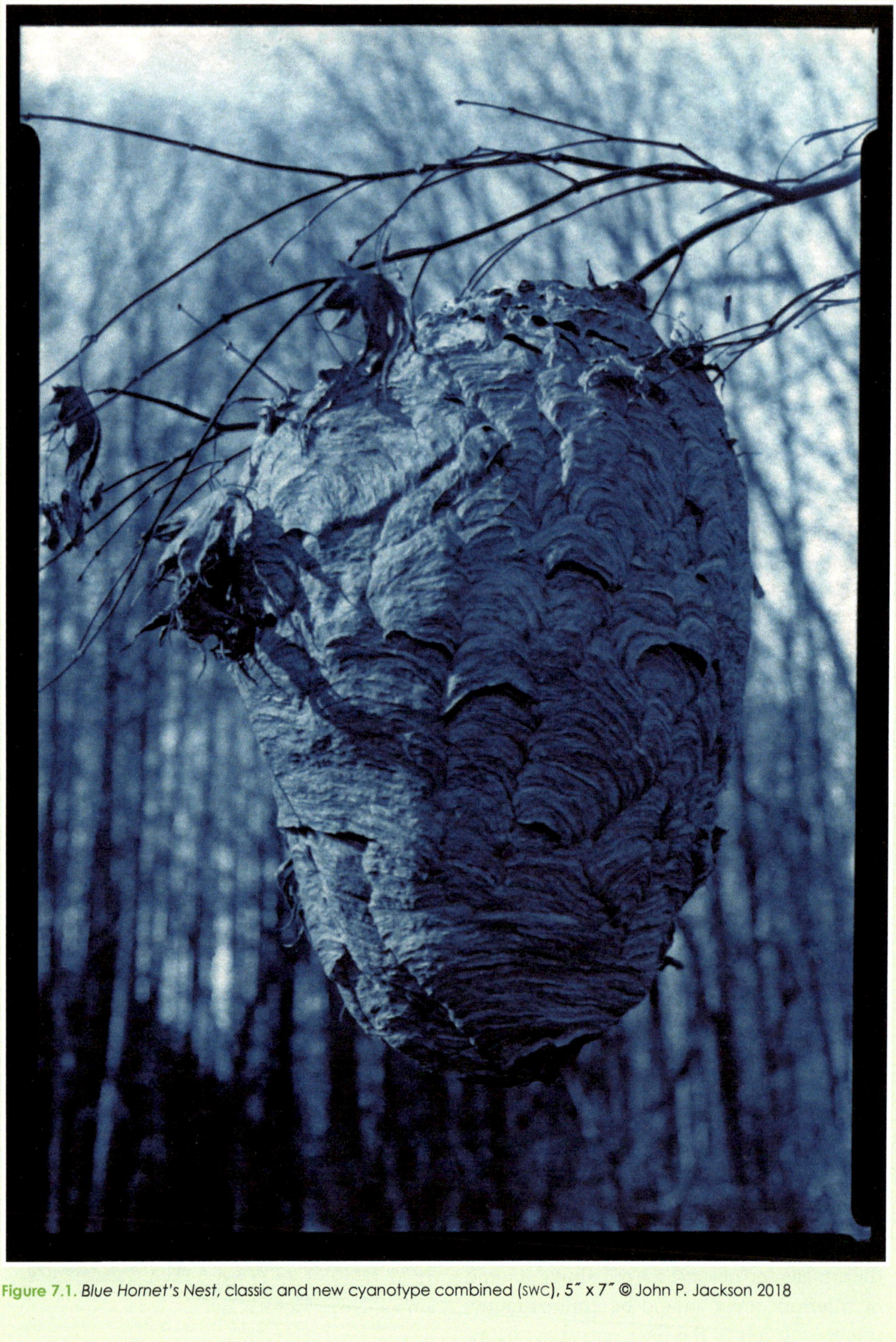

Figure 7.1. *Blue Hornet's Nest*, classic and new cyanotype combined (SWC), 5″ x 7″ © John P. Jackson 2018

Chapter 7

Further Formula Exploration

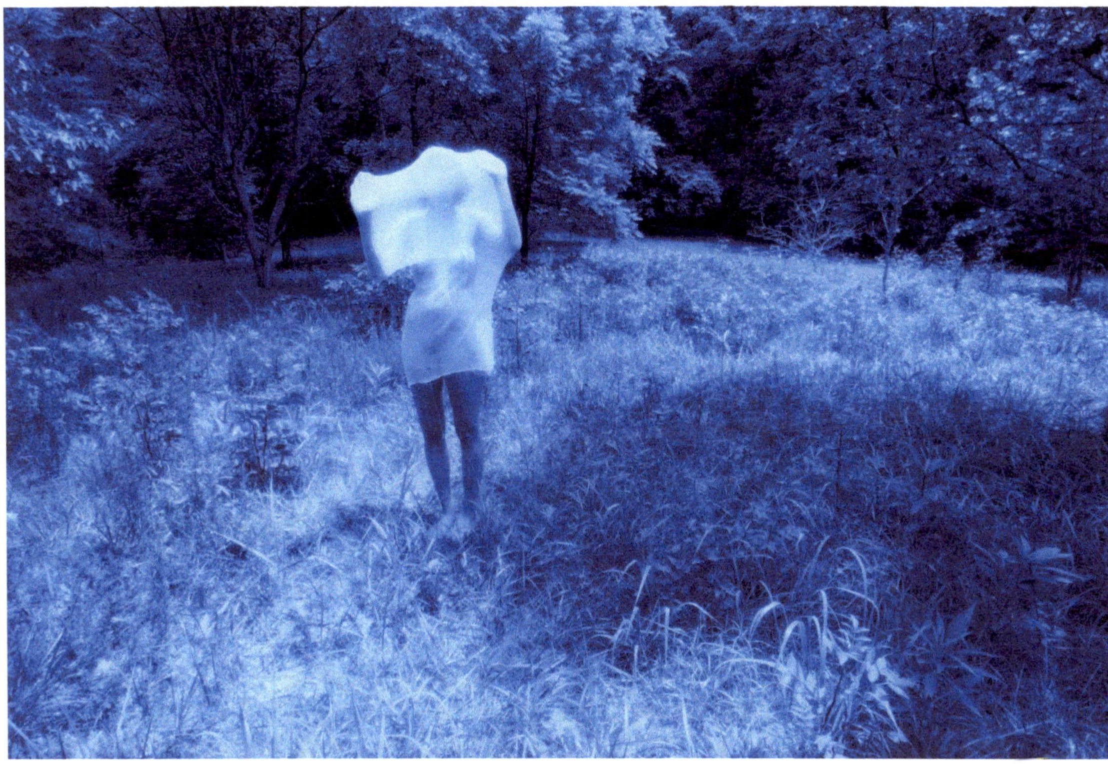

Figure 7.2. *Kristen Veiled*, classic and new cyanotype combined (SWC) © Sam Wang 2018

In 2017 when Sam Wang, John Jackson, Sandy King, and I began testing cyanotype formulas and sharing information back and forth, Wang was using a 50/50 mix of classic cyanotype and new cyanotype which he called CC/NC (and we jokingly called SWC). It wasn't anything he had read about doing; he was just thinking outside the box and responding to his intuition to do so. One of the hallmarks of an artist, which Sam is, is to experiment and explore even when something isn't necessarily "done," and to be comfortable with failure should it so arise.

Whereas new cyanotype completely replaces ferric ammonium citrate with ferric ammonium oxalate in combination with potassium ferricyanide to create ammonium ferric oxalate/ammonium ferricyanide in situ, Wang's hybrid formula was using ferric ammonium oxalate in *addition* to ferric ammonium citrate and potassium ferricyanide. After Wang used the combined formulas for a while he found it to be the best of both worlds: much faster than CC, not as grainy as CC, and having a *long* exposure scale (ES) with softer, less contrasty highlight detail.

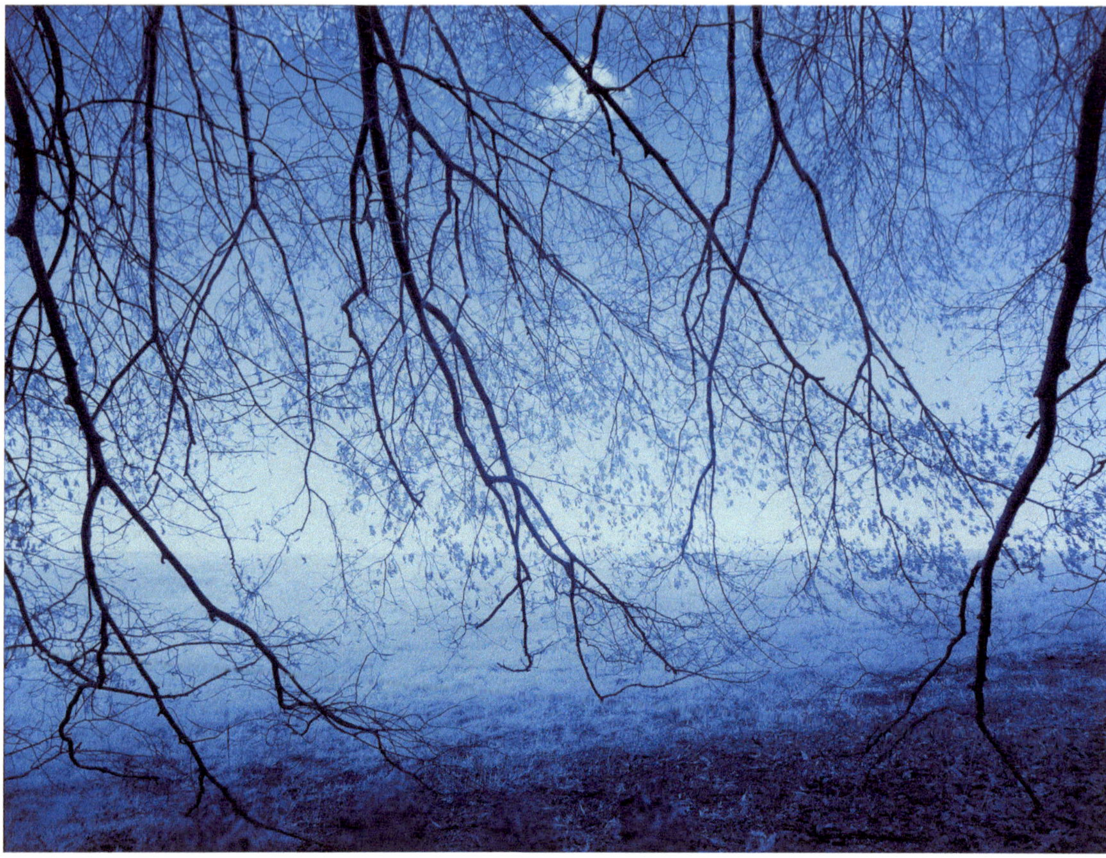

Figure 7.3. *Spring Breeze*, classic and new cyanotype combined (SWC) © Sam Wang 2018

At first I didn't pay much attention because I thought as most people do, "Why not use one formula or the other, classic or new?" Typically cyanotypists gravitate toward one formula or the other and don't have both mixed up and on hand. Often, once the faster exposure time of new cyanotype is experienced, there's no turning back.

However, in reviewing all my research, there were actually three formulas in the literature before 1950 that had ferric ammonium oxalate (FAO) and ferric ammonium citrate (FAC) in combination with potassium ferricyanide (PF) as Wang was doing. Two of the three formulas were probably for commercial use since they mixed up to 8 liter amounts.[1] One was probably for "home use" in the typical liter amount. It seemed therefore to be on the commercial radar but not the amateur or professional photographer radar.

Though uncommon—several formulas out of 107 recorded—maybe Wang was on to something. At the very least, throughout the winter's testing as a number of us were experiencing bouts of dendritic splotches with new cyanotype, Wang was merrily making print after print without coming across this malady.

I finally gave the SWC formula a try and found it to be quite fast, on par with new cyanotype which means much faster than classic cyanotype (my exposure time is 5½ minutes, Wang's is less). The Dmax was healthy. What seemed slightly different about the SWC formula is the detail from midtones to highlights. The midtones have good tonal separation and the highlights are softer and less contrasty. There is a long trailing off of the highlights that looks like fog on a step wedge, but it is not. In looking at the exposure scale it was

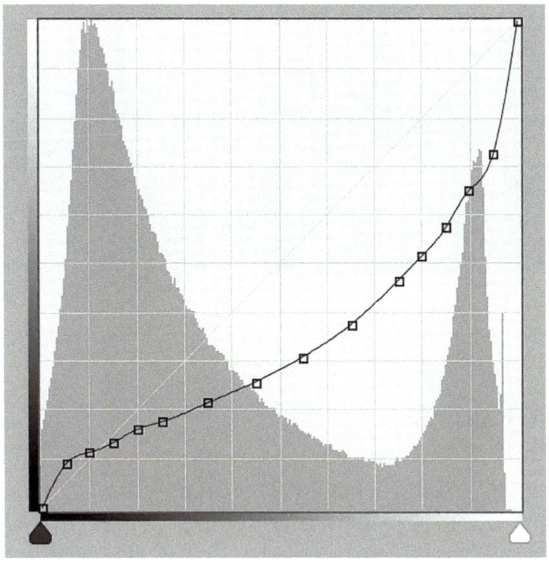

Figure 7.4. Note the gradual curve approaching the straight line with minimal adjustment required for the process.

apparent that this hybrid formula might need its own curve. Sure enough, the calibrated curve in comparison with other cyanotype curves is very gradual and quite close to the straight line. Hence I have included its own curve in the **Digital Negatives for Cyanotype** chapter. What's not to like about a deep Dmax and highlights that print forever?

This is not to say that combining CC and NC is ground-breaking practice better than using either CC or NC alone. Each formula has its benefits and drawbacks. The point in sharing other formula practice is to encourage further exploration into the cyanotype process.

For example, John Jackson did extensive testing with SWC using both 20/10 and 15/10 CC in equal proportions with NC and noted that using more FAC affects both Dmax and ES. More FAC gives a longer ES, a greater Dmax, and slightly different contrast. In Jackson's opinion, there is enough difference to be a tool to use in print control.

Download and read the MSDS sheets at ScienceLab.com before using all chemicals. Exercise caution in the handling of all photographic chemicals. Use of any such chemicals constitutes some risk, and some are poisonous. The workplace should be well ventilated. Chemicals should be mixed only in the manner described. Avoid contact between the chemicals and eyes, skin, clothing, and furniture. Do not eat or drink while using chemicals. Keep them away from pets and children. Wear protective eye wear and gloves if necessary. In short, be mindful of all safety procedures for yourself and others.

Wang's formula in a nutshell

You don't have to follow these directions if you already have classic cyanotype and new cyanotype solutions at hand. Just mix classic cyanotype's Solution A plus Solution B in equal proportions as usual, and then mix this combined amount with an equal amount of new cyanotype. In other words, 1A CC + 1B CC + 2 NC. It's that simple.

Solution A:

20 g ferric ammonium citrate (5 tablespoons)
100 ml distilled water

Mixing Solution A:

1. Measure out 20 g of ferric ammonium citrate and add it to 75 ml distilled water and stir until dissolved.
2. Add water to make 100 ml total volume.
3. Store in a brown eyedropper bottle and label the bottle "Solution A."

Solution B:

10 g potassium ferricyanide (2 teaspoons)
100 ml distilled water

Mixing Solution B:

1. Measure out 10 g of potassium ferricyanide and add it to 75 ml distilled water and stir until dissolved.
2. Add water to make 100 ml total volume.
3. Store in a brown eyedropper bottle and label the bottle "Solution B."

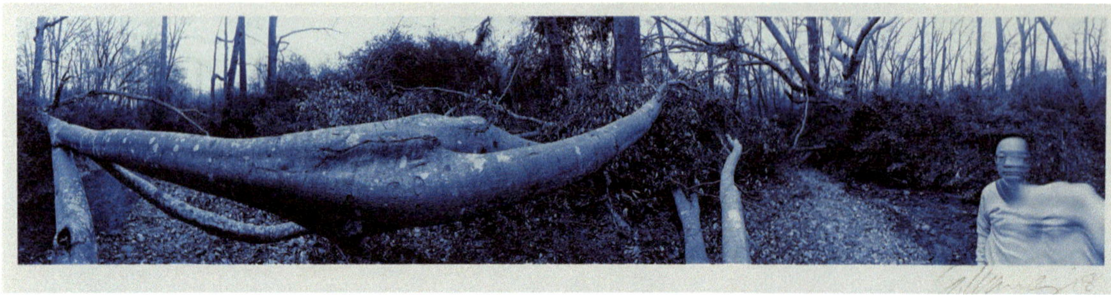

Figure 7.5. *Self Back When*, classic and new cyanotype combined (SWC) © Sam Wang 2018

Solution C:[2]
10 g potassium ferricyanide (2 level *tea*spoons)
30 g ferric ammonium oxalate (2 level *table*spoons)
0.1 g ammonium dichromate (optional)
Distilled water to make 100 ml

1. Under low room light, place 10 g of potassium ferricyanide in a 100 ml beaker and add 20 ml water. Heat until about 160°F. Keep hot.
2. Dissolve 30 g ferric ammonium oxalate (FAO) in a separate 100 ml beaker in 30 ml distilled water and heat to 120°F until dissolved. Add the 0.1 g ammonium dichromate. [Note: dichromate may be omitted, but it helps with contrast and shelf life of the sensitizer. If dichromate is omitted, the solution last 6–12 months; if included, 2–3 years.[3] If there is no scale accurate enough, make a 20% solution by mixing 5 g ammonium dichromate in 20 ml distilled water, and then add ½ ml of this to the solution.]
3. While hot, add the potassium ferricyanide solution to the ferric ammonium oxalate solution and stir vigorously.
4. Let set in a dark place to cool to room temperature and crystallize, about 2 hours or overnight. It takes a while so don't rush it.
5. Filter the solution through a coffee filter. Weigh the crystals and observe the remaining volume of solution; there should be approximately 15 g crystals left in the filter, and about 62 ml formula remaining. If this is the case, you have done everything correctly. The crystals can be thrown away.
6. Add water for a total volume of 100 ml.

At time of use:

1. Mix one part Solution A, one part Solution B, and two parts Solution C. The new cyanotype solution should always equal the amount of combined Solution A and Solution B. 2 ml (about ½ teaspoon) of combined solution are enough for an 8″ × 10″, 4 ml (scant 1 teaspoon) for an 11″ × 14″, but mix a few ml extra to account for initial brush absorption. [Note: there are 5 ml in a teaspoon.] Wang adds 1 drop of 40% citric acid to every ml of solution to prevent highlight fog so this is an option if you have fog issues (I did not).

Wang uses an LED unit and his exposures are around 2 or so minutes. He develops his prints in a heaping teaspoon of citric acid per 500 ml water, for 2–10 minutes, followed by a 5- to 30-minute water wash.

Quick 'n' easy hybrid formula

This is for someone who doesn't keep both formulas on hand and primarily uses classic cyanotype. All you have to add to your practice is to mix up a 15% ferric ammonium oxalate Solution C, which will require purchase of only one extra chemical, ferric ammonium oxalate. The neat thing about this quick 'n' easy method is that you can explore varying the amount of the three solutions quite easily to see how Dmax, printing time, and color are affected. Watch how much FAO you add because the formula has a tendency to solarize. Exposure times before solarization occurs can be as short as several minutes. The solarization produces an odd etched edge effect, where adjacent to the solarizing there

Figure 7.6. Note the odd edge effect of the quick 'n' easy hybrid formula

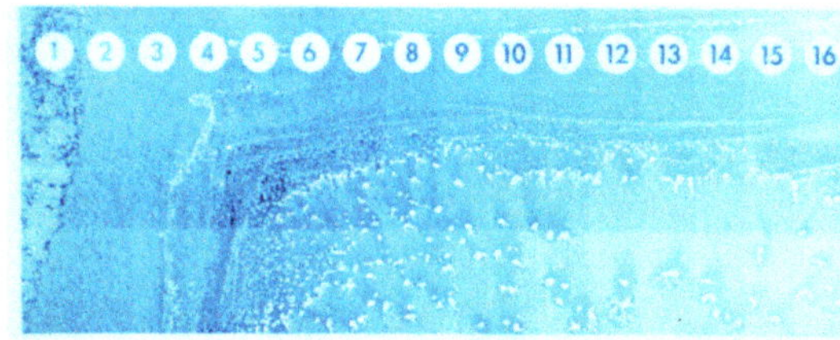

Figure 7.7. Fiasco Wall formula experience. This is a picture of the 250 FAO/250 PF/1000 ml water formula that was way too concentrated and crystallized and washed off the surface, leaving a pale blue ugly print.

is a line of darker blue; this might have creative possibilities.

Solution A:

Mix the 20% ferric ammonium citrate solution as directed in this chapter.

Solution B:

Mix the 10% potassium ferricyanide solution as directed in this chapter.

Solution C:

15 g ferric ammonium oxalate (1 tablespoon)
100 ml distilled water

1. Measure out 15 g of ferric ammonium oxalate and add it to 75 ml distilled water and stir until dissolved.
2. Add water to make 100 ml total volume.
3. Store in a brown eyedropper bottle and label the bottle "Solution C."
4. At time of use, either mix equal parts of Solution A, B, and C or else vary the amounts as desired. For instance, mix equal parts of A and B and add mere drops of C. 2 ml (about ½ teaspoon) of combined solution are enough for an 8″ × 10″, 4 ml (scant 1 teaspoon) for an 11″ × 14″, but mix a few ml extra to account for initial brush absorption. [Note: there are 5 ml in a teaspoon.]

Wall's formula

In E. J. Wall's 1924 *Photographic Facts and Formulas* there were two formulas, one a 250 g FAO + 250 g PF in a liter of water. Although this formula is similar to an FAO/PF-only new cyanotype, there was no complicated mixing procedure involved where solutions had to be heated up, combined, and left to precipitate. I thought, how great, so I mixed up the liter amount and it was a disaster, as can be seen in the step wedge above. The solution was *way* too saturated, crystallized all over the paper surface, and produced a pale print; completely unusable. I thought to myself, is it *possible* Wall even *tried* this or just repeated someone else's formula? For one, a single solution like this mixed up into a liter amount is equivalent to double the chemistry of a two-solution mix combined at time of use. It made me realize the reason Dr. Ware used heat and precipitation to get the two chemicals to combine and behave with one another before coming to a final mix.

The other formula in Wall's book was a combination formula of FAO/FAC/PF like Wang was doing. The formula was 166 g *brown* ferric ammonium citrate (the equivalent of about 197 g green), 133 g potassium ferricyanide, and 33 ferric ammonium oxalate in a liter of water. Again, I tried the full strength formula and it behaved similarly, way too saturated a formula, so I diluted it with an equal amount of water and it worked like a charm. In other words, it is best to mix up to *two* liters of combined solution, not one. Otherwise it produces a pale, scratchy, slow to expose, low Dmax print. Trust me; I know this because I fell for this same mistake twice.

What is the benefit of this formula over classic or new cyanotype? It is faster than CC by about ⅓ stop, slower than SWC and NC by about 1 stop, but the exposure scale and Dmax rival both SWC and NC and the tones are smooth. How long the combined solution will last, your guess is as good as mine, but in this user-friendly 100 ml amount

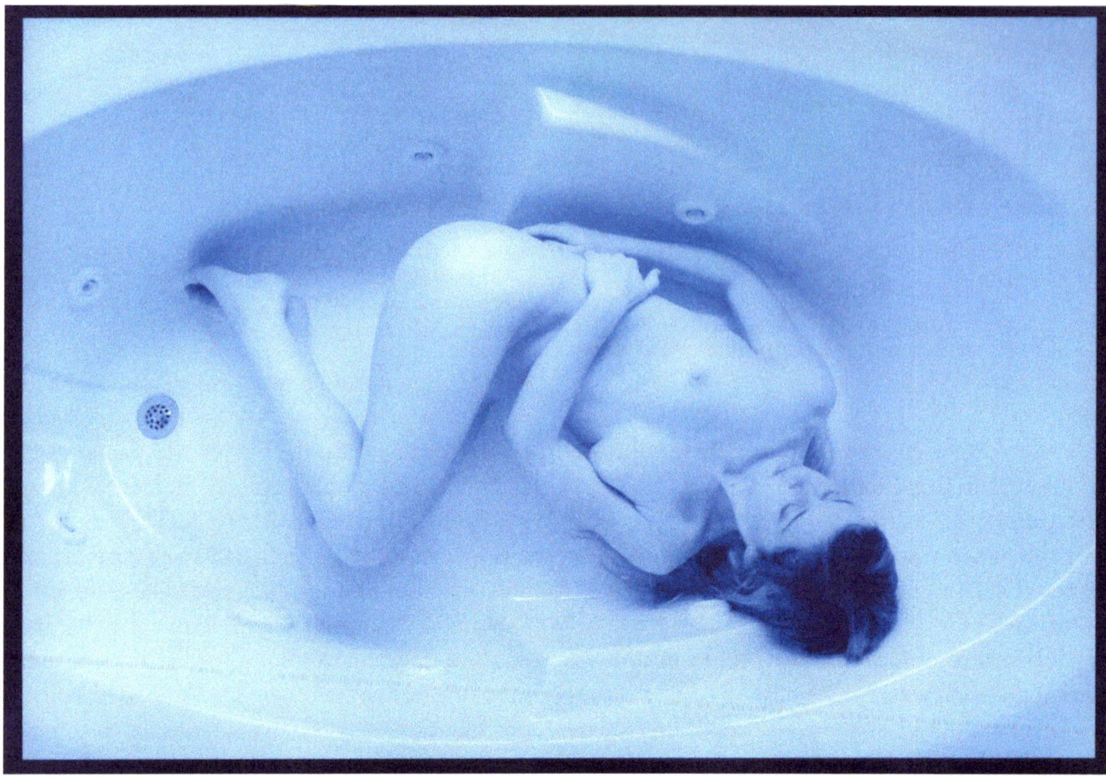

Figure 7.8. *Jessica in Tub*, classic and new cyanotype combined (SWC) © Sam Wang 2018. Note the delicate highlight detail.

you'll probably use it up before it goes bad. The numbers are rounded slightly up, but it is somewhat like using a 20/14 two-solution cc with an added 4 g ferric ammonium oxalate to bump up the speed, Dmax, and smoothness.

10 g ferric ammonium citrate (7 level teaspoons)
7 g potassium ferricyanide (1½ level teaspoons)
2 g ferric ammonium oxalate (½ teaspoon)
Water up to 100 ml

1. Measure out the three chemicals and add them one by one to 75 ml distilled water and stir until dissolved.
2. Add water to make 100 ml total volume.
3. Store in a brown eyedropper bottle and label as a single solution mix.
4. At time of use, measure out 2 ml (about ½ teaspoon) of solution for an 8″ × 10″, 4 ml (scant 1 teaspoon) for an 11″ × 14″, but mix a few ml extra to account for initial brush absorption. [Note: there are 5 ml in a teaspoon.]

Endnotes

1. Wall, E. J. *Photographic Facts and Formulas*. Boston: American Photographic Publishing Co., 1924, p. 259; Neblette, C. B. *Photography, Its Principles and Practice*, 4th ed. NY: Van Nostrand Co., Inc., 1946, p. 699.
2. Ware, Mike. *New Cyanotype, Siderotype Workshop Notes*. Buxton: self-published, 2009, p. 6. http://www.mikeware.co.uk/mikeware/downloads.html
3. Ware, Mike. *Cyanomicon II, History, Science and Art of Cyanotype: photographic printing in Prussian blue*. Buxton: self-published, 2016, p. 210. http://www.mikeware.co.uk/mikeware/downloads.html

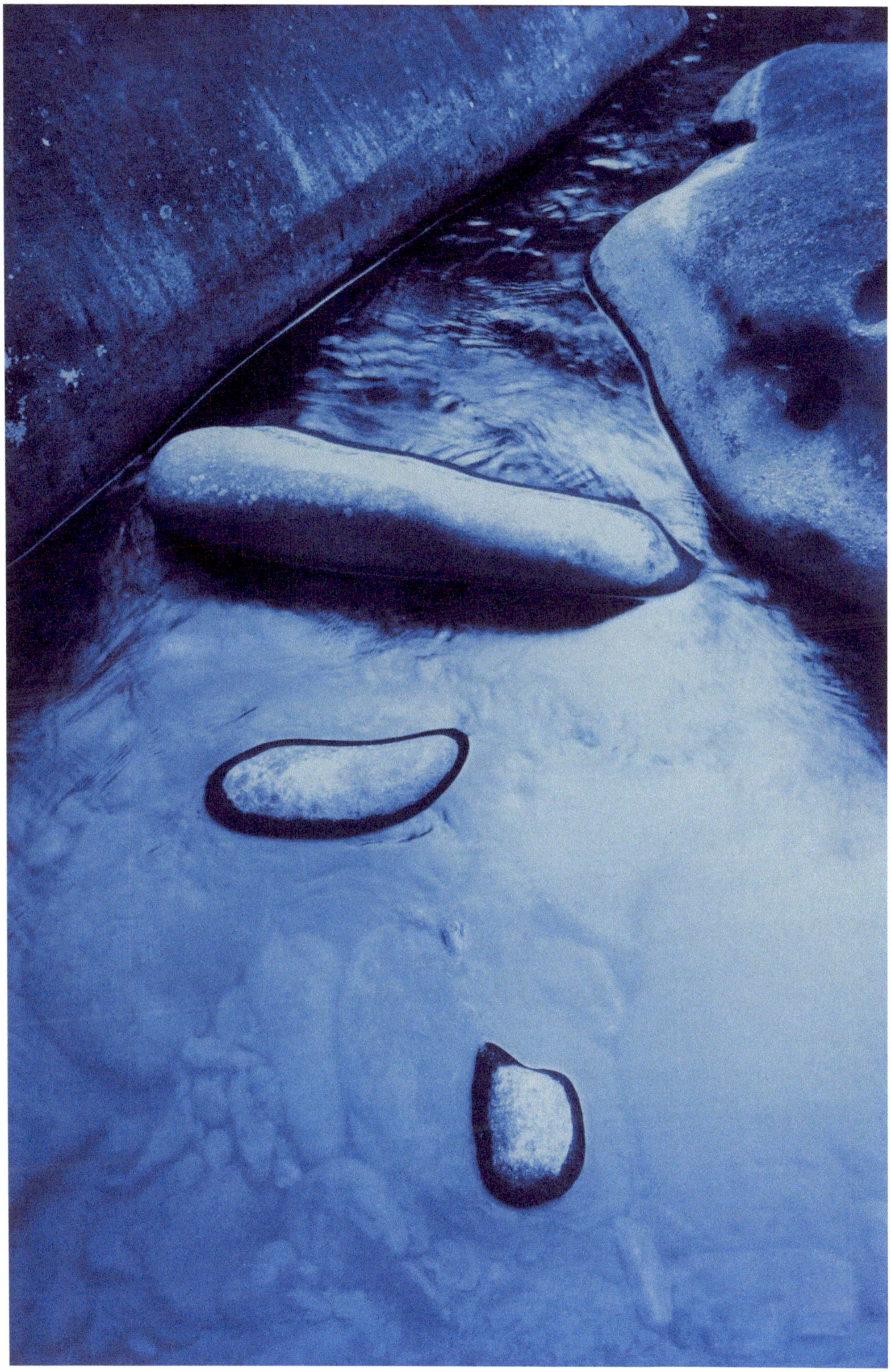

Figure 7.9. *Three Pebbles*, classic and new cyanotype combined (SWC) © Sam Wang 2018

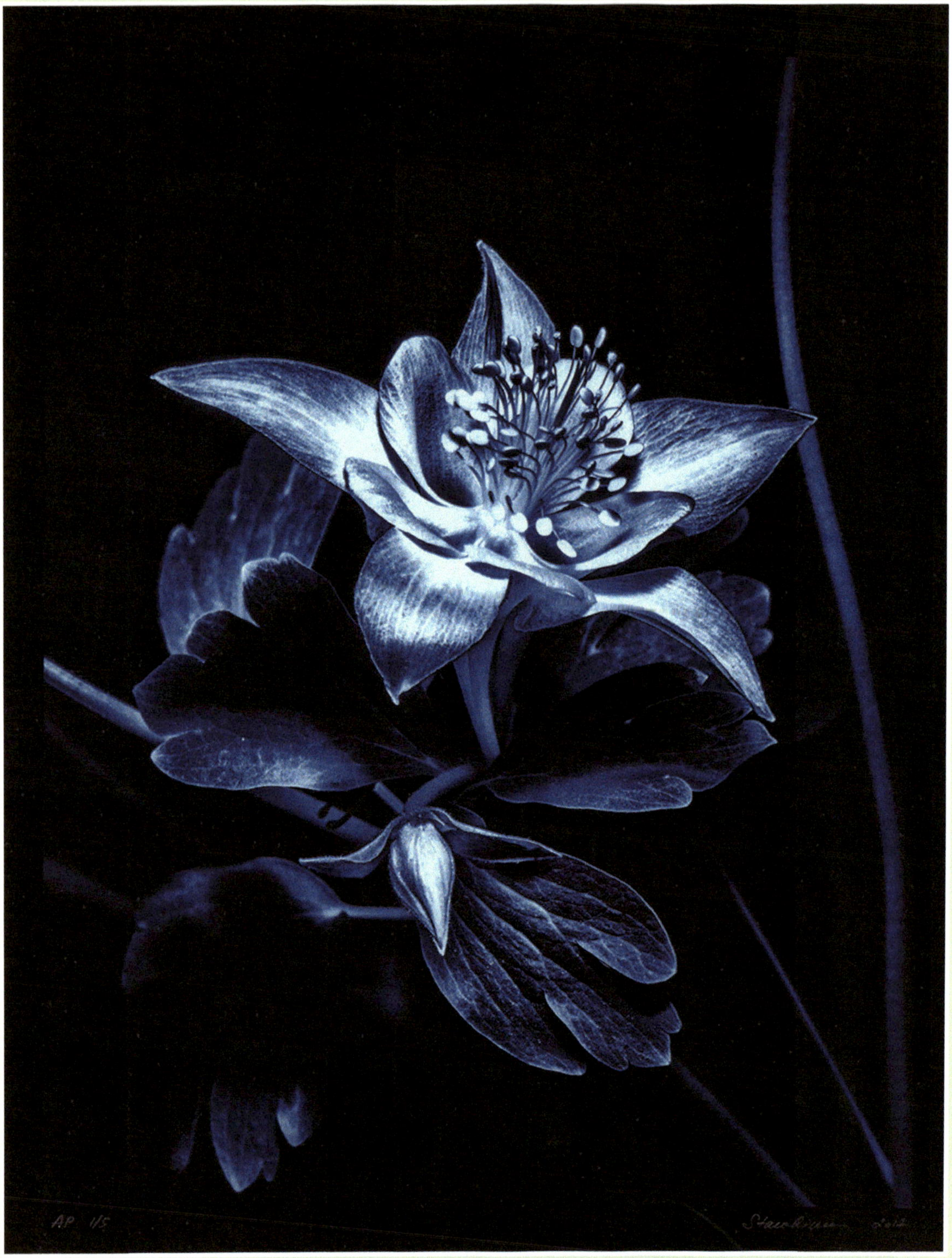

Figure 8.1. *Aquilegia*, classic cyanotype, 11″ x 14″ © Eugene Starobinskiy 2017

Chapter 8

Making the Cyanotype Print

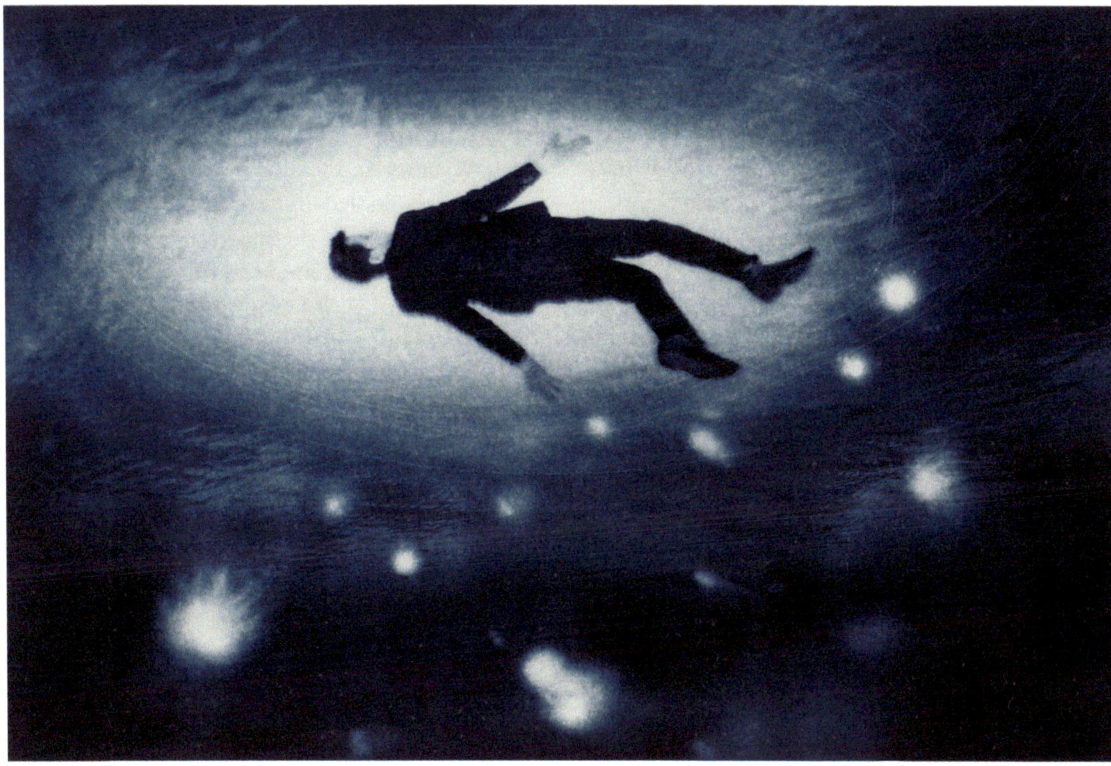

Figure 8.2. *Rising*, from the series *Out of The Blue* (Part I of *A Million Suns*), collaged cyanotype with digital drawing, 11.5″ x 17″ © Jonah Calinawan 2013

If the cyanotype solution is mixed and ready to use, a standard printing time (SPT) has been determined as outlined in the **Digital Negatives for Cyanotype** chapter, and an appropriately curved negative has been created according to that chapter also, it is now time to coat, expose, and process a print.

Brushes to use

Cyanotype is very user-friendly when it comes to brushes. Cheap foam brushes from the hardware store, natural hair hake brushes, synthetic fiber brushes—all work. My favorite for classic cyanotype is the 2″ Connoisseur 150 series hake brushes because they are absorbent and springy. Since they have a stitched ferrule I can soak them after a coating session in plain water for hours without worrying about a ferrule rusting. For combination or new cyanotype I prefer a 2″ synthetic bristle brush because it is coated more thinly to prevent pooling of chemistry and resultant splotching and crystallization (see **Troubleshooting Cyanotype**), and unless you brush very carefully, strokes can be visible. The synthetic fibers are softer and more refined.

Coating the paper

1. Note: If you live in a dry climate or want to experience smoother prints, pre-humidify the paper by creating a makeshift humidity chamber of a tray filled with water, a screen on top, and another inverted tray on top of this to retain the moisture. Leave the paper in this chamber until limp and cool, overnight, or even permanently. Pencil in the size of the negative on the paper. For an 8″ × 10″ use an 11″ × 14″ sheet. For clean white borders, tape off the area with low tack masking tape and coat within the taped borders.
2. For classic cyanotype, measure out ½ teaspoon of Solution A and ½ teaspoon of Solution B into a cup and mix together. This is enough for an 8″ × 10″ on 11″ × 14″ paper. For new cyanotype, a single solution that requires a thinner layer, measure out ½ teaspoon of solution for the same size print (perhaps extra to charge a dry brush). Rule of thumb is it is *about* 1 drop of solution per 2 square inches, and a ml contains 20 drops. I recommend adding 1 drop of 5% Tween per ml of new cyanotype solution to help the paper absorb and to perhaps increase Dmax.
3. Pour in a line along one edge of the area needing to be coated and immediately begin brushing the solution horizontally and vertically until it covers the area completely and evenly.
4. Hold up the coated paper vertically by one side and let the solution flow downward to even out the brush strokes. Immediately turn the paper 90° before any solution drips off the paper onto the counter and let it again even out downward. Repeat this on all four directions until the solution evenly covers the entire surface with no puddles or areas of drips. A small piece of paper towel can sop up the excess solution at the edge if necessary because that excess solution "blooms" back into the print area and creates marks. Classic cyanotype takes a while to sink into the paper fibers, so it is a good thing to even it out with this technique. New cyanotype absorbs a bit more quickly into the paper, and since you use less solution it is important to smooth out all brush strokes at the offset before holding the paper vertically for that extra solution to even out.
5. Once the paper is evenly coated, hang it or lay it flat in the dark to dry (hanging minimizes pooling and splotching with new cyanotype).

Tips

- Some double-coat cyanotype for deeper blue. I have never found this necessary and, in fact, I found it pooled more and created splotches that bloomed into the image area with edges of white when not coated carefully. It is definitely not necessary with new cyanotype. If you want to try it, let the first coat dry for 20 minutes and then coat again.
- In case you do not have an adequate dark area, I have left classic cyanotype out in regular light bulb type room light for two days (not sun!), developed it in water, and have seen no evidence of residual exposure. Test your dimroom light by doing the same. Put a quarter on top of a small piece of coated paper and leave it out in your room light for the time you want to test—always make it overkill—and then see if the paper comes clear in a five-minute development. That will tell you how safe your light is in the room.[1] Be sure to test both classic and new cyanotype with this method because new cyanotype is more sensitive.

Drying the paper

Depending on the climate, coated paper takes ½–1 hour to dry. I expose the paper as soon as it is dry to the touch but still somewhat limp and cool, about ½ hour. Pre-humidifying paper so it is limp and cool and coating and exposing as soon as it is dry to the touch produces a much smoother, less grainy print; it is best practice for both classic and new cyanotype, and especially so for new cyanotype because it greatly minimizes NC's splotching issues.

How long does coated paper last?

I have no problem with using coated paper for a couple days, but paper that is much older is slower to expose, prints less contrasty, and the Dmax or darkest blue is not quite as dark. As one author said, "Paper will keep for weeks but the blue is never so bright as when the paper is freshly prepared."[2] It is best practice to be consistent with this and coat and expose within the day, but paper doesn't magically go off in 24 hours. One well-known cyanotype printer coats a bunch of paper at once and uses it for a month. In the summer he merely stores the coated paper in the fridge.[3] Minimizing light, temperature, humidity, and air allows coated paper to last longer, but watch that the paper has remained bright green and not turned blue-green[4] which will result in a fogged, dull print.

Coating washi and tissue-thin papers

Tissue-thin papers require a different coating technique. They soak up more chemistry in general. With some, the solution soaks right through the paper to the coating surface below. You cannot just pour solution onto the paper in a line on one side like with other papers. Coat these papers either one of two ways:

1. Pour a line of cyanotype solution on a level piece of glass.
2. Pick the paper up by the edges and form it into a u-shape.
3. Touch the bottom of the u-shape to the solution starting at one edge and then lay the entire width of the paper down in the line of solution.
4. In an even movement, slowly drop the paper down flat, watching that the solution soaks up all the way to the paper edges. This takes a bit of practice to know how much solution to puddle on the level piece of glass that will soak the entire sheet.
5. Let the solution even out for a while and then keep flat to dry.

or

1. Fill a cup with solution that is wide enough to fit the brush.
2. Lightly dip the end bristles of the brush to load it with solution.
3. Brush a horizontal swath of solution on the top edge of the area needing to be covered, the entire length of the area.
4. Load the brush again and brush another swath of solution overlapping and parallel to the previous coat.
5. Do this brush by brush until the entire area is covered.
6. Take the brush and do the same with vertical strokes from one side of the print to the other.
7. Once the area is evenly coated let the solution even out and then keep the paper flat to dry, which may require a paper towel underneath to absorb excess chemistry.

Coating vellum papers

Vellum papers such as Wyndstone, Opalux, Vidalon, and Pergamenata, when brushed with solution, spring up into a scroll that is impossible to stretch out flat. There are two ways to work with this stubbornness. One is to pin or tape down the paper on all four sides before coating and leave it to dry pinned or taped in place. The other is to cut a much larger sheet of vellum than necessary and coat a smaller center of the sheet; this way you can easily cut off the cockled edges and the paper will in general lie flatter.

Exposing cyanotype

Cyanotype prints can be exposed either in an indoor exposure unit or outdoors in the sun. Exposure units are perfectly predictable and consistent but about two stops slower (4× the exposure time) than the sun and they require an initial investment of $$; the sun is not so predictable but ¼× the exposure time of an indoor exposure unit and it's free. With care in observation, and keeping a detailed log of exposure results at different times of the day, weather, and season, you can print fairly predictably outside in all seasons, including extremely cold weather. See

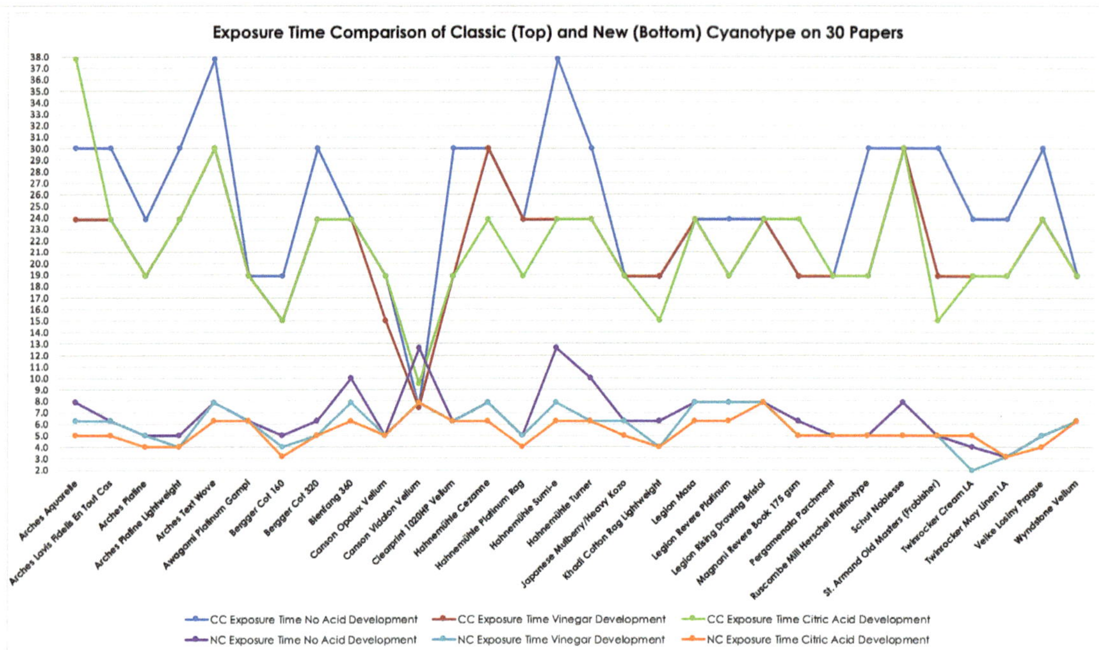

Figure 8.3. Note how short the exposure times, measured in minutes, for new cyanotype (bottom three lines) are in comparison with classic cyanotype (top three lines). Note also that exposure time decreases slightly for each process when an acid is used in the water bath, so it is important to calibrate the time based on the acid used. Prints were exposed in an Edwards UVBL unit.

the **Ballpark UVBL Exposure Times** chart pp. 96–97 for some time suggestions (all times are for classic cyanotype 10/10 and new cyanotype 200 ml dilution).

If you have read the **Digital Negatives for Cyanotype** chapter and found the correct exposure time for the paper you are using, under UVBL there is nothing to do but place the negative, emulsion to emulsion, on the coated and dried paper and expose it for the determined time.

If you are printing outside in the sun, it will be of great benefit to expose one step wedge print first to determine printing time for that moment, given different weather and seasonal UV amounts (see the **Digital Negatives for Cyanotype** chapter for how to determine standard printing time). Otherwise, a guestimate for exposure between 10AM and 2PM during late spring to early fall sun is 5–15 minutes for classic cyanotype and 1–5 minutes for new cyanotype.

In order to print by inspection, you will need a hinged back contact frame that allows checking of the exposure without disturbing the negative registration. Make sure the contact frame makes a very snug contact with the negative/paper sandwich so when checking exposure you will not move the negative out of registration. You can do this by adding sheets of watercolor paper inside the contact frame. Do not use cardboard because it does not bend well enough to peel back to check on the image.

1. Expose until the borders of the print that are not under the negative are completely solarized to a lighter silver-blue. This color is much different between CC and NC; with CC it is a dark, reversed navy. With NC it is a bright silver blue-white, and on some transparent vellums the deepest shadows become invisible.
2. Bring the contact frame inside away from bright sunshine or window light, and open the back of the contact frame carefully and on one side only.

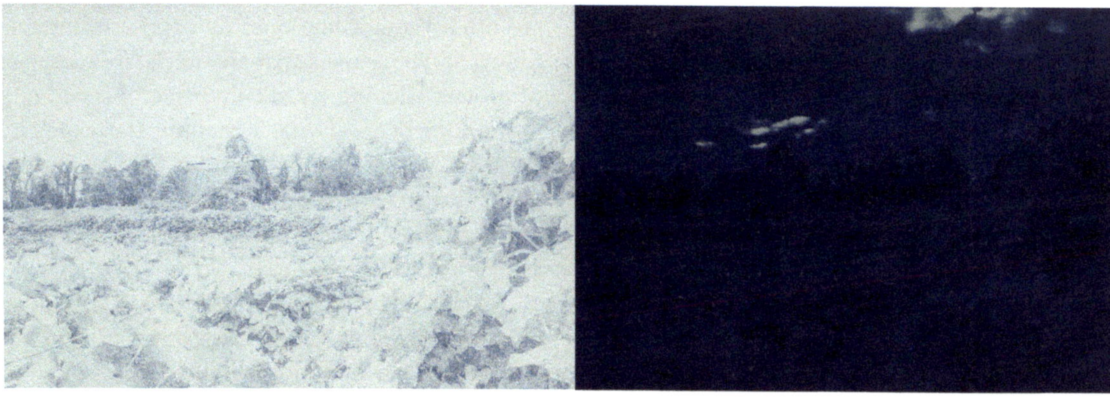

Figure 8.4. This seems really obvious, but this is a side-by-side comparison with under- and over-exposure. What might not be so obvious is how the highest detail in the clouds still did not print after gross overexposure, showing how lowering the ink density on a negative is sometimes warranted for classic cyanotype.

3. Look at the deepest shadows in the image area; they, too, should be solarized and lighter, like the borders. Look at the highlights of the image. They should look overexposed, with only specular highlights green. Anything still bright yellow-green will wash out to paper white in general. The print as a whole should look quite overexposed. Err on the side, always, of overexposure because you can always develop longer to remove some print density but you can't develop shorter as it risks not removing all the iron from the print.
4. Process the print, dry, and assess. Record the paper name, cyanotype dilution, exposure length, time of day, date, weather (bright sun, cloudy, etc.) and any other pertinent exposure information in a notebook, and or on an edge of the print itself (best).

Exposure tips

- There is no problem exposing prints one by one and saving the processing steps until the last one has been exposed, provided the exposed prints are kept in the dark while awaiting processing.
- To understand how many steps are lost and then regained during processing and drying, print a step wedge and put a pencil mark where there is detail in the highest highlights, just before maximum white. Then note where that detail is when the step wedge is processed and dried.
- To understand how much exposure is necessary to get maximum blue but not so much that the darkest blue begins to lighten, overexpose a step wedge and put a pencil mark where the first darkest dark step begins to lighten and solarize. After processing and drying, note if there are any steps that are slightly lighter in the darkest area. The step just below that first slightly lighter step indicates the best exposure time.
- Papers differ in exposure time, exposure scale, color of blue, and darkness of blue. Some papers are faster and some are slower. The speed can be affected by the alkalinity of the paper and also how much the paper absorbs chemistry. I find the more absorbent the paper is, the longer the exposure scale. The color of blue can range from turquoise (greener) to royal (purpler) blue.
- "Apparently a portion of the ferrous salt first formed by the reducing action of light reacts further with a portion of the Prussian blue to form the white insoluble ferrous ferrocyanide. After the print is washed thoroughly and dried, the ferrous ferrocyanide is oxidized gradually by the air to form further amounts of ferric ferrocyanide (Prussian blue). This explains why the cyanotype darkens after being subjected for a considerable period to the action of the air."[5]

Ballpark UVBL exposure times

The following table lists the exposure times I calibrated for classic (CC) and new (NC) cyanotype on a number of papers. I tested a 10/10 classic cyanotype formula and a 200 ml new cyanotype formula. If using classic cyanotype with more FAC, e.g. a 20/10, or a more concentrated new cyanotype solution (100 ml or 150 ml instead of 200 ml), your exposure times will probably vary, but once one paper is tested, it will be easy to extrapolate from this chart how other papers will respond.

A few interesting observations surfaced with the data. Exposure times generally decreased with acid development. Papers that are slow to expose are often the alkaline ones and those unsuitable for cyanotype. Papers that are quick to expose are often less alkaline, or alkaline-free, thinner, and more absorbent. [Note: I include the exposure times for no acid and sulfamic acid for CC and no acid and vinegar for NC for comparison purposes only. I would not recommend those developments as mentioned further on.]

As always, your results may vary (YRMV): UVBL units differ somewhat, and water supplies differ drastically, in which case merely use the charts to clue you in to slower versus faster papers. As we all know, nothing is cut in stone with cyanotype!

PAPER NAME	CC no acid	CC vinegar	CC citric	CC sulfamic	NC no acid	NC vinegar	NC citric	NC sulfamic
Arches Aquarelle	30.0	23.8	37.8	30.0	7.9	6.3	6.3	5.0
Arches Cover	37.8	30.0	30.0	9.5	6.3	6.3	7.9	5.0
Arches Johannot	37.8	37.8	30.0	30.0	12.6	7.9	10.0	5.0
Arches Lavis Fidelis En Tout Cas	30.0	23.8	23.8	23.8	6.3	6.3	6.3	5.0
Arches Platine	23.8	18.9	18.9	9.5	5.0	5.0	5.0	4.0
Arches Platine Lightweight	30.0	23.8	23.8	23.8	5.0	4.0	4.0	4.0
Arches Text Wove	37.8	30.0	30.0	30.0	7.9	6.3	7.9	6.3
Arnhem 1618	30.0	23.8	30.0	37.8	6.3	7.9	7.9	6.3
Awagami Platinum Gampi	18.9	18.9	18.9	15.0	6.3	6.3	6.3	6.3
Bee Watercolor	30.0	30.0	30.0	30.0	12.6	7.9	10.0	7.9
Bergger Cot 160	18.9	15.0	15.0	9.5	5.0	4.0	4.0	3.2
Bergger Cot 320	30.0	23.8	23.8	15.0	6.3	5.0	5.0	5.0
Bienfang 360	23.8	23.8	23.8	18.9	10.0	10.0	7.9	6.3
Bockingford	23.8	23.8	23.8	7.5	7.9	7.9	6.3	6.3
Canson Bristol Recycled	37.8	30.0	37.8	23.8	12.6	7.9	7.9	6.3
Canson Edition Etching	23.8	23.8	18.9	18.9	7.9	6.3	6.3	5.0
Canson Heritage Watercolor	37.8	30.0	30.0	30.0	10.0	7.9	7.9	6.3
Canson Mi-Teintes	30.0	30.0	23.8	23.8	10.0	7.9	7.9	7.9
Canson Mix Media	37.8	23.8	23.8	30.0	10.0	10.0	6.3	6.3
Canson Montval	37.8	23.8	23.8	18.9	6.3	6.3	4.0	4.0
Canson Opalux Vellum	18.9	15.0	18.9	18.9	5.0	5.0	5.0	5.0
Canson Pro Layout Marker	37.8	18.9	18.9	15.0	7.9	6.3	6.3	5.0
Canson Universal Sketch	23.8	11.9	11.9	9.5	12.6	10.0	10.0	7.9
Canson Vidalon Vellum	7.5	7.5	9.5	15.0	12.6	10.0	7.9	7.9
Canson XL Watercolor	30.0	23.8	23.8	18.9	7.9	6.3	7.9	16.3
Clearprint 1020HP Vellum	30.0	18.9	18.9	7.5	6.3	6.3	6.3	6.3
Fabriano Artistico	37.8	23.8	23.8	23.8	12.6	12.6	10.0	7.9
Fabriano Tiepolo	37.8	37.8	37.8	23.8	10.0	12.6	10.0	6.3
Fabriano Tiziano	18.9	18.9	18.9	18.9	6.3	6.3	6.3	6.3
Fabriano Unica	37.8	37.8	30.0	15.0	6.3	6.3	6.3	5.0
Fluid 100 HP	30.0	30.0	30.0	23.8	12.6	10.0	7.9	1.9
Garzapapel Drawing	23.8	18.9	15.0	11.9	4.0	3.2	3.2	2.0
Garzapapel Ink-Pen	37.8	23.8	30.0	23.8	2.5	3.2	5.0	2.5
Hahnemühle Bamboo	23.8	23.8	23.8	18.9	7.9	6.3	6.3	5.0
Hahnemühle Cézanne	30.0	30.0	23.8	18.9	7.9	6.3	7.9	6.3
Hahnemühle Lanaquarelle	37.8	30.0	23.8	18.9	7.9	7.9	7.9	27.9
Hahnemühle Platinum Rag	23.8	23.8	18.9	7.5	5.0	5.0	5.0	4.0
Hahnemühle Sumi-e	37.8	23.8	23.8	18.9	12.6	10.0	7.9	6.3
Hahnemühle Turner	30.0	23.8	23.8	23.8	10.0	7.9	6.3	6.3
Indigo 100% cotton	30.0	23.8	18.9	15.0	6.3	5.0	5.0	4.0

PAPER NAME	CC no acid	CC vinegar	CC citric	CC sulfamic	NC no acid	NC vinegar	NC citric	NC sulfamic
Japanese Mulberry/Hvy Kozo	18.9	18.9	18.9	18.9	6.3	6.3	6.3	5.0
Khadi Cotton and Plant Paper	37.8	23.8	15.0	9.5	7.9	6.3	5.0	5.0
Khadi Cotton Rag	18.9	18.9	18.9	11.9	7.9	7.9	6.3	5.0
Khadi Cotton Rag Lightweight	18.9	18.9	15.0	15.0	6.3	5.0	4.0	4.0
Legion Bamboo	30.0	30.0	23.8	15.0	6.3	6.3	6.3	5.0
Legion Coventry Rag	37.8	37.8	37.8	23.8	7.9	7.9	7.9	5.0
Legion Domestic Etching	37.8	30.0	30.0	30.0	7.9	6.3	6.3	6.3
Legion Folio	30.0	30.0	30.0	15.0	7.9	7.9	7.9	6.3
Legion Goyu	37.8	23.8	23.8	23.8	10.0	10.0	7.9	10.0
Legion Lenox 100	37.8	30.0	37.8	23.8	12.6	7.9	7.9	6.3
Legion Masa	23.8	23.8	23.8	18.9	7.9	7.9	7.9	6.3
Legion Parchtone	30.0	23.8	23.8	18.9	6.3	6.3	6.3	6.3
Legion Revere Platinum	23.8	18.9	18.9	15.0	7.9	7.9	7.9	6.3
Legion Rising Barrier	37.8	37.8	37.8	37.8	12.6	12.6	12.6	10.0
Legion Rising Drawing Bristol	23.8	23.8	23.8	18.9	7.9	7.9	7.9	7.9
Legion Stonehenge	30.0	23.8	23.8	18.9	7.9	7.9	7.9	6.3
Legion Stonehenge Aqua	30.0	30.0	23.8	23.8	6.3	10.0	6.3	5.0
Legion Stonehenge Colors	37.8	23.8	23.8	37.8	7.9	12.6	7.9	7.9
Legion Thai Kozo	37.8	37.8	30.0	23.8	12.6	12.6	12.6	10.0
Magnani Pescia	30.0	23.8	23.8	9.5	6.3	7.9	6.3	5.0
Magnani Portofino	37.8	23.8	23.8	23.8	5.0	5.0	5.0	5.0
Magnani Revere Book 175	18.9	18.9	23.8	18.9	6.3	5.0	5.0	5.0
Magnani Vergata	30.0	23.8	18.9	15.0	2.5	2.5	2.5	2.5
Pergamenata Parchment	18.9	18.9	18.9	18.9	5.0	5.0	5.0	5.0
Rives BFK 280 gsm	18.9	18.9	15.0	15.0	5.0	5.0	5.0	5.0
Rives BFK Hvywt 175 gsm	37.8	30.0	30.0	23.8	4.0	4.0	3.2	3.2
Rives de Lin	30.0	30.0	30.0	30.0	7.9	7.9	6.3	6.3
Rives Lightweight 115 gsm	30.0	18.9	18.9	18.9	6.3	6.3	5.0	6.3
Ruscombe Mill Chateau Vellum	37.8	23.8	23.8	15.0	7.9	7.9	5.0	4.0
Ruscombe Mill Herschel	30.0	18.9	18.9	9.5	5.0	5.0	5.0	5.0
Saunders Waterford	37.8	30.0	30.0	23.8	4.0	4.0	3.2	2.0
Schut Noblesse	30.0	30.0	30.0	18.9	7.9	6.3	5.0	5.0
Somerset Satin	37.8	37.8	37.8	37.8	7.9	7.9	7.9	6.3
St. Armand Colors	37.8	23.8	18.9	11.9	7.9	5.0	6.3	5.0
St. Armand Old Masters	30.0	18.9	15.0	9.5	5.0	5.0	5.0	5.0
St. Armand White Rag	23.8	23.8	18.9	15.0	6.3	5.0	4.0	5.0
Strathmore 500 Bristol	30.0	23.8	23.8	30.0	7.9	6.3	7.9	5.0
Twinrocker B-Buff T	18.9	18.9	18.9	15.0	3.2	2.0	2.0	2.0
Twinrocker Biscuit HT	18.9	18.9	11.9	6.0	6.3	5.0	3.2	3.2
Twinrocker Bleached Abaca HT	30.0	23.8	15.0	7.5	5.0	5.0	5.0	3.2
Twinrocker Bone T	30.0	23.8	15.0	7.5	6.3	5.0	5.0	5.0
Twinrocker Cream LA	23.8	18.9	18.9	18.9	4.0	2.0	2.0	5.0
Twinrocker Da Vinci T	37.8	23.8	15.0	11.9	5.0	5.0	5.0	2.5
Twinrocker Dickinson TT	18.9	15.0	15.0	11.9	5.0	5.0	4.0	4.0
Twinrocker May Linen LA	23.8	18.9	18.9	15.0	3.2	3.2	3.2	3.2
Twinrocker Oyster HT	18.9	18.9	18.9	11.9	4.0	5.0	4.0	3.2
Twinrocker White Cotton Rag TT	23.8	18.9	18.9	9.5	5.0	5.0	5.0	5.0
Twinrocker Yale T	30.0	18.9	18.9	15.0	3.2	3.2	3.2	4.0
Velke Losiny Moldau	30.0	23.8	18.9	18.9	4.0	4.0	4.0	3.2
Velke Losiny Moravia	30.0	23.8	18.9	9.5	6.3	6.3	5.0	5.0
Velke Losiny Prague	30.0	23.8	23.8	15.0	5.0	5.0	5.0	4.0
Winsor & Newton Watercolor	37.8	30.0	30.0	23.8	7.9	7.9	7.9	5.0
Wyndstone Vellum	18.9	18.9	18.9	18.9	6.3	6.3	6.3	6.3
Zerkall Book	30.0	23.8	18.9	11.9	6.3	6.3	6.3	5.0
Zerkall Frankfurt	30.0	23.8	23.8	23.8	5.0	5.0	5.0	5.0
Zerkall German Ingres	23.8	18.9	15.0	7.5	6.3	6.3	5.0	5.0
Zerkall Nideggen	30.0	23.8	18.9	11.9	7.9	6.3	6.3	5.0
AVERAGES	29.44	24.20	23.04	18.56	7.20	6.61	6.20	5.62

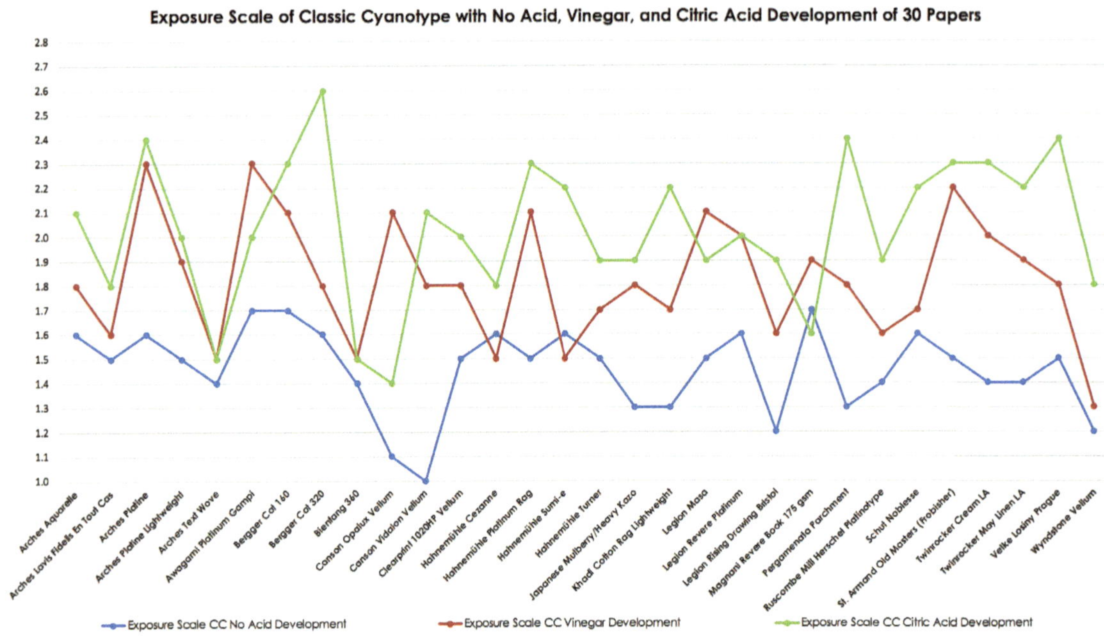

Figure 8.5. See how vinegar and citric acid extend the exposure scale of classic cyanotype.

Exposure scale

Exposure scale (ES) is a term that refers to how many steps print on a step wedge from maximum black to paper white. Only one maximum black step is included in the counting. Though often photographers include paper white in their ES, I not only don't include paper white but I only include *the last pale blue step that still has a definable edge mark*, and thus these exposure scales err *greatly* on the side of under-representation by perhaps a minimum of two steps or .2.

Exposure scales can be short or long. I have measured as low as a 1.0 ES on a paper with no acidic development to a 2.8 using new cyanotype and sulfamic acid development. This means that the former paper only printed 10 steps on a 31-step wedge (very short scale) and the latter 28 steps (very long scale). A short ES can mean a "soot and chalk" print or one that is way too contrasty with blocked up shadows and blown out highlights *if the digital negative isn't designed to match this short scale.* In general, a longer ES is easier to work with, and better able to express the entire tonal range of the image from deepest shadows to highest highlights.

Processing cyanotype

One of the reasons for cyanotype's continued and increasing popularity is that aside from only two chemicals to make up the solution, it requires only water to process a print. It is as low tech a process there is. Even though low tech, it is still important to pay attention and be consistent, especially when processing. If the cyanotype print is left in the wash water for too little time, iron left in the print will result in yellowed highlights and wreak havoc down the road, fading the highlights to gray.[6] If the print is left in the wash water too long, it fades quite dramatically. If the wash water is alkaline, the print not only fades, but also turns to a brown or gray color. Here are some suggestions:

- Test the wash water with litmus paper to make sure it is neutral (pH 7). This is usually the case with well water, but not always. City water can be much different, and vary from winter to summer.
- Watch the temperature of the wash water. Anywhere from cool to quite warm is acceptable, for example, 70°–100°F.

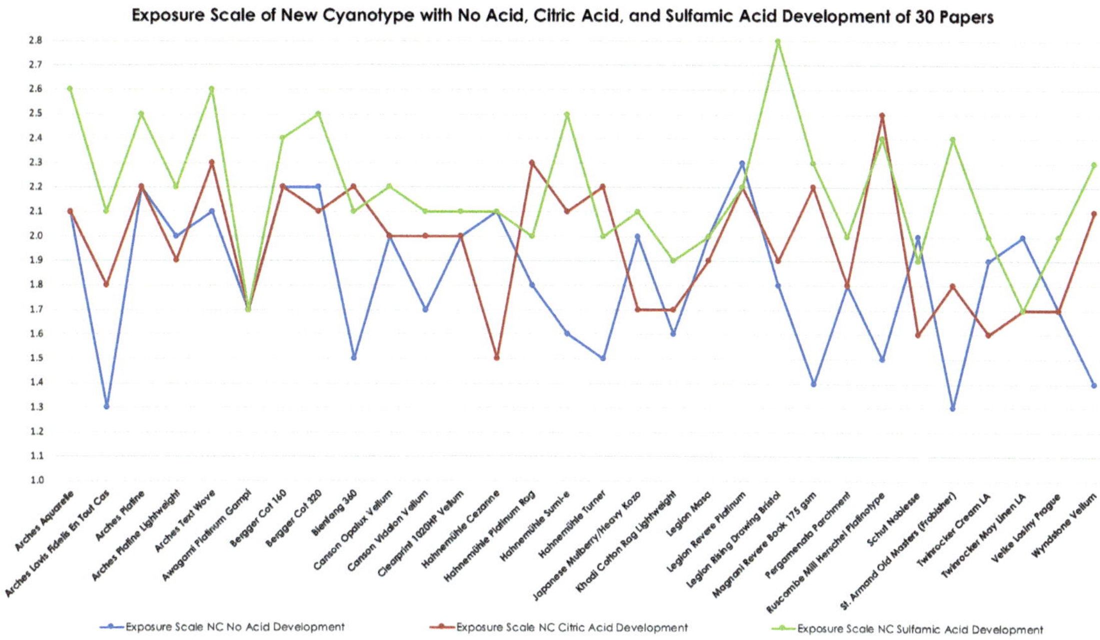

Figure 8.6. See how citric and sulfamic acid extend the exposure scale of new cyanotype.

- Keep the time of the water bath consistent. Don't pull the print too soon because it looks like all the yellow is gone, or forget the print and leave it in the wash for an hour because of the reasons discussed above.
- Keep the agitation consistent. Agitate continuously for the first few minutes, and then the print soaks face up for the remainder of the wash time, with an occasional wiggle every few minutes to avoid air bubbles which leave round ugly spots of deeper blue.

Processing with acidified water

In my previous almost twenty years of cyanotype practice I had mostly used plain water processing. I used vinegar or citric acid in the water bath occasionally, but my prints tended to end up with stained highlights. When I began the intensive paper testing I had to be systematic about all variables, including development. I chose to compare five types of development: plain water and then water acidified with vinegar, citric acid, hydrochloric acid, and sulfamic acid. I kept an Excel spreadsheet of exposure times and exposure scales of all papers and all five development techniques. I was quite surprised at what I learned. With the first ten or so papers I thought my results were anomalies, but after 136 papers put through the tests I am confident in my conclusion which is this: using some form of acidified water development is a must, and the acid of choice and amount is specific and different for classic versus new cyanotype. In an email, Mike Ware cinched this observation: "The photoproduct iron(II) (ferrous iron, Fe2+) is still bound by ligands (citrate in classic cyanotype or oxalate in new cyanotype) and this molecular complex cannot enter into and make the Prussian blue lattice with ferricyanide in this state. The Fe(II) has to be stripped of its ligands, and acid does just this, because hydrogen ions bind to the citrate or oxalate anions and pull them off the Fe(II)." If you don't take my word on it, take Ware's. Here are my suggestions:

- The stronger the acid, the longer the exposure scale, meaning more tones from maximum blue to paper white
- The use of an acid shortens the exposure time, as much as ⅔ stop

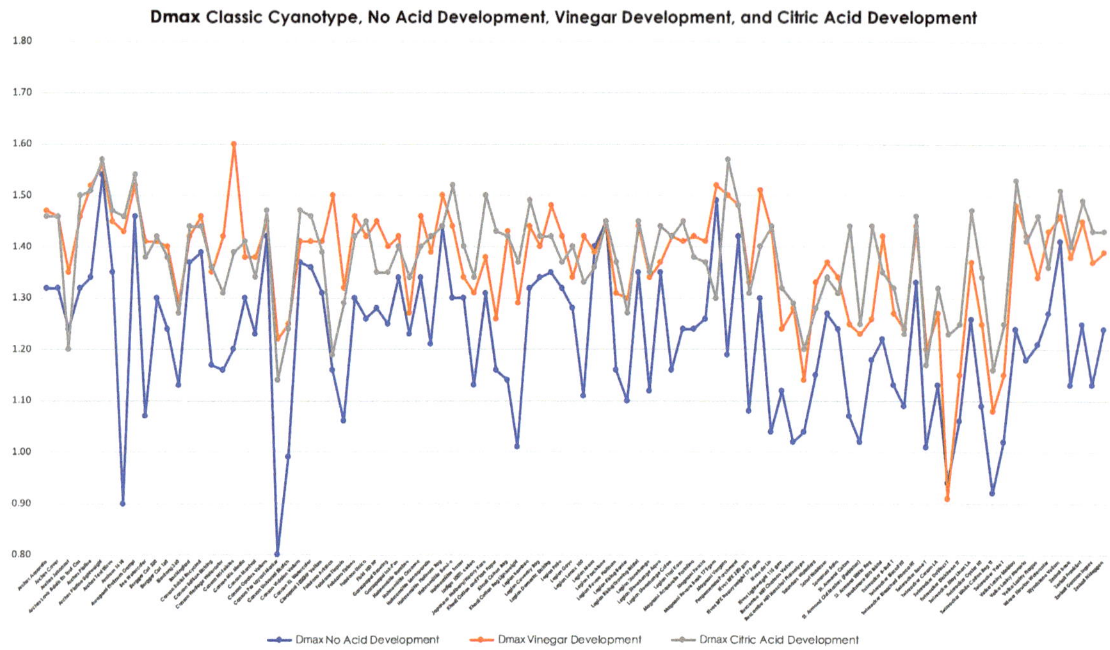

Figure 8.7. If a chart like this makes your eyes glaze over, there is one takeaway: use acidic development. The blue line representing no acidic development results across the board in a lighter, duller print. Vinegar and citric acid development are somewhat interchangeable for classic cyanotype (not new cyanotype!), with some papers performing better with vinegar in the development bath and others with citric in the development bath.

- The stronger the acid, the darker the blue/Dmax of the print; not once did the Dmax measure darkest on prints developed in unacidified water
- The stronger the acid, the less wash off occurs
- Classic cyanotype and new cyanotype differ in which acids work best
- The stronger the acid, especially with classic cyanotype, the more the highlights can stain blue
- Classic cyanotype and new cyanotype differ in how much acid can be used without staining of the highlights; for example, I can use very little sulfamic acid in a liter of water with classic cyanotype, but with new cyanotype a 1% sulfamic acid solution or 2 teaspoons in a liter poses no problem
- Classic cyanotype works best with vinegar and citric acid; new cyanotype works best with sulfamic acid; citric acid is an adequate choice, too
- With both formulas there is a sweet spot where enough acid is used to lengthen the exposure scale, shorten the exposure time, and darken the print, before staining of the highlights occurs.

Vinegar development for CC

I have found recommendations in the literature to use anywhere from full strength white vinegar (5% acetic acid) to a dilution of 1:10. How strong a solution you can use without stained highlights will have to be found out by trial and error. In my practice, 60 ml/¼ cup per liter is plenty. [Note: Ware cautions against acetic acid because it may form "insoluble colored basic ferric acetate,"[7] and though I have not seen evidence of this in my CC prints, NC prints developed in either plain water or vinegar water results in various shades of yellow to brown stain *immediately* upon immersing the print on all but the most pristine papers made for alt. This is not to say it is ferric acetate—it is probably hydrolized iron[8] which shows up most drastically on calcium carbonate buffered papers—but plain water or vinegar are not suitable developers for new or hybrid cyanotype formulas in my experience.]

60 ml white vinegar (4 tablespoons/¼ cup)
1000 ml water

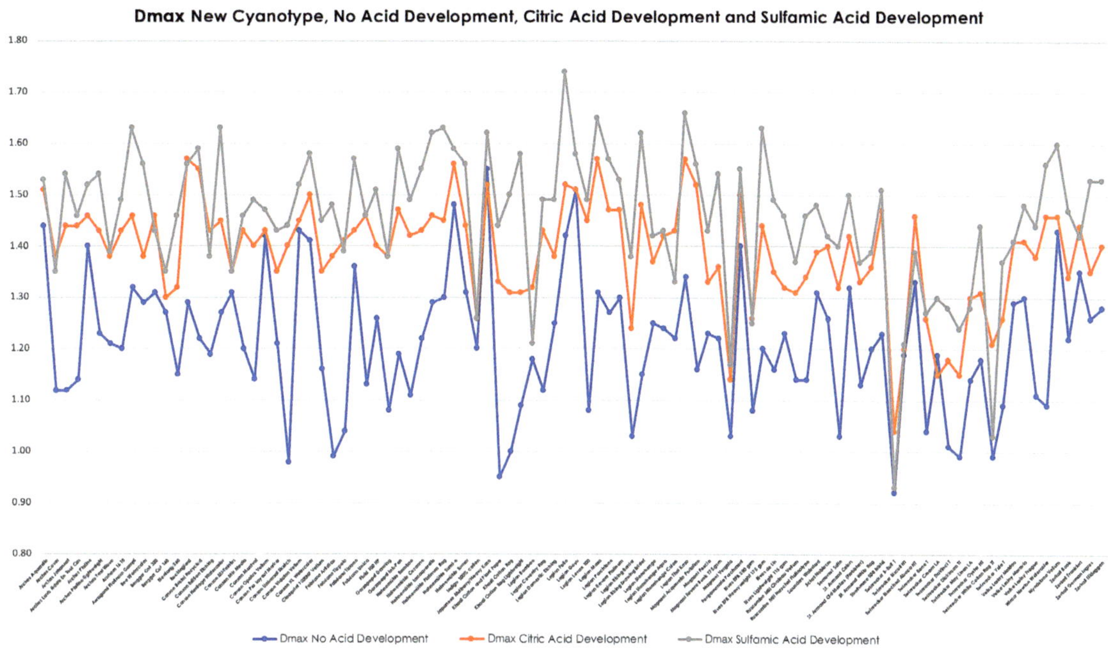

Figure 8.8. Note the comparison between the CC chart, left, and the NC chart here. New cyanotype is much more responsive to acidic development than classic cyanotype. Almost across the board sulfamic acid produces the deepest Dmax because the gray line rises to the top, but citric acid is still an acceptable alternative. Vinegar is not on this chart because it often results in stained, browned prints on any but the papers made specifically for alt (e.g. Platine, HPR, Bergger Cot 320).

1. Add the 60 ml vinegar to a liter of warm water and stir until evenly dispersed. Measure the pH before proceeding. Depending on the water source's initial pH, this is about a 4 pH. I only measure this once because my well water doesn't change, but if you are on city water that potentially changes in alkalinity with the seasons, you may want to make a practice of testing the acidified bath with litmus strips once before a printing session and adding more vinegar if necessary.
2. Remove the print from the exposure frame and immerse the whole of the paper at once in the acidified warm water.
3. You can use the entire liter of vinegar water and run perhaps 6 prints through it before it turns dark blue and should be discarded, or you can use about 250 ml/1 cup of vinegar water as a "one shot" wash, discarding after use. This minimizes any potential staining, but with the 10/10 formula I get very little washoff so the vinegar water stays good a long time.
4. Agitate the print continuously in the tray for 1–5 minutes; the water will turn blue, depending on how much wash off occurs.
5. At the end of 1–5 minutes, either throw out the one-shot developer and fill the tray with plain warm water, or transfer the print face down to another tray of warm water for 5 minutes, and wiggle the print intermittently to prevent airbells.
6. At the end of another 5 minutes discard the water and refill the tray with warm tap water again. Soak the print for another 5 minutes and wiggle the print intermittently to prevent airbells.
7. Once the print has gone through 3 washings/15 minutes, it is probably done. Some papers clear in less time, more like 10 minutes. Some take longer to clear. Take note of the time the paper

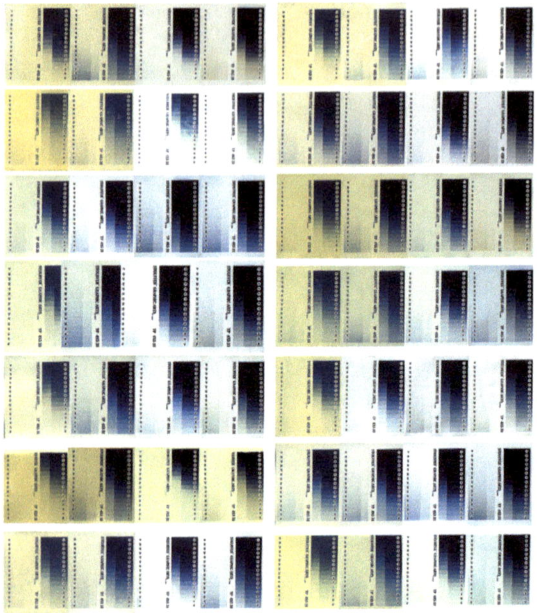

Figure 8.9. Above is a series of step wedges on fourteen papers, in two columns of seven, new cyanotype developed four ways (from left to right): no acid, vinegar, citric, sulfamic (a couple of the papers are cream colored). Notice how unsuitable plain water and vinegar water is for new cyanotype. All of these papers browned in those two development baths, even the best papers for alt which showed a slight yellow tinge. This is why I always use either citric acid or sulfamic acid development with new cyanotype, and never vinegar or plain water.

takes to clear of all yellow and use that time in the future.

8. Hang the print to dry. The print will oxidize from a turquoise to a deeper royal blue within the day, just by virtue of being exposed to the oxygen in the room air. No need to use extra chemistry (hydrogen peroxide) to hurry oxidation along.
9. When the print is fully dry and oxidized, look at the highlights. If the highlights are too white and without detail, revisit the negative and exposure time, not the development time; the development time doesn't vary unless needing to develop longer for an overexposed print. The negative could be too dense or the exposure time could be too short, but the development time should remain consistent.

Citric acid development for CC

Follow all the steps above, but use 2.5 g citric acid per liter (½ teaspoon). This amount of citric acid is about a 3 pH and a mere 0.25% dilution. Even though most acidified water recommendations are in the range of 1–5% I find staining can occur past the 0.25% amount; it may not for you, so test stronger dilutions if desired.

Citric acid development for NC

Follow all the steps above, but use 10 g citric acid per liter (2 teaspoons). This amount of citric acid is about a 3 pH and a 1% dilution.

Sulfamic acid development for CC

Sulfamic acid is recommended to be used in a 1–5% dilution for new cyanotype[9] but for classic cyanotype a mere 0.2% or *¼ teaspoon per liter of water* is the most I can use without staining of the highlights, and even with this amount, many papers stain. This is why it is not my favorite acid for classic cyanotype. This will make the water close to a 2 pH. Use it as a one-shot bath, for 30 seconds to 2 minutes maximum.

2 g sulfamic acid (scant ¼ teaspoon)
1000 ml warm water

1. Measure the water into a beaker, and carefully add the sulfamic acid to the water and stir until dissolved, being careful not to splash or spill. It is not terribly soluble so stir a while.
2. Use about 250 ml of this per print. Remove the print from the exposure frame and pour the 250 ml of acidified water onto the print and agitate continuously for 30 seconds to 2 minutes; the water will turn blue, depending on how much wash off occurs.
3. At the end of 30 seconds to 2 minutes, dump the acidified water down the drain and fill the tray with warm tap water. Agitate the print for 5 minutes in the warm water to prevent airbells.
4. At the end of 5 minutes discard the water and refill the tray with tap water again. Soak the print for another 5 minutes and wiggle the print intermittently to prevent airbells.
5. Once the print has gone through 3 washings/15 minutes, it should be done; hang the print to dry.

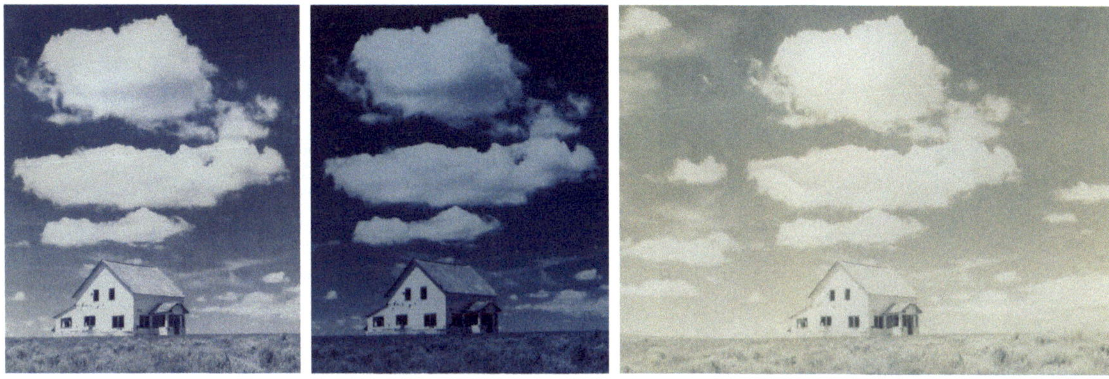

Figures 8.10–8.12. On the left two house images, note the contrast and speed change of no acid development (far left) versus vinegar development (middle). That is why it is important to calibrate your exposure time and negative density around the development you will be using. On the far right the print was left too long in the water wash. Eventually the entire cyanotype will wash down the drain and leave a pale brown or gray print if you do not time the wash carefully.

Sulfamic acid development for NC

New cyanotype does best with 1% sulfamic acid and a longer development. No need to do the bath one shot, as there is very little wash off and bluing of the water and less potential for the sulfamic acid bath to stain the paper; thus more prints can be run through the liter. It also can take a longer water wash than classic cyanotype after the sulfamic acid bath.

10 g sulfamic acid (2 scant teaspoons)
1000 ml warm water

1. Measure warm water into a beaker, and carefully add the sulfamic acid to the water and stir until dissolved, being careful not to splash or spill. It is not terribly soluble so stir for a bit. Pour the liter into a tray.
2. Remove the print from the exposure frame and put it in the sulfamic acid bath, agitating vigorously and continuously for 2 minutes. Leave the print in the bath for up to 5 minutes if desired.
3. Place the print in another bath of plain warm water and agitate intermittently for 5 minutes.
4. Discard the water and refill the tray with warm tap water again and soak the print for another 5 minutes, wiggling intermittently to prevent airbells.
5. Once the print has gone through 3 washings/15 minutes, it should be done, although some wash up to 30 minutes; hang the print to dry.

Hydrochloric acid development for NC

Vinegar and citric acid are for the most part harmless; they are in our foods, soft drinks, and vitamins. They are my preferred acids for classic cyanotype. Citric acid and sulfamic acid are my preferred acids for new cyanotype. Sulfamic acid is an industrial cleaner and should be handled with care. Even a 1% solution of sulfamic acid can be caustic to hands so use tongs when moving prints, or wear gloves if immersing your hands in the bath. It should be used with all the caution normally in place with all chemistry.

Hydrochloric has often been used in photographic processes. Now it is most used in swimming pools under the name "muriatic" acid. Muriatic acid is readily available at hardware stores. It comes in a ~30% dilution. *This is a pungent, heavy acid that should be handled with utmost care.* At the risk of being belittled, I will tell you my story. I had a small beaker of muriatic acid, perhaps 10 ml in it, and I left the dimroom for a period of time with the beaker uncovered. When I returned, my litmus strips *six feet away* from the beaker had turned red from the acid in the air. If I could make such a stupid mistake, my college level students could also. For this reason and for the reason it is corrosive to metal pipes, I do not use it, nor do I teach it. If you are a scientist and work with a fume hood, have at it. Otherwise, stay away. Sulfamic acid does not outgas and does not corrode metal.

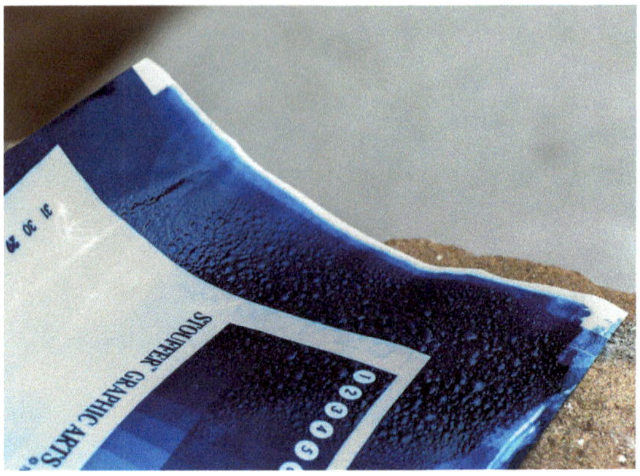

Figure 8.13. Watch using sulfamic acid on vellums; this Vidalon vellum bubbled and disintegrated in sulfamic!

There are cyanotypists and platinum printers who use it all the time in their clearing practice. I will not use it in my personal practice because I can see a white gas evaporate into the air when measuring out amounts, which means it is in the air I breathe. I do not have a fume hood in my dimroom, and most people don't either.

So with that dire warning, if you still want to test it because you are a scientist with a fume hood, the recommended dilution is a 0.1–1% solution. A mere 1 ml (20 drops) of muriatic acid per liter is enough to acidify a liter of water to a 2 pH without staining a cc print. Your results may vary; you may need more. If so, use up to 35 ml muriatic acid per liter to make a 1% dilution, especially for an NC print. It is best to use this as a one-shot bath, 250 ml/1 cup per print, discarding after use.

1–35 ml muriatic acid (1 ml = 20 drops)
1000 ml water

1. Measure the water into a beaker, and carefully add the muriatic acid to the water and stir, being careful not to splash or spill.
2. Use about 250 ml of this per print. Remove the print from the exposure frame and immerse the whole of the paper at once in the 250 ml of acidified water.
3. Agitate the print continuously in the tray for 1–2 minutes; the water will turn blue, depending on how much wash off occurs.

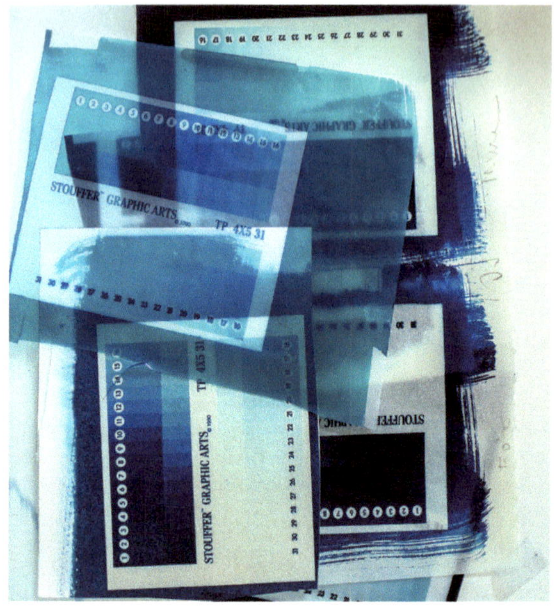

Figure 8.14. Note the lighter turquoise of the unoxidized vellum compared to the deeper blue of the oxidized (with hydrogen peroxide) prints.

4. At the end of 1–2 minutes, dump the acidified water down the drain and fill the tray with warm tap water. This time, place the print face down for 5 minutes, and wiggle the print intermittently to prevent airbells.
5. At the end of 5 minutes discard the water and refill the tray with tap water again. Soak the print for another 5 minutes and wiggle the print intermittently to prevent airbells.
6. Once the print has gone through 3 washings/15 minutes, it should be done; hang the print to dry.

Hydrogen peroxide use

A cyanotype print comes out of the development bath a beautiful soft turquoise blue. This color oxidizes to a deeper blue over 24 hours. If you want to see this change happen immediately you can add a large glug of drugstore hydrogen peroxide to the water wash just after the first acidified water bath. Amount is immaterial; you can see the print immediately darken, and if not, add more. This step is not necessary since the print darkens within 24

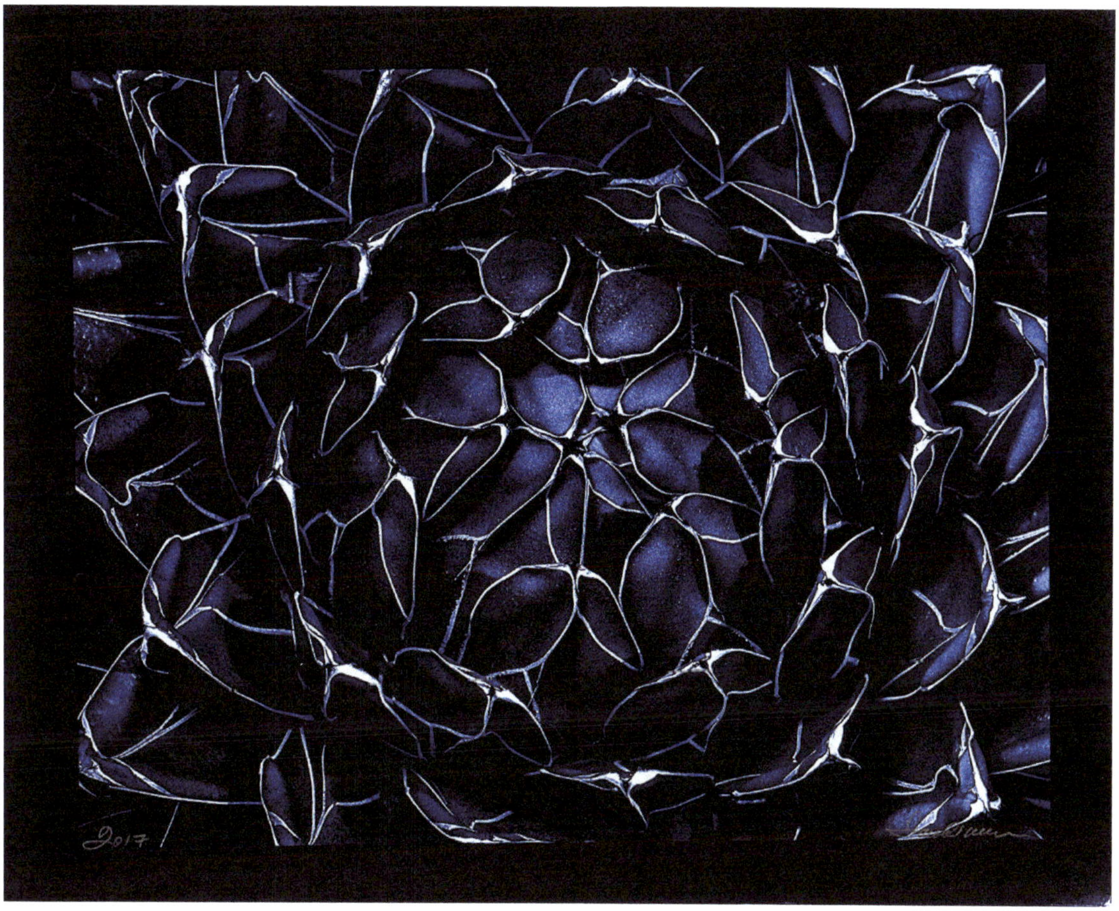

Figure 8.15. *Queen Victoria Agave*, classic cyanotype, 14″ x 11″ © Eugene Starobinskiy 2017

hours anyway. It does have a "wow" factor, though. I often add a large glug of the drugstore dilution *to my acidified water bath*; that way one glug lasts for most of the printing session instead of having to add fresh hydrogen peroxide to the water wash after each time you dump and refill the tray. [Note: add it to the acidified water bath after the acid has been thoroughly dissolved in the water; *never add hydrogen peroxide directly on dry chemistry.*]

Endnotes

1. Schaefer, John P. *The Ansel Adams Guide, Basic Techniques of Photography, Book 2*. Boston: Little Brown and Company, 1998, p. 160.
2. Harrison, Jerome. "The Chemistry of Silver Printing, Printing with Salts of Iron—Cyanotype and Kallitype" in *Photographic Times*, Vol. XXI, September 4 1891, pp. 440–442.
3. Seigel, Judy ed. *The World Journal of Post-Factory Photography*. New York: Post Factory Press, April 1998–April 2004, Issue 5, p. 28.
4. Fabbri, Malin and Gary. *Blueprint to Cyanotypes*. Alternativephotography.com, 2006, p. 19.
5. Farber, Richard. *Historic Photographic Processes, A Guide to Creating Handmade Photographic Images*. New York: Allworth Press, 1998, p. 59; quote taken from O'Hara, Charles E., and James W. Osterburg. *Practical Photographic Chemistry*. Minneapolis: American Photographic Publishing Company, 1951.
6. Worthington, J. C. and J. C. Millen. *The Photographic Primer, A Manual of Practice*. Riverton: The Riverton Press, 1897, pp. 130–131.
7. Ware. *Cyanomicon II, History, Science and Art of Cyanotype: photographic printing in Prussian blue*. Buxton: self-published, 2016, pp. 216–217.
8. Ibid., p. 176.
9. Ibid., pp. 216–217.

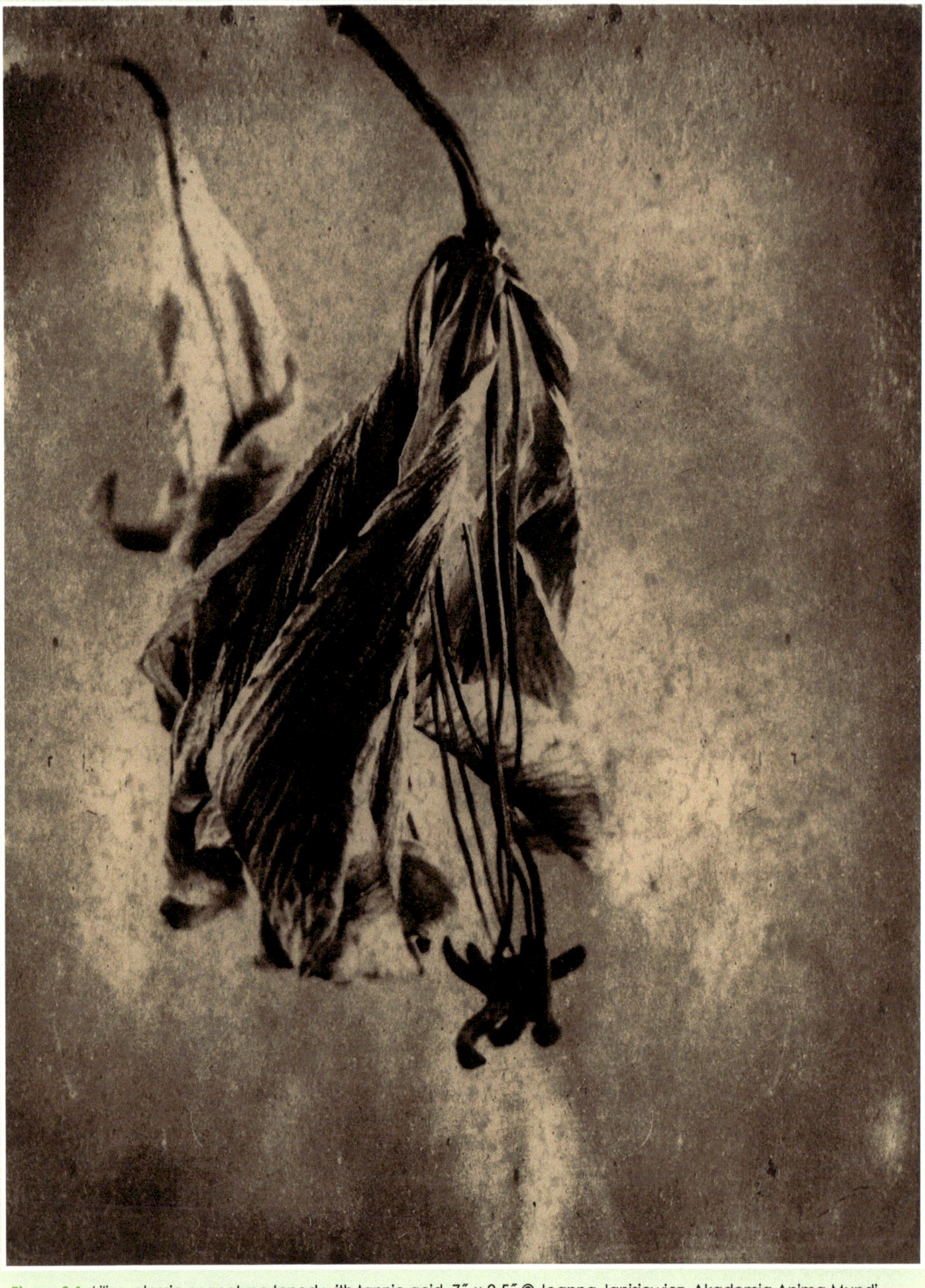

Figure 9.1. *Lilies*, classic cyanotype toned with tannic acid, 7″ × 9.5″ © Joanna Janisiewicz, Akademia Anima Mundi 2018. Joanna Janisiewicz, originally from Poland, is a teacher at Akademia Anima Mundi, an arts organization basing its work on Plato's Academy principles, with its core interest being the idea of Beauty. The art of photography (in the analog and digital darkroom) is taught along with aesthetics, elements of criticism and interpretation, history of art, drawing, calligraphy and graphic arts. For more information visit **www.animamundi.com.pl**.

Chapter 9

Toning Cyanotype

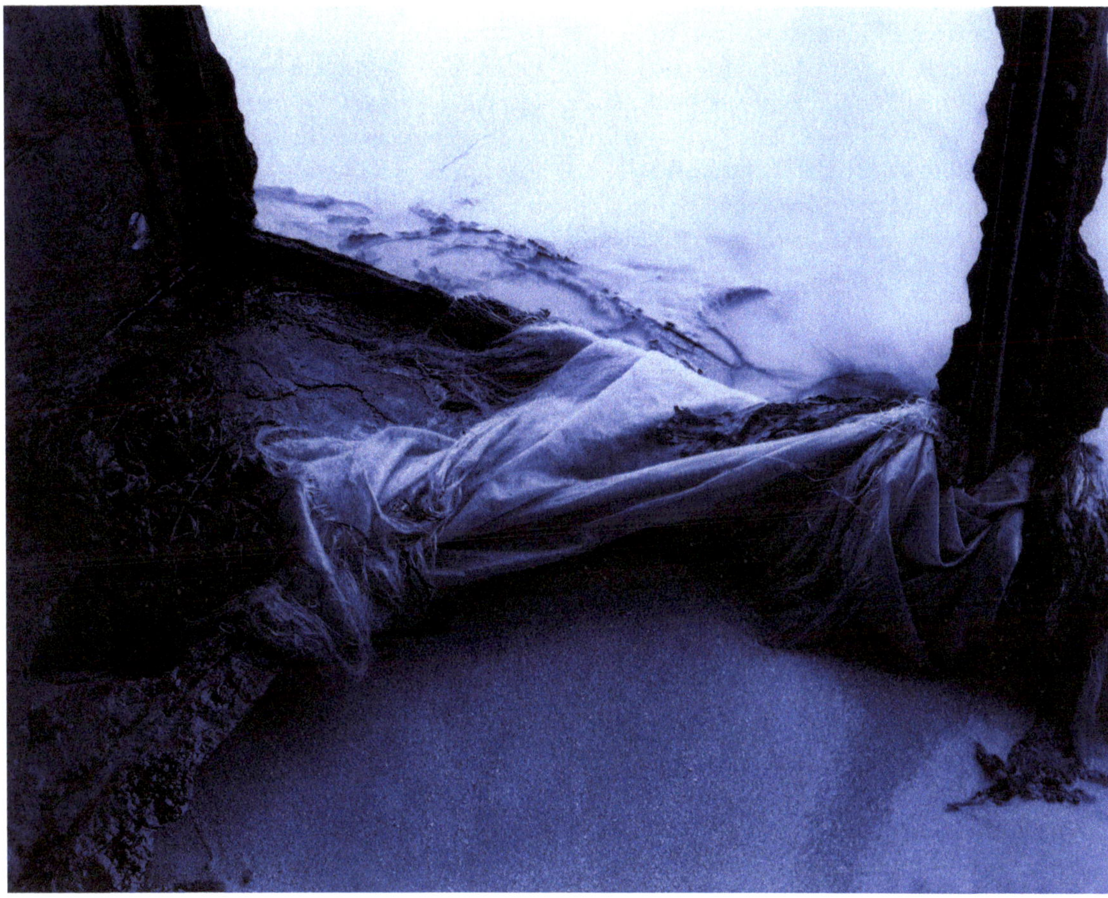

Figure 9.2. *Shrouded Blockship, Orkney*, cyanotype toned with lead acetate, 238 x 188 cm, image from a scanned 6 x 9 cm film negative © Mike Ware 2005

Toning a cyanotype is fairly simple. There are two components, an alkali (a substance with a pH over 7) and an acid (a substance with a pH under 7). First an alkali such as ammonia or sodium carbonate is used to bleach the print. Then an acid such as tannic or gallic acid is used to tone the print. This order is sometimes reversed or doubled back and forth; nothing is cut in stone. There are also one-bath toners that use only one or the other, acid or alkali, and one-bath toners made from a combination of the acid and alkali.

In 1900, author George Brown penned these cranky words: "Many of these [toners] do not come within the sphere of practical photography, for they merely enable a tone to be obtained which is very frequently only a sorry apology for one which can be obtained much more readily by some other process…it is a mistaken policy to torture the blue ferro-prussiate image with chemical reagents, with the object of imitating these other processes."[1]

What Brown didn't take into account was how much fun toning cyanotype can be! This chapter has been designed to take out the complexities of toning, by simplifying formulas and providing equivalent teaspoon measures for those who prefer the "kitchen" approach. Only the formulas I have tested have been included, and though step wedges are certainly less exciting than an actual print, they do a good job of illustrating the very subtle color differences from deepest shadows to highest highlights that result from the different toners.[2]

Choice of alkalis, called "bleaches"

Most people have alkalis around the house because they are commonly part of laundry detergents or cleaning products. The easiest to use and the least caustic alkalis are:

- Ammonia
- Sodium bicarbonate (baking soda)
- Sodium carbonate (washing soda)
- Borax (e.g. 20 Mule Team brand)
- Trisodium phosphate (TSP)

Sodium or potassium hydroxide have also been used in toning, but these chemicals are very caustic—think drain cleaner—and overkill. If you've ever gotten a drop of them on your skin, you'll see what I mean; they burn. Ammonia is a bit smelly, and if that bothers you, sodium carbonate is an acceptable substitute and a very rigorous alkali besides, although ammonia does give good purples.

Choice of acids, called "toners"

Toners can range from simple household ingredients to stuff you buy online for winemaking, basket dyeing, or from a chemistry store. No need to get chemical reagent quality. They are:

- Tannic acid powder
- Gallic acid powder
- Pyrogallic acid powder
- Catechu powder
- Strong brewed tea
- Strong brewed coffee
- Water from boiled walnut husks

Tip: buy a pound of sodium carbonate and a pound of tannic acid and you'll have the two chemicals that are easy to mix, don't outgas, and with a teaspoon of each to a liter, you will have enough to tone prints for the foreseeable future.

Quick and easy measurements

Cyanotype toning is as simple as a teaspoon of this and a teaspoon of that in tap water. No need for distilled water or any exact measures such as "up to" formulas. Just weigh or measure out the ingredient and add it to a liter of water. These volume-to-weight measures are the "kitchen" approach to toning. The weights are for a level, not heaping, teaspoon, and the grams are rounded up, no problem if using these minute amounts. If you are using larger amounts of chemistry, measuring out lots of teaspoons so being off half a gram per teaspoon matters, use a gram scale.

- 1 teaspoon sodium bicarbonate (baking soda) = 5 g
- 1 teaspoon sodium carbonate = 5 g
- 1 teaspoon borax = 5 g
- 1 teaspoon trisodium phosphate = 5 g
- 1 teaspoon tannic acid = 2.5 g
- 1 teaspoon gallic acid = 2 g
- 1 teaspoon pyrogallic acid = 3.5 g
- 1 teaspoon catechu = 3 g

Highlight staining

Staining, or coloring, of the highlights is a common problem with cyanotype toning. There are a number of possible causes and solutions. Here are ways to minimize or prevent stain.

- **All the iron has not been removed from the print.** Do a thorough 15-minute rinse of the print before toning, making sure to remove all unexposed sensitizer and residual iron.
- **The paper when it goes into the toner is still alkaline from the bleach bath.** With some toners you can use an intermediate acidified water bath (a teaspoon of citric acid or 60 ml/4 tablespoons/a glug of white vinegar per liter of water) for a few minutes between bleaching and toning to neutralize the alkali left in the paper before it enters the toning bath. However, often the alkalinity is useful to speed up the toning process.
- **The toner is mixed too strong.** Use as weak a bleaching and toning bath as possible. Some formulas in the literature call for 5% solutions which is a *lot* of chemistry, when a 0.5% will do.

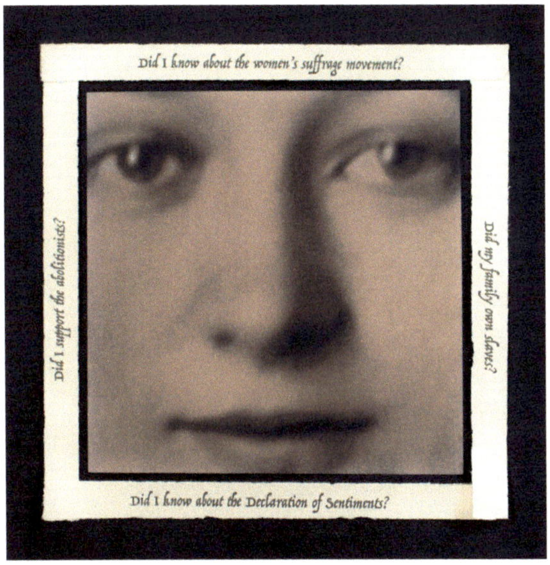

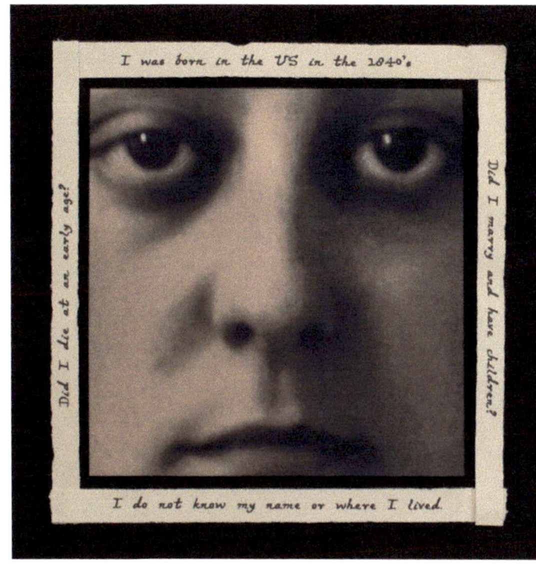

Figures 9.3 and 9.4. Left to right, *Anonymous Woman #2* and *Anonymous Woman #3*, from daguerreotypes, two coats of traditional cyanotype toned with pyrogallic acid and tea, colored pencil, on Arches Aquarelle, 8″ x 8″ © Susan Davens 2016. Susan Davens is a Maine based photographer working in 19th century photographic processes combined with digital technology. To see more of her work visit susandavens.com.

- **The toner is contaminated by being used for too many prints.** Use toners "one shot," e.g. mix the whole liter of toner, but only use 200 ml of it to tone one print at a time, and then discard the 200 ml and use another 200 ml for the next print.
- **Some papers stain more than others.** Stick to 100% cotton papers recommended in the **Papers for Cyanotype** chapter. Cellulose papers may stain more than cotton papers so if staining is a problem, see if switching papers helps.[3]
- **Paper stain may actually be exposure.** Sometimes there is very slight Prussian blue in the highlights, and when toned, it tones, too. Thus the very highest highlights are actually not stain. You can see this by looking at the back of the paper, which is probably paper white. Make sure to determine the correct exposure time for each paper before toning, and watch out for highlight fogging which will also tone.

Toning tips

- Make two step wedge prints, tone one and keep an untoned version for comparison.
- Toning can be done on an old print or on a freshly made print straight from the wash water, even though some say that the print should oxidize and age for a day or two before toning. Hydrogen peroxide in the water wash will oxidize a print sufficiently.
- Some of the powdered acids are not very water soluble, gallic acid especially. It's like mixing talcum powder and water! At time of use, mix volume to volume with rubbing alcohol before adding it to the water, e.g. 1 teaspoon gallic acid powder to 1 teaspoon rubbing alcohol.
- When presoaking a print before toning to evenly wet it, don't leave it in the water too long or the print will wash all out!
- Toning is one way to salvage overexposed cyanotypes. If you bleach a print briefly in an alkali you can restore the blue tone by soaking it in a vinegar bath, but all of the original density will not come back.
- Try selective bleaching or toning with a paintbrush dipped in dilute alkali or dilute tannic or gallic acid. With a print that has been alkali-toned only, brushing on tannic or gallic acid will immediately create brown marks; this has some creative possibilities.

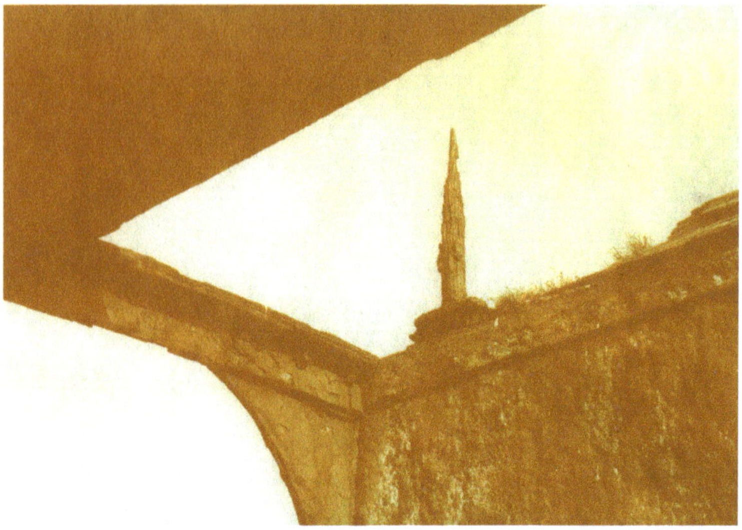

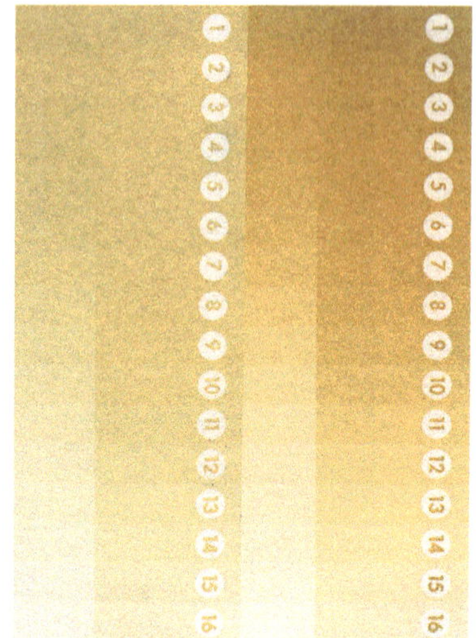

Figures 9.5 and 9.6. Left, *Stele, South Ronaldsay*, cyanotype toned with trisodium phosphate, 238 x 166 cm, image from a scanned 6 x 9 cm film negative © Mike Ware 2005. Right, a cut and paste of two step wedges, the left a full bleach in trisodium phosphate, the right a full bleach in sodium carbonate. They are similar in color but sodium carbonate tends to be slightly oranger. To get a deep orange like Ware's image, overexpose.

- Caution: tannic, gallic, and pyrogallic, but especially gallic and pyrogallic, are intense stainers, and if a grain of it gets on your countertop it will color the countertop. Be very cleanly and cautious with these powders.
- Use the empty liter hydrogen peroxide brown bottles to store toners so you can have tannic, gallic, catechu, and pyrogallic acid on hand at all times.

One-bath toners

There are a few one-bath toners in the literature. Often it is a dip in an alkali like borax, to change the tones to violet or yellow. Sometimes it is a dip in an acid like tannic to mute the blue to a browner blue. Then there are combination alkali-acid baths like sodium carbonate and tannic acid mixed into one solution.

I'm somewhat cautious with one-bath alkali toners. I wonder if the alkali merely creates a fading alkaline environment in the print, which we always try to avoid—think buffering. Besides, if you place a print back in acidified water the blue returns. A number of books mention that toners are not stable, permanent, or archival[4] or that the prints will yellow or fade within weeks.[5] A dry alkali-bleached print will actually tone quickly to brown when put in a tannic or gallic acid bath weeks later, sort of like bleaching a gelatin silver print in the darkroom and not sepia toning it until down the road. Play is always good, but forewarned is forearmed.

Trisodium phosphate toner

This tones the print a very pale chamois yellow and the resulting iron phosphate is stable.[6] If desiring a split tone look, just use a pinch and watch the print carefully, pulling it just before it looks perfect as the action will continue in the wash water. You can also put it in an acid bath and some of the tones will revert to blue if toning is not carried to completion.

5 g trisodium phosphate (1 teaspoon)
1000 ml water

1. Add the trisodium phosphate to the water and stir until dissolved.
2. Soak a dried print in plain water for a minute or two until it is evenly wet throughout.
3. Immerse the print in the trisodium phosphate solution until the color desired is reached.
4. Wash the print for 10 minutes. Hang to dry.

Figures 9.7 and 9.8. On the left is a one-bath tea toner, on the right is a one-bath coffee toner. The top step wedge for each has been toned 20 minutes and the bottom for 5 minutes. Note that a mere 5 minutes in the tea or coffee produces a more evocative color and preserves the blues in the shadows.

Sodium carbonate toner

This tones a similar but deeper yellow-orange than TSP. It can be used for just a few seconds to get a nice split tone with yellow highlights and blue shadows. It is a handy toner to clear up fog in the highlights. Note that the print will show a color change toward lavender before it changes to yellow.

5 g sodium carbonate (1 teaspoon)
1000 ml water

1. Add the sodium carbonate to the water and stir until dissolved.
2. Soak a dried print in plain water for a minute or two until it is evenly wet throughout.
3. Immerse the print in the carbonate solution until the color desired is reached.
4. Pull the print just before it looks good if wanting a split tone effect, because the action will continue somewhat in the wash water, and wash the print for 10 minutes. Hang to dry.

Variation: Substitute sodium bicarbonate or baking soda. It is useful as a mild bleach to clear the highlights and give them a faint yellow to gray-blue color. After bleaching you can return the print to an acidified water bath (a teaspoon of citric acid or 60 ml/4 tablespoons vinegar) to shift it back to blue, but all of the density will not return.

Borax or ammonia toner

This is often included as a purple toner in cyanotype kits, and borax does produce a very beautiful lilac blue color in the first few seconds of immersion but then continues to bleach to pale yellow. Same with ammonia. Both will revert to blue if placed in an acid water bath.

5 g borax (1 teaspoon) or 5–50 ml ammonia (1 teaspoon to several tablespoons)
1000 ml water

1. Add the borax or ammonia to the water and stir until dissolved.
2. Soak a dried print in plain water for a minute or two until it is evenly wet throughout.
3. Place the print in the borax or ammonia solution and tone until the desired color is reached. If lavender is the desired color, it will happen quickly so watch it closely.
4. Wash the print for 10 minutes and hang to dry.

Tea toner

I love tea toner for the evocative colors it produces. Using tea is similar to using tannic acid because it has tannic acid in it. Making your own from tea bags is easy. If you tone for a shorter time—e.g. 2–5 minutes—you'll get a split tone effect with the shadows deep blue and the highlights lavender brown.

5 tea bags
1000 ml water

1. Place the tea bags in the water and bring to a boil—remember, you're not making tea to drink here, so feel free to make a bitter, boiled brew. Let the tea boil for a few minutes and then steep after that to make a very strong tea.
2. Soak a dried print in plain water for a minute or two until it is evenly wet throughout.
3. Immerse the print in the strong tea until the color desired is reached, keeping an eye on the

Figure 9.9. One-bath tannic-sodium carbonate toner. The top step wedge was toned for 5 minutes, the bottom for 1 minute. This is a very active toner, and to preserve the Dmax, tone briefly.

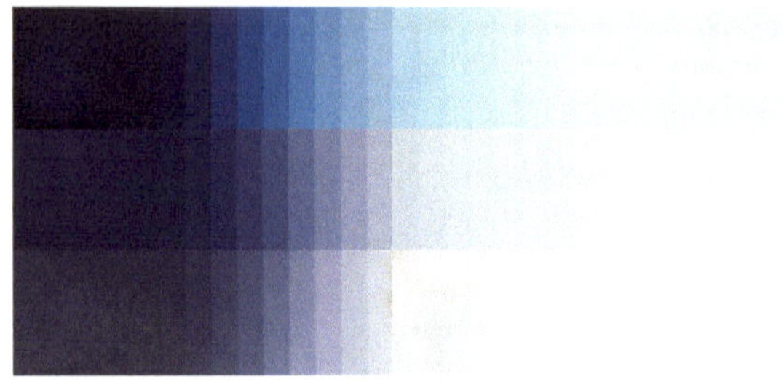

Figure 9.10. Untoned step wedge (top); 2 minute lead acetate toned step wedge (middle) versus 20 minute lead acetate toned step wedge (bottom). Note that longer lead acetate toning is not better.

highlights and the back of the paper to watch for staining.

4. Wash the print for 10 minutes, and hang to dry.

Variation: strong coffee is also a toner. If you have leftover cold coffee from breakfast, substitute it for the tea, above. This toner tones slowly and is a more yellow-brown than a red-brown. I do prefer tea.

One-bath tannic-sodium carbonate toner

This toner proceeds very quickly when mixed in warm water, and is a very handy one-bath toner, easy to mix, and uses very little chemistry.[7] I suggest mixing up a liter and using 200 ml of it one-shot, discarding after each print. Weights are immaterial; it's really a kitchen teaspoon recipe.

1.25 g tannic acid (½ teaspoon)
2.5 g sodium carbonate (½ teaspoon)
1000 ml water

1. Add the tannic acid to the water and stir until dissolved.
2. Add the sodium carbonate to the water and stir until dissolved.
3. Soak a dried print in plain water for a minute or two until it is evenly wet throughout.
4. Immerse the print in the toner until the color desired is reached. You will notice the highlights pink up immediately within seconds and by about a minute or so it is nicely split toned pinkish-brown and blue. If you tone longer, it will get more monochrome brown.
5. Wash the print for 10 minutes, and hang to dry.

Variation: use less sodium carbonate; ½ teaspoon in the liter makes this formula quite alkaline.

Variation: place the toned print in an acid bath to return the print to an acid state. The highlights will go to more of a neutral gray tone, the midtones will preserve some greenish blue, and the shadows will be back to blue, almost like a tricolor split effect.

Lead acetate toner

I hesitate to include this toner in this chapter because lead is toxic. It has to be handled with greater care than other chemicals because of the lead content, and disposed of in a manner required according to the area you live in. However, apparently lead acetate is an ingredient in hair dyes, the type that turn your hair darker over time, so go figure! The one good thing is that lead acetate toning helps with the print's tendency to fade in light, and for something so toxic, it is a one-bath toner and produces a beautiful purple.[8] Take all chemical precautions necessary when mixing this toner. The toner is often recommended to be used at a 5% solution[9] which this is. I mix a small amount, not a liter, at time of use.

12 g lead acetate (2 scant teaspoons)
250 ml water
3 ml 10% strength janitorial ammonia

1. Add the lead acetate to water and stir until dissolved.
2. Add the ammonia to the solution and adjust the pH to 7–8 by adding a ml of it at a time.
3. Tone a dry print for 2 or so minutes. Longer doesn't necessarily get better.
4. Wash for 10 minutes and dry.

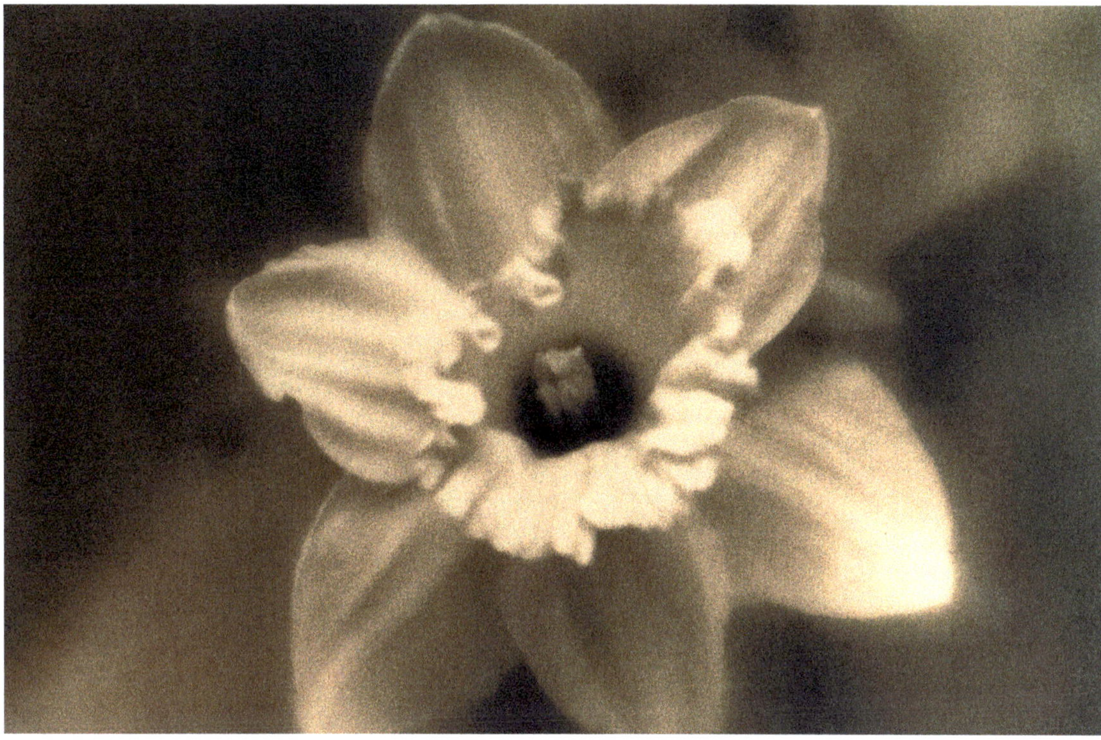

Figure 9.11. *Daffodil*, 30 × 45 cm © Guido Ceuppens 2018. The print was made with 20/8 classic cyanotype on Canson C A Grain 180 gsm paper and toned with tannic acid and sodium carbonate. Exposure was under a bank of UV tubes for 2.5 minutes. Development was in 2 changes of water for a total of 30 minutes. The print was then put in a solution of tannic acid (1 teaspoon in 1000 ml water) for 2 minutes, washed for 1 minute and briefly immersed in a solution of sodium carbonate (2 teaspoons in 1000 ml water) and again washed for 1 minute. The print was then transferred back into the tannic acid bath until the color was right. A final wash of 30 minutes followed.

Two-bath toners

Most often cyanotype toning proceeds with a two-bath process. An alkali is used to bleach the print and an acid is used to tone the print. Toning cyanotype is much less complicated than the literature would have you believe. Following is a generic toning formula that many two-bath toners follow. Once you have used this approach and understand how toning works, feel free to lessen or increase these generic amounts, lessen or increase time in the alkali bath for split toning and full toning effects, switch up the process to tone first and bleach second, or bleach-tone-bleach, or tone-bleach-tone—these three baths can be used in a multitude of ways. A few generic rules that beg for exceptions:

- As far as color, the toner of choice tends to trump all with tannic a bit more red-brown and gallic a bit more purple-brown
- As far as alkalis, ammonia and borax tend toward purple and the others toward yellow
- The stronger the alkali, the faster the bleaching proceeds
- The stronger the acid/toner, the faster the toning proceeds
- The stronger the alkali or longer the alkali bath, the more the acid colors the print
- Full bleaching can result in a dull print with a large loss of Dmax. Bleach less and tone longer
- A print always dries down darker and *duller* than it is in the water so it may benefit from a varnish to give it some gloss and depth.

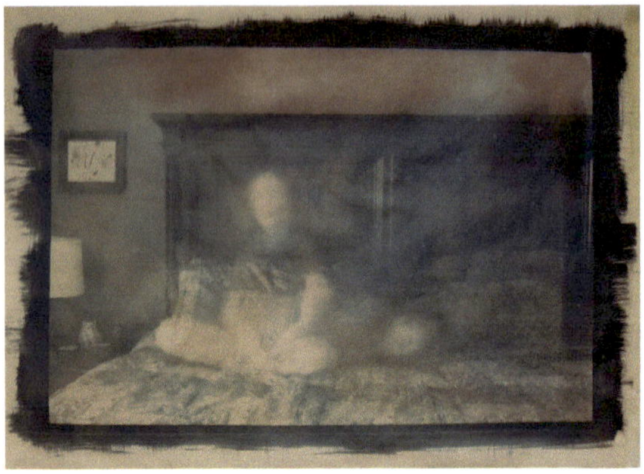
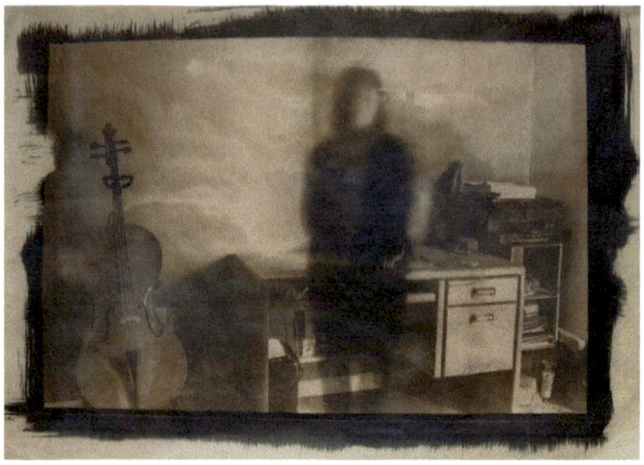
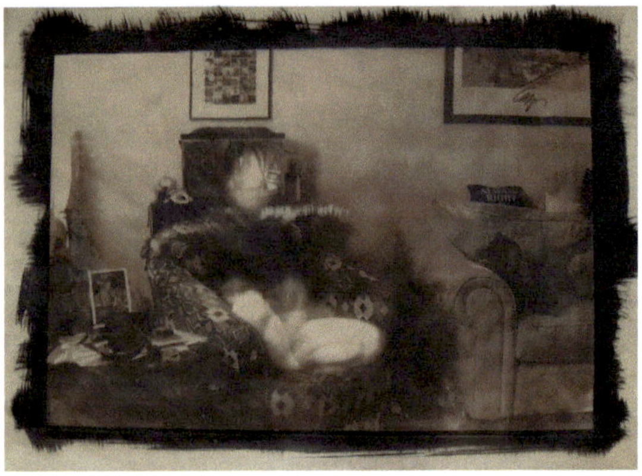
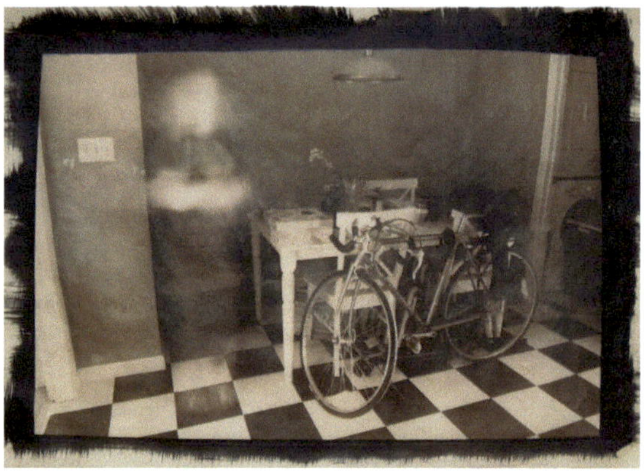

Figures 9.12–9.15. Clockwise from top left: *Waiting Series 1*, *Waiting Series 2*, *Waiting Series 3*, and *Waiting Series 5*, green tea toned cyanotypes printed on drafting vellum, 18″ x 24″ © Holly Veenis 2018. "My husband, having attention deficit disorder, becomes so extremely focused on a given task that he loses track of time which often causes him to be late. I, on the other hand, easily become anxious about being late and tend to arrive everywhere fifteen minutes early. The disparity of these two mentalities results in my spending a lot of time waiting on my husband. The *Waiting Series* is an exploration of patience that utilizes long-exposure self-portraits that are printed in processes that need an extended amount of time compared to modern day digital printing." The work of Holly Veenis is self-exploratory and typically utilizes the handmade, such as alternative processes and books, and has been exhibited nationally and internationally. She received a BA from Bridgewater College and an MFA from The Pennsylvania State University. To see more of her work, visit veenisphotography.com.

Generic two-bath alkali-acid toner

Solution A 0.5% alkali:

5 g sodium bicarbonate or sodium carbonate or borax or trisodium phosphate (1 teaspoon) or 5–50 ml ammonia

1000 ml tap water

Add the alkali to 1000 ml water and stir until dissolved. Borax and trisodium phosphate take a bit of stirring to go into solution. The others readily dissolve.

Solution B, optional:

60 ml plain white vinegar (4 tablespoons) or 5 g citric acid (1 teaspoon)

1000 ml tap water

Add the vinegar or citric acid to 1000 ml water and stir until dissolved.

Solution C 0.25–0.35% acid:

2.5–3.5 g tannic acid or gallic acid or catechu or pyrogallic acid (1 teaspoon)

1000 ml tap water

Add the acid to water and stir until dissolved. Tannic acid is fairly water soluble (and foamy) but gallic acid does not go into solution easily, so add a teaspoon or two of rubbing alcohol to the powder first, mix it into a paste, and then add it to the water. Give it a stir to keep it suspended in solution.

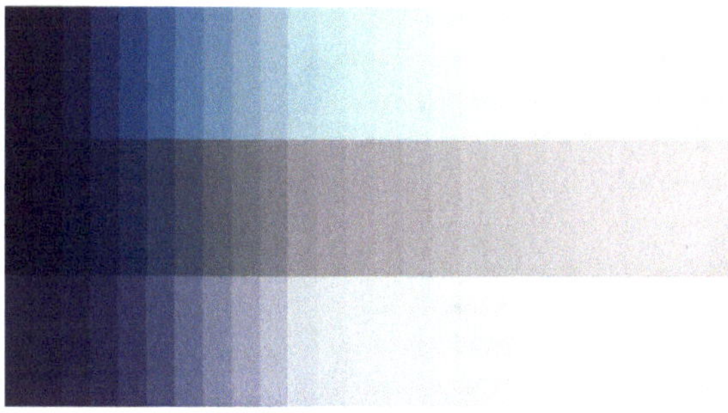

Figure 9.16. A comparison of an untoned step wedge, top, a step wedge bleached in ammonia and toned in tea, middle, and a step wedge toned in lead acetate, bottom. If lilac tones are desired, ammonia bleaching and tannic, gallic, and pyrogallic acid will lean towards violet browns but lead acetate is quite clearly violet.

1. Pour the alkali solution into one tray, the acidified water solution into another, and the acid solution into a third tray.
2. Soak a dried print in plain water for a minute or two until it is evenly wet throughout.
3. Immerse the print in the alkali bath either briefly or fully—if brief, split tones will result; if full, a monochrome print will result.
4. Optional: rinse in an acidified water bath for several minutes to remove the alkali. This helps prevent staining in the toner and keeps the alkali from contaminating the toner. Many do not do this, and, in fact, the alkalinity of the print can speed up the toning action, but staining can result.
5. Tone in the chosen acid for as long as it takes, pulling it just before it looks perfect because the toning action continues slightly in the water wash.
6. Water wash for 10 minutes to remove any acids and alkalis remaining in the paper and to bring the paper back to neutral pH.
7. Hang the print to dry. Some toners will cause the darks to bloom and bleed into the highlights if laid flat, so hanging to dry minimizes this as does the 10-minute water wash. Note that a print will dry down darker and duller than it looks when wet.

Variation: combine gallic and pyrogallic together, or tannic and pyrogallic, or gallic and tannic, for subtle color variations. Tannic is more red brown, gallic and pyrogallic more purple-brown.

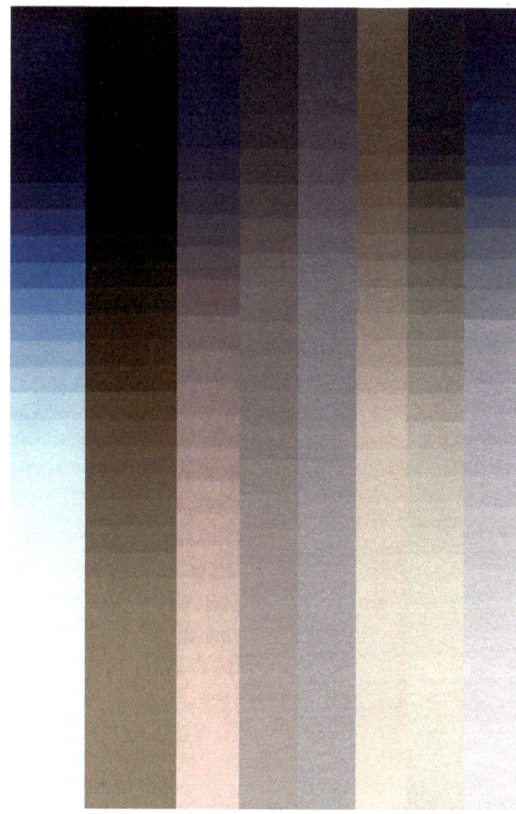

Figure 9.17. These step wedges show the colors available with toning, compared to the untoned step wedge on the left. Various shades result from purple brown to red brown to the deep chocolate brown of pyrogallic in conjunction with hydrogen peroxide. Brief or partial bleaching works well because the darks of the original print are preserved. Brief toning tends to produce the more purple browns and longer toning the more red to chocolate browns, as does stronger toners. However, it's not an exact science by any means, which is what can be so much fun!

Pyrogallic-hydrogen peroxide toner

This is a lovely toner.[10] There is no bleach involved, because both hydrogen peroxide and pyrogallic acid are acidic. It is a fairly strong toner, and proceeds quickly, especially when the toned print is put in the hydrogen peroxide bath where deep, dusky chocolate browns occur. As with all toners, it can be done partially or fully to achieve different effects from split to full toning.

Solution A:

60 ml drug store hydrogen peroxide (4 tablespoons/¼ cup)

1000 ml water

At time of use add the peroxide to the water and stir.

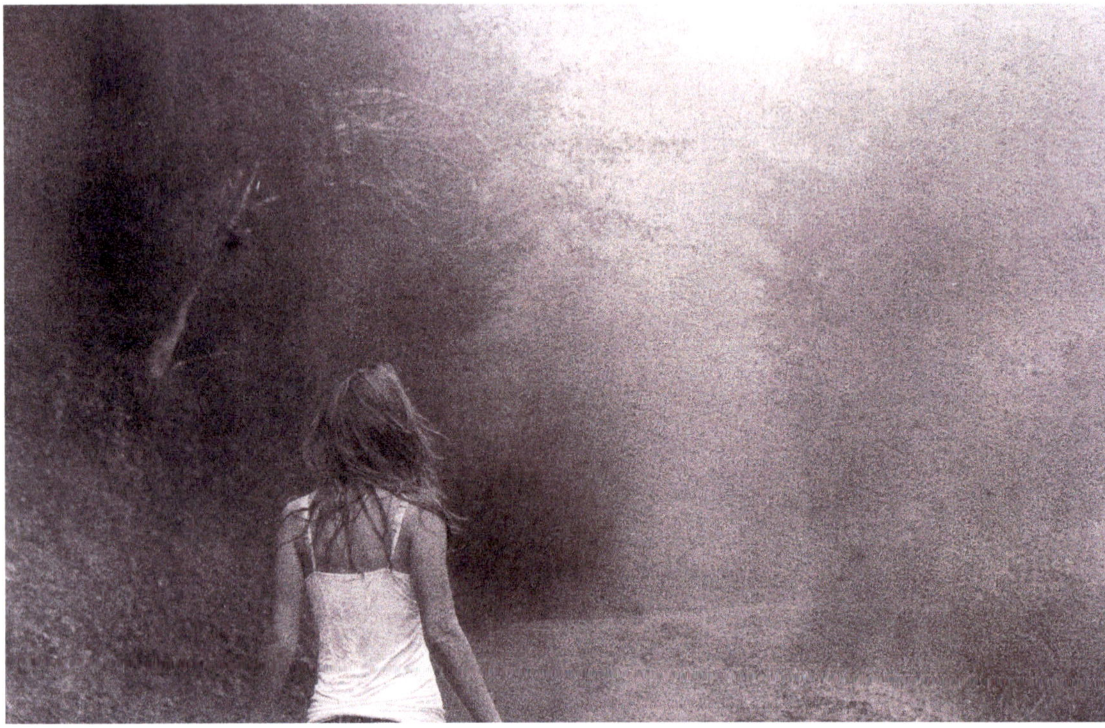

Figure 9.18. *Unattached*, classic cyanotype toned with tannic and gallic acid, 8″ x 10″ © Jessica Hays 2017. Jessica Hays' work focuses on human relationships and issues of mental health, trauma, and loss. In addition, Hays explores places of healing in the landscape and their place in contemporary culture. Hays grew up in Montana, and is currently pursuing degrees in Photography and Environmental Studies. Her work has been shown both regionally and nationally, and she has presented at conferences and events dealing with issues in art and mental health. To see more of her work, visit jessicahaysart.com.

Solution B:
12 g pyrogallic acid (3 teaspoons)
1000 ml water

Add the pyrogallic acid to the water and stir until dissolved. Pyrogallic acid is between tannic and gallic as far as solubility goes so it does require some stirring.

1. Soak a dried print in plain water for a minute or two until it is evenly wet throughout.
2. Immerse the print in the pyrogallic acid toner for 1–5 minutes. Start out with the lesser time because you can always return the print to the toner.
3. Drain the print and immerse it in the hydrogen peroxide bath. You will notice the print proceeding to tone darker, and the bath will turn brown.
4. Wash the print for 10 minutes, and hang to dry.

Endnotes

1. Brown, George E. *Ferric and Heliographic Processes: A Handbook for Photographers, Draughtsmen, and Sun Printers*. New York: Tennant & Ward, 1900 [First edition 1900, reprinted 1905, this edition n.d.), p. 13.
2. Step wedges were all printed at the same time, the same paper, the same formula: Platine paper, new cyanotype, and an overexposure of 10 minutes. You will note that on some of the step wedges the reversal of overexposure is visible in the darkest steps.
3.Said years ago on the alt list by Gilbert van Reenen.
4. Webb, Randall and Martin Reed. *Alternative Photographic Processes, A Working Guide for Imagemakers*. New York: Silver Pixel Press, 2000, p. 72.
5. Barnier, John. *Coming Into Focus, a Step-by-Step Guide to Alternative Photographic Printing Processes*. San Francisco: Chronicle Books, 2000, p. 46; Arnold, Brian. *Alternate Processes in Photography: Technique, History, and Creative Potential*. New York: Oxford University Press, 2017, p. 59; Hirsch, Robert. *Photographic Possibilities*, 3rd ed. Boston: Focal Press, 2009, p. 201.
6. Ware, Mike. *New Cyanotype, Siderotype Workshop Notes*. Buxton: self-published, 2009, p. 11. http://www.mikeware.co.uk/mikeware/downloads.html
7. Van Keuren, Sarah. *A Non Silver Manual. Cyanotype, Brownprint, Palladium, and Gum Bichromate with instructions for making light-resists including pinhole photography*, 3rd ed. Woodlyn, Pennsylvania: Munro Printing, 2005, p. 8.
8. Ware, p. 11.
9. Brown, p. 16; Crawford, William. *Keepers of Light*. New York: Morgan and Morgan, 1979, p. 166; Barnier, p. 46.
10. Enfield, Jill. *Photo-Imaging, A Complete Guide to Alternative Processes*. New York: Amphoto Books, 2002, p. 88.

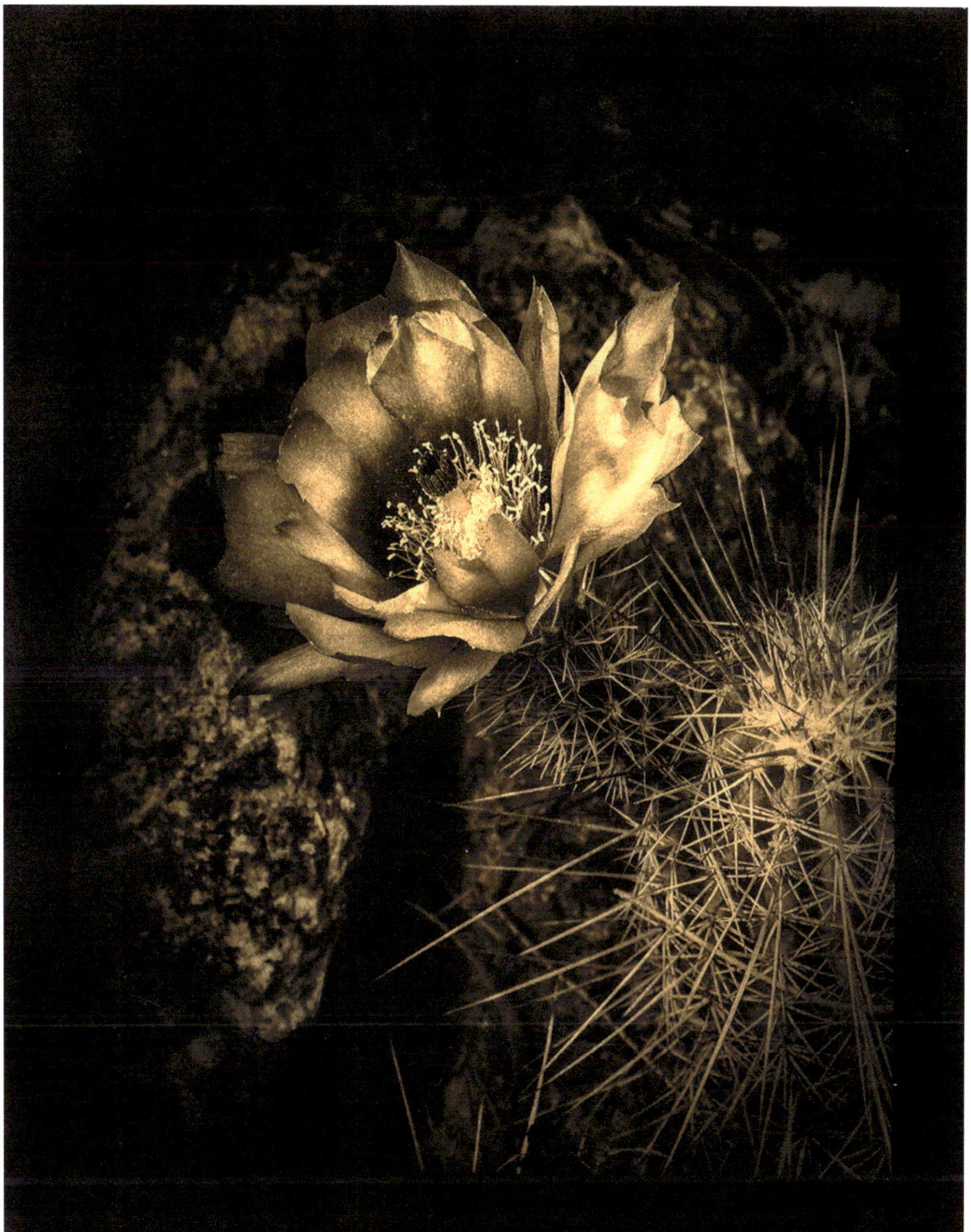

Figure 9.19. *Blooming Cactus*, tea-toned, sodium carbonate-bleached cyanotype © Eugene Starobinskiy 2017. "I first started with cyanotype because of the ease and relatively low cost of chemistry used in the process, and later on, discovered an entire universe of color/toner combinations available, and have not stopped printing in Prussian blue since. I like Hahnemühle Platinum Rag, Stonehenge and Somerset papers, but my best looking prints are on Canson XL Watercolor. I make my negatives according to Dan Burkholder's system and also adjust my image intuitively. I use the classic cyanotype formula, and tone with natural, organic toners like coffee and tea." Eugene Starobinskiy grew up in Kiev, Ukraine. His education is in the field of medical engineering. He moved to Arizona in 2002, where in 2007 he began his interest in photography. Influenced by his classes at Pima Community College and his mentor Ann Simmons-Meyers, Starobinskiy's work focuses on exploring the world around him in historic photographic processes and documenting the beauty of it by means of chemistry and light.

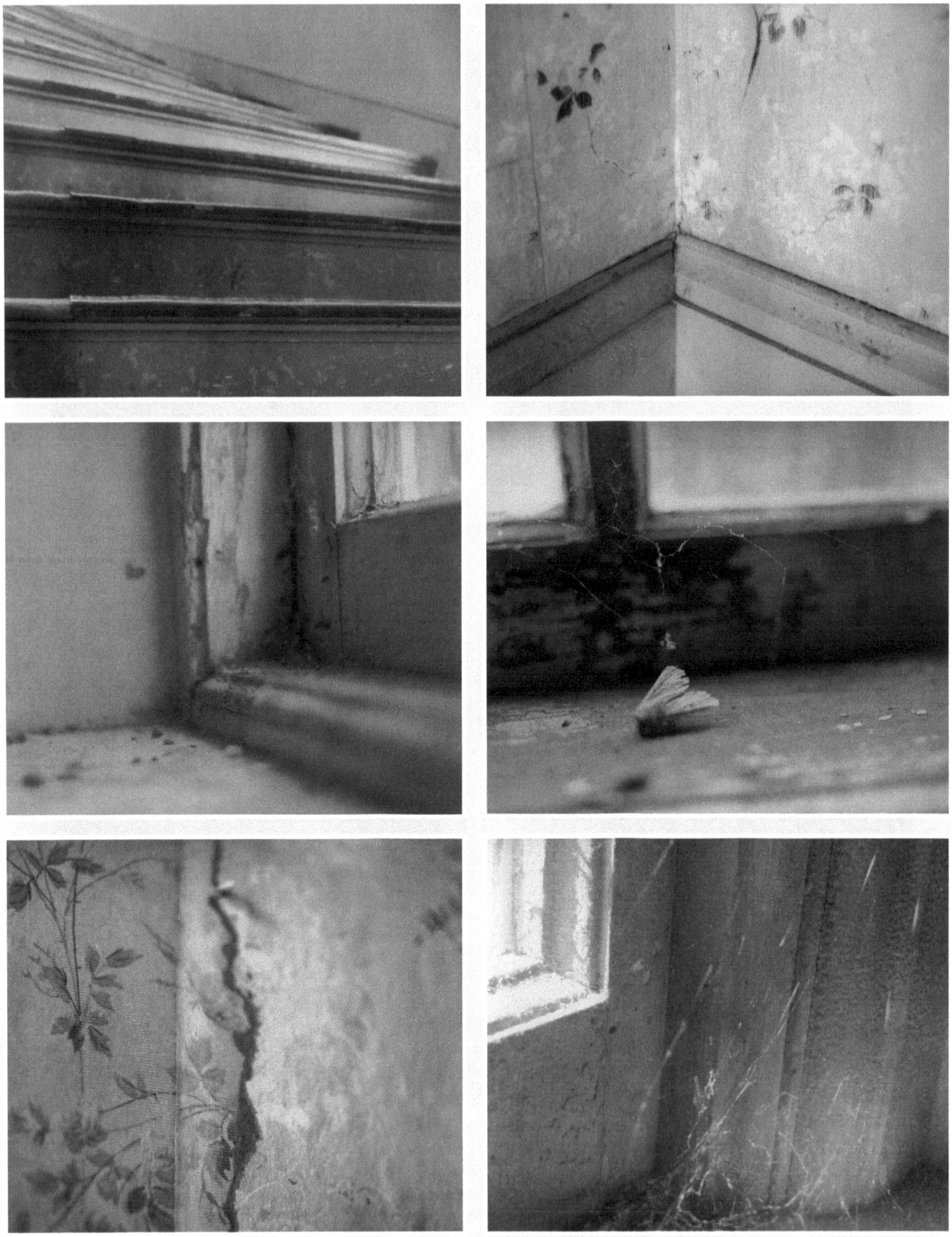

Figures 10.1–10.6. *Forsaken*, cyanotype over palladium, 7.5″ x 9″ © Julia C. Martin 2014. "The ghost town of Bannack, Montana was vacated in the 1950s and has remained there since; sitting, disintegrating, and waiting for its inevitable collapse. The claustrophobia that this series evokes allows the viewer to become trapped in the images, just as the present Bannack is trapped between its former liveliness and its imminent decay." Julia C. Martin's work deals with the themes of femininity, death, and mortality. She primarily shoots film and often prints using alternative processes. Julia received her Bachelor of Arts in Photography from Montana State University and is currently pursuing her Masters of Fine Arts at Arizona State University. To see more of her work, visit **www.juliacmartin.com**.

Chapter 10

Printing Cyanotype Over Palladium

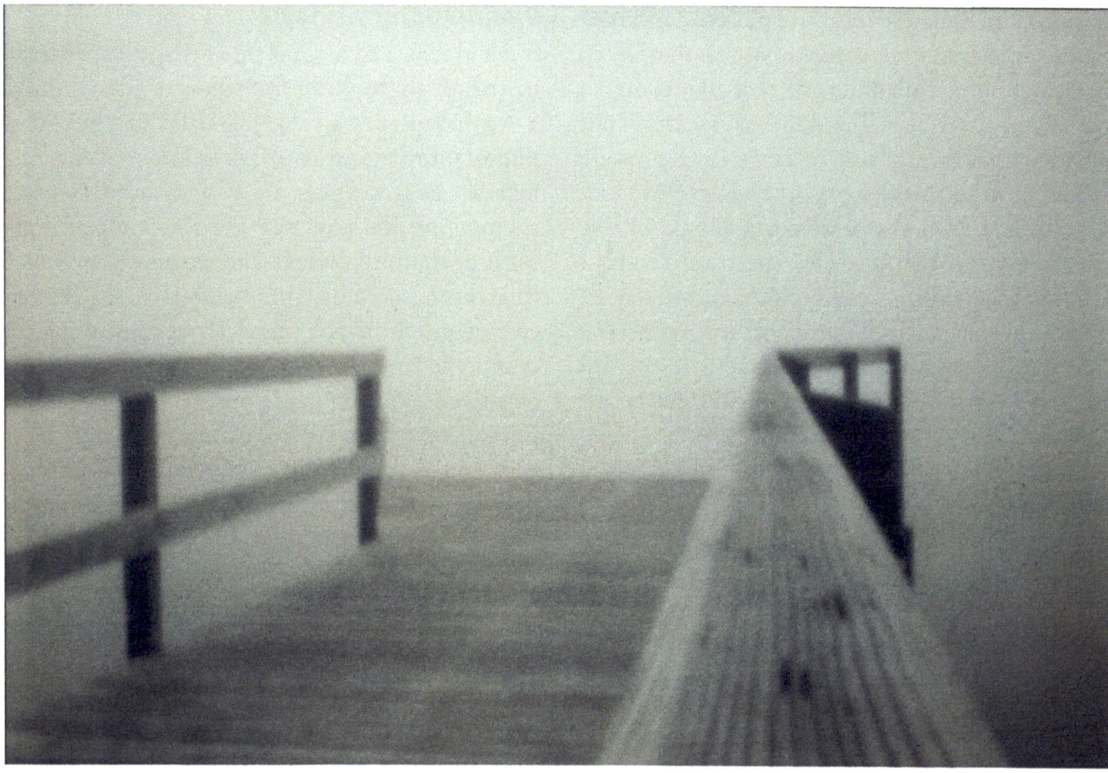

Figure 10.7. *Misty Morning #1*, cyanotype over palladium © Annette Golaz 2017

One of my favorite combination prints of all is cyanotype combined with platinum/palladium. The complementary color palette of blues and browns evokes a full color print without spelling it out completely. Creating a combination cyanotype-platinum/palladium print does require learning another process, and platinum and palladium are a bit pricey in comparison with cyanotype. Nevertheless, once you make one of these prints, you'll be hooked.

Platinum and palladium are metals that become photosensitive in combination with iron salts. The two metals are most often used together when printing. The Na2 platinum/palladium method outlined in this chapter is a platinum/palladium process because the Na2 part of the formula used as a contrast agent is a platinum salt. However, as of late, nomenclature has been called into question because the final print is almost completely composed of palladium. Unfortunately, the general public is familiar with the term platinum but not so much with the term palladium. Nevertheless, to be on the safe side, I will refer to the process herein as palladium as can be seen in the title of the chapter.

There are several ways to print a combination cyanotype/palladium print:

- Print both processes using the same monochrome palladium negative

- Print the processes with two duotone negatives made from an RGB image
- Print the processes with two duotone negatives made from a CMYK image

The look of a print made from a monochrome negative is different than one made from duotone negatives. In the single-negative process the darkest blues of the cyanotype superimpose on the darkest browns of the palladium, sort of like toning a palladium print blue. The duotone negative split, whether done from an RGB or a CMYK image, results in a more true-to-color image; blue is where blue should be, brown is where brown should be, but the absence of the full color spectrum creates a twilight-esque feel, like shooting a movie day for night. This chapter will outline all three processes in a step-by-step method with a sort of "just the facts, ma'am" approach, which should not be too difficult for a beginning palladium printer.

Supplies needed

Aside from distilled water, all that is necessary is to buy the Na2 (sodium chloroplatinate) Platinum/Palladium Kit for Digital Negatives from Bostick & Sullivan (**bostick-sullivan.com**) which comes with palladium, ferric oxalate, a 5% dilution of Na2, a developer and a clear. This kit, although close to $200 at the time of this writing, is enough for 25 8″ × 10″ prints. Sometimes the kit is packaged with potassium oxalate developer; sometimes it is packaged with ammonium citrate. The former produces warmer tones than the latter. I also recommend buying di-EDTA and citric acid for the first of three clearing baths. The first clear should always be of an acid nature. The kit may come with sodium sulfite and tetra-EDTA which will make up the alkaline clears for the second and third baths.

Paper choice

For a combination process such as this that has to go through wet processing for two different processes, it is important to have a sturdy paper, somewhere around 300 gsm in weight. Consult the **Papers for Cyanotype** chapter, but your best papers for both processes singly or together are Hahnemühle Platinum Rag, Bergger Cot 320, Arches Platine, and Magnani Revere Platinum. Palladium is a bit paper-picky like cyanotype so save yourself some trouble and stick with these top four papers specifically made for alt.

Preshrinking paper

Since this is a two-print process and paper tends to shrink slightly the first time it goes through a wash and dry cycle, just like clothes, it is important to preshrink paper before printing. This merely involves filling a tray with hot water and submerging the paper in the water for an hour for it to thoroughly soak and expand, after which time it is hung to dry completely. This should take care of any shrinkage issues. If, when printing the cyanotype layer on top of the palladium print, the negative is still slightly larger than the palladium print, take a damp sponge and lightly dampen the back of the paper to relax and expand it into the negative while the negative is taped in place on the front side and ready to go. The paper will expand ever so slightly and fit into the negative within a few minutes.

A word about brushes

The best brush for palladium printing is a 2″ flat synthetic fiber brush like one of taklon fiber. The brush hairs do not absorb as much precious chemistry as a natural hair brush does. The shape of the brush should be flat and as wide as necessary; 2″ brushes are best. At time of use, wet the brush thoroughly with distilled water, give it several shakes to get rid of excess water and then coat. After use, the bristles can be soaked in distilled water and blotted dry. Occasionally it might be necessary to clear the brush in one of the clears for 5 minutes and then rinse it in distilled water.

Developer choice

The B&S kit comes with a particular developer powder inside a liter bottle and only requires distilled water to fill up the bottle. However, if you want to make your own developers here are two formulas.

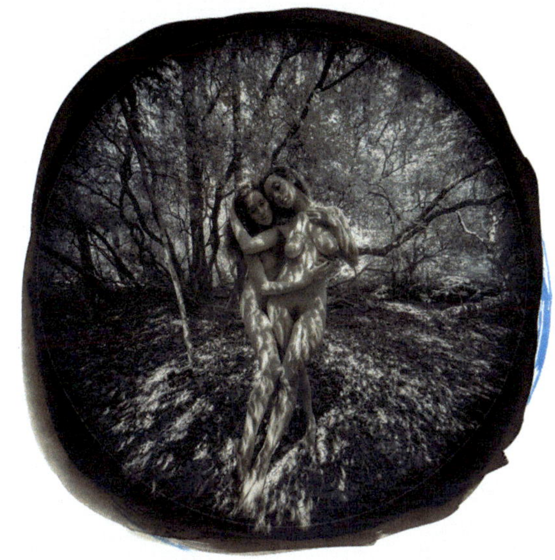

Figure 10.8. *Eclipse 2017*, cyanotype over palladium © Sam Wang 2017

Potassium oxalate developer

454 g/1 lb potassium oxalate
1400 ml distilled water

Caution: potassium oxalate is an anticoagulant and poison so handle with care. Add the potassium oxalate to the water. It's that simple. I actually keep a plastic gallon milk carton and fill it with several pounds of potassium oxalate powder. Then I fill the jug to the top with distilled water. As long as there are potassium oxalate crystals on the bottom of the jug it is a saturated solution, and at time of use I pour off liquid from the top of the jug into the tray. The developer is reused over and over so once done with a printing session, the developer is returned to the jug, and more water added to top it off if necessary. When the crystals at the bottom start waning, I add another pound of potassium oxalate to the jug. Note: for greater speed and warmth, potassium oxalate is used hot, but can also be used at room temperature. Whatever the choice, be consistent.

Ammonium citrate developer

225 g/½ lb ammonium citrate
1400 ml distilled water

Add the ammonium citrate to the water. Ammonium citrate is a very neutral toned developer; next to potassium oxalate it looks quite gray-black. It is used at room temperature, but can be heated, too, and will result in warmer tones.

Developer tips

- Palladium developers are "develop to completion" developers. This means that if the print is left in the developer for 2 minutes or for 5 minutes, it doesn't make the print any darker.
- If a developer is heated, it will increase the speed and warmth of the print. Therefore, if consistent print color is desired, it is important to be consistent with the developer temperature.
- Every developer has its own speed, contrast, and tone, so a separate printing time should be calibrated with the developer that will be used.
- All developers are reused over and over. Do not throw the solution away after a printing session, but return the developer to the bottle. They all get dark and grungy with time, and a certain amount of sludge accumulates in the bottom of the bottle, which can be filtered out, if desired, but more common is to leave it in there, and as the developer amount wanes in the bottle, to top it off with freshly mixed developer.
- The developer needs to always remain on the acid side, or else it will form insoluble iron salts in the print. Buy some pH testing strips and check it from time to time. If it becomes alkaline, add a couple spoonfuls of citric acid until the pH lowers.
- Some use developers one-shot, meaning they will pour only a small amount in the tray and discard it after each print to make sure the developer does not get alkaline.
- Developer formulas are all over the board in the literature. It really isn't an exact science. When the old literature talks of using "1 pound to 48 ounces" it equates to about a 30% solution. Just don't let it get too dilute or the print will be dull, washed out, pale, and grainy.

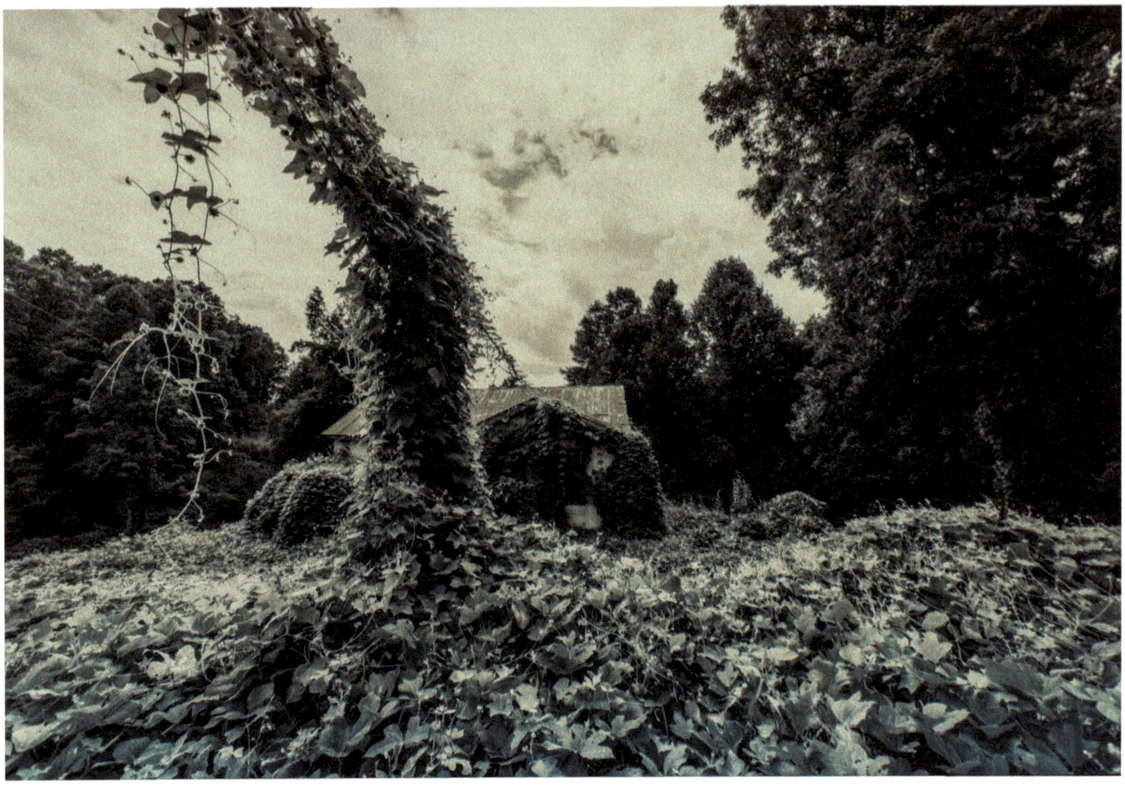

Figure 10.9. *Georgia*, cyanotype over palladium on Arches Platine © Christina Z. Anderson 2015

Clear formulas

The clearing step removes iron salts which will, if left in the print, yellow the paper over time. There are acid clears and alkaline clears. It is important to have the first clearing bath be acid no matter what to prevent insoluble iron salt formation. Thereafter the clears can be alkaline, but if the paper is left in an alkaline condition, it will wreak havoc with the cyanotype layer, so you may find it best to use only acid clears when printing cyanotype over palladium.

Clears are always used in a 3-tray, 3-bath system. Some clears do not work well in tandem with some developers. EDTA may work well with a citrate developer but not as well with potassium oxalate. If clearing is difficult, switch to a different clear. Also, some papers take a longer time to clear. Note: these are "kitchen measurements" because clears really only need a heaping spoonful of this or that, which is about a tablespoon. A tablespoon weighs right around 15 g of these chemicals. All are mixed the same, by adding the chemical to the water and stirring thoroughly.

Citric acid clear (acid)

Some papers don't clear well with plain citric acid.

3 tablespoons citric acid
1 liter water

Di-EDTA clear (acid)

Di-EDTA is an acid version of tetra-EDTA.

2 tablespoons di-EDTA
1 liter water

Di-EDTA/citric acid clear (acid)

1 tablespoon citric acid
1 tablespoon di-EDTA
1 liter water

Tetra-EDTA/sodium sulfite clear (alkaline)

This is an alkaline clear.

1 tablespoon tetra-EDTA
1 tablespoon sodium sulfite
1 liter water

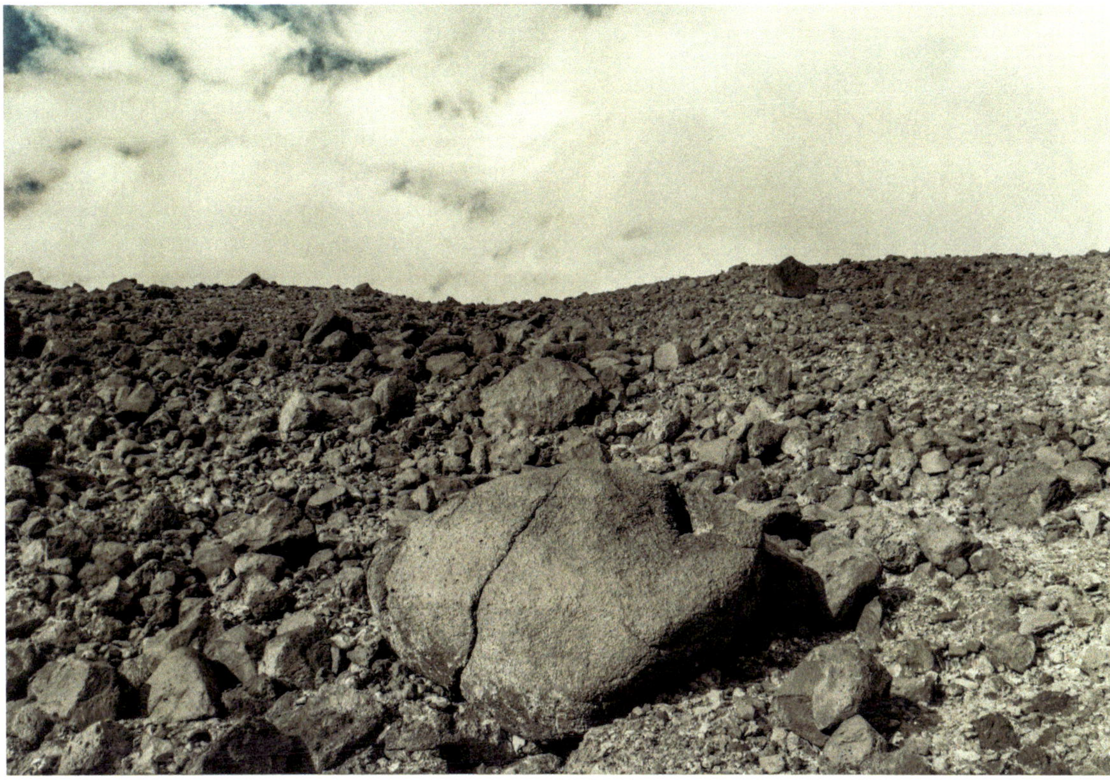

Figure 10.10. *Water from the Rock*, cyanotype over palladium on Arches Platine © Christina Z. Anderson 2016

Coating formula

Each drop will cover 2 square inches of paper. Be sure to include the borders in the calculation, though. The three solutions are mixed together in units of 6:6:1, which equals a combined total of 13 drops. Always round up to the next unit, never down. Hence:

Drops of:	8″x 10″ w/ ½ Borders	11″x 14″w/ ½ Borders	16″x 20″w/ ½ Borders
Ferric Oxalate	24	42	84
Palladium	24	42	84
5% Na2	4	7	14

One note about drop counts: palladium drops are fatter than ferric oxalate drops and thus adding 3 more drops of ferric oxalate to each unit makes a more balanced mix if measuring out solutions with eyedroppers and in small quantities. Since many palladium printers measure solutions by the ml with pipettes or some other accurate measuring device, not by the drop, drop size is not a factor.

Making the palladium print

1. Preshrink the paper so registration of layers is accurate.
2. The dimroom should be suitably humidified before beginning the coating process or bleeding may occur in the processing steps.
3. Prepare a palladium negative as outlined in the **Digital Negatives for Cyanotype** chapter, whether from a monochrome image, a color image converted to monochrome, an RGB split, or a CMYK split.
4. Mark the area on the paper that needs to be coated with faint pencil marks.
5. Figure out the square inches of the print, and divide by 2 to get the final drop count necessary.
6. In a shot glass (the rounded bottom works wonders) measure out the required amount of ferric oxalate with a dedicated ferric oxalate-only eyedropper.

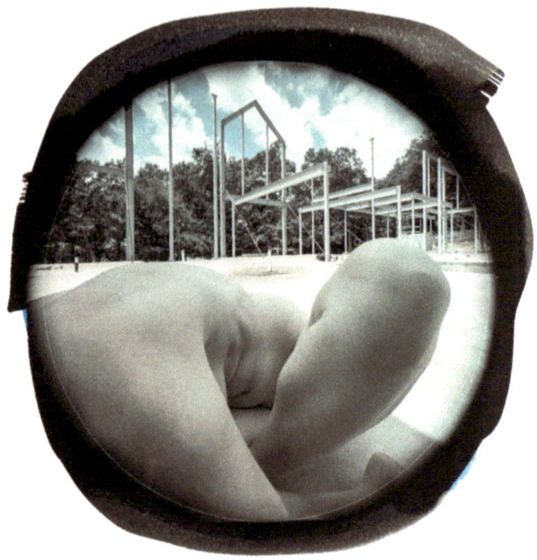

Figure 10.11. *Mansion*, cyanotype over palladium © Sam Wang 2018

7. Add the required amount of palladium drops to equal ferric oxalate using a dedicated palladium-only eyedropper.
8. Add the required number of Na2 drops with a dedicated Na2-only eyedropper.
9. Take a dedicated "mixing" eyedropper and squeeze the solution in and out until mixed.
10. Dampen the coating brush and shake excess water off.
11. Take the entire amount of solution and pour it into the center of the paper.
12. With the brush at a 30° angle, immediately begin coating from the center, moving outward to the edges of the coating area. Don't leave any area uncoated, while expanding the center "square" larger and larger until the coating area is completely covered.
13. Brush the entire area horizontally, vertically, and diagonally. As soon as the entire area is coated, brush marks are smoothed out, and it begins to look slightly matte, stop brushing and hang the paper to dry.
14. Rinse the brush in distilled water and hang to dry. For exposure, the paper should be dry to the touch but still retain some humidity.

Exposing

Exposure time is paper-dependent and developer-dependent, among other things, and thus should be calibrated with a simple step wedge as outlined in the **Digital Negatives for Cyanotype** chapter. Palladium is expensive, and an overexposed palladium print can not be corrected by a longer wash time like an overexposed cyanotype print. It is important to get it right.

Exposure can range anywhere from 2 to 5 minutes in the sun, and 4 to 20 minutes under UVBL. I use a 9-minute exposure in my UVBL box. The image when ready will look lavender-gray in the shadows and yellow in the midtones and highlights but it will darken in the developer instantly.

Processing

1. Set up one tray for developer and three trays for 1 liter each of clear. The first clear is acid; the next two are alkaline *unless you notice the alkalinity fading the cyanotype. If so use a final acid clear.* Once 10 8″ × 10″ prints go through the first tray, discard and mix fresh. Same for the next two alkaline trays; discard the first alkaline tray (tray no. 2) after 10 prints and move the second alkaline tray (tray no. 3) to tray no. 2's spot and mix a fresh tray no. 3.
2. Take the print out of the contact printing frame and have it ready to go because this part of processing goes quickly.
3. Pour the developer into a tray and tip the tray slightly so that all the developer is on one side of the tray, and put the print in the tray face up. Immediately un-tip the tray so that the developer washes over the entire face of the print in one fell swoop.
4. Agitate the developer continuously for 5 minutes and then drain the print and move it directly into the first acid clearing bath.
5. Clearing is for 5 minutes in each tray, continually agitating, for a total of 15 minutes.
6. Wash the print for 30 minutes and hang to dry. When dry it is ready to be coated with cyanotype, dried, and printed as normal in the cyanotype process.

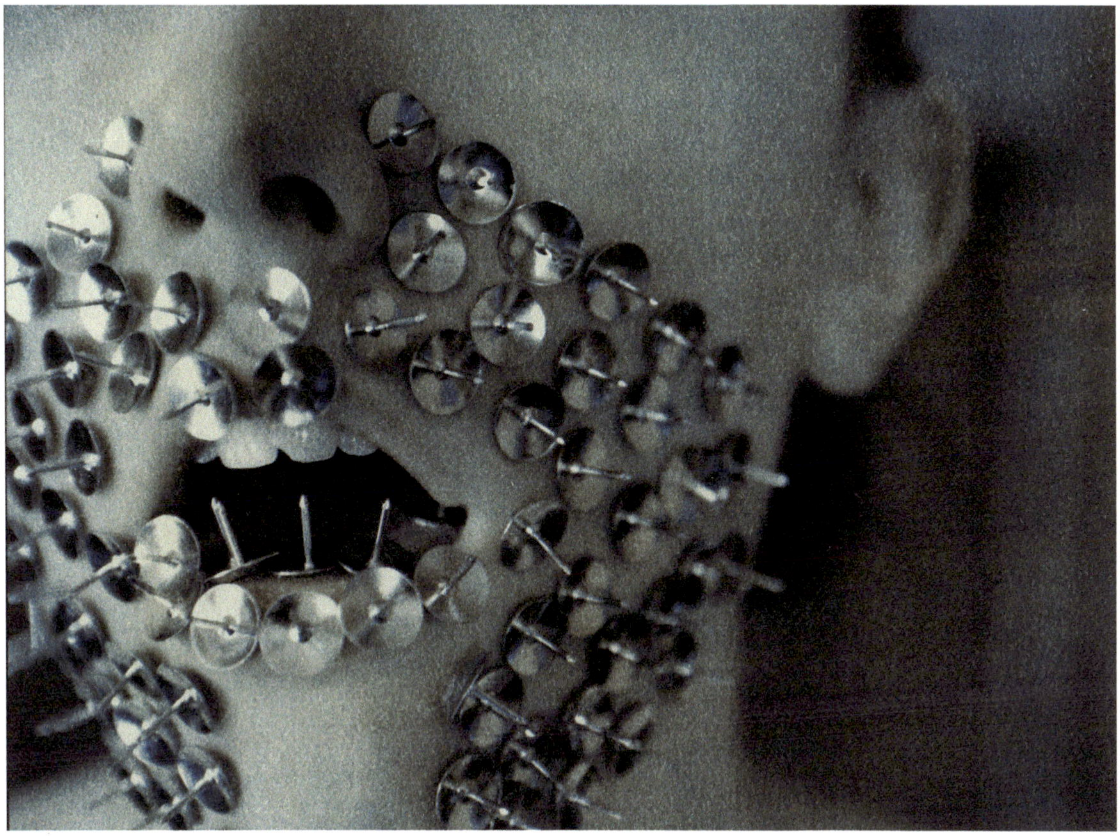

Figure 10.12. *Speaking Of*, classic cyanotype over palladium made with duotone negatives, 5″ x 8″ © Jessica Hays 2017

Printing with the palladium negative

The palladium layer is always printed first because the palladium developer and clear chemistry will bleach the cyanotype layer as will extended washing.

The palladium negative is often quite a bit denser than a negative for classic cyanotype, so the cyanotype layer will print mostly in the shadows and give the print more "drama" without affecting the highlights much. It is almost as if the palladium print has been toned blue. If using new cyanotype, the negatives for palladium and NC are similar and thus the blue layer will superimpose over the entire print with more of a noticeable blue tone throughout. The choice to make is whether to stick with the standard printing times for both processes or whether to vary either time by 25% less, for instance, to have one or the other process dominate. This takes finesse and probably a test print.

1. Preshrink the paper so registration of one printing on top of the other will be accurate.
2. Prepare the negative for palladium printing.
3. Expose and process the palladium print and hang to dry.
4. When the print is completely dry, coat it with the cyanotype solution and expose either as normal for new cyanotype, or with an increase in exposure if using new cyanotype.
5. Process the cyanotype as normal and hang to dry.
6. If the cyanotype is too dominant, a longer wash will decrease some of the blue color. If the cyanotype is not dominant enough, it can always be printed again.

Printing with RGB duotone negatives

See the **Digital Negatives for Cyanotype** chapter to create the duotone negatives. Essentially, an RGB color image is split into three color separation negatives; the two negatives that print warm colors, G which prints magenta and B which prints yellow (RGB negatives print their opposite or CMY so red(R) prints cyan (C), green (G) prints magenta (M), and blue (B) prints yellow (Y)!) are superimposed on top of each other as two layers in Photoshop with the top one set at 50% opacity so that both contribute to the image equally. The layers are flattened to yield a single negative that is used to print the palladium. The remaining R negative, which normally would print C or cyan, is used to print cyanotype.

1. Preshrink the paper so registration of one printing on top of the other will be accurate.
2. Prepare the duotone negatives as outlined in the **Digital Negatives for Cyanotype** chapter. On the warm negative, apply the palladium curve. On the cool negative, apply the cyanotype curve.
3. Coat, expose, process, and dry a palladium print using the warm negative. I would choose to do this and the cyanotype layer at their full standard printing time for your first try.
4. When the print is completely dry, coat the print with cyanotype and dry.
5. With the cool negative, expose the cyanotype layer as per normal and develop, wash, and dry.

Printing with CMYK duotone negatives

In this method an RGB color image is converted to CMYK. With CMYK, CMYK prints CMYK, not its opposite as in RGB (I know, confusing). Thus the C negative is used to print cyanotype and the K negative which normally prints black is used to print the palladium. The M and the Y negatives are discarded.

1. Preshrink the paper so registration of one printing on top of the other will be accurate.
2. Create CMYK negatives as outlined in the **Digital Negatives for Cyanotype** chapter. On the C negative apply the cyanotype curve. On the K negative apply the palladium curve.
3. Coat, expose, process, and dry a palladium print using the K negative. I would choose to do this and the cyanotype layer at their full standard printing time.
4. When the print is completely dry, coat the print with cyanotype and dry.
5. With the C negative, expose the cyanotype layer as per normal and develop, wash, and dry.

Troubleshooting palladium

- The blacks bleed into the highlights in the developer and/or clear—usually caused by too low humidity and the paper dries too quickly after coating so the solution doesn't sink into the paper enough. Sometimes it can be that the developer has excess metal in it and needs to be filtered out with a coffee filter. Sometimes the developer is too cold (keep everything at room temperature or warm the developer). Sometimes there is excess metal in the coating solution, but humidity is the usual culprit. Buy an inexpensive cool-mist humidifier and keep the dimroom humidified to at least 50%.
- Bronzed shadows—always the same three causes singly or in combination: too little solution coated with too little humidity and too much exposure.
- Minute black specks in the print—somehow there is contamination that got into the workflow, whether paper manufacturing impurities, specks of precipitate around the threads of the ferric oxalate bottle, a contaminated brush, undissolved palladium crystals in the coating solution, iron in the water supply. The answer to this problem is sort of inconclusive and it is probably the bane of a palladium printer's existence.
- White blobs or spots—usually a water drip before development, but it can be fingerprints or oils on the paper, or irregular paper sizing or paper fibers.
- Contrast issues in the print, provided the negative has been curved and made correctly—too high is usually too low humidity or too much Na2 or the

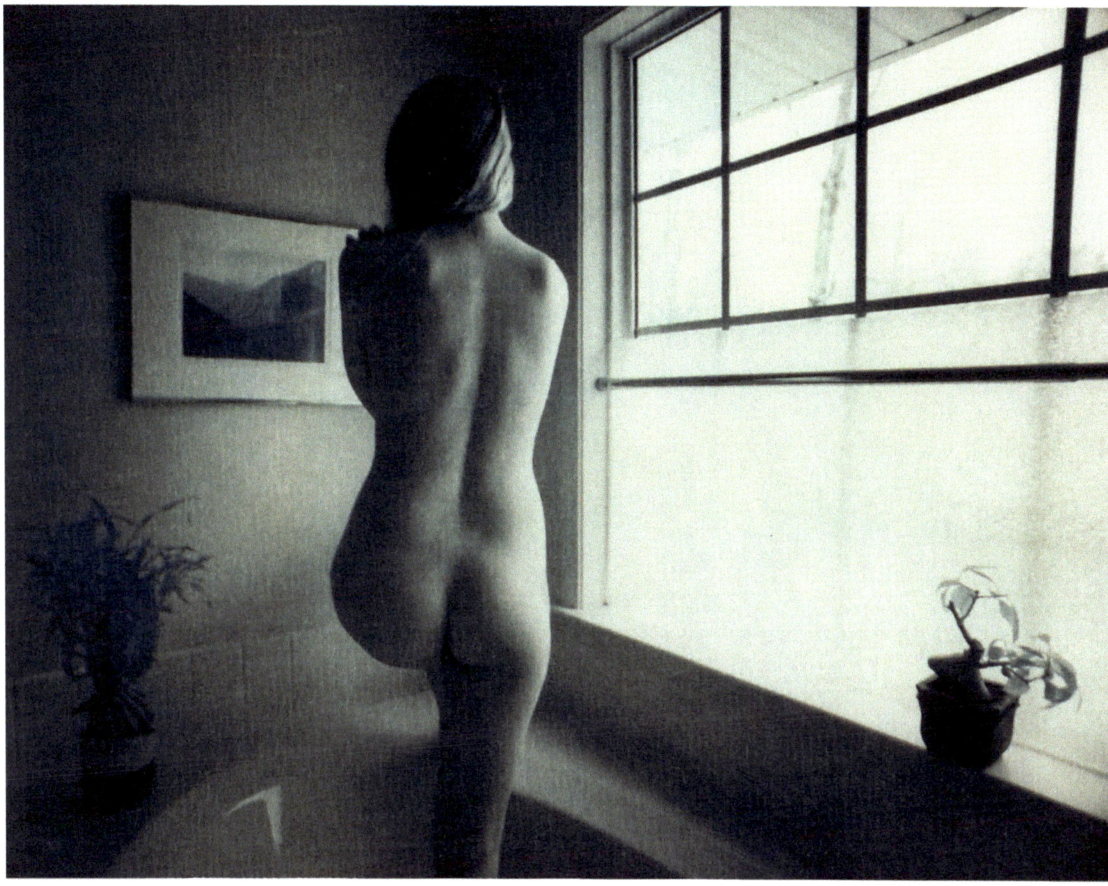

Figure 10.13. *Blue JM*, cyanotype over palladium on kozo paper © Sam Wang 2018

developer is too cold. Too low can be too little Na2, the ferric oxalate is old, the developer has gotten too alkaline and iron hydroxide stain has happened (keep developer between 4 and 6 pH), the developer is too dilute. Fog which looks low contrast (non-image color, usually a gray) can be from heat, light, or chemistry.

- Color is inconsistent from print to print—color is dependent on developer choice, developer temperature, paper choice, humidity, heat, paper sizing, contact frame temperature. If any of these factors is inconsistent so will be the color.
- Graininess—too cold or dilute a developer.
- Streaks in the print—usually not enough solution used to coat the paper, especially with more absorbent papers or when humidity is low and the solution absorbs too quickly into the paper before it has time to be coated evenly, but watch brush contamination, and make sure to rinse and clear it after use.
- Yellow stains in the highlights—this is common with too little clearing time, too cold clearing baths, or improper matching of clear to developer (tetra-EDTA works well with ammonium citrate and not as well with potassium oxalate). It is iron hydroxide left in the paper and once dry it will not come out and the print is ruined. It can also result from too alkaline a developer or too alkaline a first clear bath, so use the acid clear recommended and test the developer with litmus paper occasionally. Less likely is the water supply which doesn't allow the iron hydroxide to clear, or unsuitable paper that doesn't release the iron salts easily. It boils down to better clearing practice all the way around and perhaps having to resort to using distilled water.

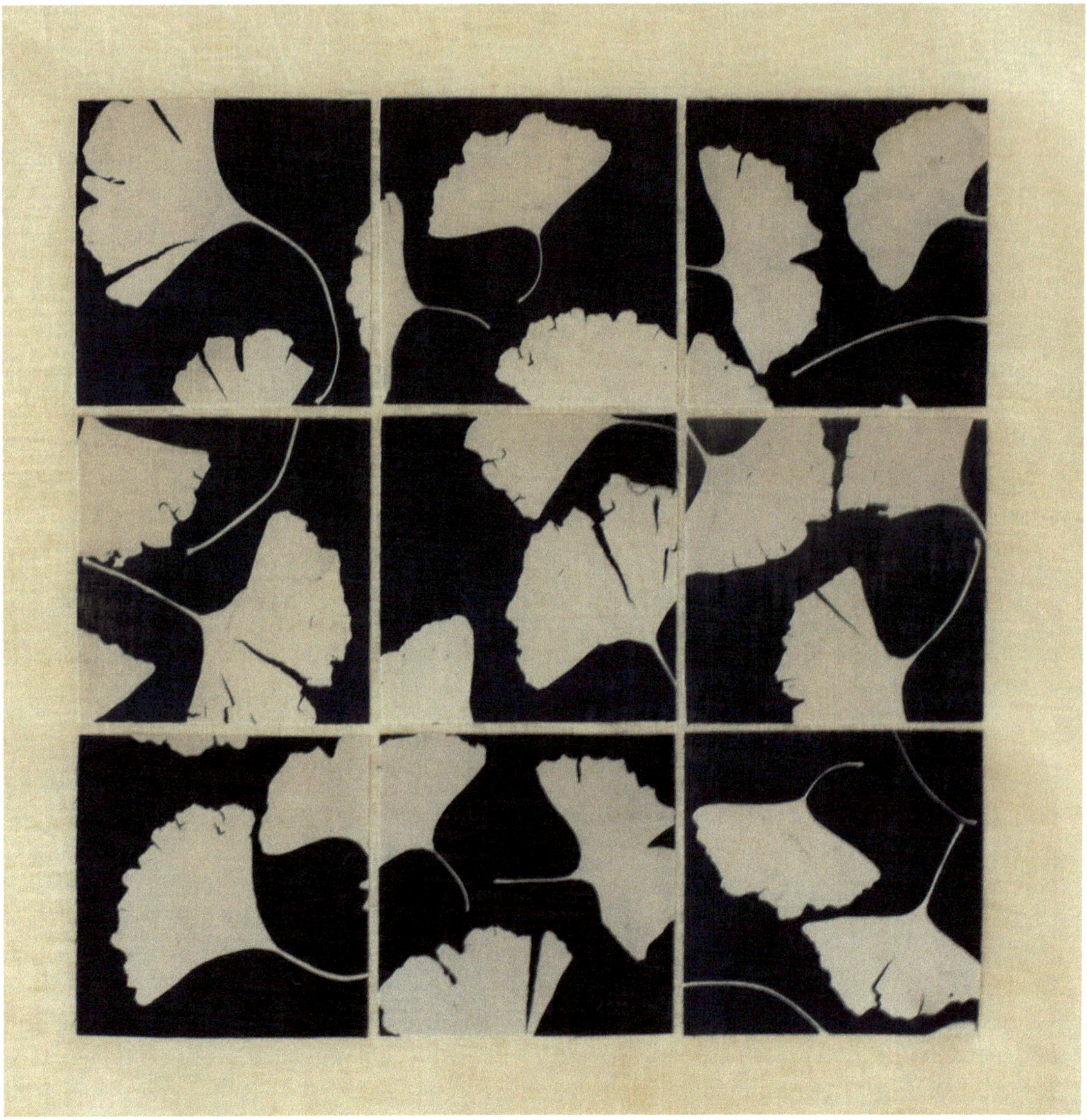

Figure 11.1. *Gingko Leaves*, cyanotype photograms on linen, 20″ x 20″ © Carol Byron 2018. "These images are part of an ongoing series of botanical photograms. I love experimenting with different processes and surfaces. I use different types of fabric and also Japanese tissue paper. The excitement for me is the unpredictable result you get and to see the shadow images these plants leave behind." Carol Byron is a fine art photographer from Naperville, Illinois and a graduate of the College of DuPage Photography Program. She has bodies of work in both traditional and alternative process photography, specifically lumen prints, cyanotypes and salt prints. She is a member of the Fleeting Moments Artists Collaborative and has exhibited in several photography exhibitions in the Chicago area and across the country. To view more of her work, please visit **carolbyronphotography.com**.

Chapter 11

Printing Cyanotype on Alternate Surfaces

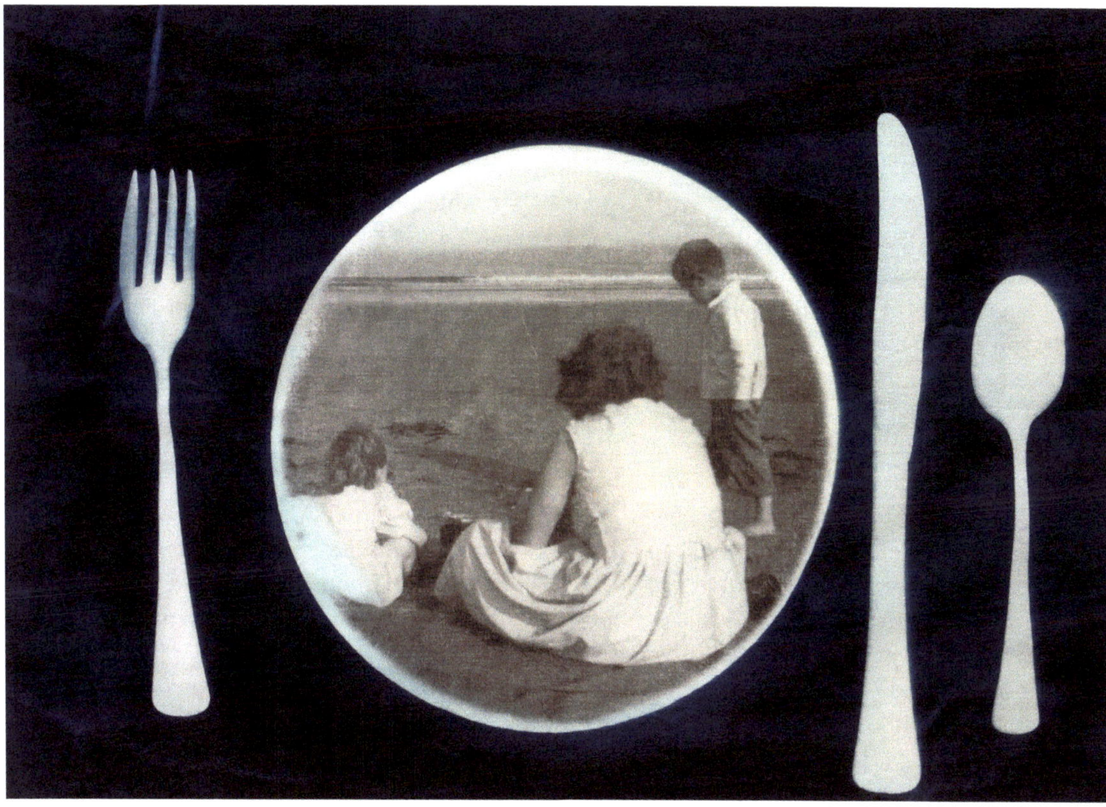

Figure 11.2. *The Beach*, classic cyanotype on fabric and image transfer, 16″ x 12″ © Vicki Reed 2015

Cyanotype can be printed on just about anything. The most common surface aside from paper is fabric, and then perhaps glass and wood, but I've also seen it printed on porcelain, eggshells, metal, bark, and bone. This chapter will discuss methods for printing cyanotype on fabric, glass, and wood.

Cyanotype on fabric

Printing on fabric has been part of cyanotype's history from the late 1800s, and continues to this day with several companies selling readymade sensitized fabric sheets or squares. Cyanotype will print well on any natural fiber fabric like cotton, silk, linen, rayon, old cotton sheets, t-shirts, gauze, unbleached muslin, and painter's canvas. The tighter and finer the weave of fabric the deeper the blue and the finer the detail.

Wool, although a natural fabric, can be problematic because of the lanolin inherent in the fibers which prevents absorption of the chemistry.[1]

Silk, if it is not PFD or "prepared for dyeing"—free of surface treatment[2]—will need a wash to remove the silkworm's natural gum "sericin": take 10% of the fabric's weight of sodium carbonate and combine it with 40% the fabric's weight in water. Simmer the silk for 30 minutes in this solution, allow to cool and then rinse *very* well so no longer alkaline, adding a splash of vinegar to the final rinse.[3]

Synthetic fabrics are not good choices because the fibers don't absorb the solution well and then after exposure the image washes off, leaving a pale, grainy print.[4]

Before committing to a lot of one fabric, take a piece of it and expose, process, dry, and then tear it in half and wash one half to test it.[5]

Wash, rinse, and dry any fabric before coating to remove factory sizing, starch, or who knows what else fabrics are coated with, or else brown stains may occur.[6]

Cyanotype formula for fabric

The quantity of cyanotype solution needed for soaking large pieces of fabric is much greater than a typical 100 ml formula provides. It may be that as much as a gallon of solution is needed for a large bed sheet and that is like mixing 40 small bottles! It is best to think differently about mixing up the formula, and use volume measurements instead.

Potassium ferricyanide is much heavier than ferric ammonium citrate so a much smaller volume is needed of it. Given equal weights, ferric ammonium citrate is about 3× the volume of potassium ferricyanide, or, said another way, potassium ferricyanide is about ⅓ the volume of ferric ammonium citrate. It's not quite this tidy, but it is close enough. If you've read the **Classic Cyanotype** chapter you will by now know how forgiving the formulas are, and how you could almost throw together a handful each of the two chemicals into a bucket of water and it will result in a great fabric print. However, consistency is always a good idea so you know how to repeat what great results you've gotten.

A lot of people store single-solution mixes to sensitize fabric. The formula below can be done either way, kept in separate containers or combined into one solution, and it can easily be quadrupled to make half a gallon of sensitizer. One-solution formulas will last for months. This formula gives a range of FAC:PF from 10% FAC:10% PF to 20% FAC:10% PF. As said in the **Classic Cyanotype** chapter, a 10% FAC:10% PF is faster than a 20% FAC:10% PF and plenty dark; it happens to be my favorite ratio.

Fabric formula

Mix one cup of this formula for every yard of fabric.[7] The solution will last up to 6 months just fine[8], stored cool and dark. Test a long-stored solution on a small piece of fabric before sensitizing larger pieces. Fabric is best if coated, dried, and exposed within the week; otherwise highlights may fog.

25–50 g/6–12 tablespoons/⅓–¾ cup ferric ammonium citrate
250 ml/8 oz/1 cup distilled water
25 g/2 tablespoons potassium ferricyanide
250 ml/8 oz/1 cup distilled water

1. Measure out the ferric ammonium citrate and add it to the 250 ml water and stir until dissolved. FAC goes into solution right away.
2. In a separate container add the potassium ferricyanide to another 250 ml water and stir until dissolved. PF takes a bit of stirring to go into solution, so make sure it is completely dissolved before dipping fabric in it.
3. Either combine both solutions into one or keep the solutions separate and at time of use combine them in equal proportions.

Coating fabric

Caution: wear gloves while coating, and do not touch fingers to the fabric because the skin oils will leave fingerprints.

If the fabric pieces are small, there is no difference coating fabric than coating paper. Both can be done with a brush. However, the solution may soak through the cloth and stain the table, so protect the table with glass, Plexiglas, or even plastic garbage bags. If the fabric is wrinkled, iron before brush-coating.

With larger pieces of fabric too big to table coat, better to soak them in a bucket of solution, wring out excess, and hang with clothespins on a clothesline to dry. It'd be great to do this at night in a garage, with plastic bags on the floor to catch the drips.

When hanging fabric to dry, hang it from all four sides so that it dries evenly and the solution doesn't drip to the bottom of the piece of cloth

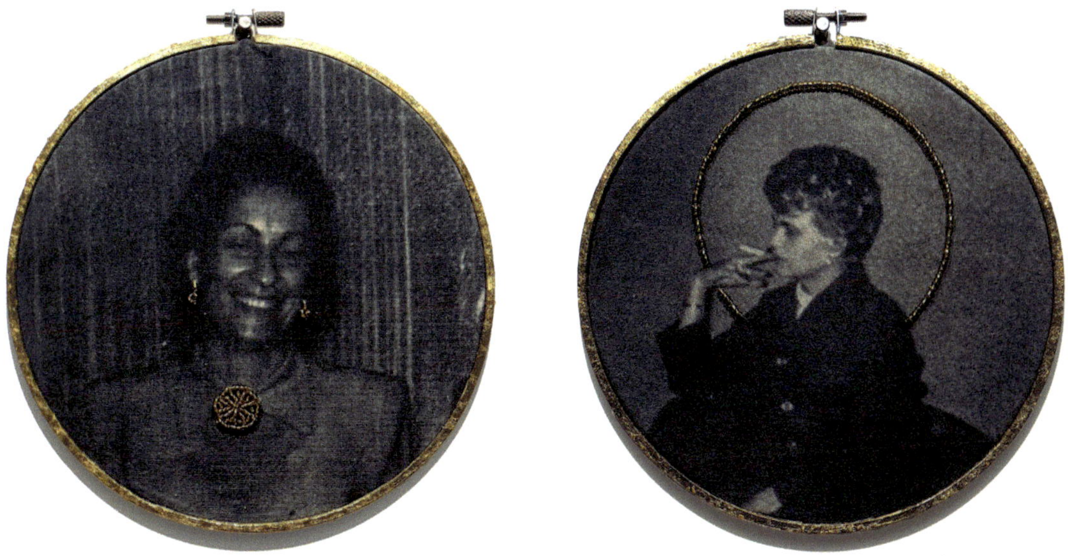

Figures 11.3 and 11.4. Left, *At the Altar of a Stranger II*, 8″diameter; right, *At the Altar of a Stranger VIII*, 7.5″ diameter, classic cyanotype on cotton, toned in hydrogen peroxide and coffee, with hand embroidered gold beading, framed in a gold leafed embroidery hoop © Ellie Cooper 2018. "The intent of *At the Altar of a Stranger* is to make the forgotten people held within the images important again. The source images have been pulled from the artist's own collection of vintage photographs gathered from various antique and thrift stores. I began collecting out of a sentimental sympathy for the discarded people within the images, wanting to adopt them and give them a new importance. Being found in these areas of commerce is the result of people thoughtlessly abandoning these images, meaning that no one cared about the people within them enough to hold onto their likeness. Outside of the context of passed down family history these images lose their meaning, lose their importance. My goal is to highlight these people using iconography from art history that signals notability to viewers. Using historic techniques and symbols from early photography and the Romanesque and Gothic periods of painting, the viewer will recognize the importance of not only the people within the images but the significance of acknowledging and remembering our past as it relates to our present and future." Ellie Cooper obtained her BFA in Photography at the Savannah College of Art and Design. She specializes in handmade traditional darkroom and alternative processes images. To view more of her work, visit Cooperphotographic.com.

and result in an uneven coating.[9] This is especially necessary for silk. Once the coating is dry the fabric can be ironed and stored in light-safe plastic bags.

Exposing fabric

Exposing fabric is similar to exposing paper, but can take longer than paper, especially if the fabric is heavy and has absorbed a lot of emulsion.[10] Expose some small test strips and test times first. Exposure times can range as short as 5 minutes in direct summer sun to as long as an hour at other times and seasons. Test exposures outside starting at 5 minutes and every 5 minutes thereafter for a total exposure time of one hour. Silk exposes in minimum time, cotton sheeting average time, and cotton t-shirts maximum time. Press the fabric with an iron before exposing if there are wrinkles.

Processing fabric

Processing fabric is the same as processing paper, with a first acidified water wash of several minutes followed by a final water wash for a total of 10–15 minutes' wet immersion. In the second water wash a glug of hydrogen peroxide can be added to hasten oxidation and darken the blues. For small pieces of fabric, a liter of acidified water is plenty. For large fabric pieces, make a bucketful.

60 ml white vinegar (¼ cup) or 2.5 g/½ teaspoon citric acid

1000 ml/1 liter water

1. Add the vinegar or the citric acid to a liter of warm water and stir until evenly dispersed.
2. Remove the fabric from the exposure frame or out from under the glass and immerse the whole of the fabric at once in the acidified warm water.

3. Agitate the fabric continuously in the tray for 1–5 minutes; the water will turn blue, depending on how much wash off occurs.
4. At the end of 1–5 minutes transfer the fabric to another tray of warm water for 5 minutes, and wiggle intermittently to prevent airbells.
5. At the end of another 5 minutes discard the water and refill the tray with warm tap water again. Soak the fabric for another 5 minutes and wiggle the print intermittently to prevent airbells.
6. Once the fabric has gone through a total of 3 washings/15 minutes, it is probably done. Hang the fabric to dry. There should be little to no blue running off of it.

Toning fabric if desired

Toning is as simple for fabric as it is for paper, so all the formulas in the **Toning Cyanotype** chapter will work, but in larger quantities if toning larger pieces of fabric. A mere bucket of water with a tablespoon of Arm & Hammer washing soda or TSP[11] will bleach the fabric, and then a bucket of strong brewed tea (depending on how much fabric, 10 tea bags to a liter of water steeped for at least 10 minutes) will tone. Having tannic acid on hand is *so* much easier than brewing large quantities of tea, so do yourself a favor and buy a pound of it. ½ cup of tannic acid powder to a gallon of water is enough to tone a couple yards of fabric.[12]

Laundering cyanotype fabric

You have to be very careful with washing cyanotype; any alkaline detergents or phosphates in the detergent will bleach the cyanotype to yellow. Test every detergent before using, because once the yellow happens, you can't get all the blue back even with an acid wash. However, you could tone it to brown if desired.

- Cyanotype-safe soaps are Dove, Ivory, Joy, and Palmolive, and liquid laundry soaps such as Clout, Tide, Wisk, and Woolite.[13] Use less detergent, under ¼ cup for a large load, or even a mere half teaspoon in a sink to wash one shirt.[14]
- For stains, RIT puts out a Stain and Grease Remover No. 90 that works.[15]
- Silk should not be washed with soap, merely rinsed in a few changes of water acidified with a little lemon juice.

Tips

- Don't just use white or natural cloth; use patterned cloth, colored cloth, etc.
- Boiling fabrics is a great way to get rid of fabric sizing.[16]
- Keep fingers off the material; fingerprints leave their mark.
- Coating fabric is easiest if you soak it in the solution, but if the fabric is bulky, thick, and stiff, you can also apply the solution with a foam applicator, brush, or sponge.
- When doing fabric photograms, get a large piece of cardboard and put the fabric on top and then pin everything to it. Or, polystyrene foam or insulation board can be used and pinned to with sequin and bead pins from the sewing store.[17]
- If the design elements are not in direct/flat contact with the fabric it'll produce a "ghosted" edge which can be quite beautiful.[18]
- Fabric takes 5× as much solution and exposure times are longer.[19]
- To create a makeshift light-tight bag to store fabric, wrap it in plastic and then aluminum foil.[20]
- If a fabric image fades, put it in the dark, give it a hydrogen peroxide rinse,[21] or soak it in diluted vinegar.

Sources for cyanotype fabric

If coating fabric seems like too much work and mess, there are places that sell pre-coated fabric: Jacquard Products (**jacquardproducts.com**), Blueprints on Fabric (**blueprintsonfabric.com**), and the Cyanotype Store (**cyanotypestore.com**).

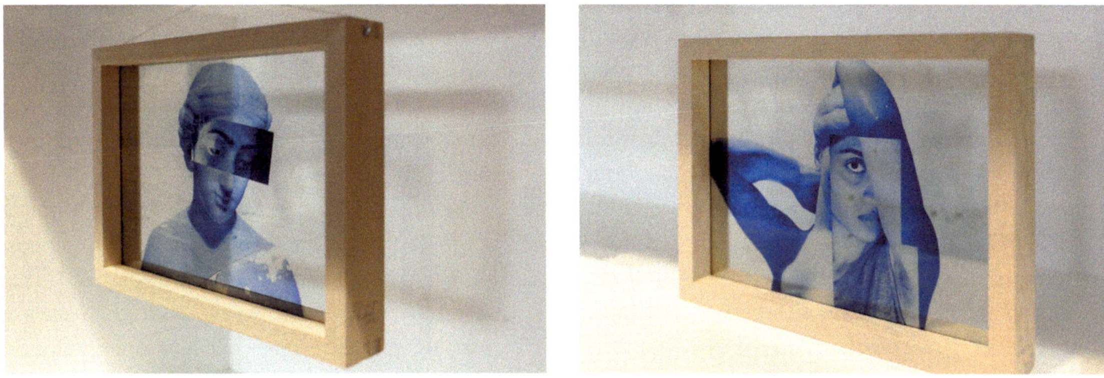

Figures 11.5 and 11.6. *Human Condition*, cyanotype and gelatin on glass, 11″ x 14″ © Kyeong Ran Jang 2016. "One pack of Knox gelatin (about 20–28 g) is added to distilled water to make total of 100 g solution. The gelatin is dissolved in a water bath. The cyanotype mixture is mixed with the gelatin mixture in equal parts. The gelatin is now about 10% of the solution, and the cyanotype mixture has been diluted to 50%. Painter's tape is used to mark the borders of the print before the chemistry is applied to the glass. There are many different techniques to coat the glass, such as using a scoop to pour on the chemistry or using a brush; however, brushing did not produce optimal results. Adding a little bit of alcohol into the mixture helps with the coating process. The sensitized glass is dried using a fan until the chemistry is dry to touch, and is exposed for 15–22 minutes under UV light. The prints are washed in distilled water for several minutes and dried.

"Late Enlightenment German philosopher Hegel thought that the role of art is to bring truths about humans, represented through ideal human form (often divine), found especially in ancient Greek sculptures of gods and heroes. Hegel believed that the gods in Greek sculptures were beautiful because of their perfect physical shape, which embodied spiritual freedom not marred by marks of physical frailty or dependence. However, Hegel believed also that while the Greek sculptures may portray the perfect human form, they were deprived of the dimension of a human being and will always fail to speak to us in a truly human manner. Drawing from Hegel's idea of art and beauty in relation to human nature, I photographed sculptures in museums and combined them with photographs of human physique in order to bring more dimensionality to the great works of art. In a sense, it is another display of human desire to reach ideal beauty. The fragile surface represents, in both the sculptures and the human body, that they will break apart if not well cared for. Even though glass may be stronger than paper, it is more likely to be damaged if it encounters a destructive event, much like the human body."

Kyeong Ran Jang is a PhD student in neuropharmacology at Emory University, and a photographer. During her childhood, she photographed her daily life and used it as a primary source of communication with her family in Korea. She became particularly interested in the intersection of art and science in photography, especially in the darkroom and later through alternative processes, which require similar experimentation with chemistry as her work as a research technician. Her work focuses on historical processes on non-traditional surfaces.

Cyanotype on glass

Printing cyanotype on glass is much trickier than printing it on fabric; fabric is absorbent, glass is not, and the surface has to be carefully cleaned, subbed with gelatin, coated with sensitizer, and processed in such a way that the solution will stay adhered to the glass and not frill or peel off. If I know anything about printing on glass it is from Galina Manikova (see the **Contemporary Cyanotype Artists** chapter) whom I "met" on the alternative processes listserv. Much of the following Manikova graciously shared with me via email.

Purchasing the glass plate

Purchase glass from any glass store and have them sand all edges for you. Otherwise you can use an Arkansas stone (a whetstone found at hunting supply stores) and abrade all edges, top edges, bottom edges, and then directly along the outer edges—perpendicular. This provides a tooth for the edge of the coating material to adhere to as well as prevents cuts.

Cleaning the glass plate

This should be done extremely carefully. Although many first clean the glass with detergent and water, Manikova has dispensed with this step because the alkalinity of detergents can wreak havoc with cyanotype further down the road.

1. Make a pumice out of whiting powder and rubbing alcohol or methylated alcohol and scrub the glass well (this is what Manikova

uses). You can also use Bon Ami glass cleaner, a mild pumice that will not scratch the glass but will remove any silicone coating that the manufacturer might have put on the glass for shipment. Alternatively, you can make your own pumice mix of equal parts Rottenstone (a ground up rock grit finer than pumice), denatured alcohol and distilled water, and put it in a squirty bottle. Shake before use.[22] Squirt this mixture on the glass and use dry, soft, lint-free cotton cloths such as old t-shirts, three in succession, to first polish the glass aggressively, then remove the cleaner, and then buff the glass dry. To verify that the glass is clean, breathe on the surface of the glass and examine the fogged glass for any streaks or imperfections. Any grit, no matter how small, must be removed or it will cause the coating to wash off or mar the final image.

2. Take rubbing alcohol, methylated alcohol, or Everclear and clean the glass completely of any remaining dirt, grit, grease, or fingerprints and let dry for 2–3 minutes before one last brush off of any remaining lint, dust, etc., before coating. Note that this whole time, hands should be free of grease, hand cream, etc., and the surface of the glass not marred with fingerprints.

Subbing the glass plate[23]

Manikova says, "The thinner the gelatin coating, the better. It needs to be free of any bubbles. If one makes many thin layers by coating in different directions, all the small imperfections and small holes will be eliminated, dissolved and covered by the next layer of thin gelatin. Both room temperature, air humidity and even the mood of the person at that particular moment of time will influence the results. This demands a lot of patience and understanding of the goal."

1. Add 30–60 g gelatin to 1 liter of cold water (3–6%). Let the gelatin bloom in the water a bit, and then heat until the gelatin is dissolved, to about 140°F.
2. Warm the cleaned glass plate. When warm, pour a first coat of gelatin on the glass and let it cover the glass plate and drip off the edges. Let dry.
3. If the gelatin solution has cooled, heat it again, and coat the glass plate with another layer of gelatin in a different direction. Let dry.
4. Do the same one more time with a third layer of gelatin.
5. Some add formalin to the gelatin in the amount of 5% of the total coating amount. Some add 3 ml glyoxal 40% per 100 ml gelatin.[24] This hardens the gelatin layer. Manikova uses a separate formalin bath because once you add formalin to the gelatin you cannot reuse the gelatin.
6. Once the plates are coated, place them on a cold horizontal slab to cool, and as soon as the gelatin sets up, let the glass dry. Manikova notes that gelatin takes about a *month* to harden. "If the gelatin coat is fresh, it might slide off or water might find its way under the coating during development. Hence, making the coated plates long before use is important. Drying everything properly leveled is also important, unless you dry it in a vertical position and turn around for each new coating. The room should be dust free as possible and with good ventilation. It can be done outside as well, if the weather allows. There is no need to do this part in the darkroom. It is worth building a drying closet for this part, easily made of old cardboard boxes or similar as one needs to let them sit there for about a month."

Coating the glass plate with cyanotype

Everything is to be held warm: glass plate, gelatin/cyanotype, even brushes, if you are using them. Add the 3–6% gelatin solution to the cyanotype solution in the proportion of 1 part gelatin to 2 parts cyanotype. The gelatin makes it adhere well to the gelatin layer on the glass and also makes it hardier during development. Let it dry.

Exposing the glass plate

Greater exposure is required on glass than on paper to get the translucent image to have enough density. Try 45 minutes for classic cyanotype.

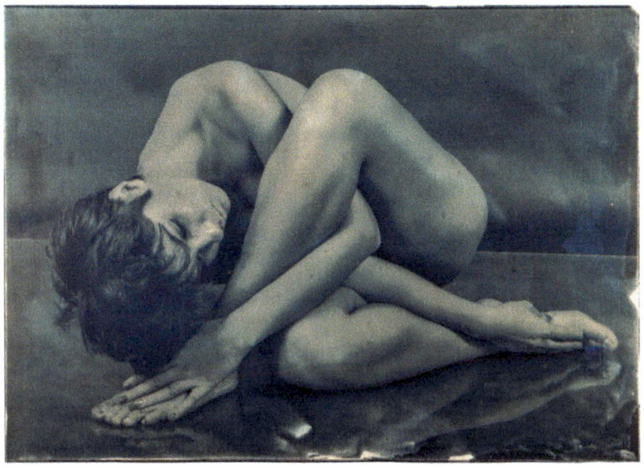

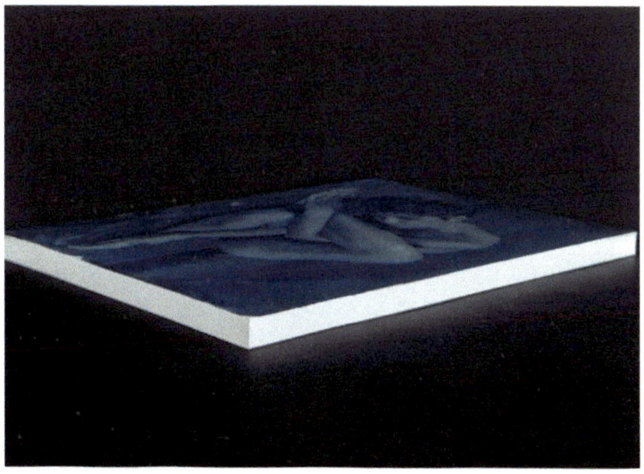

Figures 11.7 and 11.8. *Moldavian Nude*, cyanotype on Forex, 11.7″ x 16.5″ x 0.75″ © Salvatore Abate 2018. Abate first primes the Forex with acrylic paint. Then he mixed 3 ml FAC (20%), 3 ml PF (8%), 6 ml gelatin (20%), 6 ml 96% alcohol, and 6 drops of formalin together into one solution. This is poured on the Forex and evened out. When the layer is completely dry it is exposed and developed normally. "With my work I try to mix old historic photographic processes with modern materials. Forex is a white rigid PVC foam sheet. My first tries were failures because I used cyanotype solution with gelatin and formalin without primer on Forex. Now the cyanotype is so firmly attached that when you put the print in water, even if you leave Forex in the water upside down for a long time, it will not flake off." To see more of Abate's work, visit **www.salvatoreabate.com**.

Processing the glass plate

Development is in acidified *cold* water so that the gelatin does not soften. No running water! Manikova says, "I usually just place a plate carefully into the bath with cold water, without moving it too much. I take out the plate before changing the water and placing the plate back into a new water bath carefully. The last water bath is acidic. Then oxidizing by air and fixing in the sun works more gently, and should never be forgotten.

"If everything is done properly there will be no need for an acrylic after coating. I like the surface of the gelatin/emulsion mixture much better. If done right, it will be very strong and scratch resistant after about a month of drying in a dry and hot atmosphere. It will not be sensitive to humidity after that. It can be even washed in cold water if it gets dusty. One can also make a citric bath to freshen them up once in a while, if fading occurs. It takes another month to harden the final print, but after that it sits very well on the glass, and it will be not easy to wash it away, not even with hot water and a scrub. Otherwise I find it very satisfying that any unsuccessful print can be easily removed using hot water, and the glass plate can be used again."

Printing cyanotype on OHP

Cyanotype can be printed directly onto Pictorico OHP transparency film to make a sort of "poor man's" glass print or lantern slide. The OHP has a layer of ceramic coating that absorbs the solution well with no bleeding or smearing.[25] When initially coating, the ceramic will grab onto the solution and leave a white mark, but keep brushing thinly until it evens out. Lift it vertically and from all four sides so the solution runs down (watch holding it because it is slippery!). If the OHP is laid down flat it will puddle and be ruined. Expose and process as you would a normal print; new cyanotype looked great at 7 minutes and overexposure produces a somewhat odd shimmery purple on the surface.

Printing cyanotype on wood

Wood is a breeze compared to glass but brings its own set of problems. For one, the cyanotype solution will sink into the wood and stain it, so therefore it is best to size the wood with gelatin or PVA. Two, the highlights of the image are no lighter than the wood itself, so take that into consideration.

Michaels or other craft stores sell birch plywood pieces and birch is quite light with minimal wood grain so that might be your best bet for a first foray.

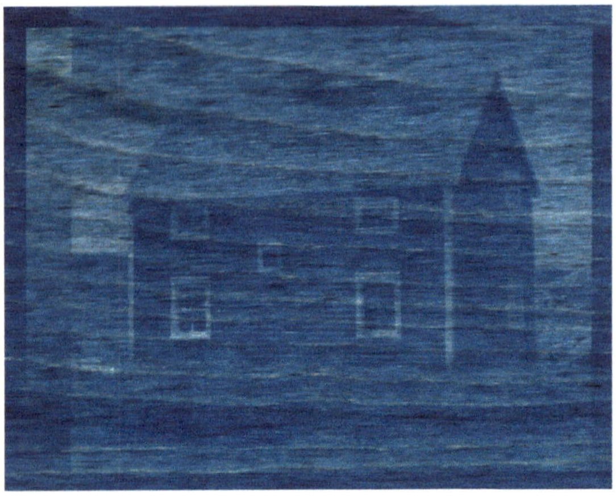

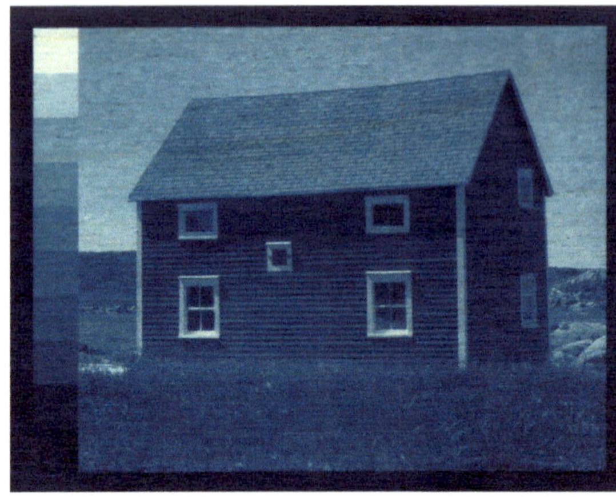

Figures 11.9 and 11.10. Test prints, 20/8 cyanotype on raw 3 mm birch plywood, 4″ x 6″ © Peter Friedrichsen 2018. The prints were exposed for 20 minutes under UVBL. Development was 5 minutes in 0.5% oxalic acid, and 5 minutes water rinse. The image on the left is unsized. The image on the right is sized with one coat full strength Gamblin PVA size and one overcoat of a 7.5% tapioca starch sizing solution. See a complete how-to in the **Contemporary Cyanotype Artists** chapter under Friedrichsen.

Check the wood for imperfections before buying. There are also cradled wood panels in birch or basswood that come attached to the frame and they are smooth and sanded. They are reasonably priced.

1. If the edges of the wood are splintery, sand them smooth.
2. Clean the wood thoroughly with plain water and let it dry completely.
3. Size with Gamblin PVA full strength, two coats, or use gelatin: add 30–60 g gelatin to a liter of cold water and let the gelatin bloom for 10 minutes.
4. Heat the gelatin and water slowly until it reaches around 140°F or thereabouts. No need to be exact.
5. Brush the wood with the warm gelatin in a thin coat and let the gelatin dry thoroughly.
6. Coat the wood with either the plain cyanotype solution or a gelatin/cyanotype solution, up to 1 part gelatin to 3 parts cyanotype. Let the layer dry thoroughly.
7. Expose and process the wood print as normal, but err on the side of overexposure and longer development because wood is absorbent and you want to make sure to remove any iron salts so the print doesn't fog with time.

Endnotes

1. Brown, Ruth. *Cyanotypes on Fabric. A blueprint on how to produce…blueprints!* East Yorkshire: SC Publications, 2016, p. 13.
2. Ibid., p. 12.
3. Ibid.
4. Ibid; Fabbri, Malin and Gary. *Blueprint to Cyanotypes.* Alternativephotography.com, 2006, p. 42.
5. Brown, p. 78.
6. Hewitt, Barbara. *Blueprints on Fabric, Innovative Uses for Cyanotype*. Loveland, Colorado: Interweave Press, Inc., 1995, pp. 19–20.
7. Hewitt, pp. 21, 24; Brown, p. 17.
8. Hewitt, pp. 27–28.
9. Newman, Thelma R. *Innovative Printmaking, The Making of Two- and Three-Dimensional Prints and Multiples*. New York: Crown Publishers, 1977, p. 114.
10. Hewitt, p. 53.
11. Ibid., p. 65.
12. Ibid., p. 67.
13. Ibid., p. 68
14. Ibid., p. 69.
15. Ibid., p. 94.
16. Howell-Koehler, Nancy. *Photo Art Processes*. Worcester, Massachusetts: Davis Publishing Inc., 1980, p. 67.
17. Brown, p. 32.
18. Ibid, p. 31.
19. Antonini, Marco, Sergio Minniti, Francisco Gomez, Gabriele Lungarella, and Luca Bendandi. *Experimental Photography: A Handbook of Techniques*. New York: Thames & Hudson, Inc., 2015, pp. 136–145, 206–207.
20. Brown, p. 23.
21. Brown, p. 78.
22. Mark Andrews on the alt list.
23. An APUG thread talks about Kuraray R-1110 silated 100% hydrolyzed PVA, which you coat on glass and dry overnight, then bake for 30 minutes (250°F) and let cool. The cyanotype print is then coated, exposed, and processed as normal. After processing and drying, the plate is again baked for 10 minutes. The product is mixed in a 10% solution and heated to 200°F to get it to dissolve. See **www.kuraray-am.com**. Others coat the cleaned glass with a layer of Silane Glass Treatment, available from Bostick & Sullivan.
24. James, Christopher. *The Book of Alternative Photographic Processes*. New York: Delmar Cengage, 2016, pp. 220–222.
25. Ware, Mike. *Cyanomicon II, History, Science and Art of Cyanotype: photographic printing in Prussian blue*. Buxton: self-published, 2016, pp. 102–103. **http://www.mikeware.co.uk/mikeware/downloads.html**

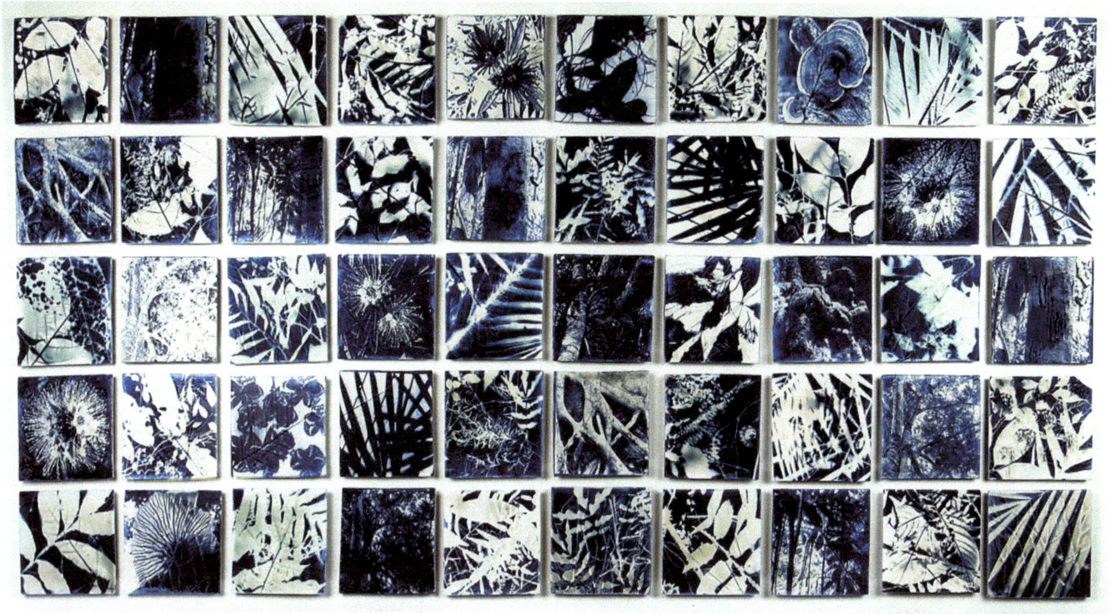

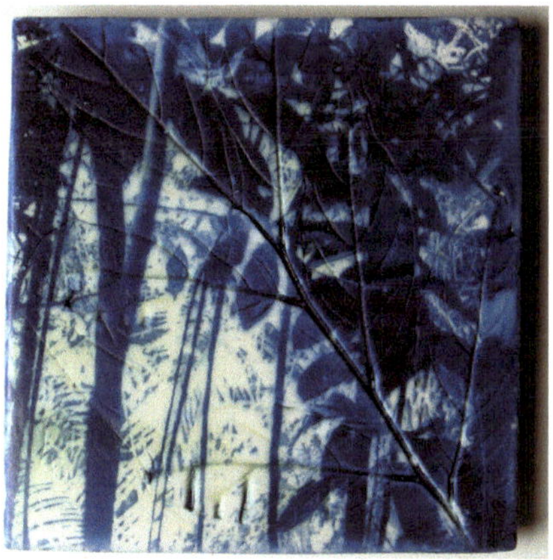

Figures 11.11 and 11.12. *A Meditation on Light*, 50 tiles 14 cm square each, installation 1.6 m x .7 m © Mollie Bosworth 2013. "The cyanotype prints were printed on handmade, textured, vitreous porcelain tiles. The tiles were subbed with a hardened gelatine layer before coating with cyanotype chemicals. Images were sun-exposed using digital transparencies or actual plant material. The tiles are sealed with clear satin acrylic.

"I came across the cyanotype process several years ago when looking for a way to transfer images onto my porcelain work. I've mostly used the new cyanotype formula, mixing it to the directions given and adding the .1% potassium dichromate. Mostly, I use the sun for exposure. When first starting on the process of cyanotype on clay, I experimented on porous bisque but had trouble washing out all the chemical. My bisque was at 1000°c and I may have been successful if it was fired higher rendering the clay less porous. Not having the facilities to do lots of temperature trials, I decided to develop the image onto vitreous porcelain and I followed advice for techniques of cyanotype on glass. I created a subbing layer of gelatine hardened with chrome alum. I poured it on evenly when warm. When it was set, the chemical was brushed on, dried and exposed. The image requires careful washing and limited careful soaking.

"For my porcelain tile artwork, each piece is hung separately. I glued a small square of plywood on the back of each tile and had a concealed eye hook for hanging. The plywood was 2 cm in from each edge, so it was invisible from the front. I liked the floating look that this gave the pieces.

"I've been an artist working in ceramics for over 30 years and started my cyanotype adventure only in the last 5 years. After working in the spare bathroom, then in a corner of my pottery studio, I now have two separate studios on my property and work between them. As I work with sun exposure, I can work on ceramics on wet days.

"*A Meditation on Light* was inspired by the qualities of intense sunlight filtering through the local rainforest and the minutiae of the forest floor. I explored the contrasts of intense light and shade and the revelation of positive and negative patterns through digital images and direct photograms. The combined work reflects the visual intensity of the rainforest."

Mollie Bosworth is a visual artist working in fine porcelain and cyanotype processes, with a studio based in far north Queensland, Australia. Her work often explores complex surfaces that reflect her rainforest environment. She has exhibited widely during her 30 years of arts practice. Her work can be seen at **www.molliebosworth.com**.

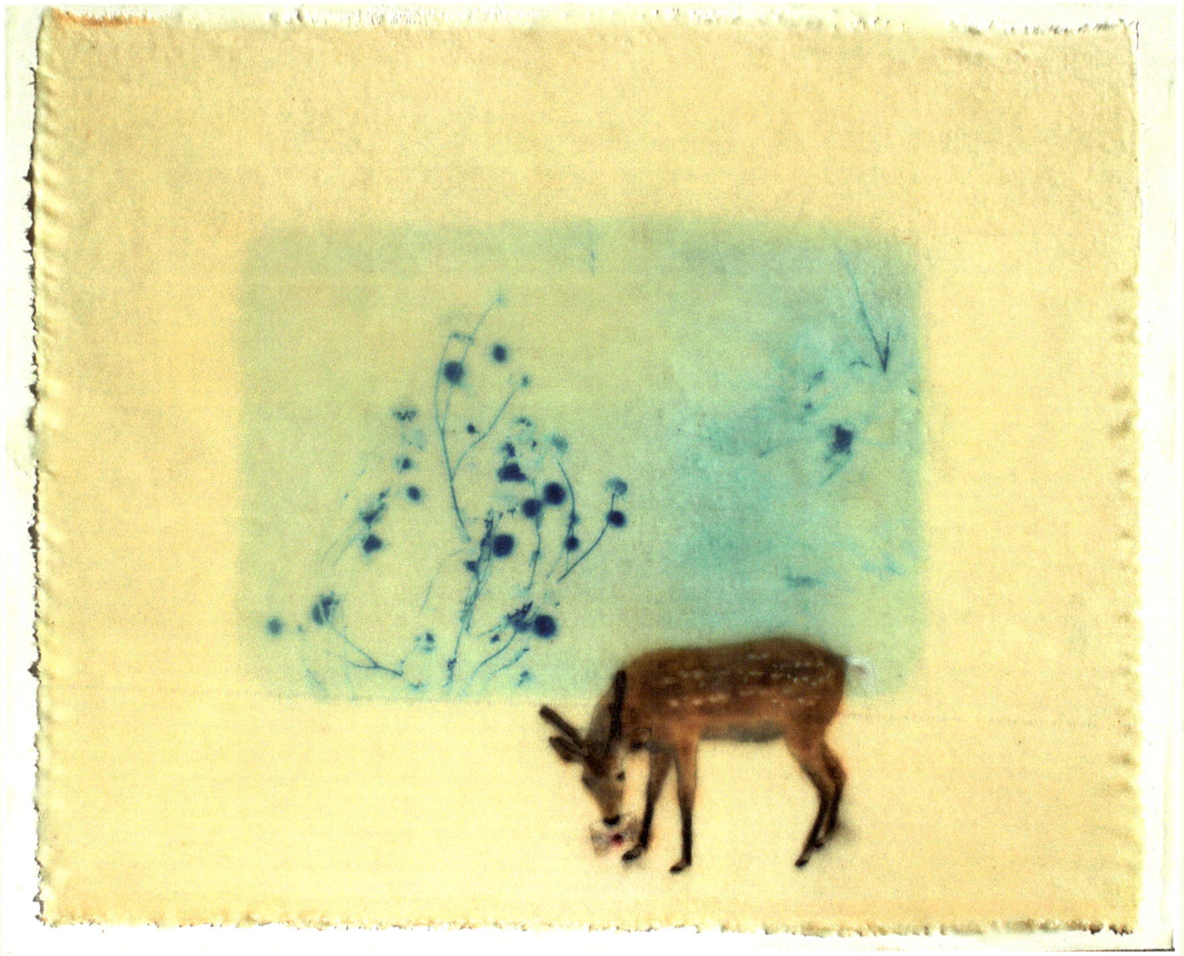

Figure 12.1. *Transference*, cyanotype on fabric and encaustic with mixed media on wood © Kathryn Murphy 2018.

"During the 11th and 12th centuries in Nara, Japan the deer that inhabited the forest were believed to be messengers of the gods. Every moment spent tuning in and connecting to nature can open our hearts to the transference of wisdom that is available to those willing to receive. When I start to feel out of tune with my own inner creativity, I take a hike in the woods or a trip to the lake. This always refreshes me and keeps me centered. Having a connection to the natural environment is at the core of all of my work. Even a quick pinch of the geranium plant that I keep nearby and the grounding aroma that follows is sometimes enough to keep me going. I believe it is this connection that opens up the pathways to my inner perception and imagination.

"I was drawn to cyanotypes right away because they are user friendly and I have the ability to print them in my home studio. Most of my work revolves around my interaction and perception of nature, so using a process that involves printing with the sun fits in conceptually with my art practice.

"I prefer to print on 100% cotton fabric. The softness of the fabric combined with the deep blue of the cyanotype is very appealing. I print my all-ink negatives on an Epson P800 at -30% ink density. I use the classic cyanotype 10/10 formula. Whenever possible I print using the sun; otherwise by UVBL. Development is 5 minutes in vinegar water for the first wash, and then 2 minutes in plain water. My process involves layering the cyanotype with wax onto a wooden panel.

"The key to printing on fabric starts with the brushing on of the cyanotype solution. I place the fabric on a piece of glass and then pour a small amount of cyanotype in the center of the fabric. I then very lightly begin to spread the solution out. If you press too hard the solution tends to 'pool' beneath the fabric and it is hard to spread evenly."

Kathryn Murphy is an artist, nature enthusiast and collector of rocks. Her work inquires into the essential interconnection between the heart, the mind and the natural environment. Her constructed scenes dance lightly between the subconscious realm and reality. She is a mixed media artist who has embraced alternative photographic processes into her art practice. A native of Montana, Murphy currently resides in Bozeman, Montana, where she received her BFA in Studio Arts, Graphic Design and a minor in Photography. To see more of Murphy's work, visit **bluemothcommittee.com**.

Chapter 12

More Creative Ideas for Cyanotype

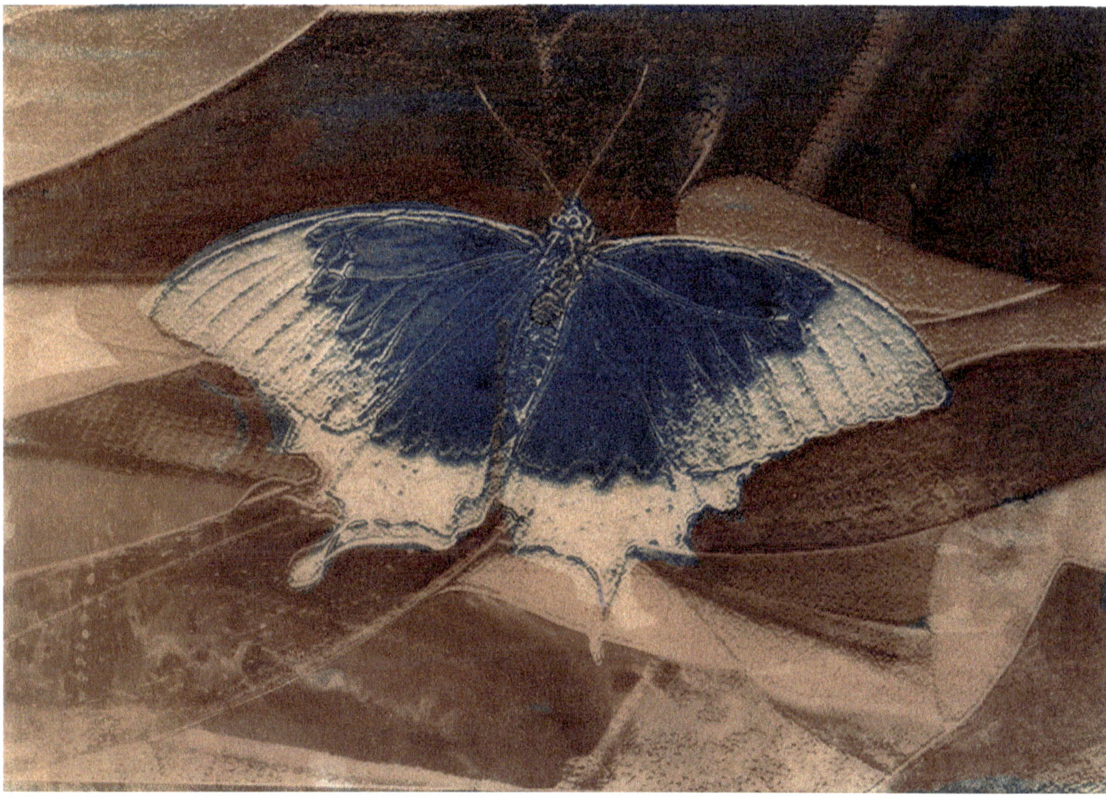

Figure 12.2. *Butter Fly*, digital artwork laser printed on acetate and coated and printed in traditional cyanotype, over-printed in Van Dyke brown, 27.5 x 19 cm © Robyn Davis 2010. Robyn Davis first experimented with cyanotype and Van Dyke brown while at university. Her enthusiasm increased after purchasing *The Experimental Photography Workbook* (Anderson 2012). Inspired by this book and her personal creative studies, she began experimentation with alternate photography processes. Davis travels extensively throughout Australia collecting inspiration and ideas and continues to push the boundaries of the process.

Printing on alternate surfaces such as fabric, glass, or wood is one way to push the boundaries of the cyanotype process, but there are other creative ways to use the process, and because the chemistry is so cheap you can experiment with abandon without breaking the bank. This chapter only touches the surface of how the chemistry can be used, but hopefully it will provide an impetus to discover new ways not even thought of today.

Printing cyanotype over gold leaf

Dan Burkholder has an excellent video outlining how to gold leaf platinum/palladium prints which applies equally well to cyanotype. The cyanotype is first printed on a transparent vellum paper. After processing and drying, it is painted on the back with a particular adhesive specifically to adhere gold leaf, and then sheets of gold leaf are burnished onto the adhesive. A final spray varnish both transparentizes the print and protects it. Today there is no need to

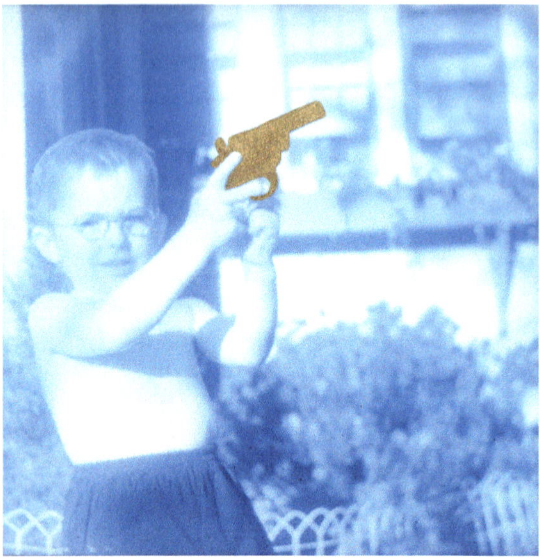

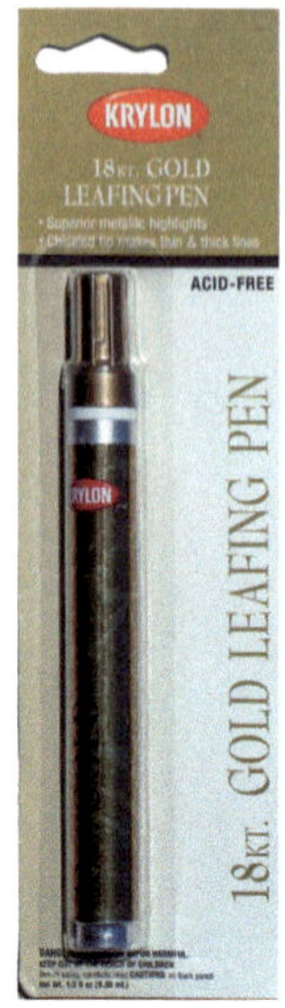

Figures 12.3 and 12.4. *Tommy Get Your Gun*, gold leaf over new cyanotype on Hahnemühle Platinum Rag © Christina Z. Anderson 2018. Krylon makes a very handy and easy to use gold leafing pen under $10. Note that gold leaf is opaque.

buy genuine 24k gold leaf ($$!) because imitation gold leaf sheets are readily available as are a variety of other "metals" like silver, copper, palladium, and pewter. Note: there is also an 18k gold leaf *pen* that can be used to draw on the surface of a print. It writes opaque and is good for detail work. In large areas it leaves noticeable strokes, so that is where gold leaf sheets should be used.

Supplies

Wyndstone, Opalux, or Vidalon vellum paper
Low-tack tape like Scotch Safe Release, Frog Tape, or Kleenedge
Rolco or other brand water-based adhesive
Gold, silver, copper, etc. leaf sheets
Krylon Kamar varnish
Soft brush for the adhesive
Burnisher or spoon

1. Cut a much larger sheet of vellum than necessary, tape the sheet down to the table, and coat a smaller center of the sheet. Let it dry in place taped down.
2. If necessary, use an iron on low or a dry mount press before exposing to flatten the paper.
3. Expose the print for 15–20 minutes CC (5–6 minutes NC) under UVBL. Erring on the side of overexposure is fine to make sure to preserve all the detail in the highlights.
4. Process the print in acidified water as normal.
5. Dry the print between two pieces of blotter paper weighted down with heavy books.
6. Iron or dry mount press the print before coating with adhesive.
7. Turn the print face down and tape the area to be coated so the adhesive stays within the print area.
8. Brush a liberal layer of the adhesive and let it dry. It should be tacky.
9. Take the sheet of gold leaf and place it carefully on top of the adhesive.
10. Burnish it in place with your fingers, the back of a spoon, or a burnisher.
11. Hold the print up to light and see if any areas are missed. If so, take another small bit of the gold leaf and fill that area in. Once this is done, let dry.
12. Outdoors, spray the front and the back side of the print with a light layer of Krylon Kamar Varnish (non-yellowing, archival, removable by conservators). Let dry and spray on another coat, both front and back. The print will be adequately protected and also become more transparent. When the print is complete, trim off any cockled edges.

Sources for gold leaf supplies

www.freestylephoto.biz
www.goldleafcompany.com
www.gildedplanet.com

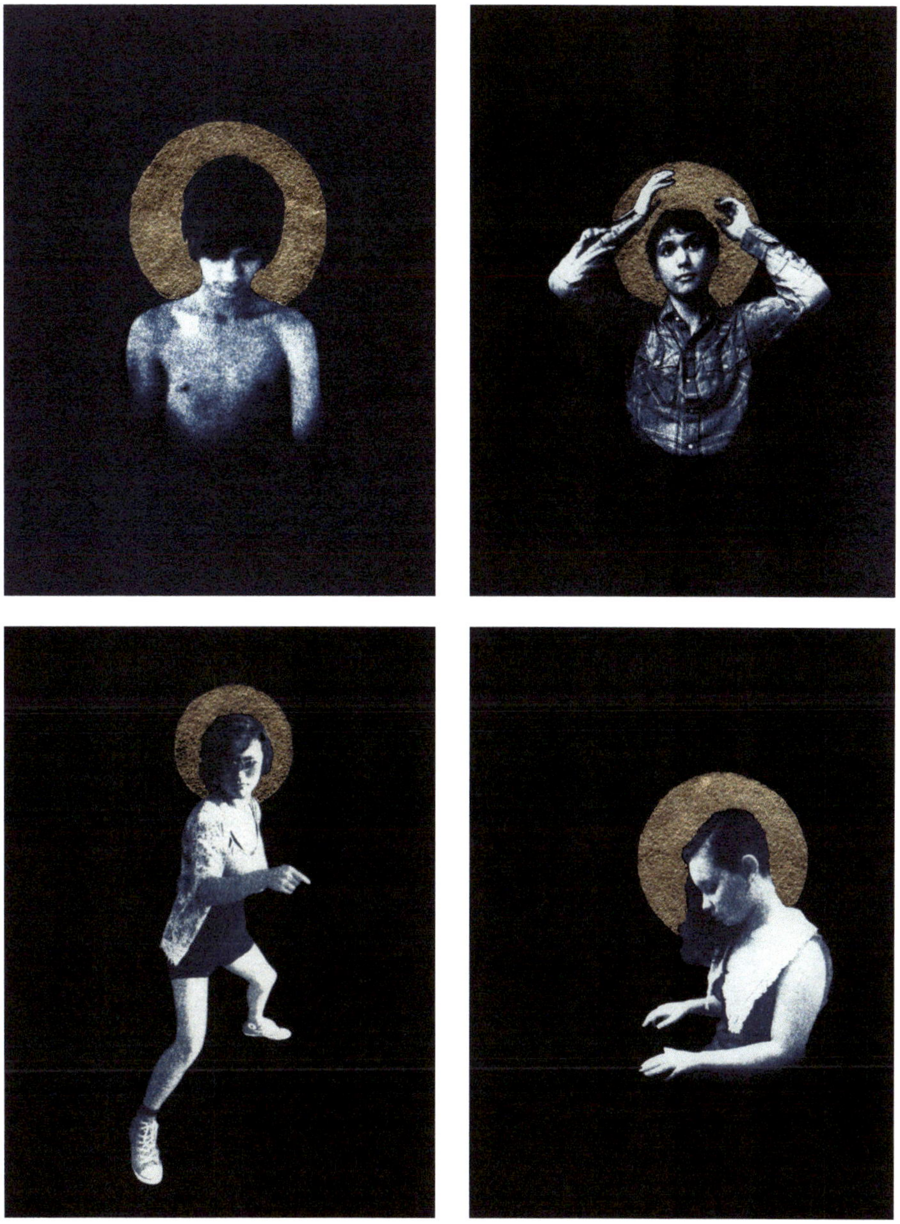

Figures 12.5–12.8. Clockwise from top left: *Wesley in the Living Room, Oliver at the Table, Edith at the Piano, Vivian in the Yard,* cyanotype and gold leaf, 5″ x 7″, from the *Reliquary* series © Alaina Hickman 2018. "Over the past couple years my family has dealt with a great deal of loss. One of the most difficult was that of our family home. I was unprepared for the emotional impact that leaving behind this physical space would have on us. Leaving this sanctuary was like being torn from a womb. This loss was most evident in family photographs. They became painful reminders of what we had left behind. I created this body of work to help me cope, heal, and move on. I removed the backgrounds in the photographs to symbolize our foundation that had been removed. I printed the photographs in cyanotype, allowing maximum exposure so the negative space became a deep, dark blue, symbolizing the heavy void left by the loss of our home. However, I found that by isolating the subject in this deep void, it brought simplicity and peace to the image and forced the focus to be on what remained. I took that further and painted gold leaf halos on the subjects to emphasize their significance, symbolize sanctity, and to be representational of their saviorhood to me personally. This body of work began as a memorial to what I had lost but became a tribute to what I have." Alaina Hickman is a maker of things, mother of four, perpetual student, and a self-proclaimed memory hoarder. She received her Associate's degree in photography from Metropolitan Community College and is the recipient of numerous awards including both a Midwest and National SPE Student Scholarship. Her work has been featured in several exhibits in the Omaha and Lincoln area as well as traveling exhibitions. Hickman is greatly influenced by the shared experiences between her physical and emotional self. Her work explores a combination of themes including motherhood, mortality, and the passage of time.

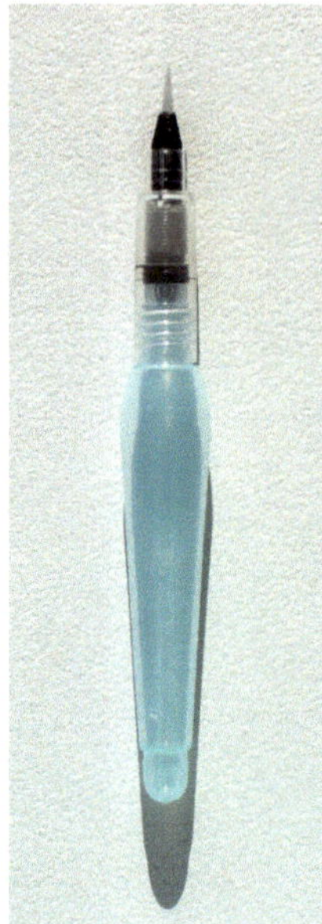

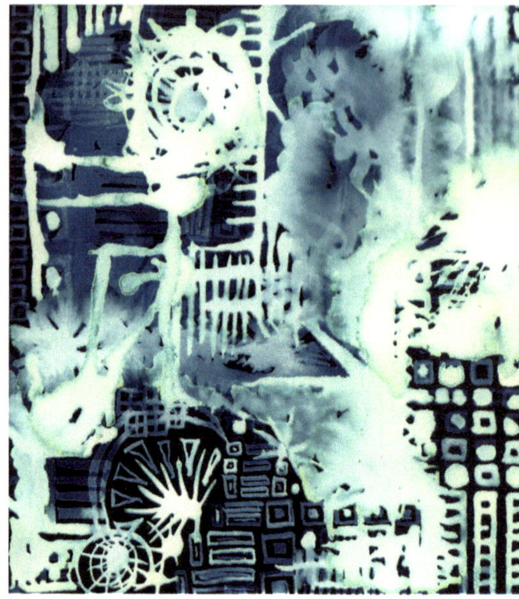

Figures 12.9–12.11. Far right, *City* © Mary Anderson 2018. On the left is a stencil design used with the potassium oxalate writing solution to create a white design on a cyanotype. When using stencils it is best to spray the back of the stencil first with a thin layer of repositionable stencil adhesive spray so that the stencil will adhere to the paper without gaps. Let the spray dry. Then, take a sponge or a sponge brush and lightly saturate it with the potassium oxalate solution. Gently dab the stencil design with the solution. It takes a minute for the lightening to appear. Once it is as white as you want it to be, rinse the design under running water with the stencil still in place to remove all the whitening chemistry. Blot dry, and then remove the stencil. This eliminates the chance of the whitening solution bleeding into the surrounding blue area.

Middle is a Pentel Aquash™ water brush which behaves like a fountain pen. The barrel of the pen is refillable and the tip is soft yet firm. Pressing and squeezing will modulate the liquid flow. Right is an example of a design drawn with the Aquash pen. The pen can be used to do all sorts of designs when filled with the potassium oxalate solution. Coat a piece of cyanotype paper and expose it out in the sun until it is maximum blue, process and dry it. Then draw whatever desired, keeping in mind that the solution does not lighten the cyanotype right away so resist the urge to keep adding more solution so that it puddles. Once the design is complete, rinse it face down quickly in warm water to remove the potassium oxalate solution and prevent further whitening, blot the face of the print, and let it dry.

Writing on a cyanotype print

There were quite a few mentions in the historical literature of writing on a cyanotype print with some form of bleaching agent that would make white writing on a blue background. This was done with potassium oxalate—a chemical any platinum printer has on hand—or sodium carbonate, with a little gum Arabic added to give the solution body so it wouldn't run. The potassium oxalate reduces the blue ferrous ferricyanide to white ferrous ferrocyanide[1] and it is quite white.

You can use a fountain pen or calligraphy brush and dip it in the solution, or better yet, get a fillable pen and in place of ink, fill with the bleaching chemistry. Pentel makes an Aquash™ Water Brush in different nylon tip sizes—fine, medium, bold, and flat—that has a handy, squishy, screw top barrel in which you can put the bleaching chemistry. Squeezing the barrel more or less will affect the flow of the bleach, from fine lines to puddles if desired. You can also stencil designs with a sponge stencil brush lightly saturated with solution (too much and it will bleed under the stencil), and pounce the stencil design into the print with the bleach.

The strength of the potassium oxalate solution quoted in the literature seems somewhat immaterial, from 3 to 17%. If you have potassium oxalate developer for platinum/palladium printing lying around, just take a tablespoon of that and add a few drops of gum Arabic to thicken. Otherwise, mix this:[2]

5 g potassium oxalate or 5 g sodium carbonate (1 teaspoon)
1 g gum arabic powder (½ teaspoon) or ½ teaspoon gum arabic solution (optional)
100 ml water

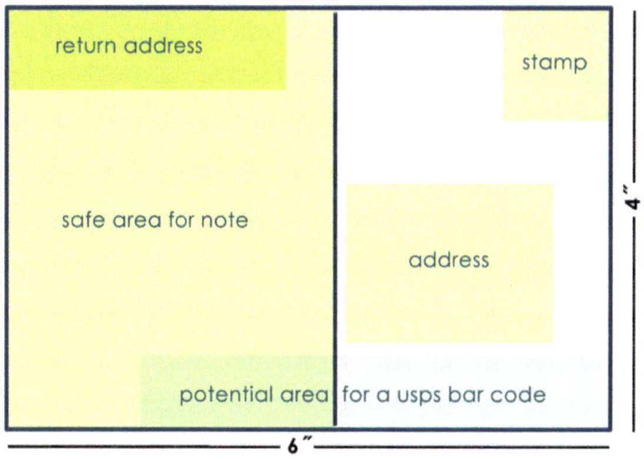

Figure 12.12. When making your own postcards be aware of the typical placement of addresses, writing, bar codes, and stamps on the back.

1. Add the potassium oxalate to the water and stir until dissolved.
2. Add the gum arabic to the solution and stir until dissolved.
3. Store in a 100 ml bottle.
4. At time of use, either dip a brush, fountain pen, or fill a pen with the solution and write on a dry print.
5. When done, dip the print in water quickly *face down* and wash off the solution so it doesn't bleed into the print. Transfer the print to a second tray of clean water, wash the print a few minutes longer and dry.[3]

Postcards

Postcards have a long tradition in cyanotype and they are an easy way to send artwork through the mail. Here are some things to keep in mind: a postcard has to be at least the weight of an index card, so any 300 gsm watercolor paper will work. Standard postcard size is 4″ × 6″, which can be sent with a postcard stamp, but postcards can range from 3.5″ × 5″ to 6.125″ × 11.5″. Follow the template here for where to keep areas free on the writing side of the postcard. If you are not into making your own, both Hahnemühle and Strathmore sell readymade postcards made from watercolor paper. What a great way to keep your family in the loop with your art.

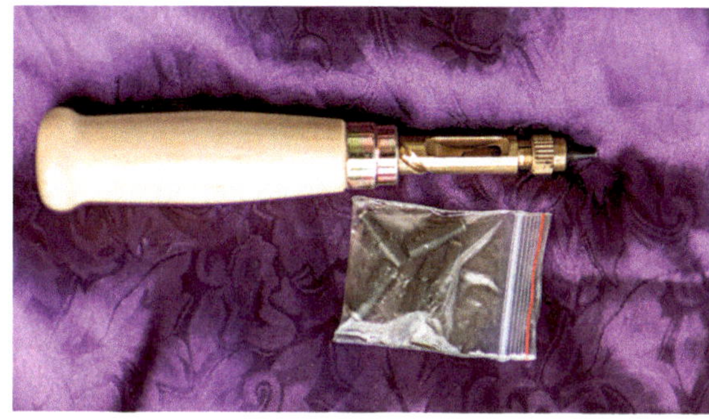

Figure 12.13. An Online River press punch comes with various sizes of hole punches. When the punch is pressed down it rotates the "drill" to make the clean hole.

Punching a cyanotype print

Onlineriver.com makes a very sharp push-punch hollow point drill that comes with assorted sizes of punches. When the punch is pushed down, the drill spins and cuts a round hole. It has a lot of possibilities for creative uses, e.g. for shining light through a print (think Amy Friend/**amyfriend.ca**), placing the print over colored backing paper, or making clean holes for sewing. The drill bit is quite sharp so place the print on top of cardboard or a towel so as not to mar furniture below.

Handcoloring

Handcoloring is perfect for adding color to a cyanotype print. Watercolors, watercolor pencils, oils, pastels, Prismacolor pencils, charcoal, and conté crayons can all be used.

Watercolor pencils

The benefit to using watercolor pencils is precision. The pencil point can be sharpened to get into little teeny areas of color. Then the addition of water softens the stroke marks and blends color effectively. It's probably the easiest way to handcolor, almost like paint-by-number.

Supplies

Faber-Castell, Staedtler, Caran d'Ache, Derwent, Prismacolor watercolor pencils
Brushes of various sizes to wet and blend the pencil; especially recommended is a medium-

size calligraphy brush which ends in a thin point and a very small round, pointed synthetic fiber brush to get into the smallest areas
Q-tips
Cotton balls
Toothpicks

1. Wrap a couple toothpicks with minute pieces of cotton ball and have them handy. These will be used for cleanup.
2. Pencil in the desired area with color. Color softly. Keep pencil tips sharp while coloring. Remember that when the color is moistened it will be more saturated so draw within the lines very carefully and evenly, and avoid extending color into areas you don't want to color.
3. Clean up any unwanted marks with a white eraser before wetting. If necessary, whittle the eraser to a point for the tiny areas.
4. As soon as the print is colored and cleaned up, take the calligraphy brush, wet it, and squeeze it out so it doesn't drip. The belly of the brush should be filled with water and the tip twisted to a fine point. Have other size brushes ready to go.
5. Lightly moisten all areas of color with the wet brush and blend the color carefully, rewetting and cleaning the brush when switching colors.
6. If necessary, take a toothpick wrapped in cotton, either wet or dry, to remove any unwanted color.
7. If the color will not remove, dry the print, pencil over the area of color bleed with either white or the color for that area, and then lightly rewet.
8. When the print is finished, hang to dry.

Photo oils

The benefit to photo oils is that the tubes are small so you can get many colors for a reasonable price. The colors are also more concentrated. The detriment is that photo oils are geared for smooth, hard-surfaced b&w paper on which the oils are easy to spread around and blend and mistakes are easy to clean up with either the Marlene solution (trichloroethylene) or the PMS (turpentine) included in the kit. Because cyanotype is printed on watercolor paper it is not quite so easy to move around the paint and clean up areas, so it is important at the offset to be much more careful when coloring and to first size the print with Gamblin PVA or acrylic medium to make the surface smoother.

Supplies

Gamblin PVA size
Cotton balls, Q-tips, toothpicks
Marlene solution (trichloroethylene), Arista cleaner, or mineral spirits
Extender (if needing to thin the colors)
Wax paper or freezer paper for a paint palette and to protect the work area
Kleenedge or some other low-tack tape, if desired, to protect the edges of the print
White vinyl eraser to clean up small areas
Needle to poke holes into dried pigment tubes
Needle-nose pliers to remove stuck tube caps

1. Size with a layer of Gamblin PVA or acrylic matte medium and dry. Tip: in the 1850s it was common practice to lay the print face down on plate glass and burnish the back of the print with an agate burnisher, rubbing hard, so that the pebbly paper surface on the front of the print flattened and smoothed out.[2] This would make the surface harder and less prone to mottle when color is applied. Something like a large spoon could be used, or buy a burnisher at an art supply store.
2. Cover the work area with freezer paper, waxed side up, to protect it if desired, and protect the print if desired by covering the borders with low-tack tape.
3. Prepare a makeshift paint palette of a double thickness of freezer or wax paper to use as a disposable palette. Make this about 4″ × 6″ at least, large enough to mix colors on.
4. Prepare the paints. Squeeze out pea-sized or smaller amounts of colors onto the palette. It is easy to replenish the palette with more paint, but once the colors are squeezed out, they will dry out and be unusable within a day.
5. Mix colors on the palette; use the extender to weaken them if desired.
6. Start the color application. Be subtle. If the color is too dark, weaken it with a cotton ball or Q-tip.

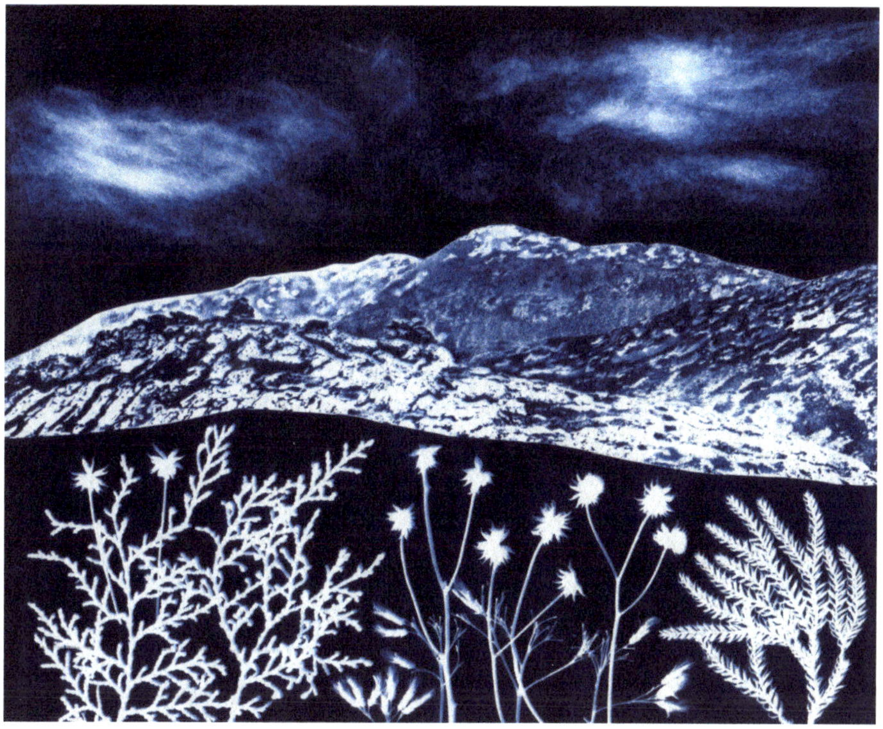

Figure 12.14. *Magic Mountains*, traditional cyanotype cliché verre, 4¼″ x 7½″ © Suzanne Izzo 2018. *Magic Mountains* began with a piece of glass onto which I rolled watercolor paint using a brayer. The result was a cliché verre negative. There was no attempt to produce mountains or anything else, but I liked what I saw on the glass and let my imagination take over. So as not to risk damaging the painted glass, I contact printed it to lith film, making both a positive and a negative. This version is from the positive and so replicates the image as seen on the original painted cliché verre negative."

7. The handcoloring mantra: Go from larger to smaller, top to bottom (if possible), messy to clean. Translated, this means to color the skies and background larger areas first. When the background larger areas are complete, work on the next smaller areas and on down, until the last elements that are handcolored are such things as gold rings, lips, eyes, fingernails, etc. at which point cleanup of paint in unwanted areas begins.
8. Painting is begun with larger tools and finished with smaller tools: large brushes to small brushes, cotton balls to Q-tips to toothpicks wrapped in teeny bits of cotton. To make the latter, wet the end of the toothpick, pick up a smidgen of cotton, and wrap it around the moistened toothpick by spinning the toothpick between your fingers.
9. Color with variety. Avoid using areas of unmodulated color; it looks amateurish.
10. Perfect cleanup is the secret to good handcoloring; investigate the print with a loupe, scrutinizing it closely for any bit of color out of place. Take a toothpick with cotton, dip in the PMS or Marlene solution, and remove the stray color as best as possible. Clean up eye whites, teeth, and highlights.
11. Dry the print in such a way that dust will not collect on the surface and dry into the paint. This can be accomplished by placing the print inside a dust-free plastic bag or box.

The wet cyanotype process

There are quite a few cyanotypists nowadays doing different variations of this process, either exposing while the cyanotype coated paper is still wet or exposing a dry, sensitized paper and wetting it while the image is exposing with water or various chemicals to get a watercolor effect.[4] A lot of this work is photogram work with actual objects laid on the wet paper out in sunlight.

The paper is coated with cyanotype solution as normal. At time of exposure it can be still wet or a spray bottle with water or mild chemistry like vinegar is sprayed on the paper. The paper is exposed to the negative or objects placed on top. Often the print is overexposed for hours or even a couple days, and then developed as normal. The water adds texture and flow to the image, changes the colors, and each print is unique.

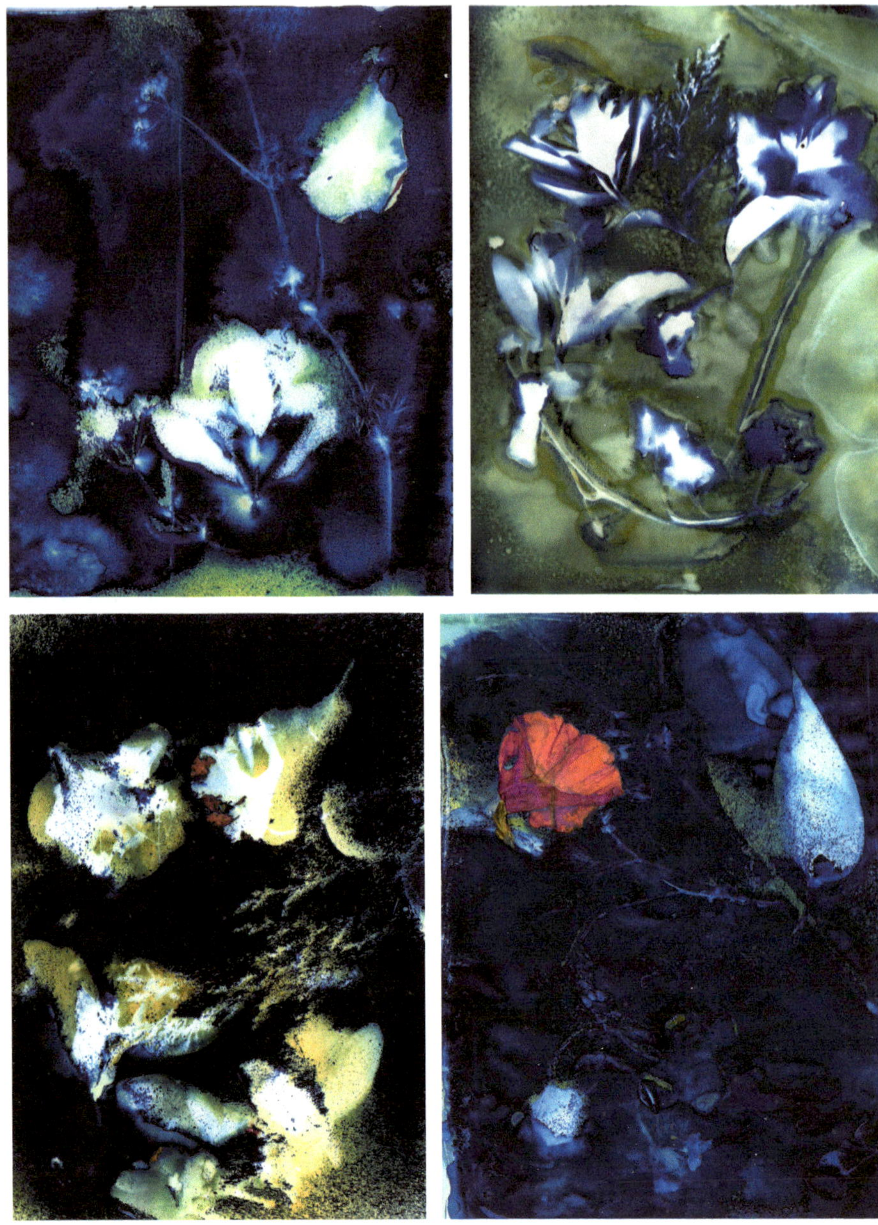

Figures 12.15–12.18. These images, all © Barbara Murray 2018, illustrate the wet cyanotype process where exposure is done while the solution is still wet on the paper and chemistry is also often added while the print is exposing. Clockwise from top left: "For this image I brushed Solution B on the paper and dried it, then exposed with various botanicals for about 30 minutes. Then I applied Solution A on top of and around the edges of the existing items. I sprayed lightly with bleach and exposed for about an hour. As I was moving the items to another location some of the items moved around, leaving ghost images. I then applied more Solution A with a brush and exposed for another hour. For the next image the assembly was sprayed with ammonia and coffee and then after an hour of sun a 50% solution of bleach and water and exposed for another hour. For the bottom right image, the same initial process as the top left was followed, but after the first exposure the surface was painted with Solution A and sprayed with ammonia. After an hour sunlight exposure some items were moved around the paper and more Solution A was painted on. The flower in the corner is the original flower which stuck to the image. For the bottom left image, the initial process followed that of the images top left and bottom right, except the sprays of bleach and ammonia were applied before the first exposure to sun. After an hour in sunlight I painted on Solution A around the edges. After 15 minutes the entire assembly was rearranged and then resprayed with coffee and vinegar, and continued to expose for another two hours."

Barbara Murray (Cardiff by the Sea California) is a multidisciplinary artist working in printmaking, film photography, and alternative photographic processes. Much of her work involves crossovers between these media, exploration, and experimentation. Her work has been shown in numerous group exhibitions and she has just completed a book of film and digital photographs of the places and things that make her hometown unique. You can find more of her work on her website at **heartthread.space** and on Instagram **@cardiffbabs**.

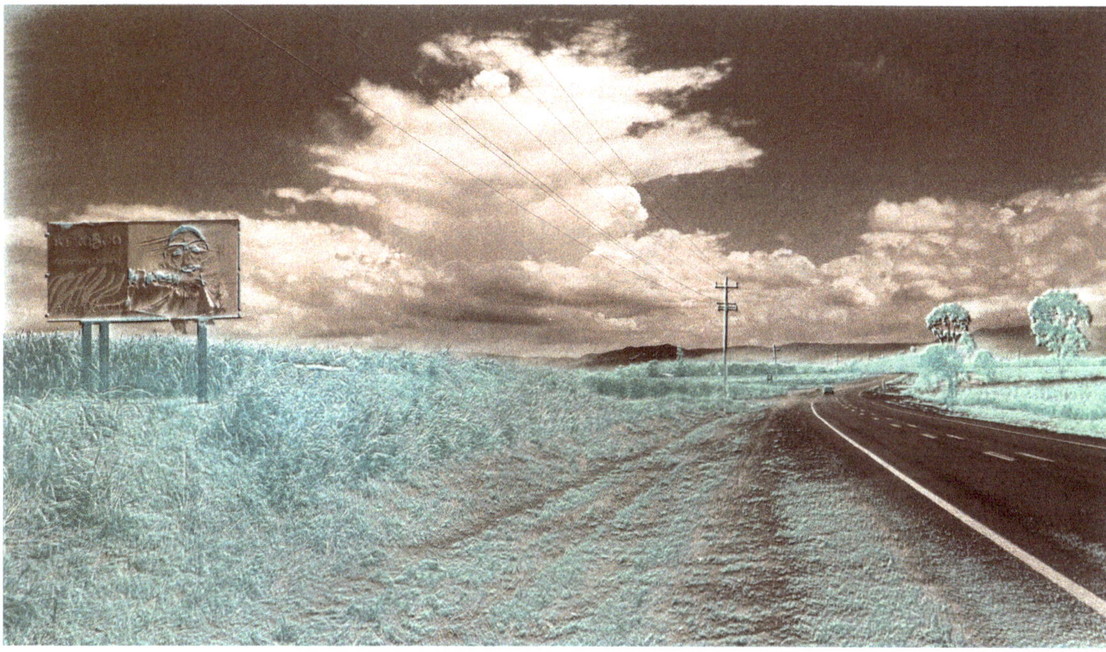

Figure 12.19. *Hot Summer Day*, salted paper over cyanotype on Hahnemühle Platinum Rag © Christina Z. Anderson 2017

Combination printing

Cyanotype is often used in combination with other processes. This book goes in depth about combining cyanotype with palladium, but it would be impossible to cover every combination process in depth without having the book end up at 500 pages! Nevertheless, it is still important to mention how the process is being used today, with a few image examples, to spur further creative exploration.

Cyanotype is perfect for a foundation layer under tricolor gum or tricolor casein. It serves as the blue layer in a tricolor print. If you are already set up to gum print, it is a no-brainer. The cyanotype layer is printed first on shrunk and unsized paper, processed, and dried. Then the print is sized with either gelatin or PVA and the gum or casein layers are printed on top. See *Gum Printing, A Step by Step Manual Highlighting Artists and Their Creative Practice* for a full how-to.

Cyanotype can be combined with salted paper to produce very unusual results. The two processes war with one another. When cyanotype solution is coated on a salted paper print there is an immediate bleaching and mottling. It is next to impossible to print cyanotype over salt unless you are going for very experimental results. That is because cyanotype chemistry contains a silver bleach, potassium ferricyanide. You can print salt over cyanotype, though, and even though they aren't happy bedfellows, the end result is intriguing. The way to make it work is to print a very high contrast cyanotype so that there is only cyanotype remaining in the deepest shadows of the image. Once the cyanotype is finished, it is salted, dried, and then sensitized with silver nitrate. The salt layer should be ½–1 stop overexposed, and then processed as usual, with a final hydrogen peroxide wash to bring back the blue. See *Salted Paper Printing, A Step-by-Step Manual Highlighting Contemporary Artists* for a full how-to.

Cyanotype can be printed with Vandyke brown. The same scenario applies here as with salted paper, because Vandyke brown is a silver nitrate process.

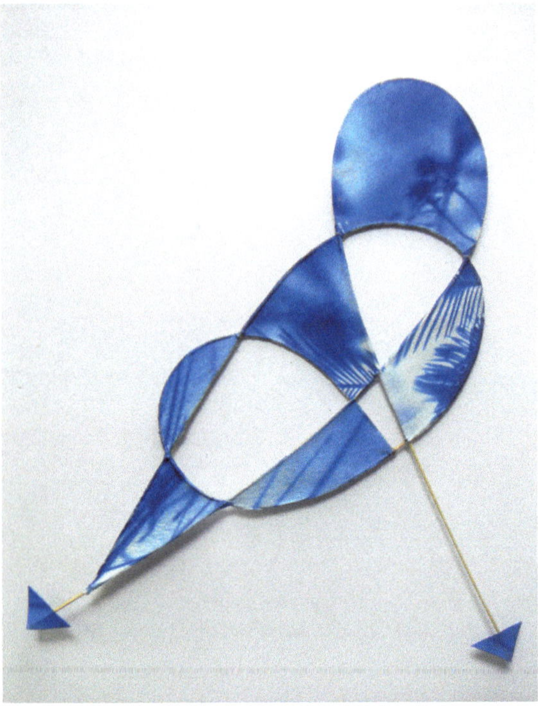 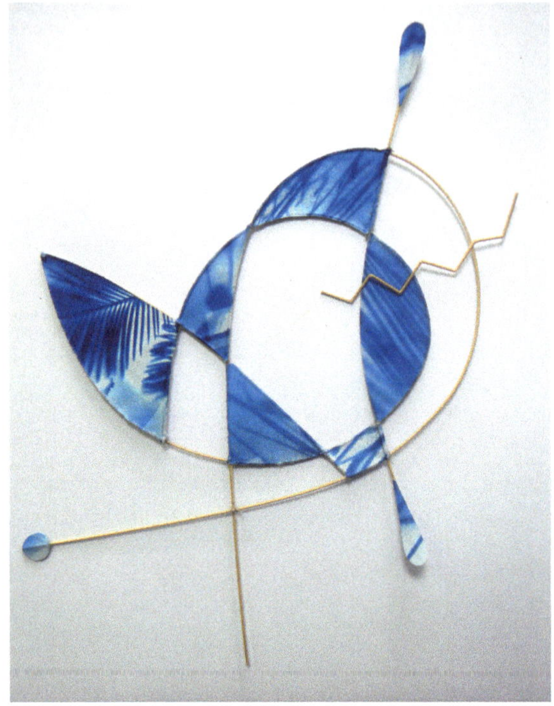

Figures 12.20 and 12.21. Left, *Navigation Without Numbers 7 (homage to Wynn)*; right, *Navigation Without Numbers 8 (homage to Wynn)*, classic cyanotype on Hahnemühle Sumi-e paper intensified with hydrogen peroxide, approximately 11″ x 14″ © Melanie Walker 2018. "I use classic 50% FAC:35% PF mixed 1/1 at time of use. Sometimes I double the amount of FAC to 100 g to increase sensitivity and speed up exposure times. I have also mixed all ingredients together and made a combined solution. I have also obtained reasonable results coating with FAC only and then coating with PF after exposure. Recently I have been using a Ryonet silkscreen exposure unit with an exposure time of 20–25 minutes. Sun is always a first choice when possible with exposures ranging from 7 to 20 minutes depending on time of year. Generally I use a short still water rinse which is placed in a recycle bin and then followed with a 10- to 15- minute running water rinse. Sometimes I tone with a sodium carbonate bleach followed by green tea toner. Matcha gives beautiful, almost orange tones."

More creative ideas

The creative possibilities for cyanotype are endless. Melanie Walker makes cyanotype kites as well as puppets. Paula Riff prints on unusual "food" papers. Suzanne Izzo scrunches paper and makes "skiatypes" as well as prints from clichés verres. Katsuhiko Inagaki makes cyanotype negatives for inside camera exposures, as does John Beaver (see p. 59). Robyn Davis does multiple manipulations before arriving at her final print. Jocelyn Mathewes sews fabric cyanotypes onto paper cyanotypes. Steffani Frideres attaches metal *milagros* to cyanotyped fabric prints. Keith Gerling tones and scrapes and slurries casein and cyanotype together. Cyanotype is alive and well in creative practice and this book is merely the tip of the iceberg of the ways the process is used now and will be in the future.

Endnotes

1. Kosar, Jaromir. *Light Sensitive Systems: Chemistry and Application of Nonsilver Halide Photographic Processes*. New York: John Wiley and Sons, 1965, p. 33.
2. Wall, E. J. "The Iron Salts I" in *American Photography*, Vol. XVI, November 1922, pp. 677–688.
3. Vogel, Dr. E. *Practical Pocket-Book of Photography. A Short Guide to the Practice of All the Usual Photographic Processes for Professionals and Amateurs*. London: Swan Sonnenschein & Co., Ltd, 1896, pp. 192–193; Brown, George E. *Ferric and Heliographic Processes: A Handbook for Photographers, Draughtsmen, and Sun Printers*. New York: Tennant & Ward, 1900 [First edition 1900, reprinted 1905, this edition n.d.], p. 96; Tennant, John. A., ed. "The 'Blue Print' and its Variations" in *The Photo-Miniature*, Vol. 1, No. 10, January 1900, p. 495; "Titles on Blue Prints" in *Popular Photography*, Vol. II. Boston, Massachusetts: F. R. Fraprie, December 1913, p. 104; Bolle, Oscar. "Prints in Turnbull's Blue" in *The Photographic Times*, Vol. XXIX. New York: The Photographic Times Publishing Association, December 1897, pp. 579–580.
4. Blacklow, Laura. *New Dimensions in Photo Processes*, 3rd ed. Boston: Focal Press, 2000, p. 104; Enfield, Jill. ***Jill Enfield's*** *Guide to Photographic Alternative Processes, Popular Historical and Contemporary Techniques*. Burlington, Massachusetts: Focal Press, 2014, p. 70.

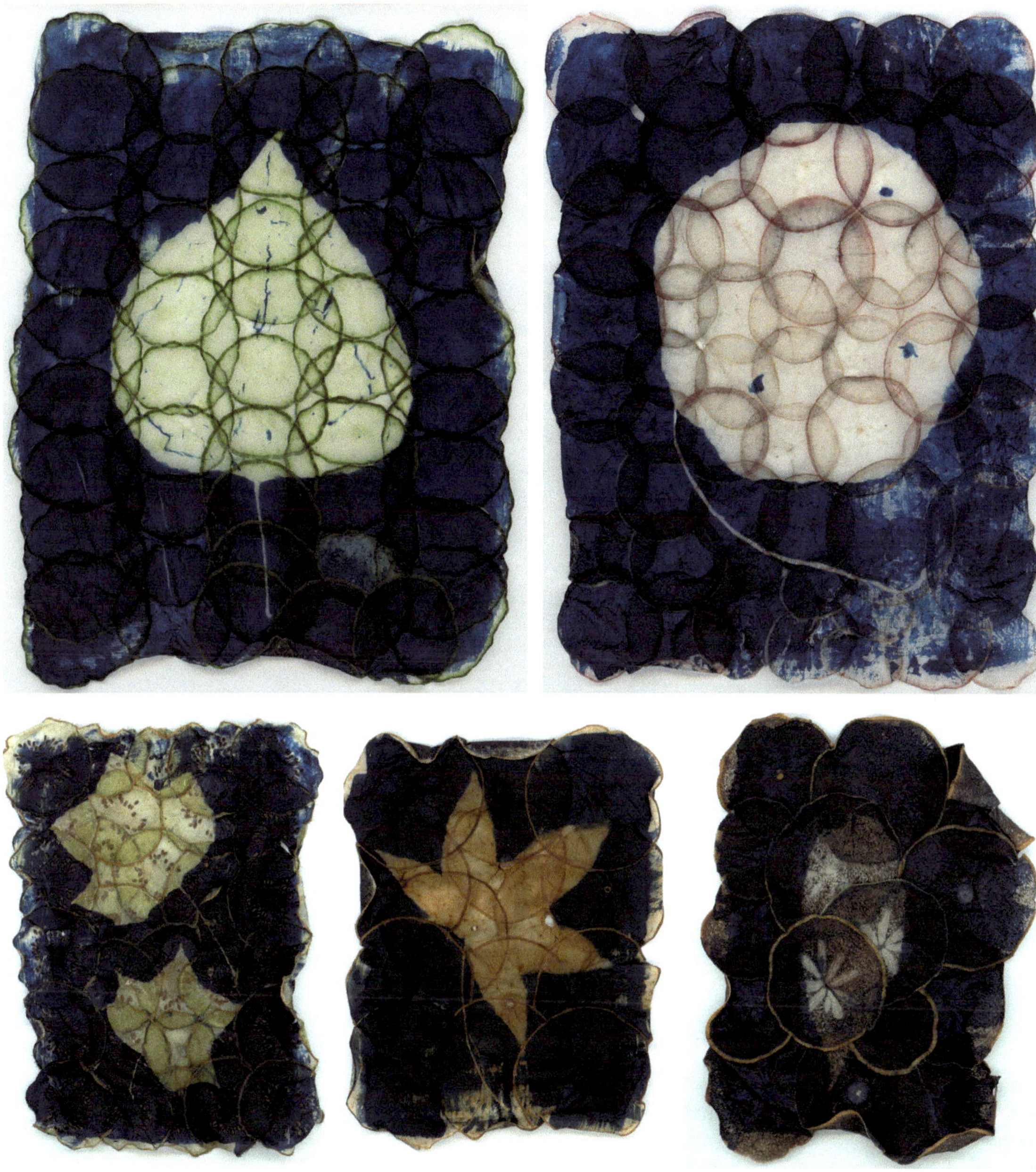

Figures 12.22–12.26. Top to bottom, left to right, *Cucumber, Radishes, Kiwi, Orange Peel, Persimmon*, from the *Earthly Delights* series, approximately 5″ x 7″ © Paula Riff 2017. "I am always contemplating the possibilities of coating different natural surfaces and my recent projects involve using different kinds of botanicals, leaves and grasses, so when I came across these beautiful Japanese vegetable and fruit 'papers' I wanted to experiment and see how these natural edibles would hold up under the cyanotype process. *Earthly Delights* is the beginning of an ongoing project using fruits and vegetables and other potential natural findings as substrates for cyanotype and other processes. The most wonderful surprise that occurs when making these is the sculptural shapes they transform into as they dry. They also become incredibly translucent, like stained glass windows when held up in the light." Paula Riff's first career did not involve taking pictures. After college, she lived in Tokyo, Japan for several years and upon her return became an interpreter for Japanese production companies in Los Angeles. She switched careers while landing an internship at the Los Angeles County Museum of Art in the photo department. She also worked at the California Institute of the Arts, taking photos for their publications. Although Riff owns digital and film cameras her new work finds her camera-less, coating her own papers and making photograms.

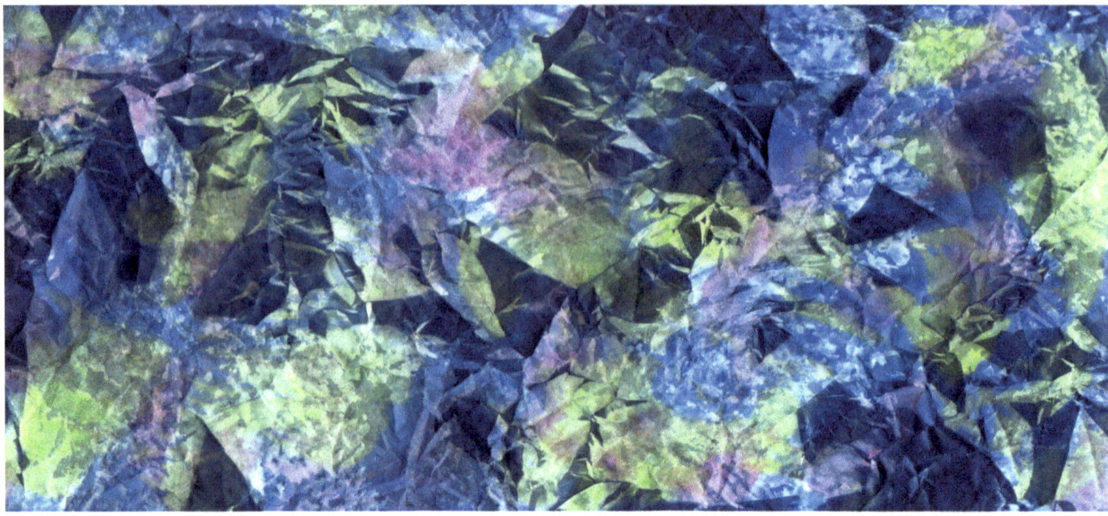

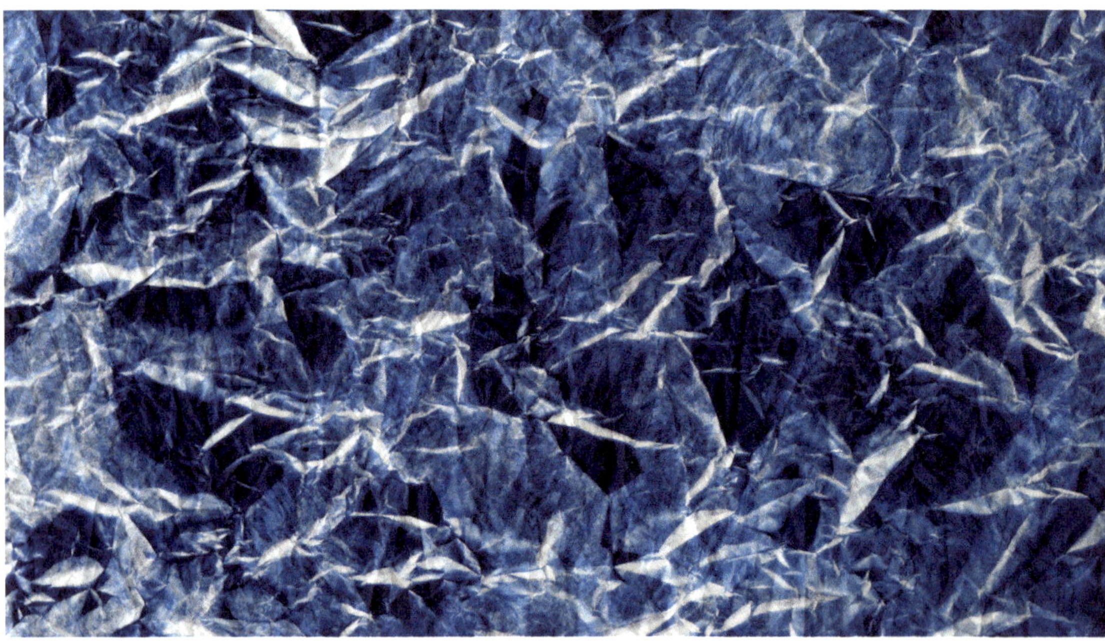

Figures 12.27 and 12.28. Top, *Iridescence #2*, classic cyanotype skiatype with gum bichromate, 12″ x 5″ © Suzanne Izzo 1999; bottom, *Soaring*, classic cyanotype skiatype, 6″ x 11″ © Suzanne Izzo 2018. "Skiatypes are made on paper which, after being coated with emulsion, is crumpled to produce a surface of peaks and valleys. In the sun, the peaks receive the most exposure, slopes less, and deep valleys none. After development, the resulting cyanotype will range from deep blue to the color of the paper support. In *Iridescence* I often add color in certain areas of the print. After this layer has been developed, the print is coated again with cyanotype emulsion and re-exposed either crumpled or covered with a random-pattern negative to break up the colored areas. In *Soaring*, the crumpled cyanotype paper was exposed on both sides so that the darkest blue areas on the one side show through the paper to be seen as light blue areas on the other side. The appeal of this technique is that it produces unexpected but interesting results which allow your imagination (and that of other viewers) free rein. Since it is impossible to crumple two pieces of paper exactly the same, each image is truly unique."

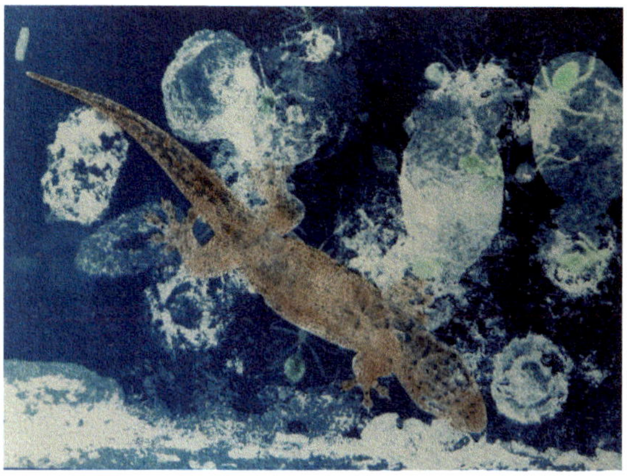

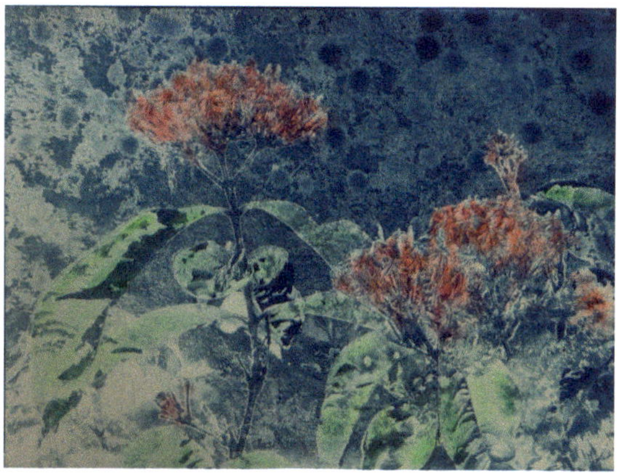

Figures 12.29. –12.30. Left, *Gecko*; right, *In Cape York*, digital artwork printed, crushed, photocopied, methylated alcohol poured over paper, more crushing, laser printed on acetate and coated in traditional cyanotype, exposed in sunlight, and hand colored, 27 x 20 cm © Robyn Davis 2018

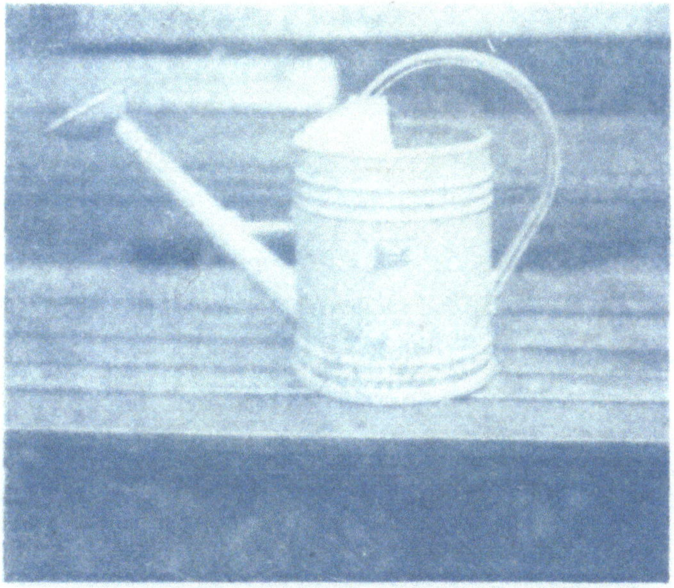

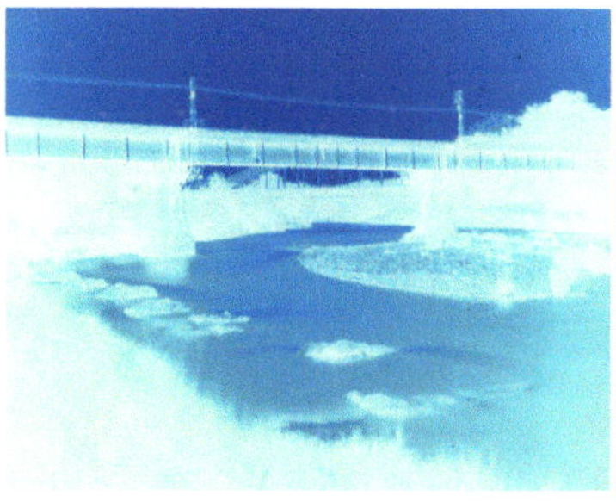

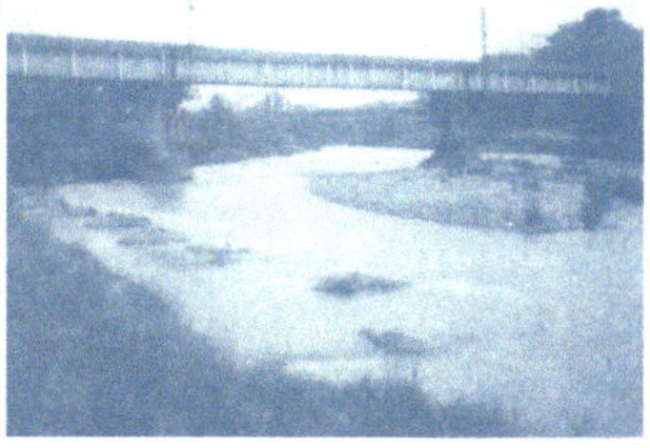

Figures 12.31 and 12.32. Top, *Watering Pot*, cyanotype negative-positive process, 55 x 68 mm; bottom, a negative and the resulting positive for *Railroad Bridge*, cyanotype negative-positive process, 89 x 127 mm © Katsuhiko Inagaki 2018. "To prepare a cyanotype negative, cut transparency film (Pictorico TPS-100) to film size. Classic cyanotype solution is brushed on the film, then dried for an hour. The dried film is set in the camera. I used a PENTAX 6 x 7 with a Super-Multi-Coated Takumar 105 mm/2.4 lens. Typical exposure is 10 hours/f/2.8 for a sunny day. The exposed film is soaked in a 2% citric acid solution for 1–2 minutes, rinsed with tap water for several minutes, and dried. To print the cyanotype from the cyanotype negative, I follow the standard process using the classic cyanotype sensitizer brushed on watercolor paper (Arches 300 gsm). Since the cyanotype negative does not block blue light, I use black light bulbs (UVBLB) from which visible light is filtered. A typical exposure is 30 minutes."

Katsuhiko Inagaki is an experimental physicist with expertise in condensed matter physics. In 2014, he was appointed as Associate Professor of Physics at the Asahikawa Medical University, Hokkaido, Japan. He also devotes himself to developing physics education apparatus. The cyanotype negative-positive process is firstly intended for use in a laboratory class to learn optics and photosensitive phenomena. The work shown here was produced during his improvement of the process. To see his work, visit **https://500px.com/katsuhikoinagaki**.

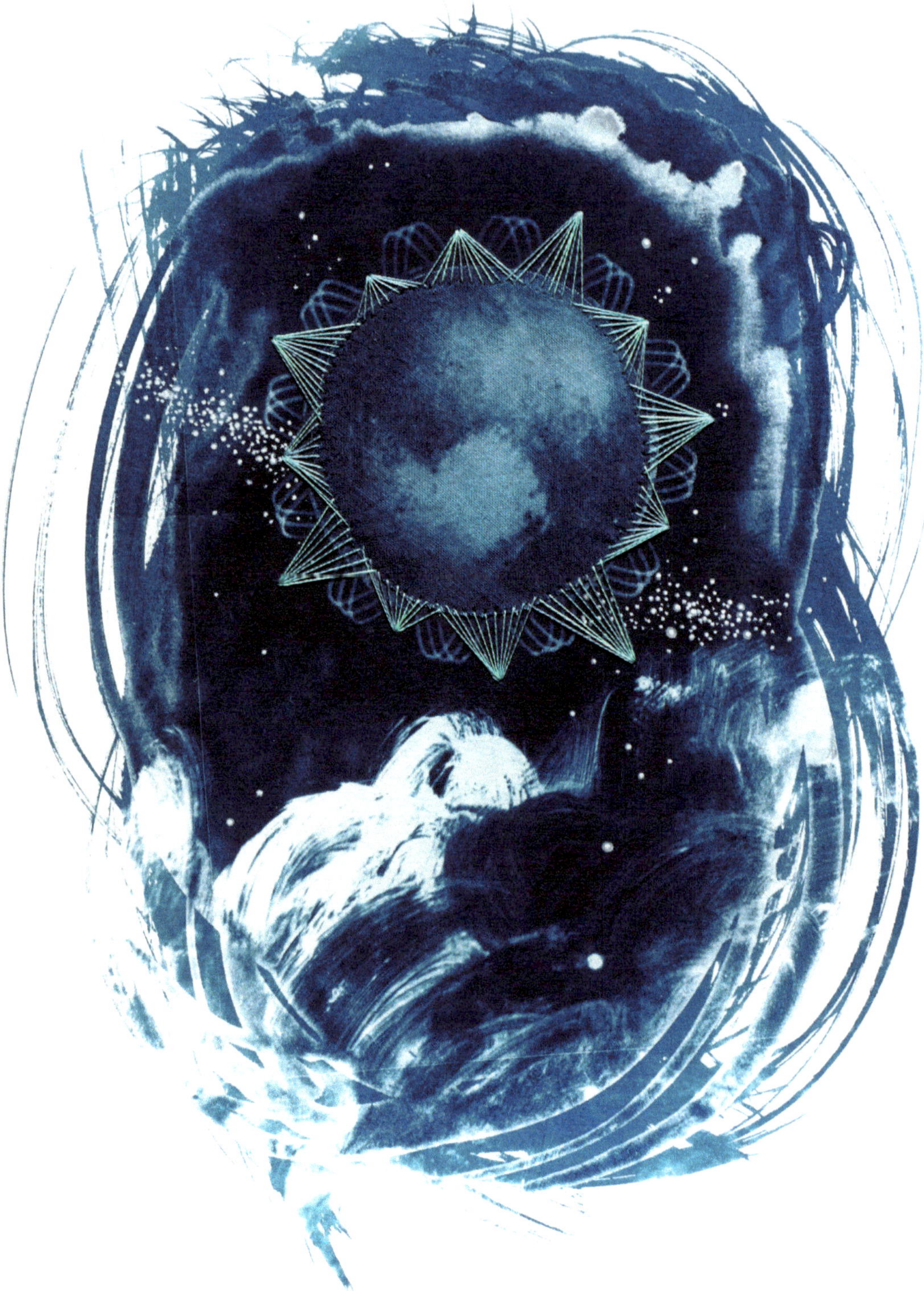

Figure 12.33. *Pluto's inversion*, cyanotype on fabric embroidered onto cyanotype on paper, 9″ x 12″ © Jocelyn Mathewes 2015. To see more of Mathewes' work, visit **www.jocelynmathewes.com**.

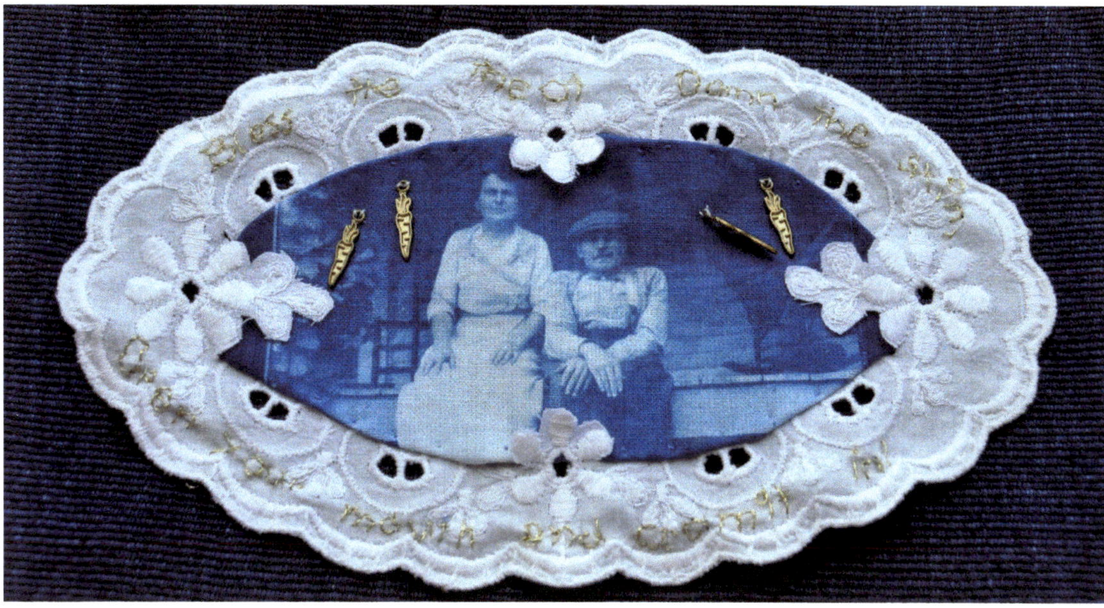

Figure 12.34. *Carrots: Bless the meat, damn the skin, open your mouth and cram it in*, cyanotype with embroidery and charms, 6″ x 4″ © Steffani Frideres 2018. "I combine historical nursery rhymes and or children's songs and cyanotype prints on cloth from old glass plate negatives from the early 1900s. Each photographic image is stitched and embroidered onto found lace doilies, linen dustcovers or other cloth items. Then using embroidery thread, I sew the rewritten nursery rhymes into the cloth. The text is chosen and altered to complement the photographic images, changing or manufacturing a false story that becomes part of the photograph. This narrative allows me to alter and assign a new story to each of the images whose original identity has been lost though time. Each piece also has a variety of charms or metal *milagros* attached which support layered interpretations of the narrative." Steffani Frideres completed her undergraduate BFA at the University of Calgary in Canada, and then her MFA from Washington State University in 1990. Currently, Frideres teaches photography and art at LSC-Tomball, Texas. To see more of her work, visit **Steffanifrideres.com**.

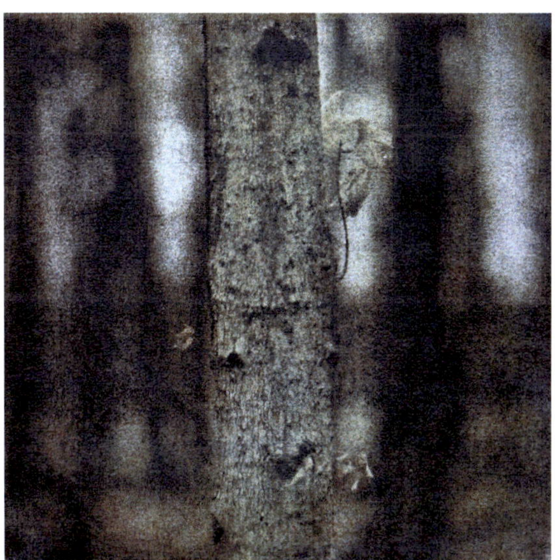

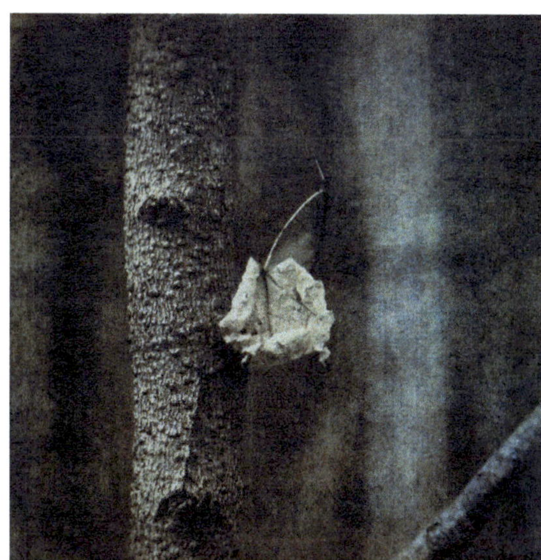

Figures 12.35 and 12.36. Left, *Remnants III*; right, *Remnants IV*, casein over cyanotype, toned/developed/distressed in a tannin-rich slurry of hardwood sawdust, chips and leaves, reprinted in cyanotype, toned, etc., 10″ x 10″ © Keith Gerling 2017. "These photographs were all taken on the same day within feet of each other in an overgrown, long abandoned farm in northern Indiana."

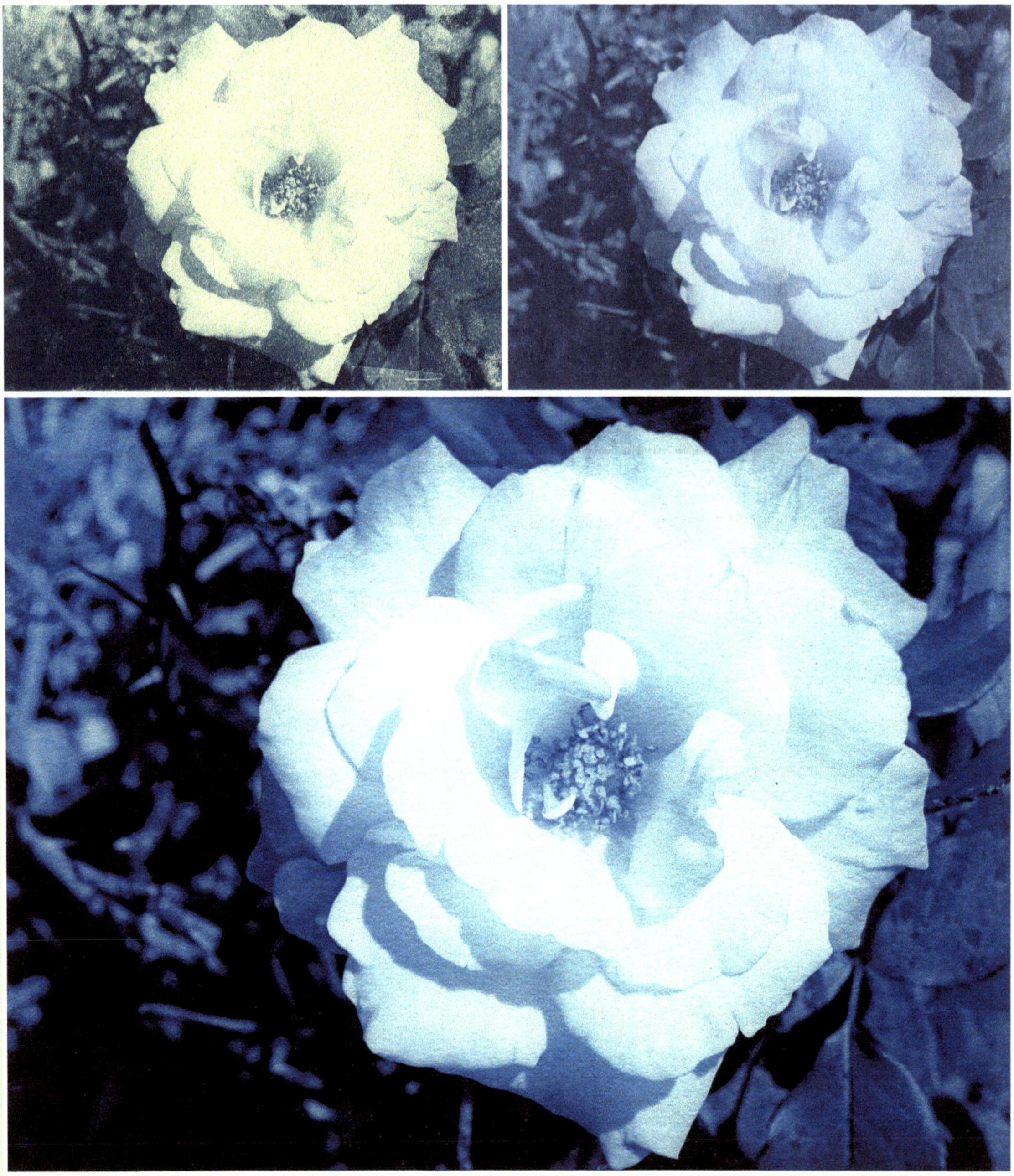

Figure 13.1. The importance of choosing the right paper for cyanotype cannot be overestimated. Here is another image I use to test papers; although it originated from the first digital camera I ever owned, a Sony Cybershot 5 megapixel point and shoot camera, the image has a range of tones from highest highlights to deepest shadows that can test the mettle of any paper. The top two papers are unsuitable papers. They print dull, grainy, washed out prints. They would need to be acidified and probably double-coated to work effectively. What a waste of time to do so when there are so many beautiful papers that don't require such machinations, such as the bottom paper.

Chapter 13
Troubleshooting Cyanotype

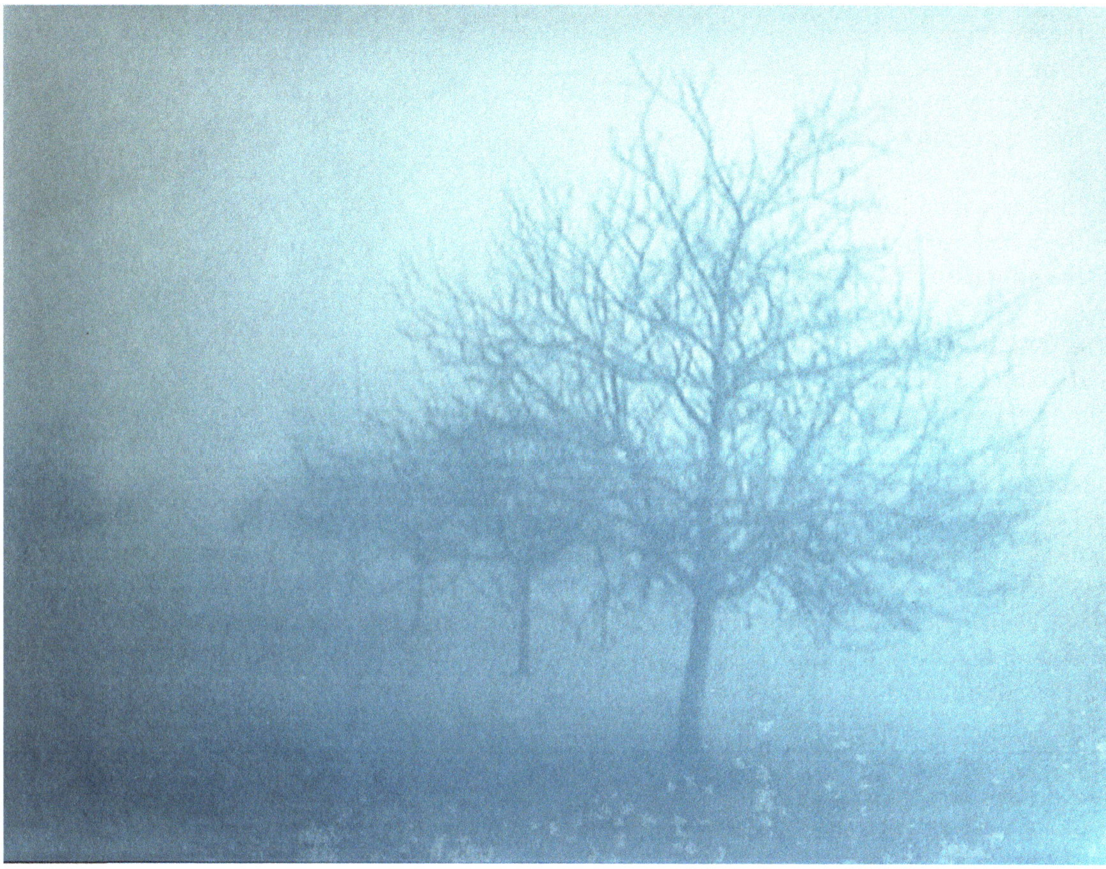

Figure 13.2. This is a beautiful image of the splotching that has plagued some new cyanotype users as of late. Image courtesy of Annette Golaz.

Cyanotype is an easy process, perfect for a beginner, but it is a process that requires some attention to detail to make an expert print. By far the most common problem encountered is a dull, washed-out gray-blue, grainy print that is slow to expose, and by far the most common cause is a poor paper choice and not using acidified water development. A multitude of sins can be covered by a good paper developed in some form of acidified water. There are, however, other issues that can crop up. This chapter addresses the most common issues and solutions.

Potential problems

- The FAC solution has turned blue
- The FAC solution is moldy
- FAC and PF when mixed turns blue
- Paper turns blue while drying
- The exposure is extremely slow
- The print is underexposed
- The print is overexposed
- The cyanotype print washes off
- The shadows bleed into the highlights
- The print is too low contrast
- The print is too high contrast

- The print is grainy
- The highlights are fogged blue
- There are lavender stains in the print
- The highlights are yellowed
- The print is mottled
- The print is dull
- The prints are inconsistent
- There are blue spots or streaks
- The print is splotchy or crystallized
- There are round white spots in the print
- The shadows are solarized
- The image is unsharp
- There are blue stains
- The print has faded

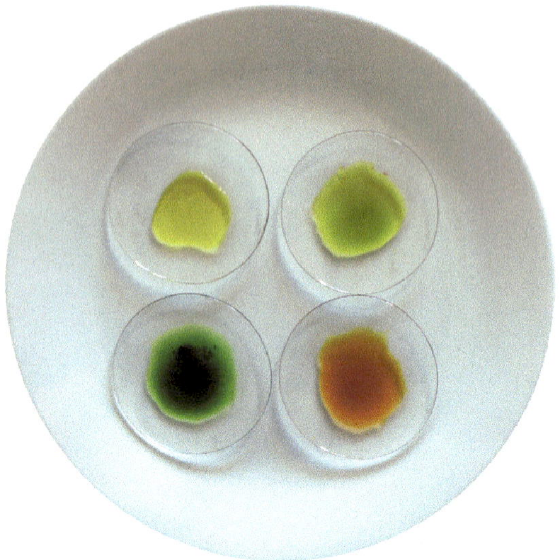

Figure 13.3. Clockwise from top left: normal cyanotype solution, then solution purposely contaminated with citric acid, sodium carbonate, and tartaric acid. Keep your coating area clean of all flecks of chemistry.

The FAC solution has turned blue

Ferric ammonium citrate is a chartreuse green (the green form) and stays that color both in powder form and in solution if stored in light safe conditions. FAC lasts a long time. It is hygroscopic, and in humid climates the powder can absorb moisture and become a sticky lump. It still works fine. However, if an FAC solution turns blue in the bottle, contamination of some sort is the culprit, whether by light or by chemistry. Light is easy to control; the solution should always be stored in a brown glass bottle in the dimroom. Chemical contamination is a harder issue to sleuth.

- If the FAC is contaminated by acid it can create iron citrate, which is blue.[1] When working with citric acid keep it away from your stock solution. I had this happen once after I had weighed citric acid crystals on my coating table. A few of the crystals were inadvertently left on the table and the next print I coated had little crystal-like spots of blue show up immediately upon coating.
- If anything other than distilled water has been used to mix the FAC solution, it can become contaminated because there are lots of minerals, including iron, in tap water. Always use distilled water to make the solution.
- The storage bottle can be contaminated, which can happen if it has been used to store other chemistry, or if the cap has a paper seal that has absorbed other chemistry. Wash the bottle with distilled water before filling with fresh chemistry. Remove any paper seal or, better, use a cap that has a plastic seal. Soak the eyedropper of an eyedropper bottle in distilled water occasionally to remove any potential contamination.

The FAC solution is moldy

FAC in solution will grow mold. The mold looks like black or brown globs of muck which streak on the paper surface when coating. FAC solution can also become sour and smelly. This is only true of FAC; PF in solution does not mold or go bad.

- Run the solution through a funnel lined with filter paper, something even as simple as a coffee filter, return the solution to the bottle, and keep using it.
- Store the FAC solution in the fridge and measure out smaller 100 ml bottles at a time.
- Add 1–2 drops formalin per 100 ml of solution if you normally have this in your dimroom.
- Add several drops of 100% thymol (10 g thymol in 10 ml methyl or rubbing alcohol) per 100 ml of the solution. This is what I use since I have it in my darkroom at all times for preserving gum Arabic. It keeps indefinitely. It is very minty-pungent. Caveat: do not get thymol solution

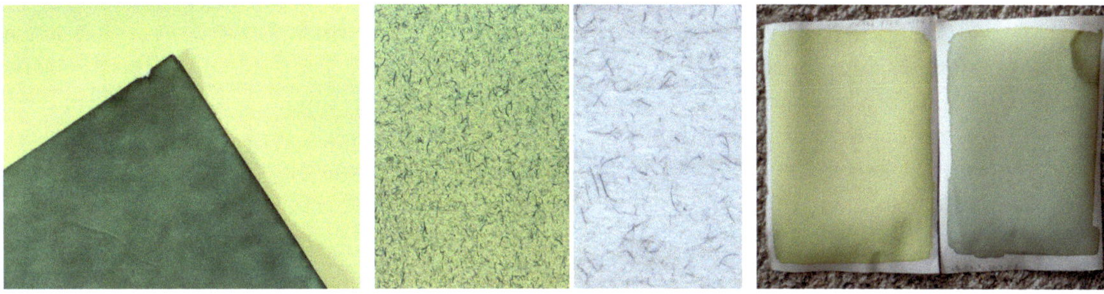

Figures 13.4–13.6. Left, a comparison between a good paper (background) and an unsuitable paper (foreground) for new cyanotype. This particular paper turned blue almost immediately and continued to darken within the hour of drying time. The resulting print was fogged blue all over, interesting because this paper is *not* buffered and works great for classic cyanotype. It's not always the buffering that causes fogging. The middle image pair is Canson Bristol Recycled coated with new cyanotype. The solution fogged immediately only on certain paper fibers and the resulting print had speckled highlights, thus an unusable paper for new cyanotype though it works fine for classic. On the right is an illustration of how alkalinity affects paper suitability; the same paper was soaked in citric acid (left) and sodium carbonate (right) before coating with cyanotype solution and it is apparent which sheet turned blue right away.

anywhere near plastics including contact lenses and food processors because it will dissolve certain plastics like that of which a food processor bowl is made.

- Add Listerine or rubbing alcohol, 5 ml per 100 ml solution.

FAC and PF when mixed turns blue

I have not seen this in my practice perhaps because I use glass bottles, distilled water, and fresh chemistry. Others have, and you can make it happen just by adding a pinch of tartaric acid to a ml of combined solution. Instant blue.

- If anything other than distilled water has been used to mix the solution, it can become contaminated. Always use distilled water to make the solution.
- The storage bottle is contaminated, which can happen if it has been used to store other chemistry, or if the cap has a paper seal that has absorbed other chemistry. Use brown storage bottles and wash them with distilled water before filling with fresh chemistry. Remove the paper seal or, better, use a cap that has a plastic seal.
- Iron contamination has happened somewhere, either from the bottle, mixing implements, or the environment. However, it is a myth that using stainless steel mixing implements is problematic.
- The FAC or PF is somehow contaminated. If this is the case the blue will happen instantly upon mixing. With poor storage, FAC may start to become ferrous, PF may start to turn to potassium ferrocyanide, and either change will result in formation of Prussian blue on mixing, without UV exposure. Buy fresh from a reputable chemistry store and see if the problem disappears.

Paper turns blue while drying

If the solution is bright green when brushing on the paper and then turns blue instantly or slowly so that upon exposure it is quite blue, it is most always a paper issue. Any hint of blue while drying indicates chemical fog.[2] One paper I tested, Canson Bristol Recycled, did the most amazing thing when coated. Only certain fibers in the paper turned dark blue. The paper then printed with fibrous blue specks. Though this paper worked great with classic cyanotype, new cyanotype highlighted the problem immediately.

- The paper is unsuitable, because it contains alkalis, buffers, alum, or some other ingredient, so choose a different paper next time. Go ahead and expose and process the print as usual if only a little blue; it will still result in an acceptable print. If the coating is *really* blue, it will result in a fogged print, especially with new cyanotype. The paper will need to be pre-acidified before use (see the **Papers for Cyanotype** chapter on how to do this) or with new cyanotype, citric acid drops can be added to the solution to alleviate this with a number of papers (see the **New Cyanotype** chapter).

The exposure is extremely slow

Classic cyanotype is slow to expose, one of the slowest alt processes of all. New cyanotype is easily two stops faster. A normal CC exposure can be around 20–30 minutes UVBL, NC around 5 minutes. If CC takes far longer, the culprit is most often paper choice.

- Buffered papers print the slowest. Switch papers.
- Pre-humidify the paper before coating; this results in a faster paper by at least ⅓ stop.
- Acidified water development contributes to faster speed. Use vinegar or citric acid for classic cyanotype and citric or sulfamic acid for new or combination (SWC) cyanotype to maximize exposure.
- Use a less concentrated FAC. A 10/10 CC formula often is faster than a 20/10 CC formula, and a diluted NC formula (mixed to 150 or 200 ml) is often faster than a 100 ml solution.
- The negative is too dense. Use a lower ink density, up to -20% for an Epson 3880 or -30% for an Epson P800.
- Pictorico Ultra was used instead of Pictorico Premium; the Ultra is denser.
- If printing with a film negative, some films have UV-blocking layers. Film negatives developed in pyro developer are also slow.
- The contact frame glass was accidentally replaced with UV-resistant glass. This happened at our university and it was weeks before we figured it out. Make sure all contact frame glass is normal, cheap, non-UV resistant.
- The potassium ferricyanide is old. I have read conflicting opinions on whether potassium ferricyanide gets old, some saying it turns into potassium ferrocyanide, which is yellow and not ruby red. Some red and orange is OK but no yellow. Buy fresh potassium ferricyanide.
- The light source is problematic. Test this by exposing a print in direct sun to compare. It probably goes without saying to check that all your UVBL bulbs are operating.

The print is underexposed

All of the reasons discussed under **The Exposure is Extremely Slow** applies here. In the olden days, when sun was weak, and photographers were printing outside during winter, etc., a print that was underexposed was intensified by immersion in ferric chloride, silver nitrate, or iron sulfate (3.5 parts to 1000 parts water).[3] Nowadays with UVBL boxes we have it made!

- See the low tech and high tech ways of finding exposure times in the **Digital Negatives for Cyanotype** chapter and be sure to calculate a standard printing time for each and every paper.
- Always err on the side of overexposure. If the print is slightly overexposed it can be washed slightly longer to lighten it, but it can never be washed less to darken it.
- Double print the paper by registering the negative on top of the pale print and reprinting.
- Use the back of the paper to make another print.
- Print another process on top.
- Wash the paper a really long time to remove most of the image, or bleach it in sodium carbonate and wash it thoroughly and reuse the paper.
- If exposing by the sun, sunlight varies remarkably and therefore so will exposure time. It is not the temperature outside that is the problem—as Dr. Ware told me once, optical photons don't feel the cold, and I proved him right by successfully exposing prints outside in -20°F weather—it is the *strength* of the sun, and winter sun between 10AM and 2PM is a lot weaker than summer sun at the same time. *With sun exposure, always do an exposure test strip before exposing your print.*

The print is overexposed

Sometimes this happens, especially with misjudging sun exposure. It can be rectified somewhat with a longer water wash (sometimes all day!) in warm to hot water,[4] but you can also resort to chemical measures by bleaching the print with a weak alkaline bath and then restoring the blue in an acid bath afterwards; the original density will not return in full. It only takes perhaps ½–1 ml of ammonia to a liter, or perhaps ¼ teaspoon of sodium carbonate to a liter, after which use a couple teaspoons of citric acid to a liter to restore the color.[5] It's a stop gap measure, not best practice.

The cyanotype print washes off

The obvious answer to this problem is underexposure (see above) but if that is not the case, look to three causes.

- The paper was coated but the coating dried too fast before the solution had time to absorb in the paper fibers—a problem in the winter, in dry climates, or using a blow dryer too soon to dry the coating. Pre-humidify the paper, or prewet it with a couple teaspoons of water to "soften" the surface of a hard-sized paper before coating, and don't use a blow dryer.
- The paper is unsuitable, perhaps too well surfaced sized. Use the techniques above.
- Too much FAC is used in the mix (e.g. a 20/10 mix as opposed to a 10/10). As was said by Channing Whitaker in 1883, "The same blue results on the same paper with different proportions of the chemicals, with excess PF slower but finer, excess of FAC is faster, but too much and whites become stained with iron and paper fogs and is redder blue which does not adhere to the paper but can be washed off with a sponge."[6] I used see wash off on Hahnemühle Platinum Rag paper in the winter months with the 20/10 cc. When I switched to 10/10 the problem disappeared instantly.
- Use 1–2 drops of a 5% dilution of Tween 20™ per ml. Tween is a surfactant which helps the chemistry absorb into the paper better.

The shadows bleed into the highlights

The number one cause of bleeding is using too much Solution A/ferric ammonium citrate in relation to Solution B/potassium ferricyanide.[7] See the discussion about this issue in the **Classic Cyanotype** chapter. As long as Solution A:Solution B is no greater than 2:1, and better yet, 1.5:1 or 1:1, this problem will be minimized or non-existent with even a hard-sized paper.

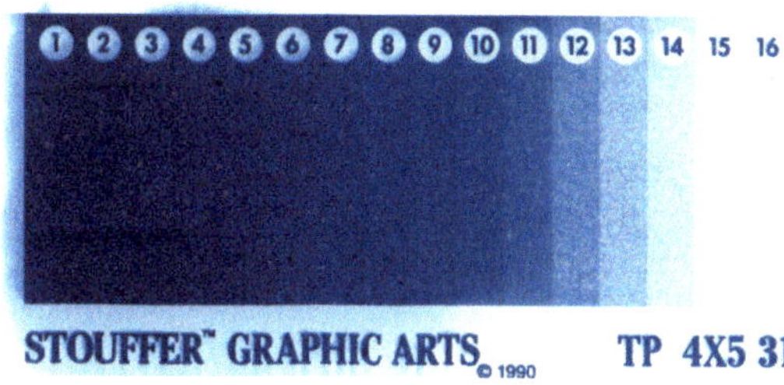

Figure 13.7. Bleeding, caused by using too great a proportion of FAC to PF in classic cyanotype.

The print is too low contrast

Low contrast can happen because of chemical fog due to an unsuitable paper, which will result in dull blue highlights, especially an issue with using NC on alkaline papers. This can be seen immediately before exposure in the hour the paper is drying: it goes from a chartreuse green to blue within the hour. Other causes are:

- A too-low ink density was chosen for the negative; use no less than a -20% for an Epson 3880 and a -30% for an Epson P800; or, use a 0% instead.
- The chemistry is contaminated. Make sure the chemistry is made with distilled water, and it is fresh and fairly clear.
- The print is overexposed and dull. Wash longer.
- The print is not processed in an acidic bath, and it will not have nice, deep darks. Use acids in the water wash.
- The coated paper has been kept too long before exposure which will result in lower contrast and less deep Dmax. Coat only as much paper as you will use in a day or so and store any unused paper in lightsafe bags and even in the fridge. This becomes more crucial in humid climates.

The print is too high contrast

Contrast is usually handled at the negative stage. Using a correct negative curve and a lowering of the ink density takes care of high contrast issues.

- Prehumidify the paper. This will allow the solution to sink into the paper better.
- Use a 15/10 or a 10/10 cc solution instead of 20/10 and you will notice an increased exposure scale (ES) and the highlights will print beautifully.

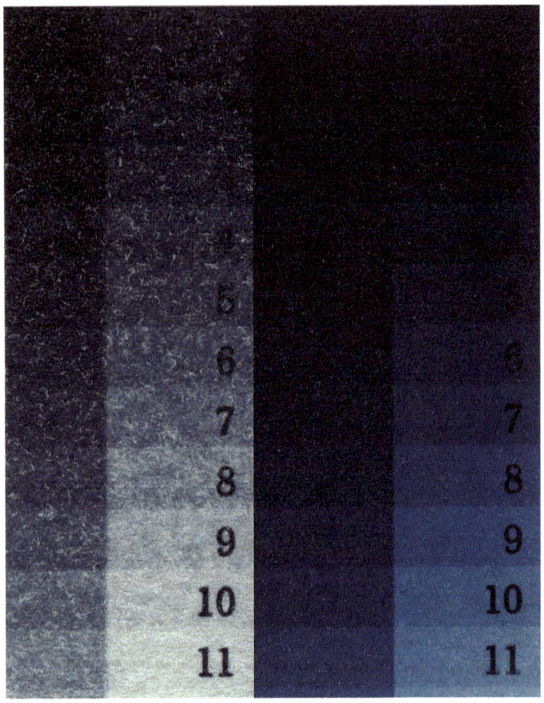

Figures 13.8 and 13.9. On the left is a side-by-side example of two step wedges, a grainy, unsuitable paper choice and a smooth suitable paper choice; grain in this case is paper-related. On the right is a cut-and-paste of unhumidified paper and prehumidified paper, all else the same. Moral to the story: choose a proper paper and humidify it.

- Look to paper choice. Papers that absorb the chemistry will print a much longer, more beautiful exposure scale. Papers that are unbuffered do the same. Thinner papers absorb chemistry better as a general rule than watercolor papers which are fairly well surface sized.
- Use a stronger acid water bath. For CC use more vinegar or citric acid. For NC use sulfamic acid instead of citric acid.
- Contrast can be somewhat reduced by delayed exposure and delayed development, but watch that the Dmax doesn't lessen to produce a less deep blue.
- Other suggested methods are to dilute the solution with water, pre-coat the paper with a 1% acid, or print in direct sun.[8]

The print is grainy

Grain can look like an all-over, unappealing graininess in the entire image, highlights to shadows, or merely minute white speckles in the deepest shadows of the print, microscopic and pin-pricky, with a certain allure.

- Choose an appropriate paper. Some mixed fiber papers pick up the solution differentially. Certain papers (eliminated from the paper chapter) print unusually grainy no matter what.
- *Prehumidify the paper and it will print very smooth tones.* A mere tray of water with a screen on top and another tray to form a lid makes a great humidity chamber!
- Use a more absorbent paper.
- It goes without saying to use an unbuffered paper.
- If the paper is soft, coat carefully, with *more* solution, brushing gently so as not to abrade the paper.
- Switch to SWC or new cyanotype. Classic cyanotype is grainier in general. Ferric ammonium citrate molecules are larger and they don't absorb into the paper as well.
- Use a 15/10 or a 10/10 classic cyanotype formula instead of a 20/10.
- Tween 20™ helps the solution absorb better; use 1-2 drops of a 5% dilution of Tween 20™ per ml.

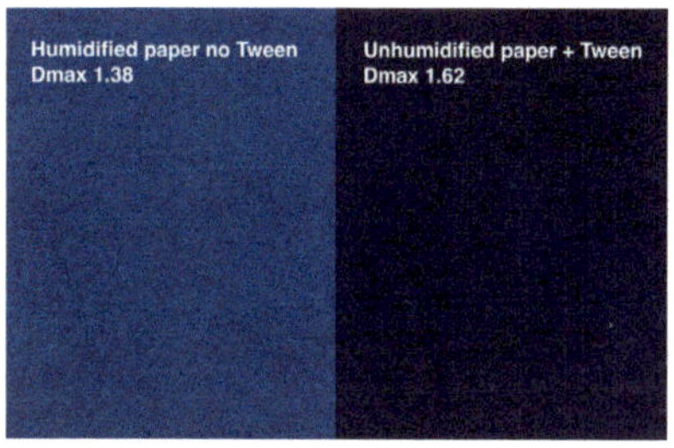

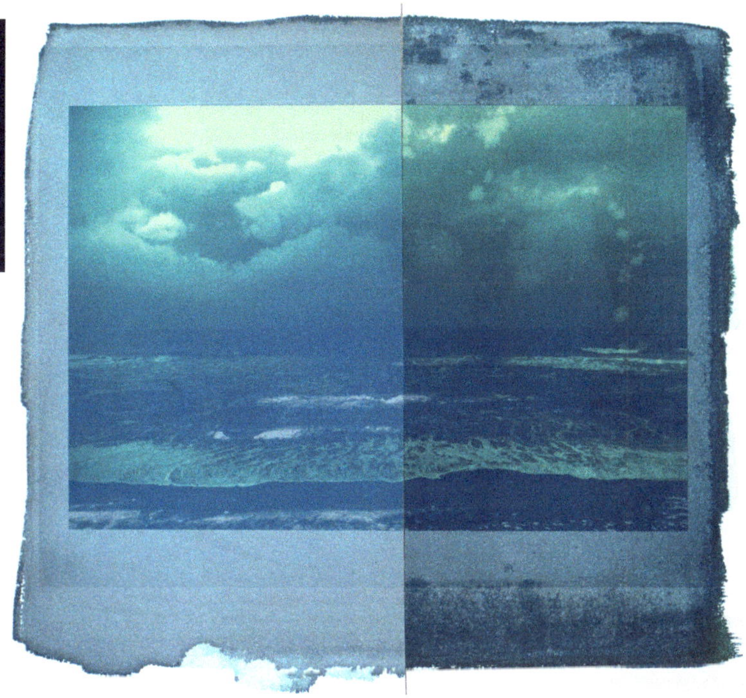

Figures 13.10 and 13.11. The image above shows how Tween 20™, which helps paper absorb chemistry even if not humidified, can sometimes result in deeper Dmax; thus the use of Tween 20™ is a win/win. On the right is a cut-and-paste of an exposed print on prehumidified paper (left) and the same on unhumidified paper, right, all else equal. This shows the benefit of prehumidification for the smoothest tones possible, a practice even more necessary with inabsorbent papers and low humidity dim rooms.

The highlights are fogged blue

This problem is readily apparent in a print that lacks brilliance and paper white.

- Make sure the negative is dense enough with a correct ink density and that it is not -20% when -10% or 0% would have been better.
- Unsuitable paper. If a paper is unsuitable the coating solution will almost always turn blue before exposure, and usually within an hour. That blueness will translate to blue fog. Use a suitable paper, to acidify an unsuitable paper (see **Papers for Cyanotype**) or to add citric acid to the coating solution (see **New Cyanotype**).
- It has been said blue fog is from an excess of potassium ferricyanide in the sensitizer[9] but I don't think we see this anymore because we don't use an excess. In any case, the 10/10 formula is fine because I never see it.
- Too much light exposure after coating and before exposure. Cyanotype has a lot of tolerance for room light, but to test this, coat a piece of paper and leave it out in room light for an extended period of time (an hour?) and then water wash it without exposure to see if there is any hint of something other than paper white.
- The sensitizer has gone bad. Usually if this is the cause it will be dark blue and noticeable while coating.
- If highlights turn blue after development, while drying, the print was not washed long enough, or the wash water was too cold; use warm water and longer wash times for difficult clearing papers and wash past the point of yellow in the highlights disappearing.
- Coated paper was kept too long before exposure, or too long after exposure before development. Usually the former is worse than the latter.
- Too strong an acid development. This is an easy thing to test: try using two full teaspoons of citric acid in the first water bath instead of ½ teaspoon and a CC print may fog blue in the highlights. An NC print will not, as it thrives in stronger acids.
- The literature says an excess of FAC has the potential to fog[10] and a print can be developed in 0.2% potassium ferricyanide and then well washed to raise contrast.[11] I have not tried this.

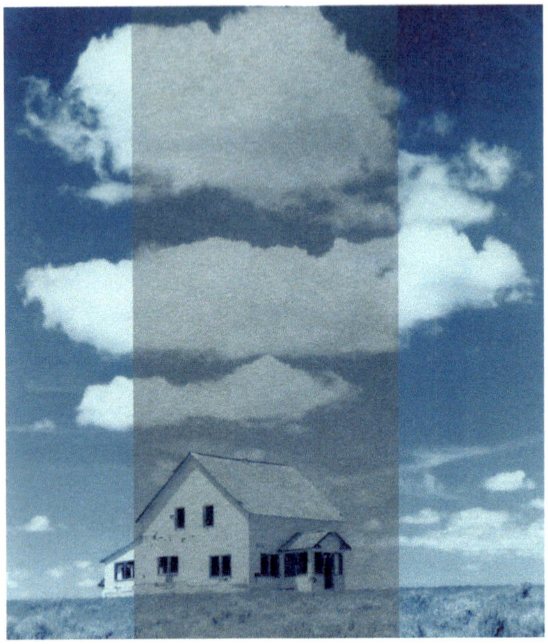

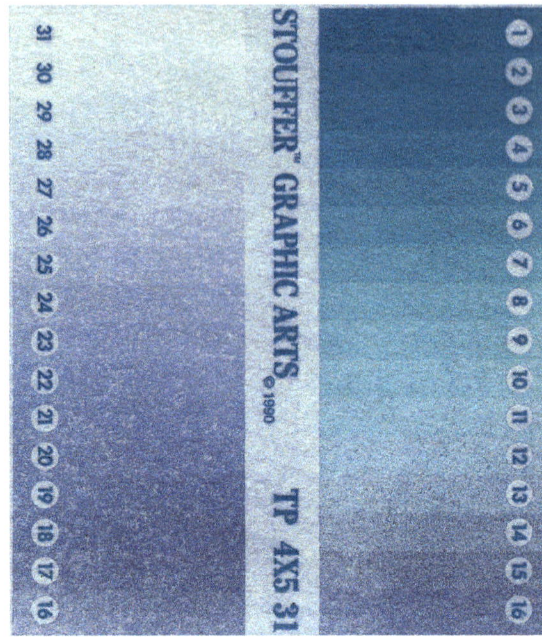

Figures 13.12 and 13.13. On the left, the strip cut and pasted into the middle of this image illustrates the result of a fogged paper and new cyanotype. On the right, an extreme instance of lavender stains due to high humidity, exposing while the print was still dampish, moisture developing in the contact frame, alkaline water wash, you name it.

There are lavender stains in the print

Lavender highlights or stains are indicative of alkalinity, moisture, or long sun exposure[12] (which can build up moisture in the contact frame). The alkalinity can originate in the paper itself or a contaminated tray or the water supply.

- Make sure the print is dry enough before exposing.
- Make sure there is no condensation of moisture in the print frame when exposing in full sun.
- Make sure the trays are clean of all alkalinity including fixer remnants.
- Make sure the paper is suitable and an unbuffered paper good for cyanotype.
- Always, have enough acid on hand and use more acid in the washing. An easy test for this is to just put a drop of vinegar on the lavender area and see if it turns blue.

The highlights are yellowed

Yellowed highlights relate to iron left in the print.

- The print has not been washed long enough. Some papers clear quickly and some take longer; keep records of this.
- Use warm water and constant agitation when developing a print. Change water as soon as it has yellowed. Only for the final minutes of the water wash does the print lie still in the water.
- Note that new cyanotype as a general rule should not be developed in plain water or weakly acidic water because most papers will turn yellow to tan brown in the highlights. New cyanotype requires a stronger acid bath such as 1% citric or 1% sulfamic.

The print is mottled

Mottling is different than splotching. Mottling is a very regular pattern of uneven tones resembling chilled flesh.

- Poor paper choice. The papers that have mottled have always been less than suitable papers. I have no idea why, but I never see it on unbuffered papers.
- Sulfamic acid development with some papers has not been carried to completion. I have seen this happen as if the acidic development didn't "do its thing" long enough and to all parts of the paper so it results in an uneven, regular

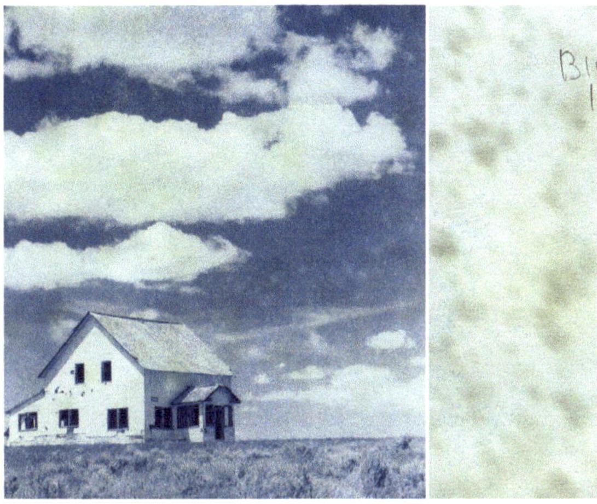

Figure 13.14. This is what it looks like on the front and back of the paper when it has not been washed sufficiently!

mottling pattern, mostly visible in the midtones to highlights.

- Too much coating solution with a textured, cold press paper. The solution settles in the hills and valleys of the paper texture and creates different densities of blue.
- A gray mottle all over the print which often appears quite rapidly indicates a paper that has not been cleared enough.

The print is dull

A dull print lacks deep darks and detailed highlights. There is overlap with this malady and a fogged and low contrast print.

- The print has not been exposed long enough to get a deep Dmax. See **Digital Negatives for Cyanotype** on how to determine correct exposure time.
- The negative has not been prepared correctly: the curve could be accidentally inverted, the colorspace is other than Adobe 1998, the chosen ink density is too low.
- A dull print that is pale and bland most often relates to an unsuitable paper. After printing 136 papers in both classic and new cyanotype with the same step wedges, the same developments, and the same image, *the number one problem for a dull print is paper choice*. A buffered paper often results in a dull, blue-gray, short exposure scale (ES) print. Consult the paper chapter for the best paper choices.

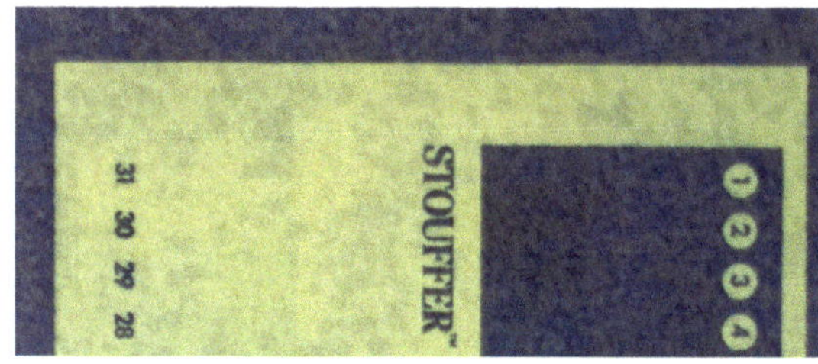

Figure 13.15. Mottling is paper-caused. This particular paper showed the uneven acceptance of the solution right after exposure.

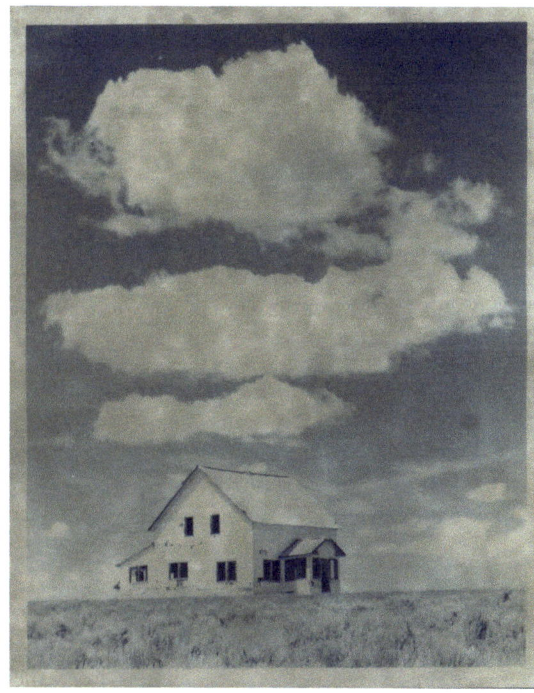

Figure 13.16. Iron left in the paper due to a very absorbent and unsuitable paper that would not release the chemistry. This same look will result from any incompletely washed paper.

- Acidic development was not used. Plain water in general will produce a duller, shorter exposure scale print. With classic cyanotype a mere vinegar or citric acid water wash will lift a print from dull to brilliant. With new cyanotype a mere citric acid or sulfamic acid bath will do the same. This has been in the literature for over a century, as Channing Whitaker said in 1897, "A few drops of a small quantity of any organic acid will make the finished print more brilliant."[13]
- Water is alkaline. If your midtones to highlights are brown and the blues are dull, worry about

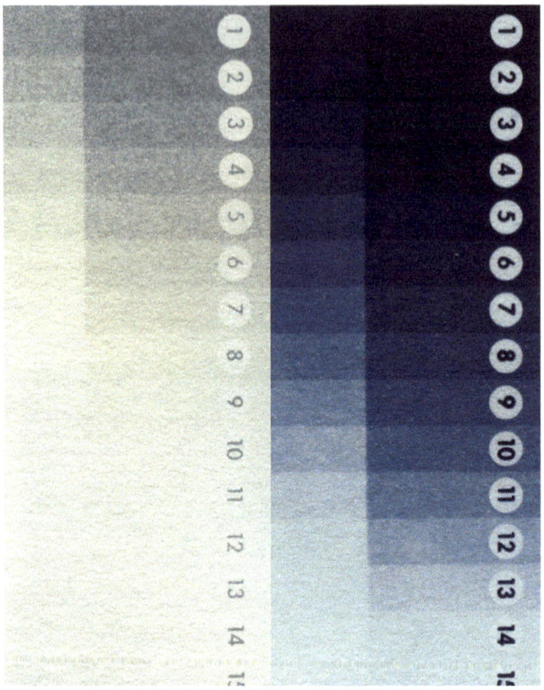

Figure 13.17. To the left in a scan of an extremely alkaline and unsuitable paper compared with a suitable paper on the right. You will note that development on an alkaline paper will happen quickly and result in this ugly bleaching effect.

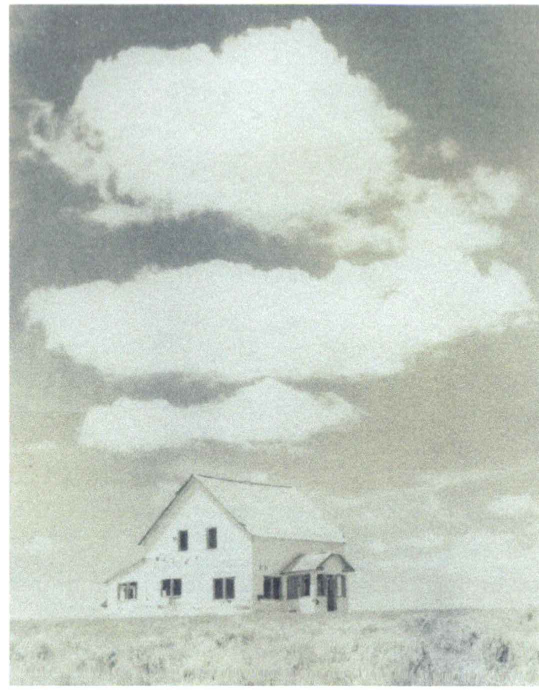

Figure 13.18. A dull, pale, yellowed print from too much washing and an unsuitable paper.

your water supply. Do a comparison print with distilled water and see if there is a difference. Alkaline water will bleach a print to a dull color. "Wash water should be free from calcareous salts, which convert the iron into carbonates which impart an ochrey tinge to the proof."[14] You can use more acid to acidify it or resort to using distilled or rain water.
- Too long a water wash. Watch the print while it is washing and don't extend the wash longer than necessary.

The prints are inconsistent

If practice is completely consistent but, all of a sudden, results are inconsistent, suspect changes in the paper at the manufacturer level, changes in the water supply at the water plant level (e.g. between winter and summer), or changes in the makeup of ferric ammonium citrate (FAC) which can be inconsistent. One author recommended to stir up the cyanotype solution before use to make sure the ingredients are always mixed in the same proportion when coating.[15]

There are blue spots or streaks

- Air bells, caused by not agitating the print well enough while it is in the water bath. The characteristic look of an air bell is that it is quite smooth and circular, like a bubble. Unfortunately, it can be really large or small, and once it's there there is no way to fade it unless you wash the print longer which in turn fades the whole print commensurately.
- Contaminated brush. At the end of every coating session soak the brush standing up in a cup of plain water; that way the chemistry diffuses out of the brush. A stitched ferrule brush doesn't mind a long soak in water. A brush with a metal ferrule, though, may rust and create problems.
- Free iron in the water can cause blue streaks when coating[16] but hopefully solutions have been made with distilled water so no tap water comes near the print before development.
- Mold chunks in Solution A will streak dark marks across the print. Brush them away to the edge of the coating area and use thymol, Listerine, or rubbing alcohol in your Solution A to prevent mold.

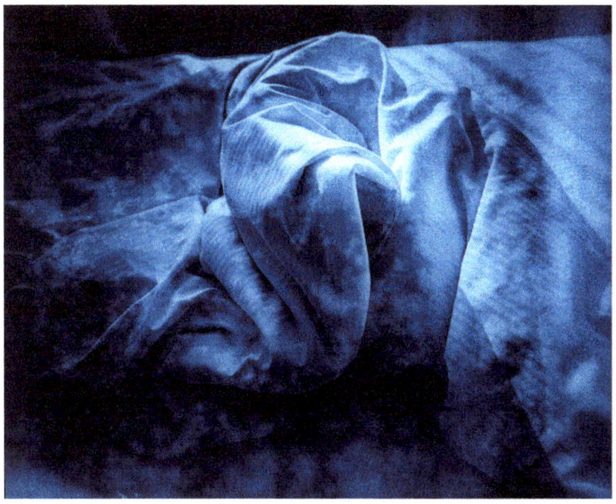

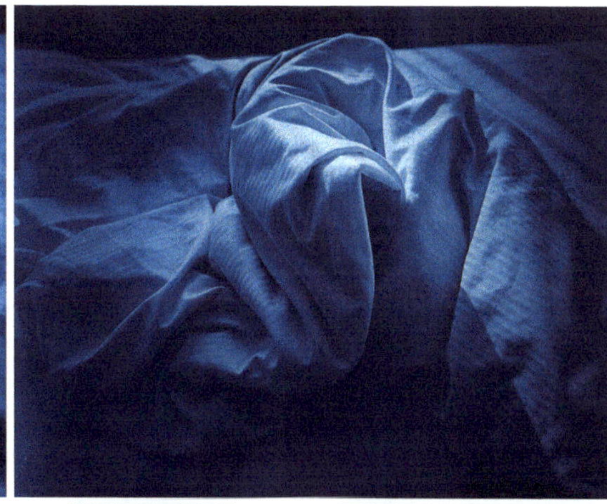

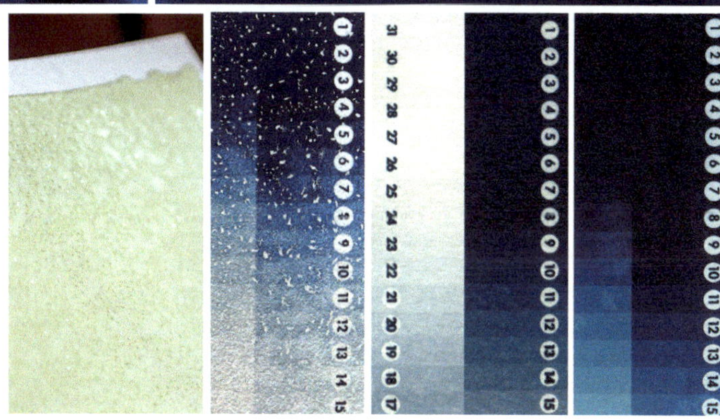

Figures 13.19 and 13.20. Splotches, mottling, and crystals: the top left image illustrates the dreaded irregular dendritic splotching. Splotches are different than crystals though causes are potentially related. The first two bottom strips show crystals from an incorrectly mixed new cyanotype, before and after exposure; the coating crystallized all over the surface of the paper, leaving sharp, white spots. The middle step wedge is the more subtle regular mottling, an overall pattern of uneven tones most apparent in the highlights of a print, and almost always a paper issue, also different from splotching and crystallization. The far right step wedge shows the irregular splotches.

The print is splotchy or crystallized

Splotching—lighter, irregularly shaped areas with branching, "dendritic" or fibrous edges—is a sporadic and frustrating issue with new cyanotype (not classic cyanotype). It can happen on a variety of papers and intermittently on the same paper, including the good ones, so it is not a result of paper unsuitability. It is not an exposure issue because it is slightly visible on coated but unexposed paper. Since the splotches are visible before and after exposure and before development, it is not a development issue. Unfortunately, at the time of this writing, there are no firm conclusions as to splotching's cause.

Dr. Ware has offered one explanation: "Splotching must be due to some non-uniformity in the paper sheet, causing a differential absorption of sensitizer, or retention of pigment, over adjacent areas. Different coating weights cause different image densities. This paper non-uniformity could have several possible origins: inadequate beating and consequent flocculation of the pulp; uneven addition of sizing agent or other additive to the stock; localized hydrophobic contamination of the formed sheet by contact; random growth of colonies of micro-organisms (especially fungi) within or between the cellulose fibres: i.e. biofouling—either at the mill or on storage."[17]

Its intermittency may suggest multiple factors operating at once; for instance, it is possible to conjure splotching at will by using a **thick coating** of a **full strength** (100 ml) NC dilution on an **unabsorbent, un-humidified** paper coated **without Tween** and **laid flat to dry**, six variables. One other variable may rest with the ferric ammonium oxalate itself, because splotching is an issue with printing out palladium and chrystotype which use FAO, too.

Though the cause is unclear, a few of us have pooled our experience on ways we have been able to minimize or avoid splotching entirely, with

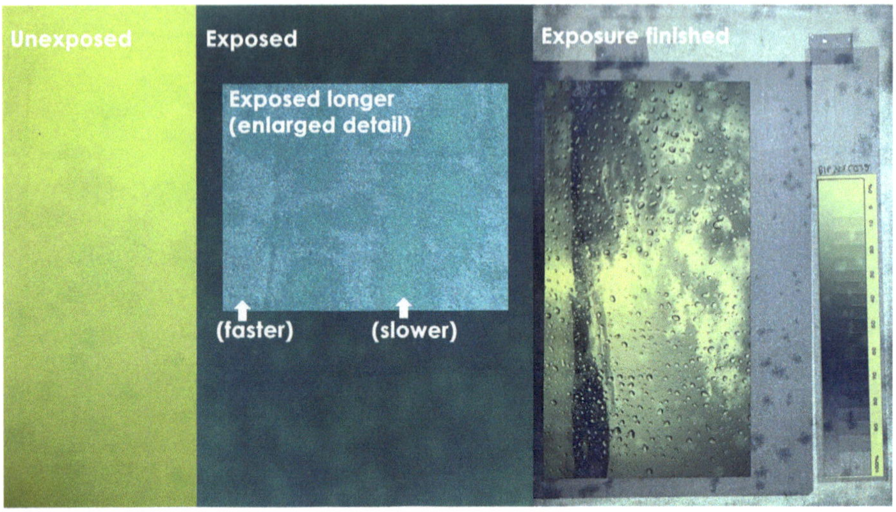

Figure 13.21. On the left and middle are enlarged details of a coated piece of paper before and after exposure, showing these splotches. Far right is the full print after exposure, with sporadic splotches everywhere. Once this happens, unfortunately the print is only destined for the trash bin.

pre-humidification of the paper being the first line of defense.

- It goes without saying to mix the new cyanotype formula according to directions, and filter out the insoluble potassium ferric oxalate, so there's no chance of it falling out of solution and crystallizing on the paper surface, creating areas of less sensitivity. [Note: splotches and crystals both result in spots on the print, but the former is dendritic-edged and the latter is sharp-edged. Although they are different phenomena, the techniques to alleviate dendritic splotching also alleviate crystals; hence the conflation of the two phenomena in this section.]
- Pre-humidify the paper. Construct a humidity chamber: fill a tray with water, place a screen on top on which the paper can rest, and invert another tray on top of that to lock in the moisture (courtesy of Sam Wang) several hours before a coating session or even overnight. Alternately, a tray of hot tap water for 20–30 minutes before coating will work but watch condensation on the inverted tray which can drip on the print. This minimizes splotching and *greatly* minimizes graininess with CC, NC, and SWC. A far smoother print results.
- Never shake the bottle of cyanotype solution before use and draw solution from the top of the bottle to avoid drawing up any potential sludge from the bottom of the bottle if there is any.
- Dilute new cyanotype to 150 or 200 ml instead of the formula's 100 ml.
- Coat with a combination of classic and new cyanotype (SWC). Splotches that are apparent after exposure often minimize during development with this formula, as if the classic part of the equation covers a multitude of sins.
- Use 1 drop of 5% Tween per ml of solution to improve paper absorption. Tween (introduced by Dr. Ware to alt pro use) is a surfactant that allows the sensitizer to penetrate the paper better.
- Shake the *measured out* coating solution well (i.e. not the bottles of solution but what's going to be coated on the paper) just before coating the paper so the chemicals are evenly dispersed.
- Keep the dimroom humid enough so the solution has a chance to sink into the paper fibers and not dry too quickly on the paper surface; 40–50% humidity is perfect. Let the paper dry in this environment for 20–30 minutes before exposure.
- Sensitize paper with a *thin*, not thick, coating of new cyanotype so there is no pooling of chemistry on the paper surface, e.g. even as low as 1¼ ml 2 ml for an 11″ × 15″! An often quoted "alt pro" rule of thumb is 1 drop every 2 square inches, but 1 drop per 3–6 square inches may be best on some papers (not washi!). Often splotches as well as crystals appear in the borders of the print where the solution is thicker. Dr. Ware in an email conversation said that if the coating

is allowed to stand on the paper without being promptly absorbed, it evaporates and can leave these crystals (possibly of potassium ferric oxalate or potassium ferricyanide). The reason new cyanotype may be more prone to this is because both FAO and PF can crystallize, whereas in classic cyanotype only PF is capable of crystallization.[18]

- *Hang* the paper to dry, don't dry it flat. That way free air circulation keeps the paper moisture even throughout the paper, and hanging prevents pooling of the solution. Be sure to wait until the paper has absorbed the solution and appears slightly matte, not wet, before hanging to dry, to avoid migration of the wet coating to the bottom of the print which will result in uneven density.
- It is possible that the citric acid recommended in the mix may help, though I have not run tests to prove this. See the **New Cyanotype** chapter.

There are round white spots in the print

- Paper that is heavily sized can "fish eye" while coating because the solution doesn't want to sink into the paper fibers. Just continue brushing and brushing until the fish eyes disappear.
- A quick fix for smaller white spots are spotting them out with Prismacolor pencils or watercolors, or even dip a brush in cyanotype solution, spot, expose, and rinse![19]

The shadows are solarized

- Overexposure. Solarization of the deepest shadows mostly turns to dark blue when the print is developed. If there are areas of lighter blue in the dark blue that look reversed, the print is overexposed and exposure should be shortened. Expose a cyanotype only as much as it needs to get the darkest blue possible, not too much so the darkest blue begins to lighten. As Channing Whitaker said, "There is a time of exposure that produces the deepest blue and with overexposure the blue gradually turned to gray but that being within several minutes of that optimum exposure is fine."[20]

Figures 13.22–13.26. (From top to bottom) White spots in a print because the coating area was contaminated with stray citric acid crystals. Initially this created blue spots in the coating. Next, water spots caused by having the print drying area too close to the water wash area. Next, hydrogen peroxide fingerprints from developing a print and then handling a coated piece of paper; luckily the fingerprints did not appear in the finished print. Any type of acidic water on fingers will do this. Next, the coating solution pooled or "bloomed" back into the print area, leaving white and dark blotches. Finally, a large round spot from an air bubble, caused by leaving a print face down in the water wash without continuous agitation.

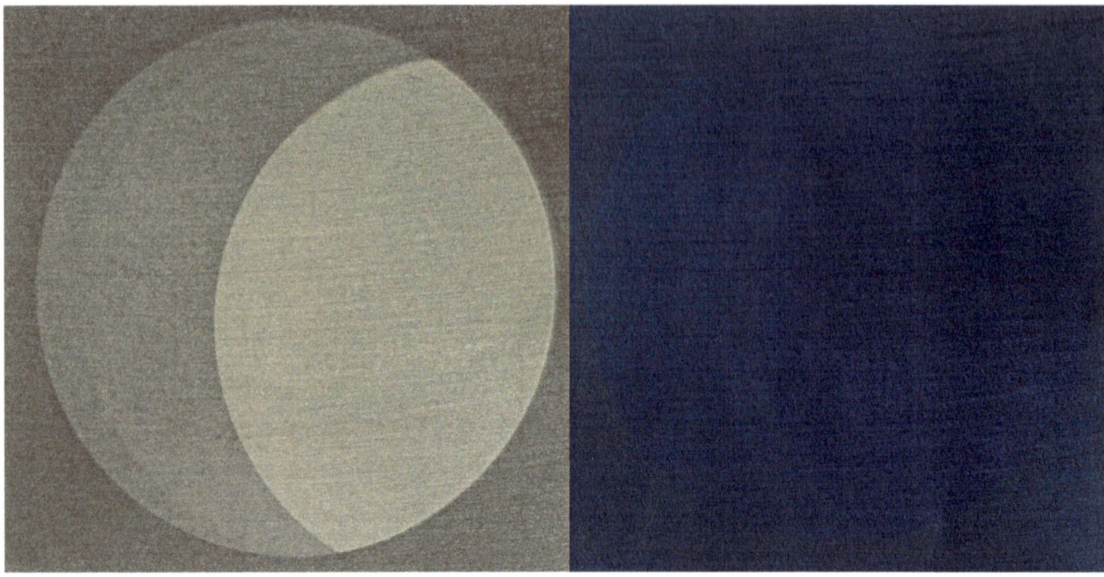

Figure 13.27. Solarizing/bronzing due to overexposure. I left this cc print outside all day with a dinner plate on top, which I moved during the exposure. The photograph on the left is before development and on the right after development. You can see the successive lightening of the blue with gross overexposure.

The image is unsharp

- Incomplete contact of the paper with the negative. I find this happens with a paper like a vellum or a delicate weight paper that wrinkles when coated. Those papers really have to be coated very carefully, coated in the center of the paper, leaving large uncoated borders, and then potentially trimmed before printing to get rid of the cockled edges. Some papers are so bad they need to be pressed flat in a dry mount press before exposing, or with a "poor woman's dry mount press" *aka* iron.
- The contact frame has lost its "oomph." Over the years use one, then two, then several pieces of paper inside the contact frame to make a tighter contact. This low tech fix works wonders as does an extra piece of felt.
- There is a chemical reason for areas of "fuzz": in the process, carbon dioxide gas is released, and it can form a bubble between the negative and the paper and cause loss of sharpness. If the contact frame has a felt backing which can absorb the gas, you probably won't see this.[21]

There are blue stains

- Cyanotype is the one process that is easy to clean up. It is so susceptible to fading with alkalis that a simple alkaline cleaner will remove stains on hands, sinks, bathtubs, clothes, etc. Use ammonia or sodium carbonate/washing soda, for instance. Another way is to mix two solutions: Solution A is 10 parts oxalic acid in 100 parts water Solution B is 12.5 parts potassium hydroxide in 100 parts water. These are mixed together at time of use to make a cleaning compound.[22] This is quite caustic. Also, if you happen to have EDTA and sodium sulfite on hand, those mixed together in solution work, too.

The print has faded

Prints fade because of sunlight, alkalis, water, and residual iron. Sunlight fading is reversible; the others are not.

- Frame the cyanotype print behind UV-resistant glass or Plexiglas and do not hang it in direct sunlight. Prussian blue (ferrous ferricyanide) converts to colorless Prussian white (ferrous ferrocyanide) in the presence of intense sunlight,

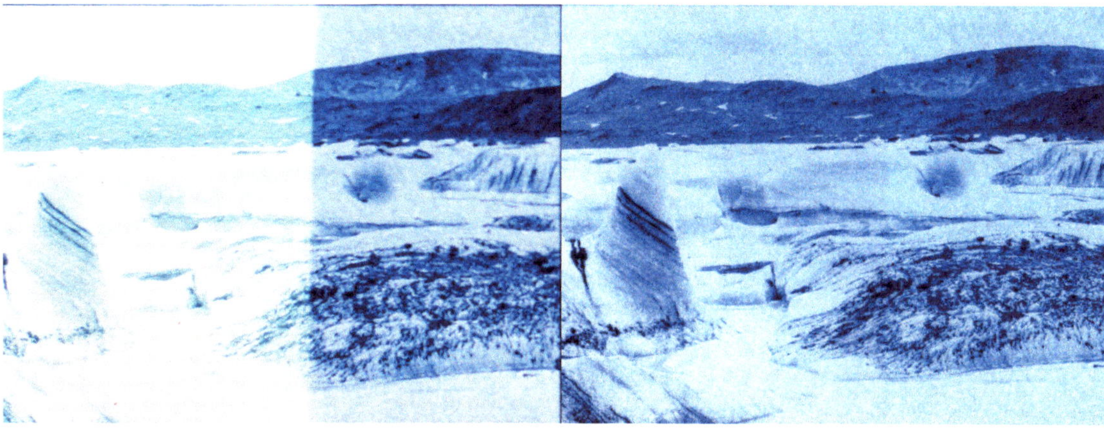

Figure 13.28. Scan of a small 4″ print left out half-covered for a full day of sun exposure, and then brought into the dark to recover. Scan on the right is the same print after two weeks in the dark.

but thankfully it reverts back to Prussian blue in dark, cool storage. If the print has faded in the sun, return it to dark storage until it regains its dark blue color. This is fun to test: put a throwaway print out in the sun, half-covered with something opaque. Within a day of full sunlight, the bleaching is quite dramatic, but after a few days in dark storage the image regains its depth. A dilute solution of hydrogen peroxide will also help the print regain its blue more quickly.[23]

- A buffered mat board has been used. Always mat cyanotypes in unbuffered mat boards.
- The cyanotype has been stored in a buffered archival box. Always store in unbuffered conditions.
- Extended contact with water such as by over-washing a print or (perish the thought) a flood where the print remains submerged in water for an extended period of time will fade a print to a pale blue-gray or yellow. Unfortunately, once the cyanotype washes out of the print it is down the drain so avoiding water contact is a must.
- Not developing and washing the print long enough so that iron is left in the paper, which will cause it to fade and the highlights to turn gray.[24] Be scrupulous with your development times and baths and don't under-wash.

Endnotes

1. Worthington, J. C. and J. C. Millen. *The Photographic Primer, A Manual of Practice*. Riverton: The Riverton Press, 1897, p. 128.
2. Barnier, John. *Coming Into Focus, a Step-by-Step Guide to Alternative Photographic Printing Processes*. San Francisco: Chronicle Books, 2000, p. 43.
3. "The Workroom-Cyanotype" in *The Photographic Journal of America*, Vol. LV. Philadelphia: Edward L. Wilson Company, Inc., February 1918, pp. 93–94.
4. R. W. K. G. "Ferro-Prussiate Dodges" *in The Photographic News, The Journal for Amateur Photographers*, Vol. XLI. London: Photographic News, November 26 1897, p. 778.
5. Tennant, John. A., ed. "The 'Blue Print' and its Variations" in *The Photo-Miniature*, Vol. 1, No. 10, January 1900, pp. 481–514; Farber, Richard. *Historic Photographic Processes, A Guide to Creating Handmade Photographic Images*. New York: Allworth Press, 1998, p. 68; "The Workroom-Cyanotype", pp. 93–94.
6. Whitaker, Channing. "Apparatus for Printing by the Blue Process" in *The Photographic News*, Vol. XXVII. London: Piper and Carter, December 7 1883, p. 359.
7. Ibid.
8. James, Christopher. *The Book of Alternative Photographic Processes*. New York: Delmar Cengage, 2016, p. 173.
9. Tennant, p. 498.
10. Whitaker, p. 359.
11. Wall, E. J. *Photographic Facts and Formulas*. Boston: American Photographic Publishing Co., 1924, p. 260.
12. Ware, Mike. *Cyanomicon II, History, Science and Art of Cyanotype: photographic printing in Prussian blue*. Buxton: self-published, 2016, p. 246. http://www.mikeware.co.uk/mikeware/downloads.html
13. R. W. K. G., p. 778.
14. Duchochois, P. C. *Photographic Reproduction Processes*. New York: The Scovill & Adams Company, 1891, p. 33.
15. Blacklow, Laura. *New Dimensions in Photo Processes*, 3rd ed. Boston: Focal Press, 2000, p. 102.
16. Barnier, p. 42.
17. Information on splotching and white spots comes from a lengthy email conversation August/September 2018 with Dr. Ware.
18. Ibid.
19. R. W. K. G., p. 778.
20. Whitaker, p. 359.
21. Ware, pp. 181–182.
22. Duchochois, p. 35.
23. Kosar, Jaromir. *Light Sensitive Systems: Chemistry and Application of Nonsilver Halide Photographic Processes*. New York: John Wiley and Sons, 1965, p. 36.
24. Worthington, pp. 130–131.

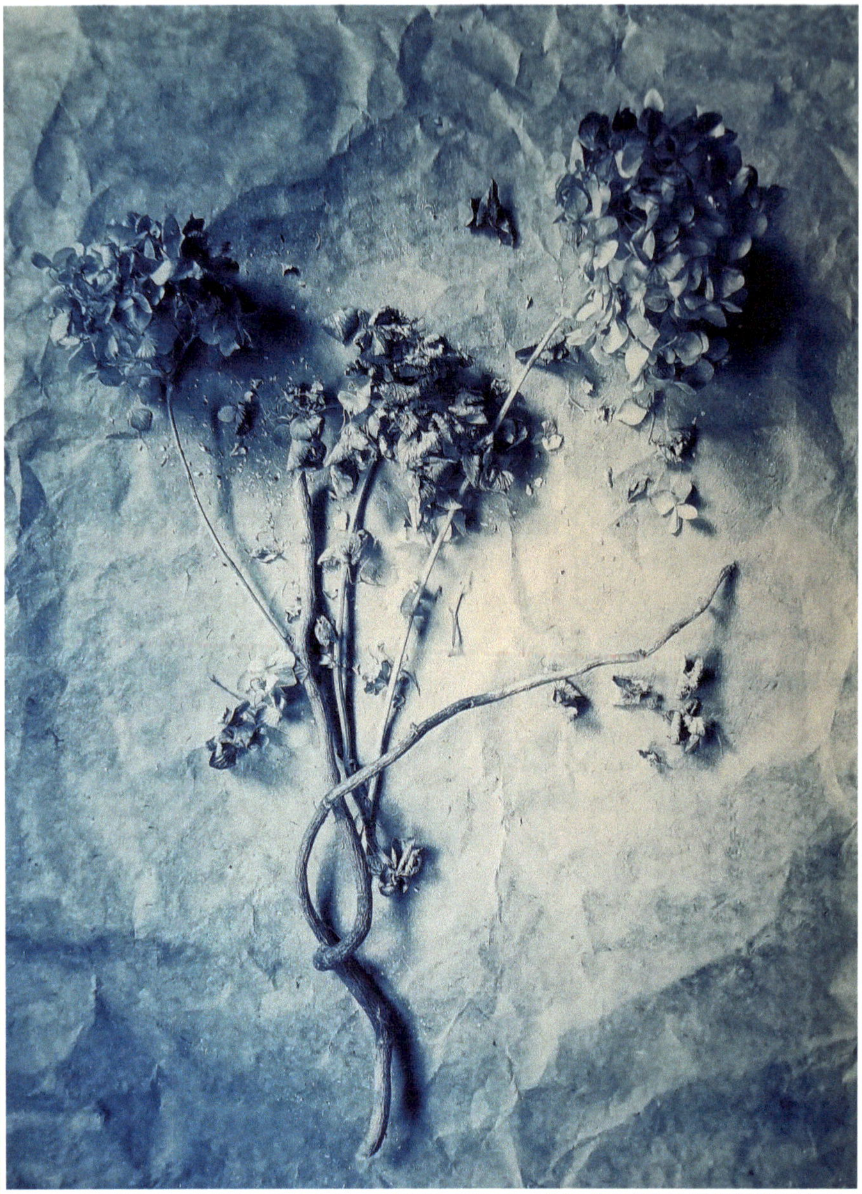

Figure 14.1. *Hydrangeas in Blue*, cyanotype on Revere Platinum, 12″ x 16″ and 9″ x 12″ © Diana Hooper Bloomfield 2018. "Inspired by a friend and fellow artist who made a piece of artwork every day for a solid year, and simply kept going after that, I, too, thought—what a great idea! I can do that! I began on February 3, 2018 with a cyanotype over platinum/palladium image. That first week—only 7 days in—seemed Herculean, but I kept going. Today, July 1, at Print/Day #125, I now feel I have my footing. I'm in a rhythm. Over these last 125 days, I've adjusted my way of working. More efficient, less worried about 'perfection,' more interested in what I can achieve, specifically when traveling or too busy with life to get anything else done, cyanotype suddenly became my friend. I have even made ribbons of color with nothing more than brushstrokes of cyanotype made with various dilutions and exposures. Over time, I started making prints with flowers from my garden. This new series, titled *The Old Garden*, is mostly tricolor gum bichromate over cyanotype. But I always start with cyanotype, and I marvel at the process. Some of the images are photograms, using the sun. Not caring about detail or tonal range, I simply wanted the deep blues and white outline of the image. All are flowers from my backyard that I see, not only from my window, but also in my mind's eye—hydrangeas and irises and climbing roses—in my grandmother's yard, all of another place and time—and as ephemeral as ever."

An exhibiting photographer for over 35 years, also an independent curator and writer, Diana Bloomfield specializes in 19th century photographic printing processes. Her work has been featured in a number of books and journals. A native North Carolinian, Bloomfield lives and works in Raleigh, North Carolina, where she received her MA in English Literature and Creative Writing. She teaches workshops throughout the country, and in her beautiful backyard studio. She is represented by Panopticon Gallery, located in Boston, Massachusetts, and the Ryan Gallery of Art Intersection, located in Gilbert, Arizona.

Chapter 14

Finishing, Framing, and Storing Cyanotype

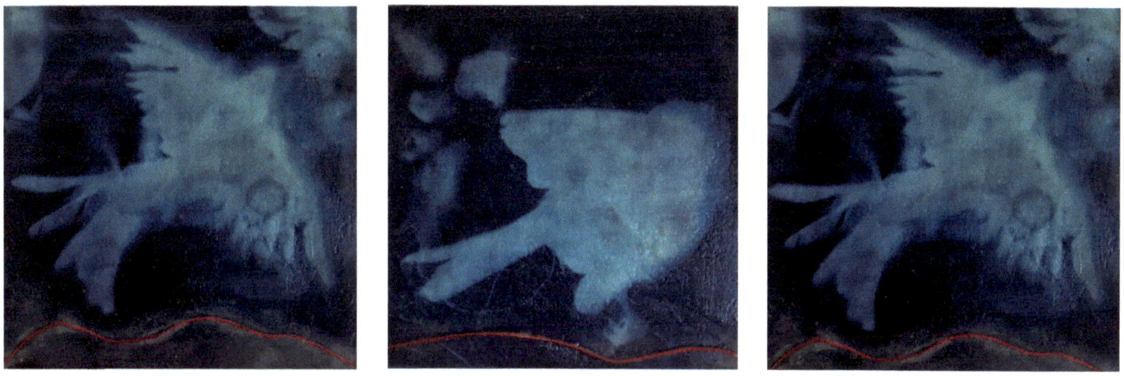

Figures 14.2–14.4. *Memento Mori: Aves I, II,* and *III*, cyanotype on fabric applied to wood panel with encaustic wax, panel edges charred (Sho Shugi Ban, a traditional Japanese wood preservation technique), red "thread of fate" and brass bells hung from each panel, 6″ x 6″ © Donna Headrick Moore 2017. "The inevitably of death is haunting, be it the usually overlooked death of birds or a mother's death from complications of Alzheimer's. This work documents and explores the invisible connection we all have with each other in life and in death." Donna Headrick Moore is a photographer and art educator from Blaine, Tennessee. She works with pinhole cameras, and historic and alternative photographic processes. She is currently the codirector of A1LabArts, a cooperative art space, in Knoxville, Tennessee. She is a founding member of the Knoxville Community Darkroom and part owner of The Big Camera, a mobile darkroom and Camera Obscura. To see more of her work visit **Donnaheadrickmoore.org**.

Cyanotype is unique in that it is alkali-sensitive not only in the making of the print—common with a lot of alternative processes—but in matting, framing, and long-term storage. Just as an alkaline paper will create a faded looking cyanotype print, and alkalis are used to bleach a print before toning, alkalis will bleach a cyanotype print within a frame or in storage by virtue of being in contact with it. This means that all storage boxes and mat boards or anything else the print comes in contact with have to be below a pH of 7.5, unbuffered, and alkaline free. Contact with an alkali will fade the print and unfortunately the fading will not regenerate. If you have a personal framer as I do, or are buying mat board from a big box store, be sure to mention this when matting and framing.

Cyanotype is also subject to a strange phenomenon in that it fades in strong sunlight, but regains its density when given a rest in the dark, almost like a living document. I thought this was a myth when I first read about it, but sure enough: I placed a print outside in direct sunlight all day, covering half of it, and leaving half of it exposed. The half of the print exposed to direct sun for that amount of time faded to almost detailless white. The print was brought inside to sit in the dark and within two weeks it had regained all density to its original. The reason for this is light changes Prussian blue to Prussian white or ferrous ferrocyanide, which oxidizes back to Prussian blue in the dark and with circulating air.

What does this mean for framed works hanging on walls? First, be sure that UV glass or Plexiglas is used in the framing of the print. Second, don't hang a print in full, direct sunlight. Third, retire the print to a dark closet for a couple of weeks every now and then if you see any evidence of fading.

A third problem is that a cyanotype is always susceptible to washing out when wet. Thus in a flood, for instance, the print will fade, and water fading, like alkali fading, will not come back.

Hopefully this latter issue you will never encounter. With minimizing sun, alkalis, and water, you can expect the cyanotype to last for as long as the paper does.

Keeping record of works

When I finish a cyanotype print, or any handmade print for that matter, I make sure to first document it before it is even matted. First I trim the edges of the print so they are straight, even, and allow for a minimum of ½″ extra paper on all sides of the image. It would be best if you could allow several inches on every side and tear the paper with a tear bar so the edges have a natural torn edge instead of a sharp cut one.

I sign and date the bottom border of the actual print (never the mat board!) in pencil; pencil is archival. Since my handwriting is so terrible I actually sign within the dark brush stroke border. When viewed obliquely the pencil is shimmery and can be read. If your handwriting is beautiful, by all means let it show in the white paper area of the print!

Then if the print is small enough to fit on a flatbed scanner, I scan it. Otherwise I take a photograph of it with my digital camera set to raw mode.

I scan at 300 ppi/16 bit/Adobe 1998, and keep the file exactly the same size as the print so I never have to wonder what its size is down the road. If it is 8¼″ longest dimension in real life, the file is 8¼″, too.

I save it as an 8 bit/300 ppi/SRGB TIFF file with no compression, provided I am not ever planning on printing a larger digital version of the print, in which case I would scan it at 720 or some other larger size so that when it goes to print it will be 300 ppi at that larger size. I choose SRGB because it works well for the web and SRGB translates more accurately to CMYK than does Adobe 1998 if it ever goes to press. TIFF and PSD files are both fine but some presses prefer TIFF.

I also save a small JPEG version of the file, same exact size, compression level 10, for digital submission to shows.

After the scan or edited raw file is done, I compare the actual print with the computer screen image. I find often I have to make a slight adjustment to the digital version in a levels adjustment to get the darkness and contrast of the print to match. I do this by Image/Adjustments/Levels and more often I find I have to set the black point at 10–20 and the white point at 240–250. If needs be, I print out a small test print to corroborate. Then a simple sharpening step of Filter/Sharpen/Unsharp Mask/Amount 150/Radius 0.5/Threshold 0 and the scan is done.

This may seem like a lot of work for nothing, but you never know when you might submit the print to a show, sell the print and no longer have a copy of it, or just need a record of work completed for some reason. In my case, for instance, I work for a research university and 40% of my contract expectation is research. My work is one part of my research. I also talked with a friend whose mother had died, leaving a lot of unexplained work in boxes. With good record keeping this problem could have been easily averted.

To that end, I keep a running visual catalog of all works created year to year. I'm lucky to know InDesign thoroughly so I do the layout for this catalog in InDesign, but there are many quick ways to make PDFs of your work, since you now have it documented.

I also keep an Excel spreadsheet *Inventory of Works*. On the Excel spreadsheet I record the title, the series it belongs to, what process the print is, the image size, the year the image was created, the mat size if matted, the frame size and type if framed, and its retail value for each and every work no matter how successful the print. When a work sells, I move it to a different sheet in the file and record who owns the print.

Varnishing cyanotype prints

After scanning or photographing a cyanotype, you may consider varnishing the print,which can protect the print from anything it might come in contact with that is alkaline. In general, cyanotype is not a process that seems to require a varnish,

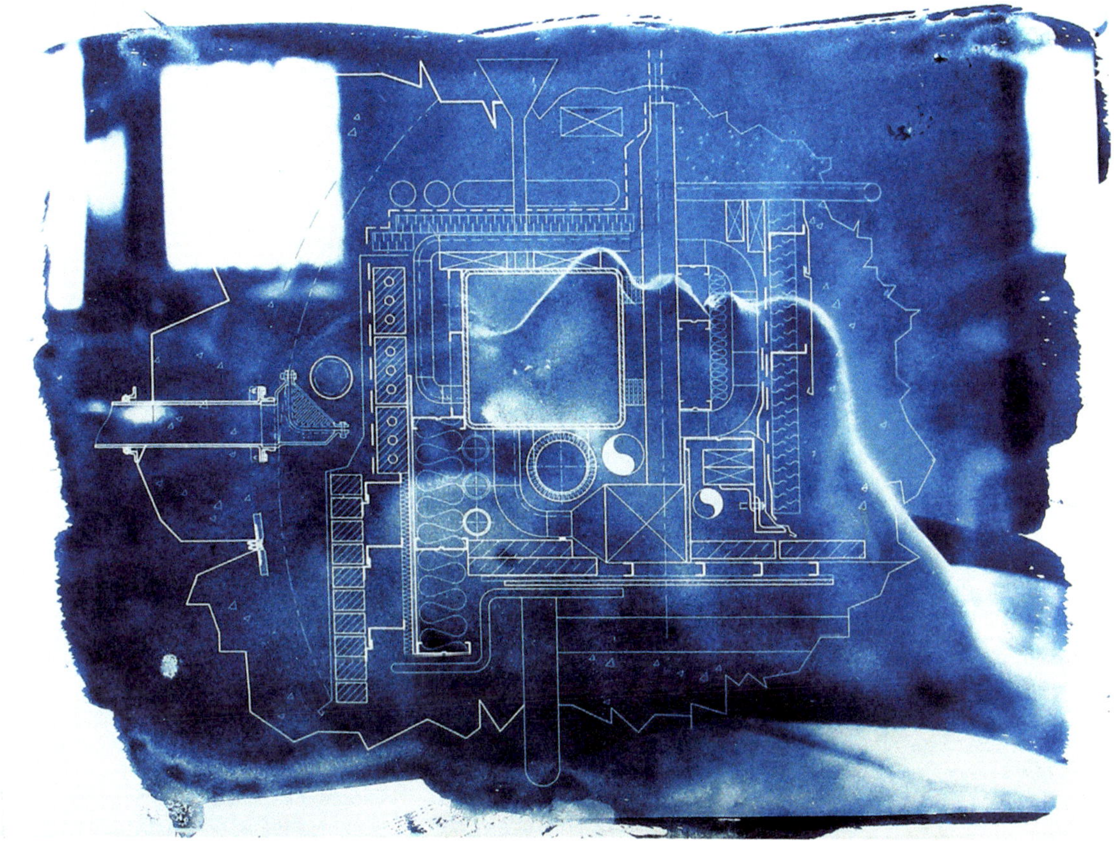

Figure 14.5. *Self Section*, cyanotype on watercolor paper, using two digital negatives, a (self) portrait and a drafted drawing of a "self section," exposed in direct sunlight, washed and dipped in hydrogen peroxide solution (diluted at 1:10), 11″ x 8.5″ © Natalia Tcherniak 2016. "We are not alone; we are not in a vacuum. We exist in a network of relationships, visible and invisible, conscious and unconscious. We connect directly and indirectly to other people, things, concepts, events, places, and everything around us. If lines were drawn to represent all the connections, they would comprise a pattern so dense, it would be solid, a Connective Tissue. An overarching theme in my work is a search for orientation. I explore methods of defining position within a Connective Tissue, creating new frames of references, mapping new readings, adding layers to the tissue. I strive for multiple readings as I explore my own psychological construction assembly and the dystopian urban environment I call home." Natalia Tcherniak works in multiple disciplines: architectural, theatre set design, performance, fine and installation art, and alternative processes in photography. Focusing on the cyanotype process, Natalia explores the historical application of blueprints in a contemporary context of urban environment and self-exploration. Her visual artwork has been displayed in galleries and public spaces in Toronto, and projects built across Ontario and British Columbia in Canada. Regular updates are posted on **blueprintjam.com**.

except a toned cyanotype which tends to dry down duller than the wet toned print. It is important to keep in mind that many varnishes are alkaline, and as soon as they are brushed on the print, the print starts turning purple. This is unfortunately the case with my favorite archival varnish called Golden Polymer Varnish that has built-in UV inhibitors. Also important to keep in mind is that cyanotype is best with free air circulation which varnish might inhibit. Nevertheless, here are some varnish suggestions that you can experiment with:

- Egg white diluted 1:1 with water
- Gum Arabic brushed on the print
- Gum Arabic and dichromate brushed on the print, exposed, and rinsed only enough to remove the yellow dichromate stain and then dried
- Gamblin Cold Wax Medium
- Dorland's Wax Medium applied with a piece of flannel, dried, and buffed
- Beeswax melted and combined with mineral spirits 1:2 or 1:3 applied with a piece of flannel to the surface of the print, dried, and buffed

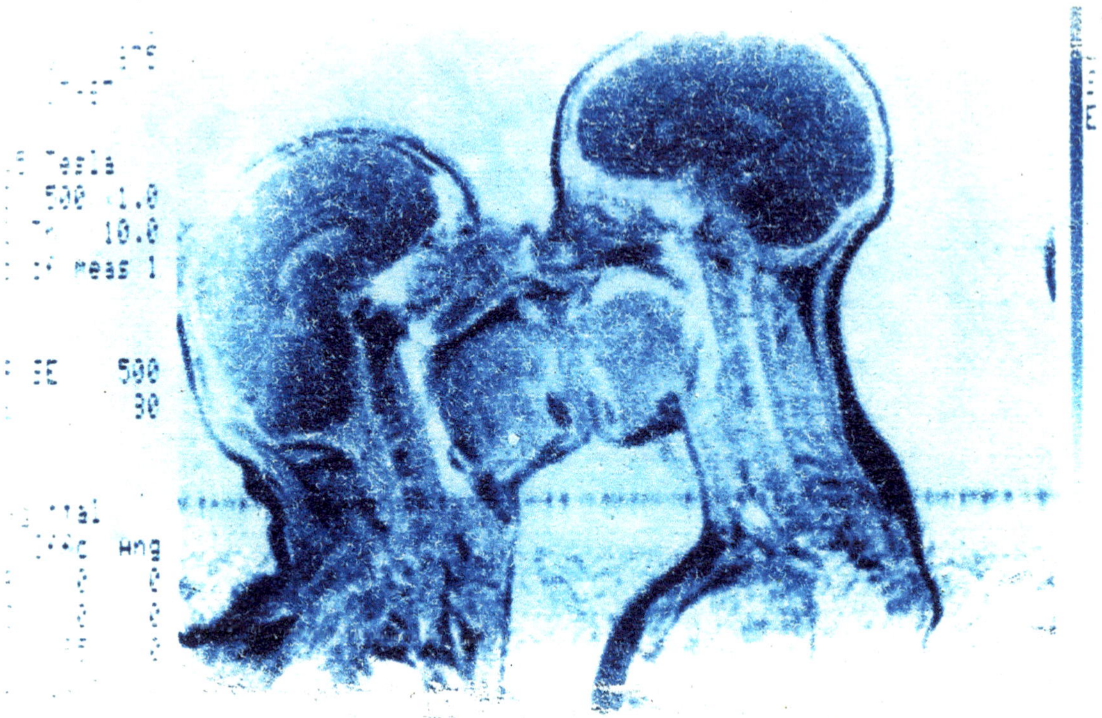

Figure 14.6. *5 minute kiss*, traditional cyanotype using MRI film, 5″ x 7″ © Russell N. Jeffcoat 2018. Russell Jeffcoat is a photographer specializing in both historic film processes and black and white photography using vintage cameras and techniques. His work is collected both nationally and internationally and is part of the permanent collections at the Ogden Museum of Southern Art in New Orleans and the Columbia Museum of Art. He currently resides in South Carolina where he teaches photography and the creative process. To see his work, visit rjeffcoat.com.

- Beeswax melted and combined with lavender oil 2:1 or thereabouts applied with a piece of flannel to the surface of the print, dried, and buffed

How to best store the print

If I am not going to mat and frame a photograph, I usually store it in its own clear archival plastic bag and in an archival photo storage box. However, cyanotype prints are better stored in a dark environment in which air can freely circulate; hence store them in boxes loose, without plastic bags.

Editioning work

Cyanotype is one process that can probably be editioned because prints can be made similarly enough to one another to constitute an edition and not a monoprint or 1/1. I remember a statistic once that said the average sales number of any one print was five. Although editioning is an artifice which has more to do with printmaking, pulling a print off a press, than it does with photography, sometimes you have to play the game because some galleries or buyers want it that way. I look at it like this: I set an edition at 15. I don't print it all at once. And if I were blessed enough to sell 15 of one image, I would gladly retire the image, because as an artist who is growing in creative practice, hopefully my future images will be as good as or better than my past ones.

Matting and framing

For an 8″ × 10″ print, I suggest a minimum of a 2½″ border on all sides, but 3″ looks best. You can bottom weight the mat—having the bottom of the mat ¼″– ½″ wider than the top—which has been

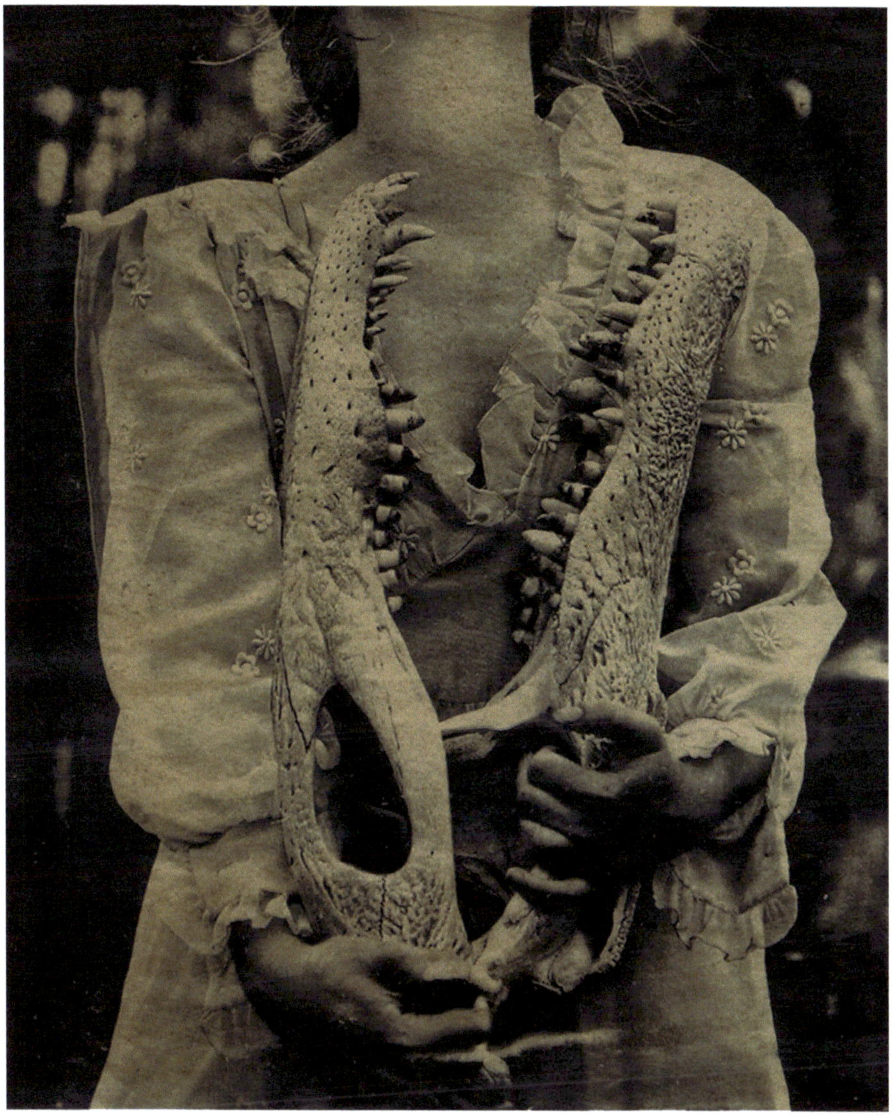

Figure 14.7. *Barney*, from the *Untamed* series, traditional cyanotype stained with Lipton tea on Japanese Kitakata paper, 16″ x 20″ © Jaime Aelavanthara 2015. Aelavanthara displays these prints unframed. Because of the tea staining and wrinkling of the fragile paper, the images hang on the wall like delicate tanned animal hides, in fitting with the series' concept.

the traditional method of matting, but modern presentation leans towards equal mat size all the way around.

Mat board color is always a matter of preference, and should probably match the paper the cyanotype is printed on, but I have standardized all my alt work matting with a slightly cream mat board such as Artique Devonshire A4906. The color complements cyanotype nicely.

Develop a relationship with a frame shop. A frame shop will have 50 or more "white" mat board samples to test. Bring in a cyanotype print and spend an hour testing each one next to the print. Each mat board has a color bias: bright whites are biased toward blue, other mat boards are biased toward green or yellow or red. Some are more neutral. The rule of thumb is this: each color will bring out its *opposite* in the color to which it is adjacent, and a warm or yellow-biased mat board will make a cyanotype therefore look more purple-blue. Once you find the color that works, make note of it and standardize with that mat board always.

Endnotes

1. Ware, Mike. *Cyanomicon II, History, Science and Art of Cyanotype: photographic printing in Prussian blue*. Buxton: self-published, 2016, pp. 181, 277.

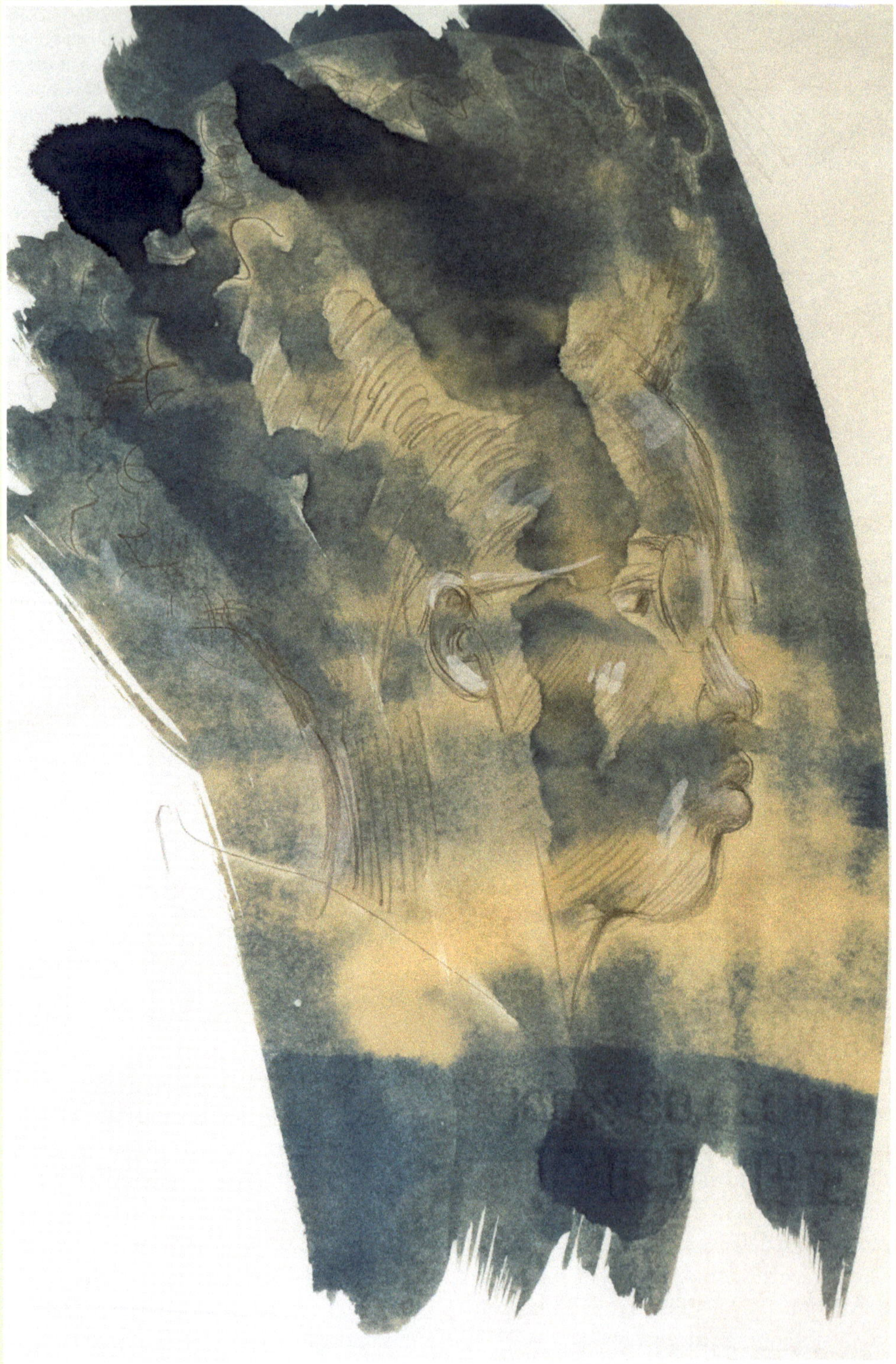

Figure SII.1. *Vicki*, pencil, cyanotype, bleached cyanotype, and white gouache, 15″ x 8.5″ © Sarah Van Keuren 2018. Sarah Van Keuren majored in art history at Swarthmore College, studied printmaking at the Philadelphia College of Art, and received an MFA in Photography from the University of Delaware. From 1980 to 2014 she taught non-silver printmaking processes at the University of the Arts (formerly PCA). She is author of *A Non-Silver Manual* available online now for free at **alternativephotography.com**. Before retirement she taught workshops across the US and in Finland. Schmidt-Dean Gallery represents her work.

PART TWO

Contemporary Cyanotype Artists

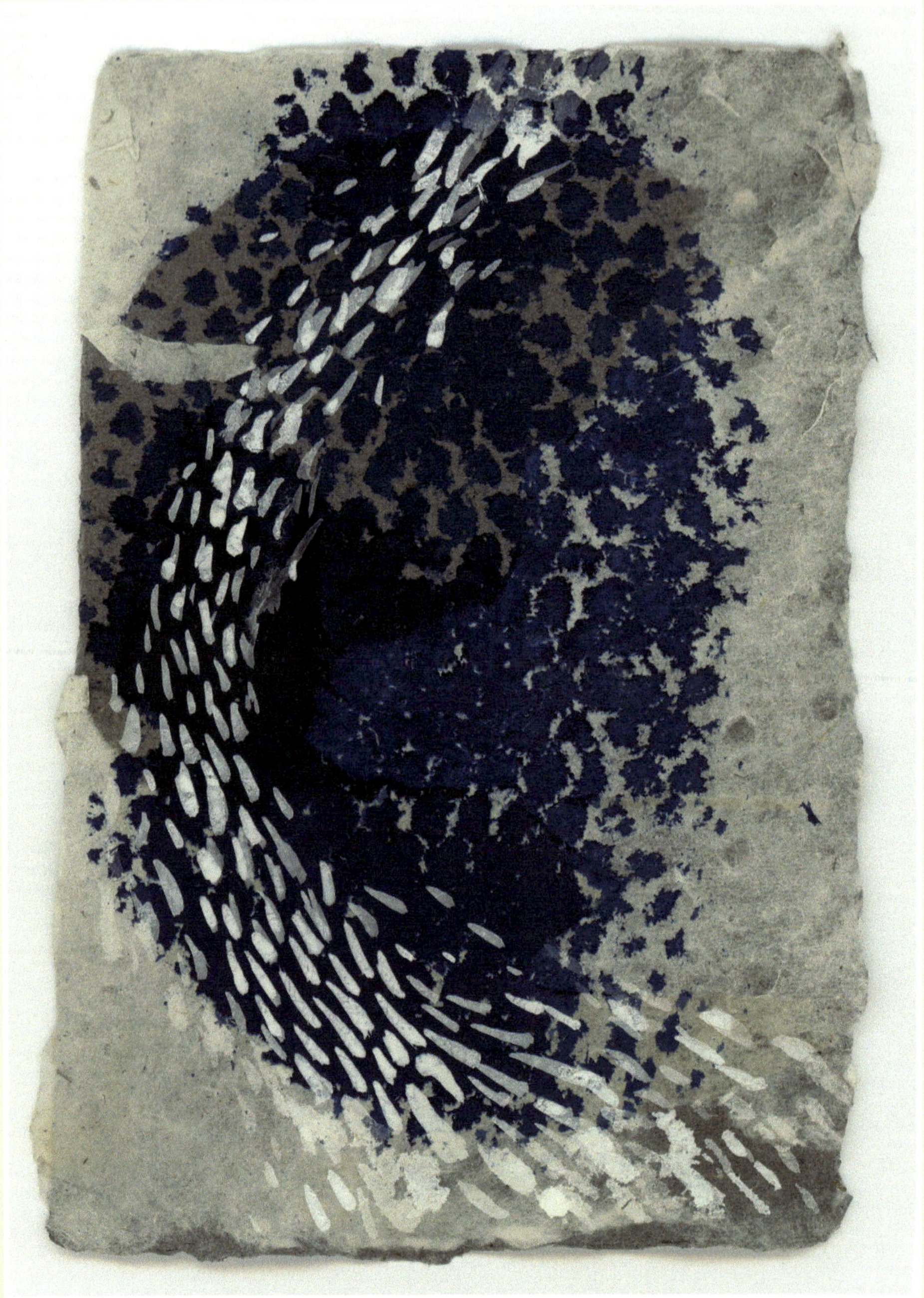

Figure 15.1. *Lamina Eddy 2*, cyanotype and gouache on paper handmade from pigmented cotton, flax and cornhusk, 6˝ x 9˝ © Lindsey Beal and May Babcock 2016–2018. "The *Lamina* series is a recent collaboration between artists Lindsey Beal and May Babcock. It focuses on experimental integration of alternative photographic processes and creative papermaking techniques. Finely beaten pulp is made from pigmented cotton, flax and corn husk, suspended in a vat of water. Sheets are formed in a Western-style dip method, pressed and stack-dried flat. Paper is then externally sized with photo-grade gelatin sizing. The papers are then coated with cyanotype chemistry and printed using photogram material or stencils. Lastly, gouache is painted on top of the cyanotype." Lindsey Beal is a photo-based artist whose work combines historical and contemporary women's lives with historical photographic processes. May Babcock is an interdisciplinary artist who creates paper, prints, and installations inspired by landscape and place. Both artists reside in Providence, Rhode Island and are educators with extensive national and international exhibition records.

Chapter 15

Contemporary Cyanotype Artists

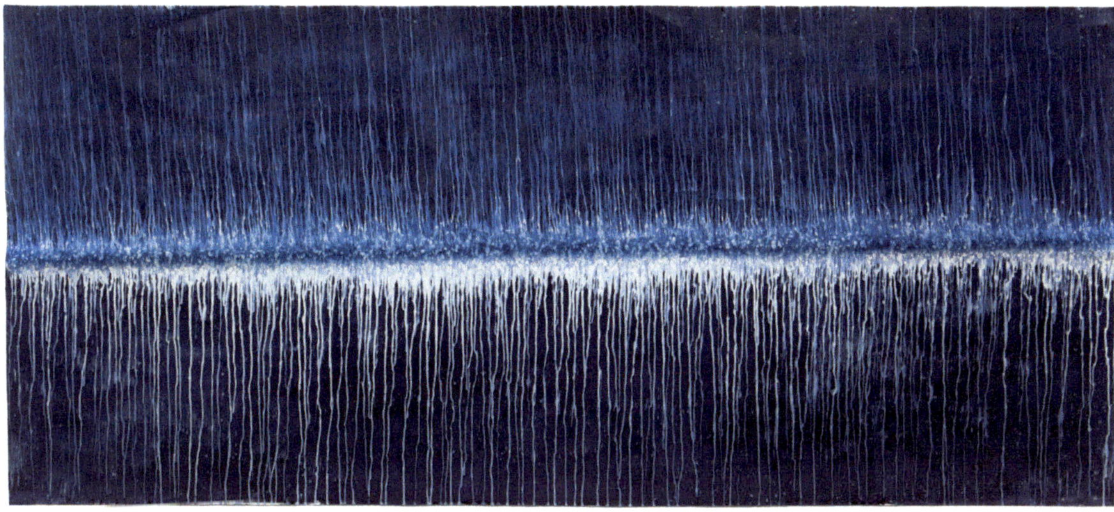

Figure 15.2. *Ecotone #04 (Bainbridge Island, WA 03.19.15, Rainstorm, One Hour and Thirty-seven Minutes)*, unique cyanotype, 42″ x 96″ © Meghann Riepenhoff 2015

Writing—or editing as the case may be—this final chapter is the fun part of creating a book. I get to see into the minds of the artists and how they think and work and make art. We all are driven to do this thing called "art" for myriad reasons, and we work in myriad ways.

Each artist was sent a questionnaire asking for an artist statement, biography, and answers to these questions about process and creative practice: *Why cyanotype, in general or in specific for this work? What is/are your favorite paper(s) for cyanotype? How do you make your negatives? What cyanotype formula do you use, classic or new? What exposure device do you use, and approximately how long do you expose? What is your development process? Do you tone cyanotypes and if so, with what toner, what formula, and how? Do you have any favorite ways of matting, framing, and/or displaying your cyanotypes? Describe any cyanotype problem(s) you have had, the probable cause, how you solved it. In what unique ways are you using cyanotype? If you are printing on alternate surfaces like wood or fabric, how? If you are doing the process in a different way, how? Do you have any cyanotype advice to share? Describe anything about your creative practice as an artist that might interest/be helpful to a reader.* Each artist could choose to answer concisely or elaborately, some questions, none, or all.

I edited out the questions, and intervened only minimally in their words so the artists could speak for themselves, in their own voice, and with their particular syntax. It is important to me that their works and their personalities are represented as one.

The reader will note the wide range of cyanotype practice from traditional to contemporary, abstract to representational, 2D to 3D, printed on various surfaces, and installed in numerous ways. We are all working with two chemicals that, when combined, make a light-sensitive substance, but often that is where our commonality ends.

Artists are presented democratically in alphabetical order in 2-, 4-, or an occasional 6-page spread, depending on the length of submitted essays and number of images submitted.

I thank these artists for taking time out of their lives to share their passion as I know it is yet one more commitment in already busy lives, and with no remuneration besides the pleasure of seeing their creative talents come to fruition in print.

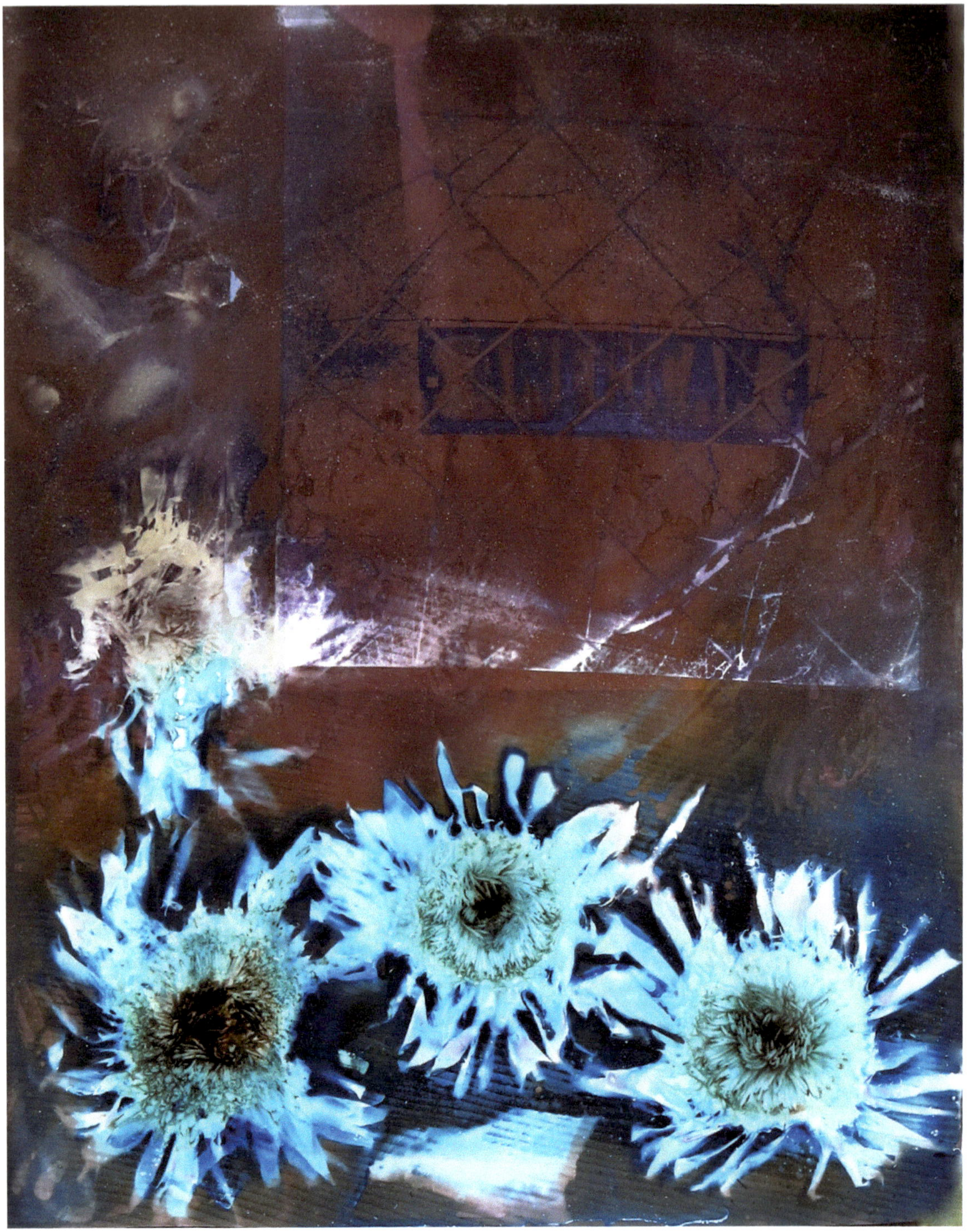

Figure 15.3. *American*, lumen print of chrysanthemums with contact printed negative, classic cyanotype, multiple toning (gold, selenium, Dektol, peroxide, Halochrome) on expired Ilford Multigrade IV WT RC Deluxe Glossy paper, 11″ x 14″ © Mariana Bartolomeo 2015

Mariana Bartolomeo

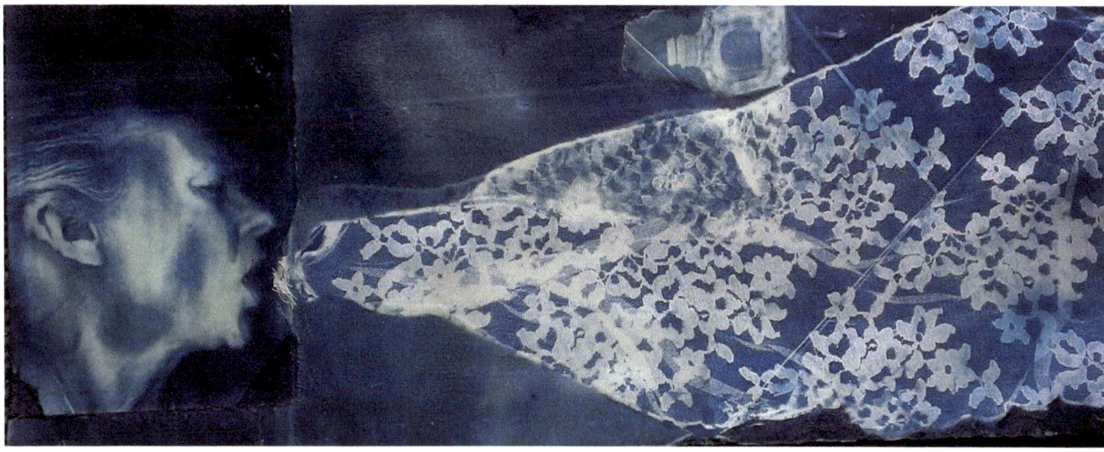

Figure 15.4. *Mi Sueño de la Noche* (from photo-narrative *Widow's Work*), classic cyanotype, double coated, contact print of a negative and photogram of lace and bottle, collage, encaustic, Rives BFK, 42″ x 14″ © Mariana Bartolomeo 2014

Process

My MFA is in Printmaking and Book Arts, so my life is paper galore. I use many different papers depending on the effects that I want to achieve. Overall, I love using large papers that can handle 2–4 coats of cyanotype chemistry: Rives BFK Heavyweight, Arches Platine (both weights), Canson XL, and Canson Bristol.

When I apply multiple coats of cyanotype I am looking for Dmax, the deepest, darkest (almost cobalt/ultramarine) blue possible...Nearly black. It allows me to make multiple, long exposures with different overlapping elements (plants, negatives, textured glass, lace, etc.). They afford me the ability to experiment as I work.

Medium weight papers such as Hahnemühle Sumi-e also produce rich, deep blues, but on projects that are less brutal. Canson Montval produces a more true cyan blue, more on the cold/acidic side of deep aqua. Japanese/Thai Unryu is a good choice when I want to work in cyanotype collage or chine collé.

If I want to contact a negative with significant sharp detail I refer back to etching/intaglio/letterpress papers such as Fabriano Tiepolo, Arches 88, and Canson Bristol. I am really searching for a 100% cotton paper that will produce image sharpness from a negative. One on my list to try is Rising Stonehenge.

Practically an anathema in printmaking, I have used cyanotype on non-rag, fiber-based paper, specifically the same Ilford Multigrade FB and RC glossy papers that I often use for lumen prints (see the print to the left). I discovered that Ilford darkroom paper produced a dreamy, almost misty image akin to dusk on the ocean. It is useless for all sense of detail, but it turns a contact printed bottle into an image that looks far away, the edges blurred and eyes squinting to clarify, with a gentle, glowing light effect that is likely very difficult to reproduce on any other paper. The only real difference I found between RC and fiber is that RC is a colder acid cyan and the fiber is a bit warmer blue. A lot of my cyanotype work is experimental. I often combine contact printing with lumen printing on various papers as in *American*.

Another way of working cyanotype is a multiple day exposure. My process is similar to "wet cyanotype" but I let the paper dry in-between repeated exposures, the use of various salts (besides just iron), and repeated developments. My effort is to create as much depth of field in an image by

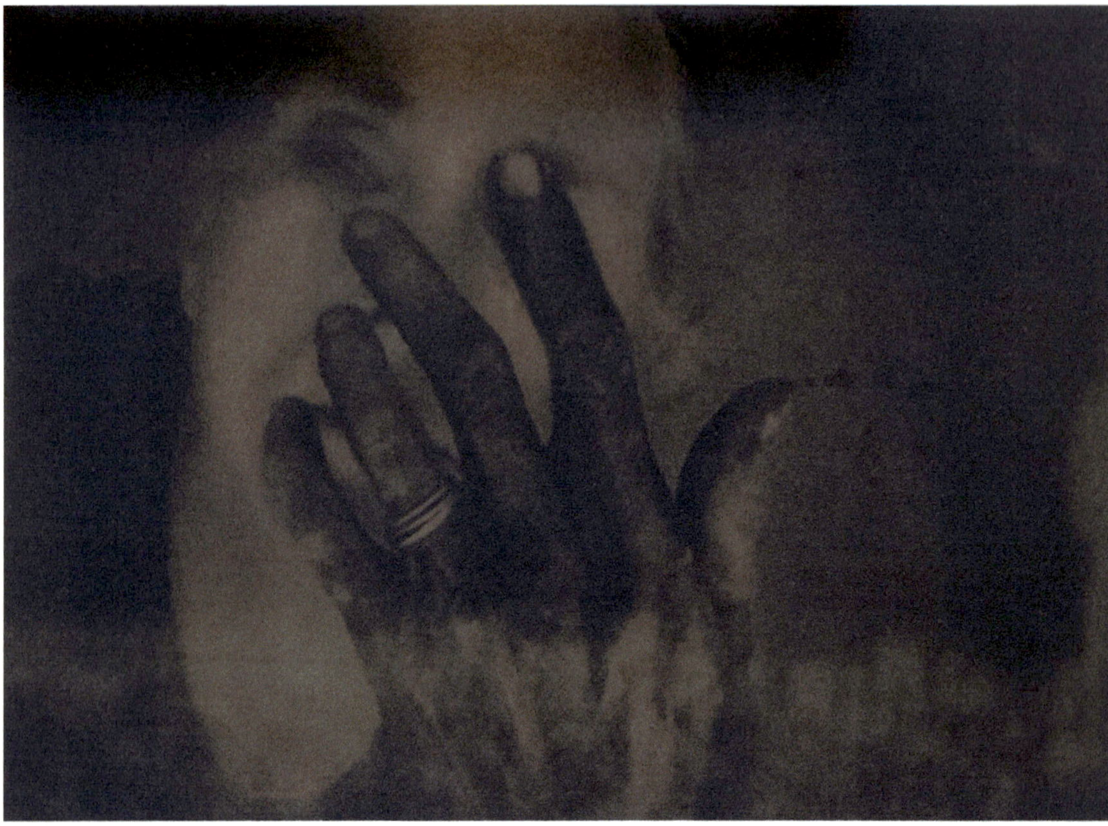

Figure 15.5. *Féin Portráid* (from photo-narrative *Widow's Work*) classic cyanotype double coated, developed in hydrochloric acid and tannin, alka seltzer, peroxide, on Arches 88 paper, 8˝ x 12˝ © Mariana Bartolomeo 2014

increasing the tonal range in stages from darkest to lightest. This is not easy at all!

One of my experiments was with Arches 88 paper which is traditionally for silk screen. It is completely unsized and highly absorbent, almost like blotting paper. First, I triple coated the Arches 88 (42˝ x 30˝) with classic cyanotype and added oxalic acid 5 g/liter. While the treated paper was drying I had the following chemistry on hand during exposure: 10% potassium dichromate, vinegar, glacial acetic acid, 10% ferrous sulfate, caffenol, 1% hydrochloric acid, household ammonia, hydrogen peroxide, magnesium citrate/carbonate, table salt, potassium chloride, potassium phosphate, in separate plastic bottles. I set these out on a bench next to the garden hose.

I have a 4´ × 4´ work table outside with a piece of plate glass on hinges attached to it. I placed the paper on the table and on it a large dried palm leaf skeleton, soaked in water to make it soft, with fruit seeds removed. I spritzed the paper with water and then lowered the plate glass to hold it in place undisturbed for an hour. Then, the paper was spritzed again, also with cyanotype solution and the glass was lowered again. This was left over night. The next day I repeated the wetting process and sprinkled the paper with magnesium powder to impart a grainy effect. On the third morning, I removed the palm and washed the paper with water and treated it with peroxide to see what I had in the way of an image. Working in full sun, I used the various toners "à la poupée" (printmaking style), washing and toning and then let it dry. I wanted a deeper blue black in the corners and some elements of the composition so I painted on some hydrochloric acid (1%), then the cyanotype solution, replaced the palm, spritzed the glass, lowered it and let it sit for another three hours. I removed the palm and washed the paper again, but this time I poured on the peroxide full strength and watched it foam up and disperse. I repeated this action 4–5 times in direct sun, then let it all sit again for another hour.

Two things happened: First the paper solarized in some areas leaving a metallic sheen, and second the paper began to break down and flake apart. Panic! I took a piece of blotting paper, laid it over the Arches 88, lowered the glass and clamped it. After two hours, I was able to slide a piece of blotter paper under the Arches and remove the top blotter. I used a bone folder (book-making tool) to press the large flakes back into the paper itself. All cotton and fiber papers contain a certain amount of glue in the plant chemistry, so it is mostly possible to repair tears or other paper problems (like falling apart) by pressing damp paper back together. I still wanted a more granular effect, so I lightly spritzed the paper with peroxide and vinegar and sprinkled areas with magnesium and sodium chloride powder. I pressed it again. I did not wash this off after it dried. Over all, I had a cyanotype lumen print that had many layers and tones of blue and was solarized. I don't have any idea if I can repeat this outcome.

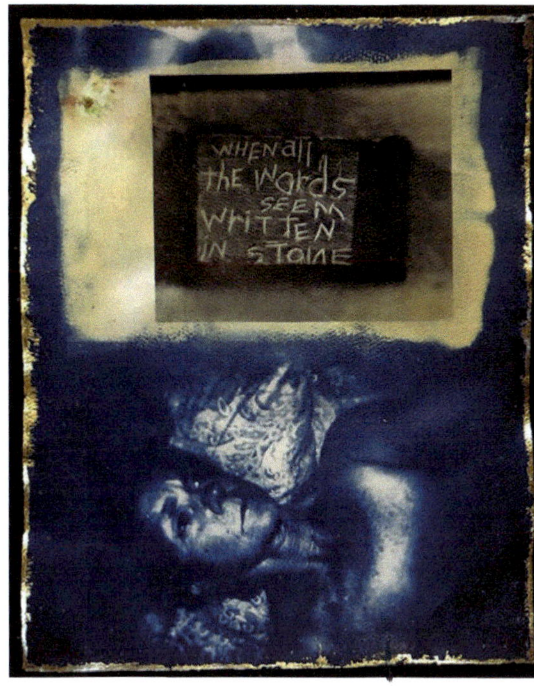

Figure 15.6. *The Stone* (from photo-narrative *Widow's Work*) contact print with two negatives, classic cyanotype, double coated, lower figure was printed first and then upper portion of the paper was bleached with sodium carbonate and re-coated to print the second negative and toned with Dektol and tannin, Canson Montval, 8″ x 11″ © Mariana Bartolomeo 2016

Creative practice

My interest in cyanotype goes back to 1960 when my father gave me my first camera, a Kodak Brownie Hawkeye and a roll of blueprint paper. He was a CPA and had a fascination with architecture. He built our house in 1947 and his own office building in 1960 and that is when I learned how to transfer a drawing to blueprint paper using wax paper, a china marker and household ammonia. I also experimented using my negatives from the photographs I took, although it was a really primitive effort and looked more like a contact sheet. But, my interest in cyanotype is an actual extension of my childhood experiences.

Artist statement

Widow's Work is a photo-illustrated, autobiographical publication about my mystical experiences related to my husband John's (also known as Don Juan) death on December 20, 2007 (The Winter Solstice and our 25th wedding anniversary) and spiritual transformation to the present. He died from lymphoma due to exposure to Agent Orange in the Vietnam Era. The text includes haiku and other poetry inspired by visions of a Dimensionless Light and our intertwined threads of existence.

Biography

Mariana Bartolomeo has been a professional artist for over 40 years. She works across mediums depending on the subject matter including: historical photographic methods, darkroom and digital processes, printmaking, narrative documentary, and handmade books. She exhibits her work internationally and is represented by Primo Piano Livingallery in Lecce, Italy. Collections include: The Cathedral Foundation, Beijing Natural Cultural Center, Gvyllm Family Trust, Center for Fine Art Photography, and Yale University/Beinecke Library Archives. To see more of her work visit **marianabartolomeo.com**.

Figure 15.7. *Teviot 4*, classic cyanotype toned with bleach and tannic acid mounted to birch plywood panel, 22″ x 15 ″ © Jennah Ward Bentley 2015

Jennah Ward Bentley

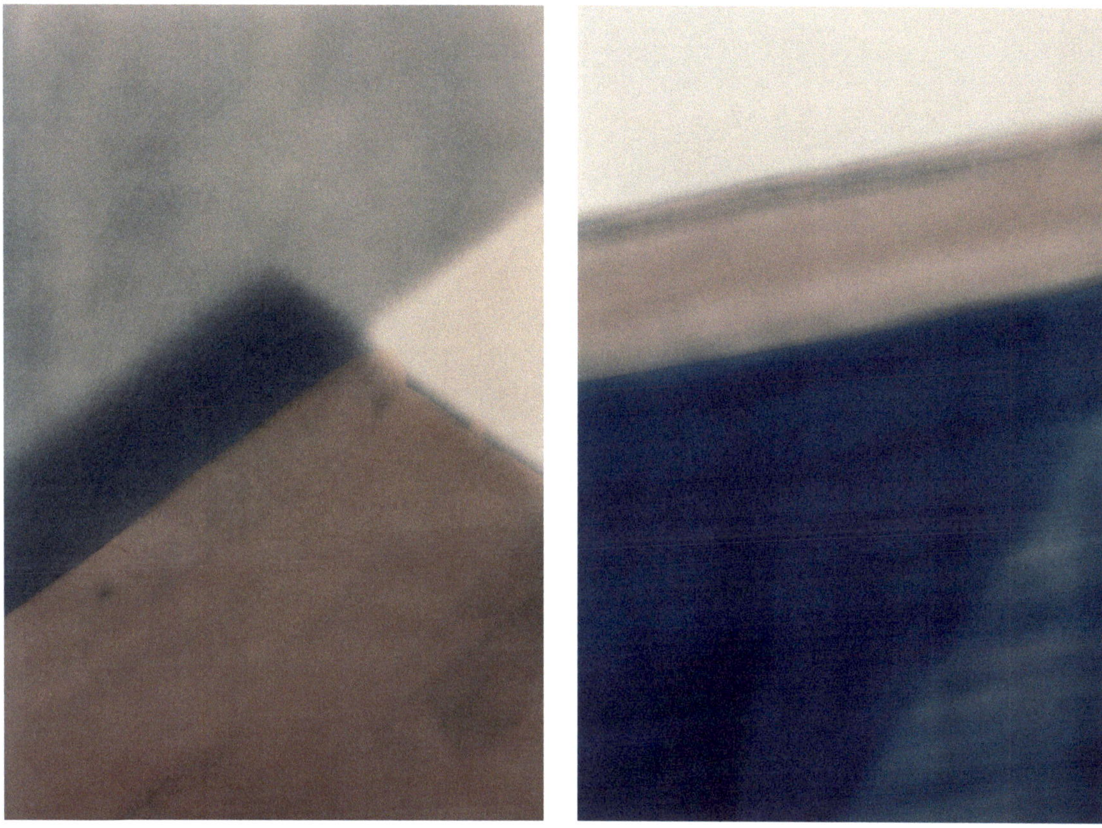

Figures 15.8 and 15.9. Left, *Teviot 10*; right, *Teviot 1*, classic cyanotype toned with bleach and tannic acid, mounted to birch plywood panel, 22″ x 15″ © Jennah Ward Bentley 2015

Process

The paper I used for this work is Strathmore 500 Series Bristol Vellum. I like the smooth, soft surface. The formula I use is the classic formula from Bostick & Sullivan. I mix equal parts of 25% green ferric ammonium citrate and 10% potassium ferricyanide and coat the paper once with a brush. The emulsion is extended past the edges of the paper so that the entire sheet is coated. Sometimes I coat the paper a second time in specific areas to emphasize variation in depth of the Prussian blue color of the cyanotype.

I don't use an exposure device per se. To make an exposure, I lay the paper on a surface in a space. The windows and doors of the space act as my aperture. The exposure times range from 30 seconds to 6 minutes. Typically, I make 2–4 exposures on each paper. With each exposure I rotate or move the paper. After exposing the paper, I fix and wash the print in a running water bath for approximately 15 minutes. I dry the prints face up on paper towels.

The *Pico* and *Helen* cyanotypes are not toned. With the *Teviot* series, I partially toned the prints with tannic acid. I used a strong bleach solution in certain areas of the prints using a wide brush, then rinsed them in a running water bath, and lastly immersed them in the tannic acid bath. While the tannic acid changed the tone of the entire prints,

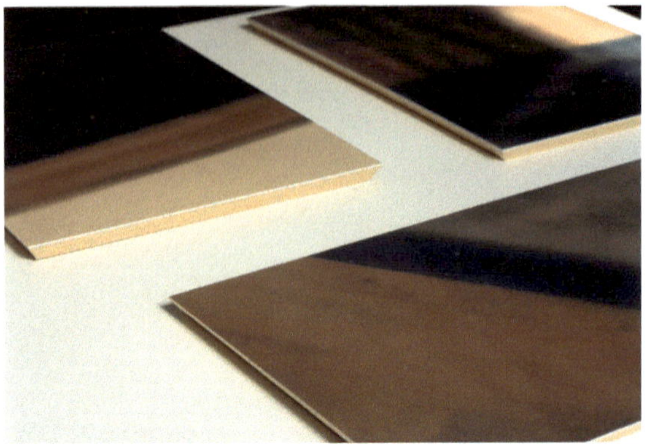

Figure 15.10. *Teviot Studio*, chamfered birch plywood panels that Ward Bentley constructs for her prints © Jennah Ward Bentley 2018

areas that were bleached also had a change in the image structure as some of the original detail was intentionally removed.

To finish the work, I mount the prints on handmade birch plywood panels. The panels extend from the wall about 1″ and are chamfered at the edges so that the image appears to float.

I had difficulty with some fine details washing out during the running water bath. I experimented with various papers and found that the smooth, soft surface of the Strathmore vellum works best to retain detail. I also had difficulty with getting really consistent results during the toning process. I don't recall that I ever got to the bottom of this problem.

Creative practice

I work within the boundaries of what defines photography. While most photographic processes utilize a camera and digital sensor or film, this work uses neither. The work is more in line with the process of a photogram, but rather than place objects on the paper, I record the light and shadow play made by the windows and doors of an interior space.

I use the cyanotype process for this work for several reasons. Technically, since the cyanotype is primarily sensitive to ultraviolet light, it suits my intention of recording the light and shadow of a specific site. Also, since I make several exposures per print and the exposures are not usually predetermined, the long exposure time of the cyanotype allows me time to move the print around in the light. Conceptually, I chose cyanotype because it was historically used to produce architectural blueprints. My work nods to architectural blueprints in that it records the light delineated by an architectural space.

Figure 15.11. *Helen 3*, classic cyanotype mounted to birch plywood panel, 22″ x 15″ © Jennah Ward Bentley 2014

Biography

Jennah Ward Bentley is a Los Angeles-based photographer. Her work has been the subject of solo and group exhibitions throughout the US and internationally. The work of Ward Bentley resides in private and public collections, including at the Albright Knox Art Gallery, the California Museum of Photography, and the Museum of Fine Art, Houston. She earned her BA at San Francisco State University in 1999, and her MFA at Art Center College of Design, Pasadena, in 2011. To see more of Ward Bentley's work, visit jennahward.com.

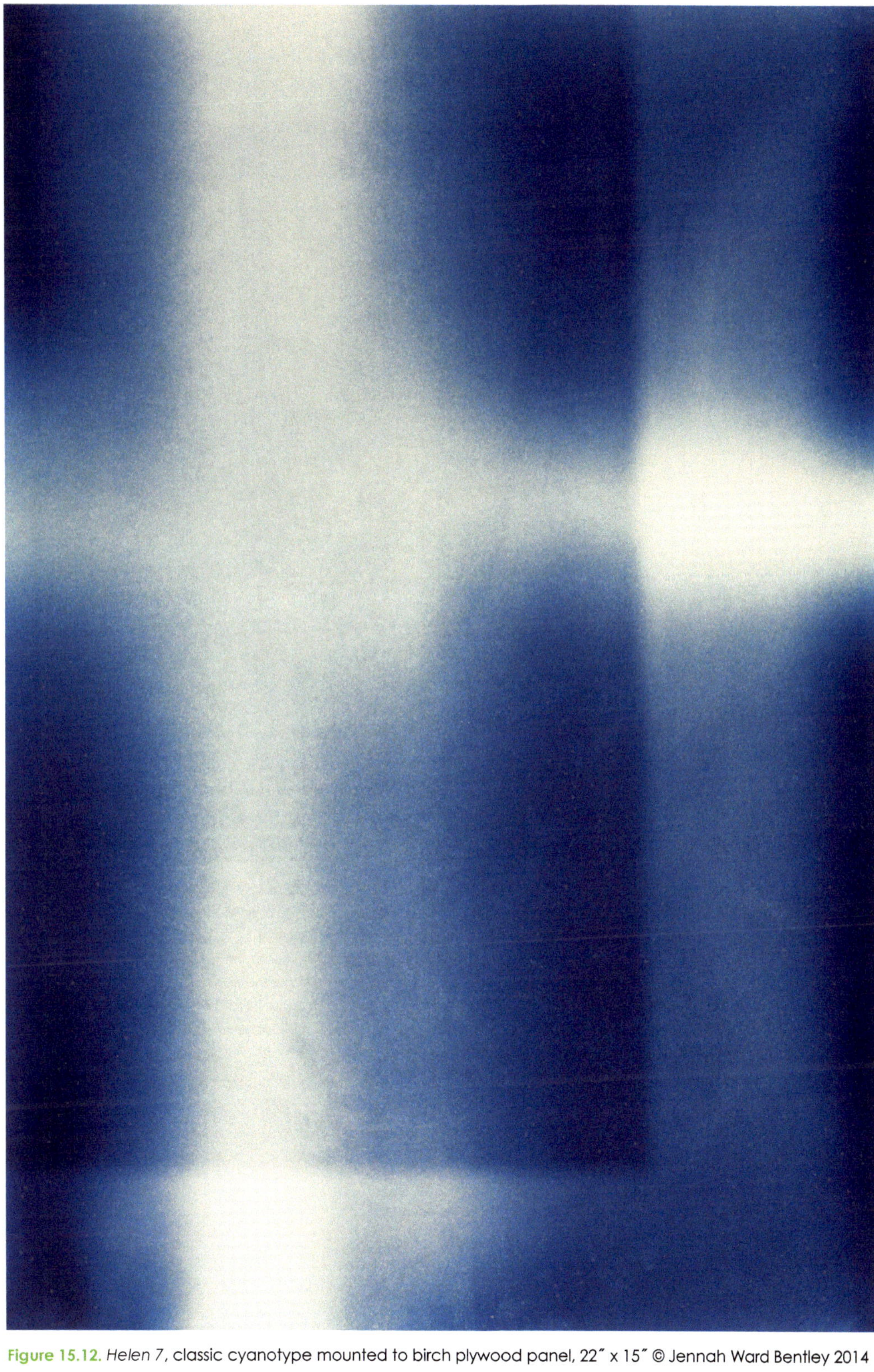

Figure 15.12. *Helen 7*, classic cyanotype mounted to birch plywood panel, 22″ x 15″ © Jennah Ward Bentley 2014

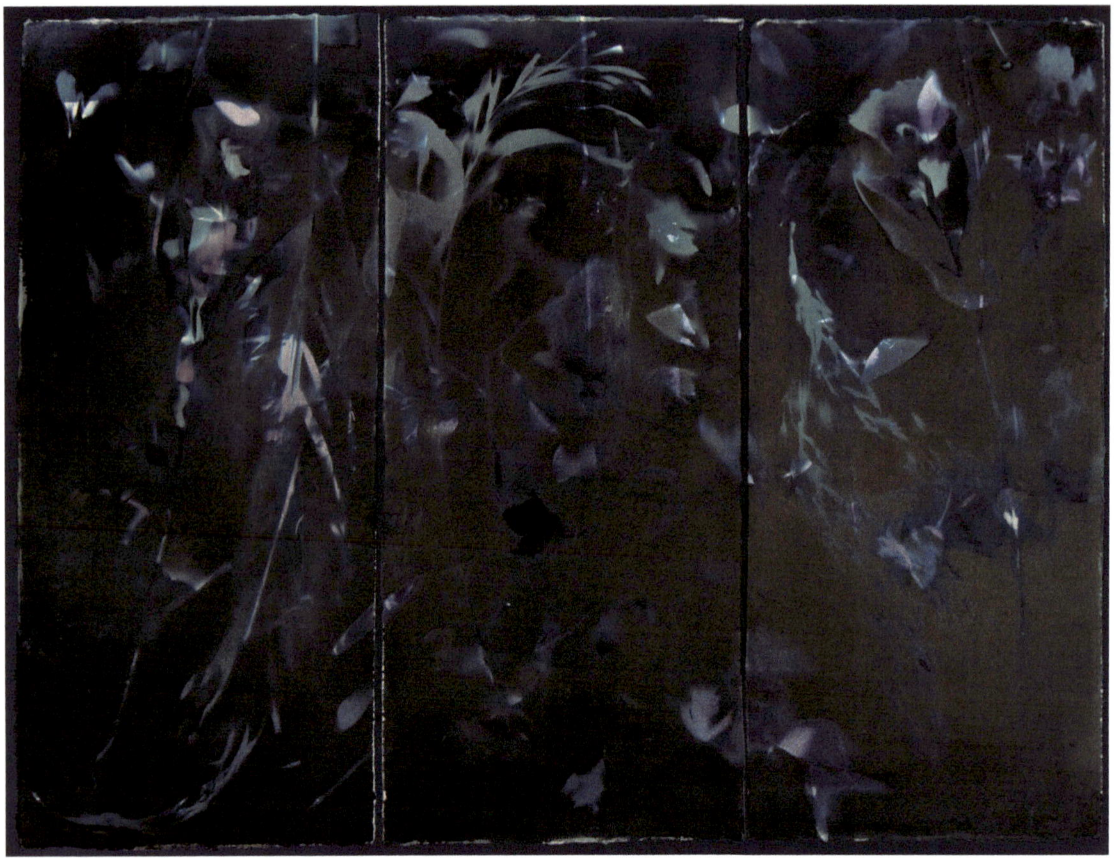

Figure 15.13. *Cuba Downpour, Triptych*, cyanotype and Vandyke brown © Laura Blacklow 2018 (author of *New Dimensions in Photo Processes*, newly revised, a must-buy). "If I start with a normal or slightly underexposed regular-strength cyanotype (processed and dried), I apply regular-strength Van Dyke over and use a normal exposure to produce a brown so dark that it looks almost black. If I start with a double-coated or dramatically (!) overexposed Van Dyke (processed and dried), I can make a slightly underexposed cyanotype on top to produce a greenish-blue. If I expose normally with cyanotype, I get faint traces of the brown, almost like a shadow, with a predominant blue print. I usually go back and forth at least four times and never size in between. However, I have noticed that fixer remover will bleach the blue print while clearing the Van Dyke. So, after fixing and clearing the Van Dyke, I do a brief wash in running tap water, then put the whole print back into hydrogen peroxide to intensify the blue without affecting the brown. Where I live and teach, I have to limit a final wash to 5–7 minutes before the city's additions to tap water start to bleach back the entire print."

Laura Blacklow

Process

Because I coat, expose, and process many times on the same sheet, I need a strong paper. I am fond of Rives BFK (which also imparts a slight green tone). I also regularly have used Fabriano Cotton and Magnani Pescia, but Pescia does start to tear after two layers and can show finger or glove prints when wet. I am pleased with recent trials on Canson Special and Strathmore 500 Series.

In the summer I use Bostick & Sullivan's pre-mixed cyanotype because it generally is slower (which I compensate for, by direct exposures in the sun) but more detailed. In the fall and in the winter, under UVBL, I use Photographer's Formulary pre-mixed cyanotype because it seems to be faster. I try to avoid mixing raw chemicals as much as possible, due to health concerns.

I have found cyanotype to be sensitive to both heat and light. Therefore, in the summer, I set up a sandwich of plate glass>coated paper>botanics>plate glass, which warms up in the sun, and I overexpose—even for hours—depending on the depth of color I want.

To develop, I first lean a piece of Plexiglas in the sink and wet it, then I suction my exposed print to it. I gently hose the print with cool running water, allowing the unexposed solution to wash off. I wash the print in a tray with running water for 3 minutes, turning it from face down to face up. I evenly slide the print into a tray of 1 tablespoon hydrogen peroxide to a liter of water, then I rock the print for about a minute while visually inspecting the change in tone. I let the print drip before moving it to the final wash, in a tray of running water, for 5 minutes. My city uses chlorine in the water, and longer wash times will bleach the print.

Sometimes I start with an overexposed Van Dyke and do a regularly exposed cyanotype on top. Sometimes I start with a deep blue cyanotype and do Van Dyke on top. If I am not satisfied, I will do a third coating and exposure and development. I am now experimenting with using both processes at the same time, such as coating Van Dyke and gently pouring cyanotype in parts while the first layer is damp.

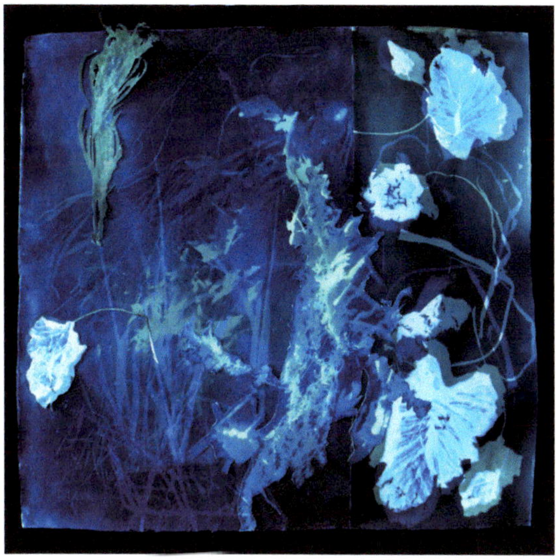

Figure 15.14. *Brigham & Women's Hospital* © Laura Blacklow 2018

Creative practice

I handmake books, and the tactile printmaking papers plus the pliable cyanotype process, are perfect for pages. I also make larger prints and collages, which I usually mount on colored archival mat board atop neutral pH foam core. I do not over-mat but I do frame with UV protective Plexiglas. Of all the historical processes, cyanotype is probably the most forgiving, so it allows me to feel free from concentrating on numerous technical details, and that enables me to experiment.

Biography

Laura Blacklow is the author of *New Dimensions in Photo Processes: A Step-by-Step Manual for Alternative Techniques*, 5th ed. (Focal Press, 2018), the recipient of a National Endowment for the Arts Regional Fellowship, St. Botolph Club Award, Polaroid's Artist Support Program, and Harvard's Rockefeller Center for Latin American Studies Grant. Blacklow teaches at the School of the Museum of Fine Arts, Tufts University. To see more of her work, visit **laurablacklow.com**.

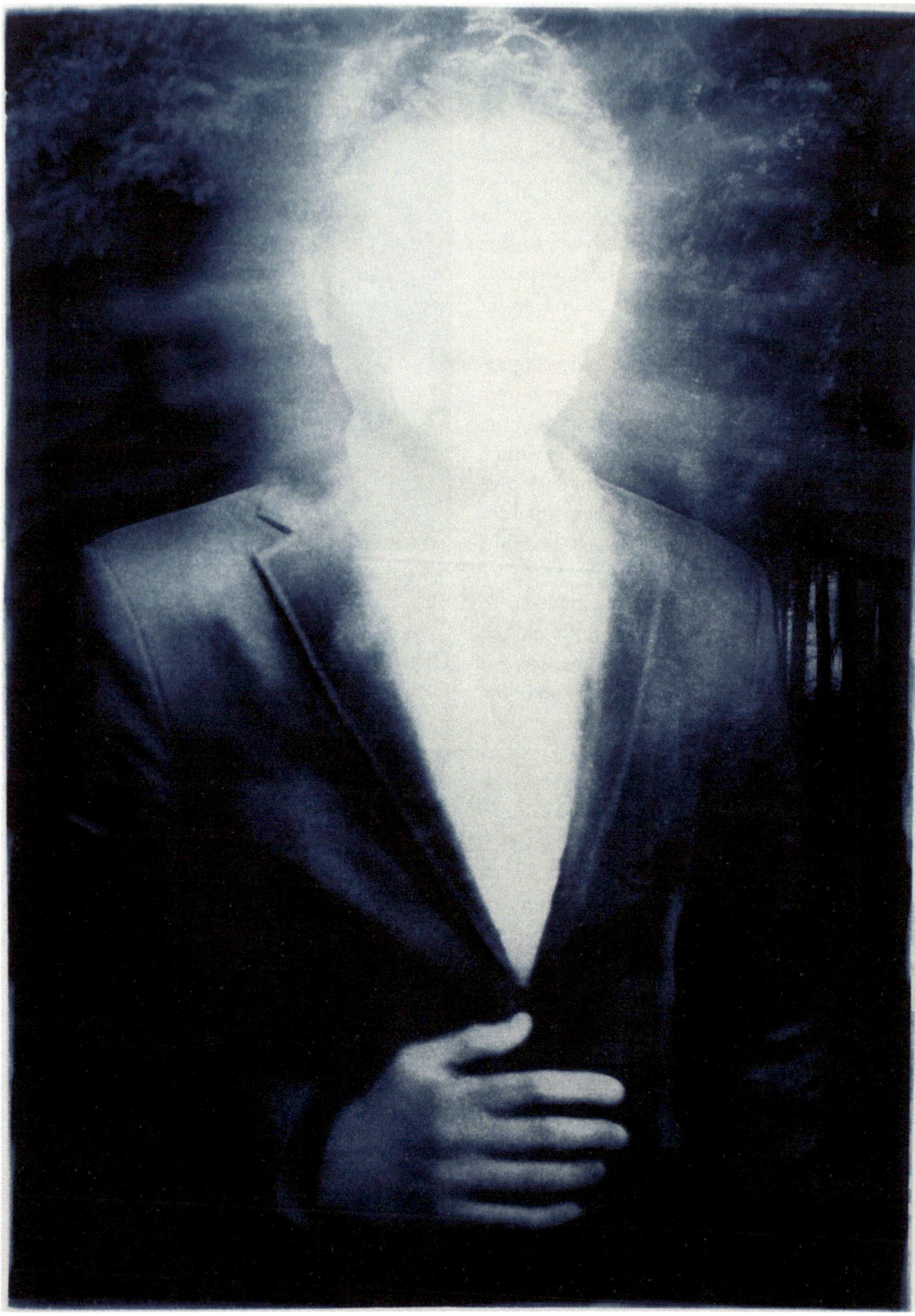

Figure 15.15. *Illuminated*, from the series *Out of The Blue* (Part I of *A Million Suns*), collaged cyanotype with digital drawing, 17″ x 11.2″ © Jonah Calinawan 2013

Jonah Calinawan

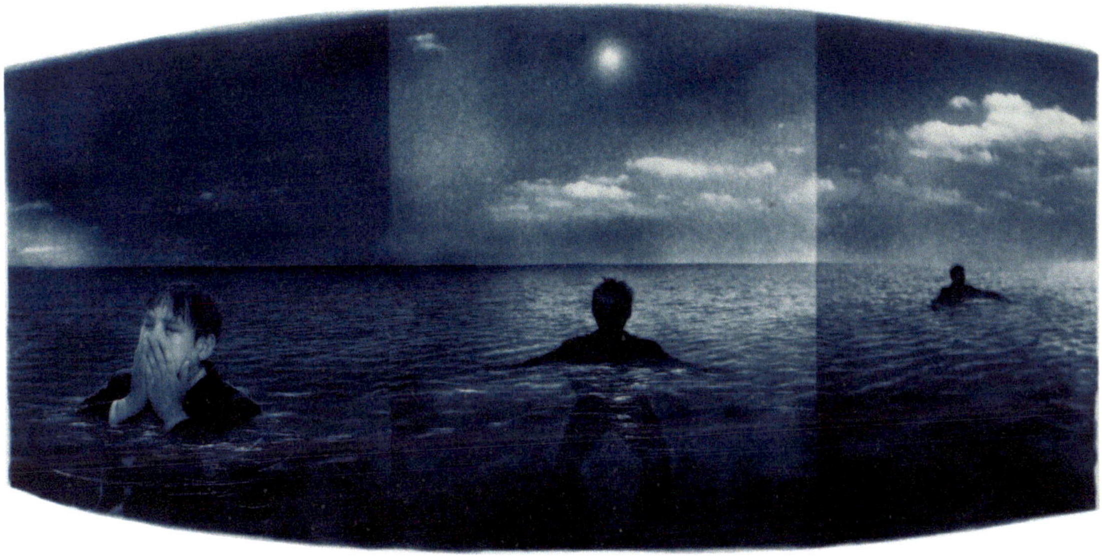

Figure 15.16. *Adrift*, from the series *Out of The Blue* (Part I of *A Million Suns*), collaged cyanotype with digital drawing, 8.3″ x 17″ © Jonah Calinawan 2013

Process

I use cyanotype for its emotional expression. I think blue is a very introspective color and that supports the intended outcome of my work in *A Million Suns*, which tells a story of a man looking for his future. To think about the future, one must look inward, to one's hopes, fears, and dreams. And I think blue, which some have regarded as a thinking color, supports my intent. I use Arches 140 lb hot pressed watercolor paper, print digital negatives using the Precision Digital Negatives system, and use the classic cyanotype formula from Bostick & Sullivan, mixed equally at time of use. I have a vacuum UV box and I expose for 12 minutes, and then wash for 10 minutes in 70° tap water. I tone with 3 g soda ash dissolved in 600 ml of water, and hand brush it in areas of the cyanotype. I hand-cut my cyanotypes and float mount them to emphasize the physicality of the print. Problems I've had: loss of contrast is one problem and it frequently means the cyanotype paper (with the dried up chemistry) is too old. I just mix a new batch of cyanotype chemistry and use it immediately. I also try to keep the water temperature the same (about 70°). If the water is too warm, I find the highlights wash away faster and the image is flatter. Of course, if I want a very clean highlight (paper white), then I might brush hot water in that area.

Creative practice

Why specialize in one color? I find that blue is a very mysterious color. All the themes that show up in my artwork—such as identity, purpose, free will, and destiny are mysterious concepts. Of all the colors, I find that blue, depending on how light or dark it is, is the most mysterious. I think of infinity when looking at a deep blue color.

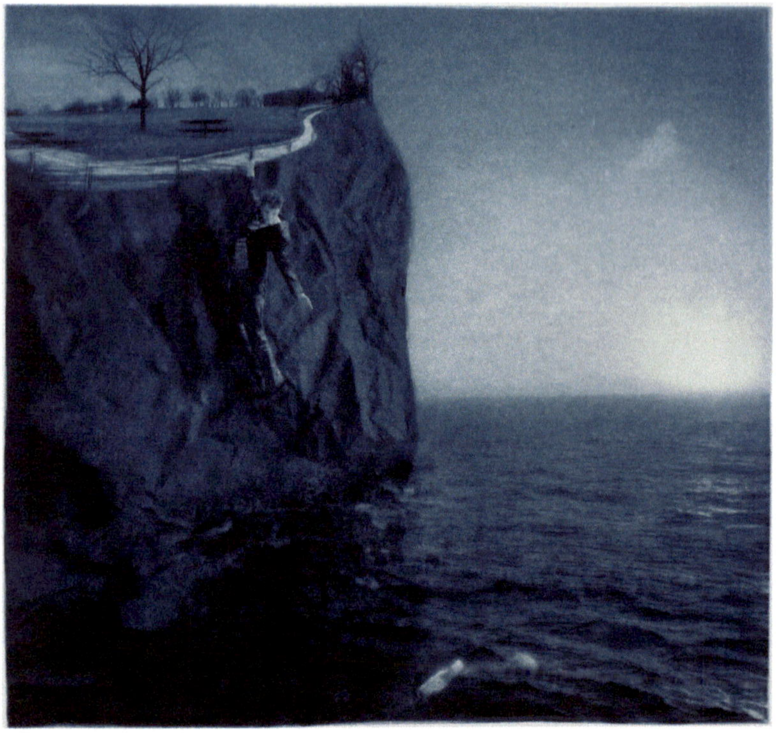

Figure 15.17. *Message in a Bottle III*, from the series *Out of The Blue* (Part I of *A Million Suns*), collaged cyanotype with digital drawing, 15.5″ x 17″ © Jonah Calinawan 2013

Blue is also a very introspective color, a quiet color. It makes you think about things. Cyanotype leads the viewer to contemplation.

Furthermore, having been an accountant for 25 years, I knew that specialization was important. I'm finding that what is true for a career is also true with art. The only way to get noticed, the only way to compete, is to specialize. Specialization is key if you want to compete.

Specialization makes me more creative. Instead of going from one technique to another technique and never being an expert at anything, by limiting my choices (like cyanotype), I get to push against those limits and hopefully come up with something original. It's an unexpected lesson that I realized along the way.

Artist statement

Merging graphic novel sequential storytelling with the stand-alone fine art photograph, *A Million Suns* tells a story of one man's search for his future. We once looked up at the stars to divine the future. How do we find this future now? Will there be a sign or a signal? Using images that are collaged, montaged, erased, and drawn on, I portray my belief in the unlimited variations of the future. Collage is like creating the future—we combine experiences and circumstances and move them around to see what fits and what makes sense. At any point, we may choose to go left or right, turn back or move forward. Each choice alters the trajectory and overall shape of our lives. There is a great mystery here: we are on a journey where there is no map—it is pieced together as we go along. Thus, we become the traveler and cartographer, the origin and the destination. For this one man's quest, his life is about to veer off its projected course, into the blue world of A Million Suns.

Biography

Born in the Philippines, Jonah Calinawan lives in the seaport city of Baltimore where he explores ideas of identity and choice in his photographic cyanotype collages. Calinawan has exhibited his work nationally and internationally. His ongoing project *A Million Suns* tells a story of one man's search for his future. To see more of Calinawan's work, visit **www.amillionsuns.com**.

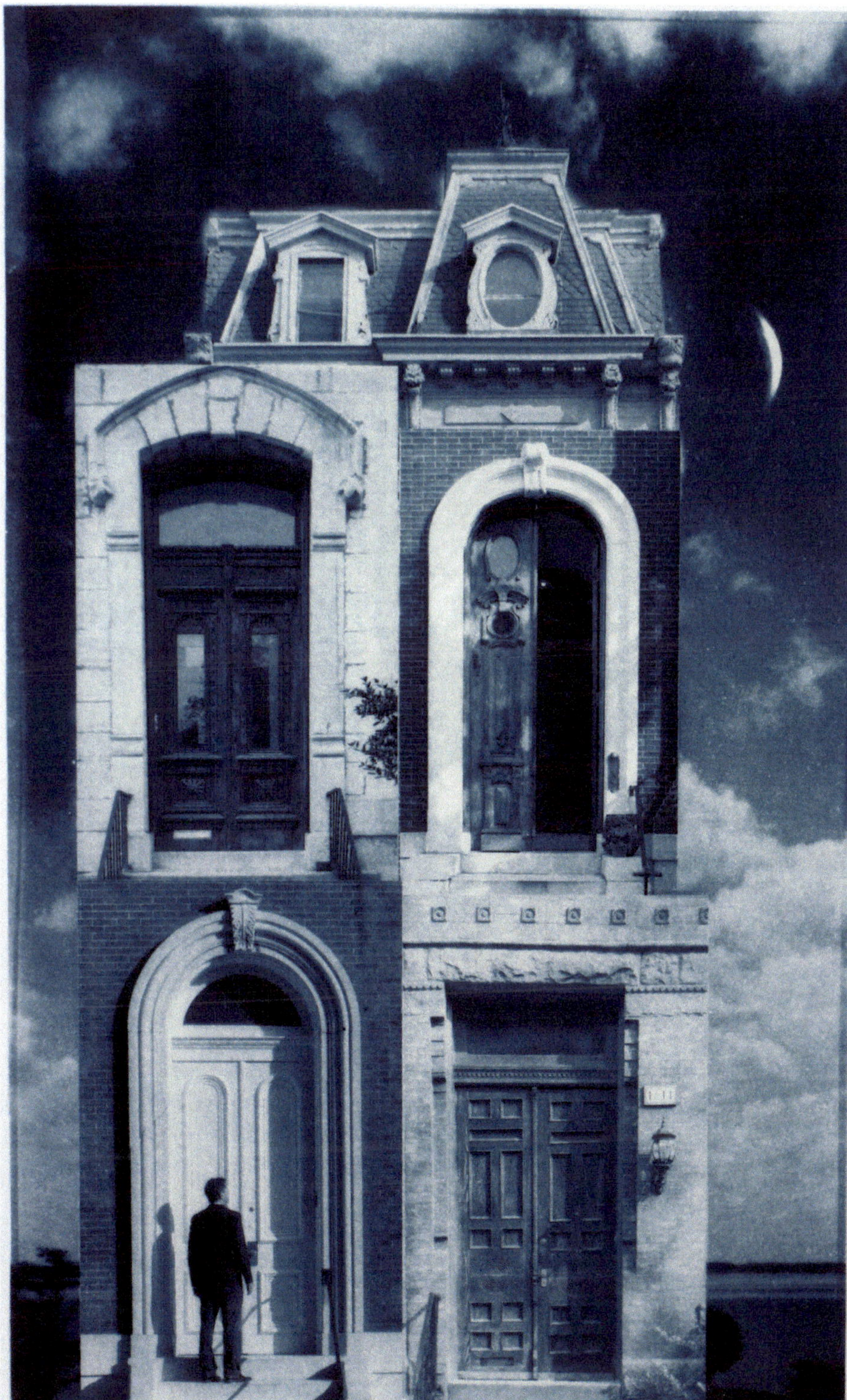

Figure 15.18. *This Old House*, from the series *Out of The Blue* (Part I of *A Million Suns*), collaged cyanotype with digital drawing, 17″ x 10.7″ © Jonah Calinawan 2012

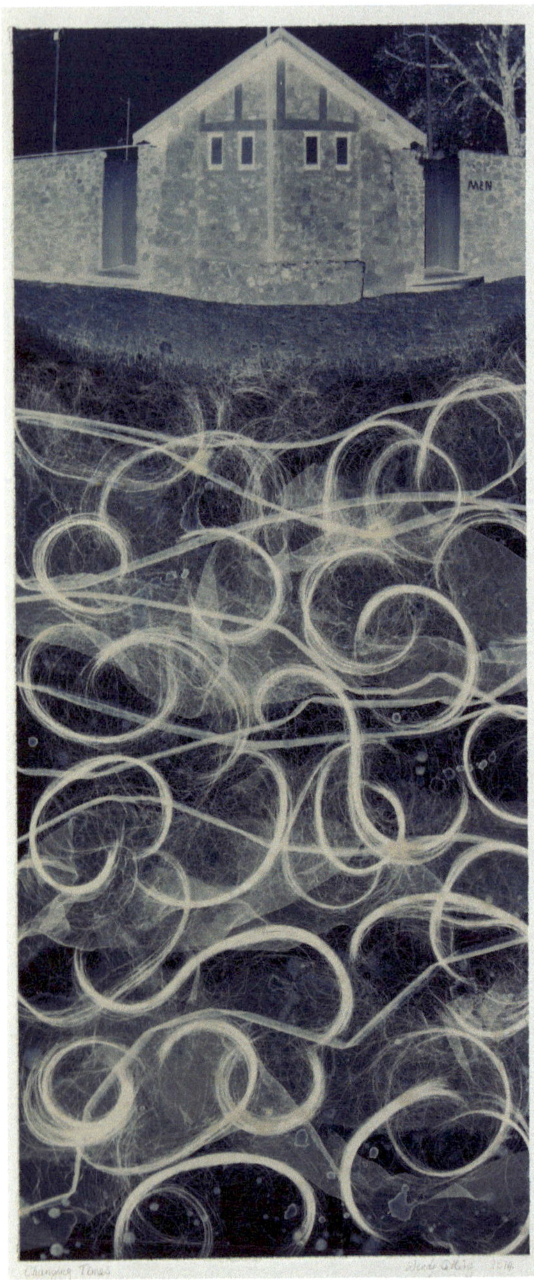

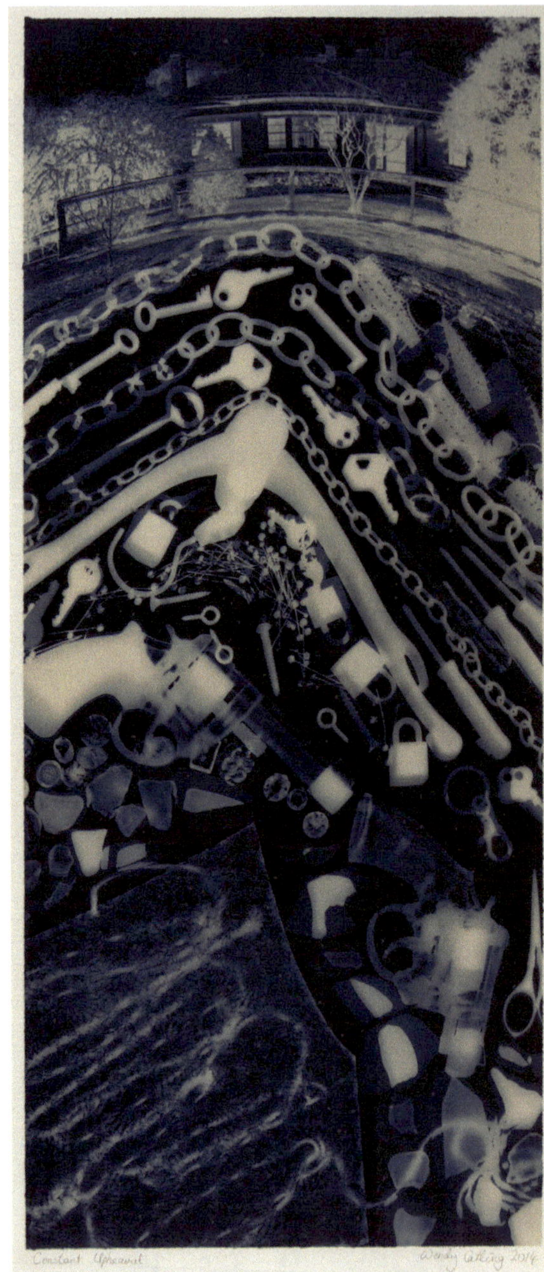

Figures 15.19 and 15.20. Left, *Changing Times*; right, *Constant Upheaval*, new cyanotype on Stonehenge, digital images distorted in Photoshop, printed onto Folex film, hand cut or scratched into the film emulsion, contact printed, along with photogram exposed found objects, hair, fabric, glass, mesh, leaf litter, glassware, insects, and water dilution, 67 x 29 cm © Wendy Catling 2014

Wendy Catling

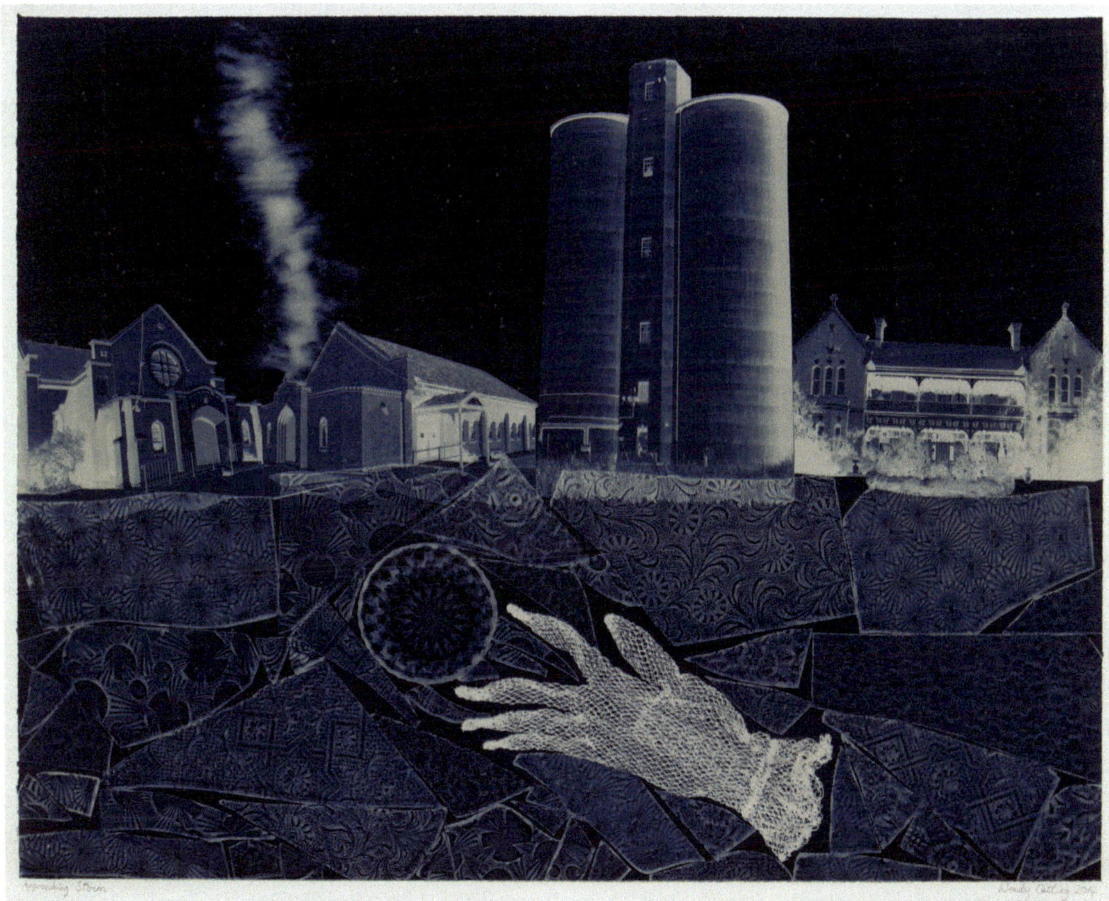

Figure 15.21. *Approaching Storm,* new cyanotype on Stonehenge, digital images distorted in Photoshop, printed onto Folex film, hand cut or scratched into the film emulsion, contact printed, along with photogram exposed found objects, hair, fabric, glass, mesh, leaf litter, glassware, insects, and water dilution, 53 x 68 cm © Wendy Catling 2014

Process

For the *Hard Light* series, I was looking for a variation of blue that expressed the appropriate mood for my ideas. I experimented with Mike Ware's new cyanotype sensitizer onto a variety of papers including Arches Platine, Arches watercolor, Hahnemühle photo rag inkjet, Kozo rice paper and Buxton paper. It seemed to me that these papers gave tones of blue that were too bright and cheerful. I settled on Stonehenge because the dark hues were best suited to the dark intensity of my memories. More recently I have been using Buxton Platinotype paper, which I order from Ruscombe paper mill in France. It responds beautifully to the sensitizer without issues, and when I need darker hues I will explore toning.

Stonehenge paper proved to be a difficult choice when paired with new cyanotype solution, which is very sensitive to additives in the paper. When I wanted to splash water onto the coated paper it created undesirable stains in the highlights. My solution to the problem was a combination of:

one drop of a 40% solution of citric acid per ml of sensitizer, work quickly not allowing the water drops to sit too long on the coated and dry paper, and sponge water drops off with paper towel.

I shoot digital images with a Canon EOS. These are contrast adjusted and distorted in Adobe Photoshop. I like to work with positive film so that my cyanotype images will print as negatives. After printing onto Folex inkjet repro film I scratch textures into the film positive and sometimes cut away areas to better integrate the contact printed image with the photogram part of my print.

My exposures are mostly done in bright sunlight between the hours of 11.00AM and 3.00PM. This time can vary depending on the time of year. I have a UV light box but for photogram prints I prefer the long, one-directional shadows of sunlight. The process of exposing prints in sunlight was also an important part of my conceptual approach to image making in the *Hard Light* series. For Australians, skin cancer is a huge problem. Working and playing outdoors in the amazing Australian light is potentially dangerous. I take great care after having four carcinomas removed. Working in sunlight I am reminded of the invisible dangers that as a child I was exposed to, so that now when I harness the ultraviolet radiation it becomes a metaphor for taking control of malevolent forces.

After exposure I move the prints continuously through three changes of water in my darkroom. Approximately five minutes overall, and then hang to dry in the darkroom. I have not explored toning yet.

My *Hard Light* series of prints have neatly torn edges and are glued to a larger sheet of Stonehenge paper. My mounting process is this: Yamato rice paste glue is brushed onto the back of the print. Both the print and the extra Stonehenge paper are dampened well with a water spray. They are allowed to absorb the water and then run through an etching press: first a sheet of 1 mm acetate cut to size with a 2 cm margin bigger than my print, the print is laid face down on the acetate, a larger sheet of paper is laid onto this. I use a paper template to position everything. The result is a "plate mark" making a subtle margin around the print. The print can then be float- or mat-mounted in a frame. It looks beautiful but is an expensive use of paper!

My best advice for working with cyanotype is to keep detailed notes of what you have done: all the materials, techniques and processes. I am not overly scientific in my approach but I have found that keeping an organized system of visual diaries documenting my experimentation with smart phone snaps and annotations has been an excellent way to keep track of and repeat successful methods, and to record accidental results, which may be useful later.

Creative practice

I have long been attracted to the color blue. After completing a Bachelor of Arts in Printmaking, I worked as a textile designer and colorist in London in the late 1980s. During this time I became skilled at hand painting tiny floral designs onto the navy or Prussian blue backgrounds, which were fashionable at the time. My daily work revolved around colors; mixing them, matching them, observing how they interacted with one another and creating moods with color.

It was not a new experience for me to be entranced by a single color. On one memorable visit to the British Museum's Egyptian department I saw the most astounding blue stone amulet. It was a bit of an epiphany moment; I could not tear my eyes from it and my fascination with blue was cemented. I love indigo dyes and was a designer of bleached denim screen print patterns for a time.

Photography has been a guilty pleasure since art school. When I first discovered and explored the photographic medium of cyanotype in 2012, my past experience as a printmaker made me appreciate the qualities of working with handmade paper. I was attracted to the cyanotype as a perfect medium for the making of unique, individual hand wrought photographic artworks. The many possible variations of Prussian blue seemed especially sublime and yet also to encapsulate a sense of melancholy that I closely identify with. Associations with blueness such as "feeling blue" or the results of violence being "black and blue" have particular poignancy for me as someone who has grown up with domestic violence in my family. Blue and blueness color my view of the world and are important in my creative output. The potential of the cyanotype medium to merge visual 'truth' with emotional and imaginative expression, along with the physical process of harnessing chemistry and ultraviolet light drive my current work.

Artist statement

For the *Hard Light* series, I have reclaimed the intense sunlight remembered from my childhood during the 1960s in a small Australian wheat belt town. Light filters through personally significant objects and photographic imagery, recreating remembered and imagined places and portraying the ghostly shadows of my past. Symbolic and expressive imagery become metaphors for family violence, social anxiety and memory, to represent the trauma of a child trying to understand and survive a violent family life. Each work in this series presents a significant place in the town of Yarrawonga, which was a literal oasis built on a lake and river in a wide, flat, dry, grain-farming landscape. Harsh truths and unhappy lives barely hidden underneath a superficially pretty surface erode the tranquility and beauty of the town. The lake, populated with dead trees standing and fallen into the strange swirling currents of the dark water, becomes a metaphor for monstrous things lurking below the visible surface. The school on the banks of the river, a safe haven for children from dysfunctional families, stands firmly for now, while murky floodwaters carve the river bank closer and closer.

Biography

Wendy Catling's work explores traces of human experience, reflecting on the anxieties that remain indelibly in the memories of victims of trauma. Catling is completing a Masters in Art Photography at Photography Studies College, Melbourne, Australia and teaches art and photography. She has shown in Australia and the United States. To see more of her work, visit www.ettamodern.com.

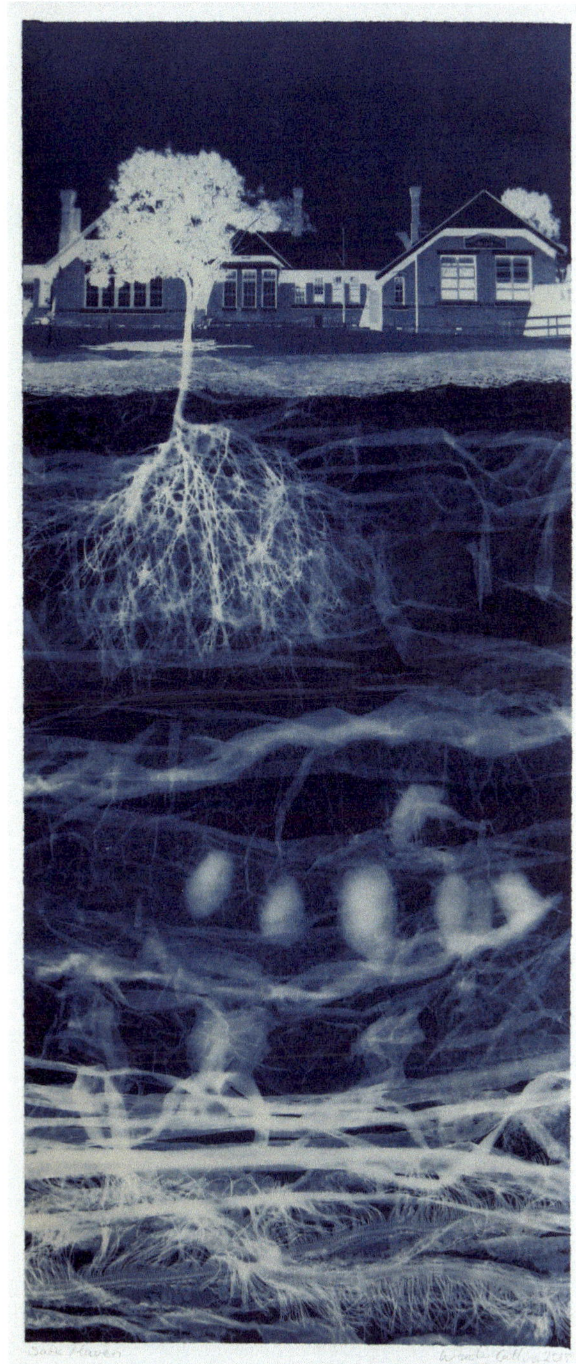

Figure 15.22. *Safe Haven*, new cyanotype on Stonehenge, digital images distorted in Photoshop, printed onto Folex film, hand cut or scratched into the film emulsion, contact printed, along with photogram exposed found objects, hair, fabric, glass, mesh, leaf litter, glassware, insects, and water dilution, 67 x 29 cm © Wendy Catling 2015

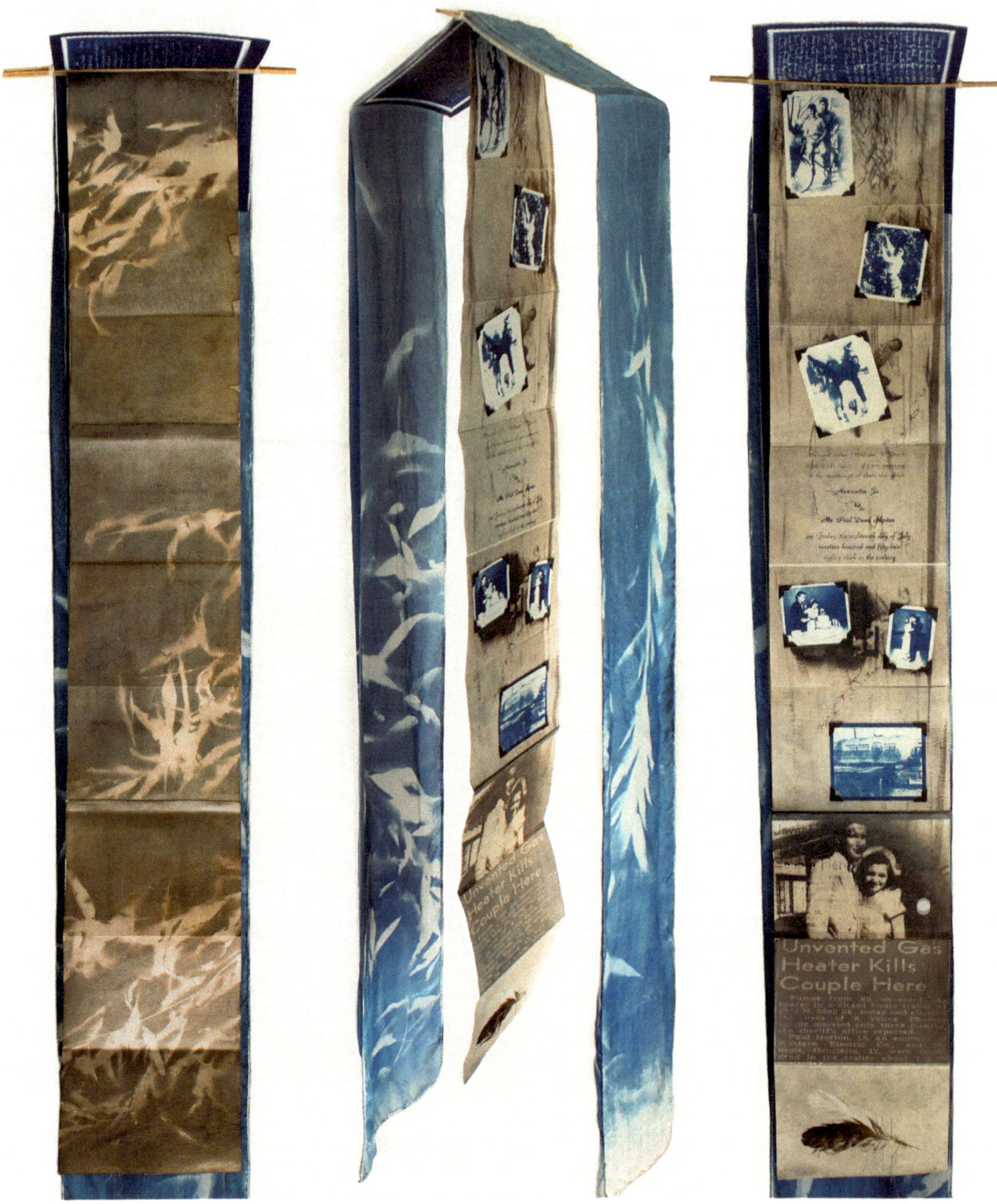

Figure 15.23. *A Short Story*, traditional cyanotype on kozo paper, watercolor paper and silk scarves using composite negatives and photograms. Photo corners were used to hold small cyanotype photos onto the accordion fold pages along with PVA glue. Parts of the kozo paper were toned in coffee and wine tannin. Bamboo stick, feathers and thread were attached to the book, 46″ x 8″ © Kimberly Chiaris 2016

Kimberly Chiaris

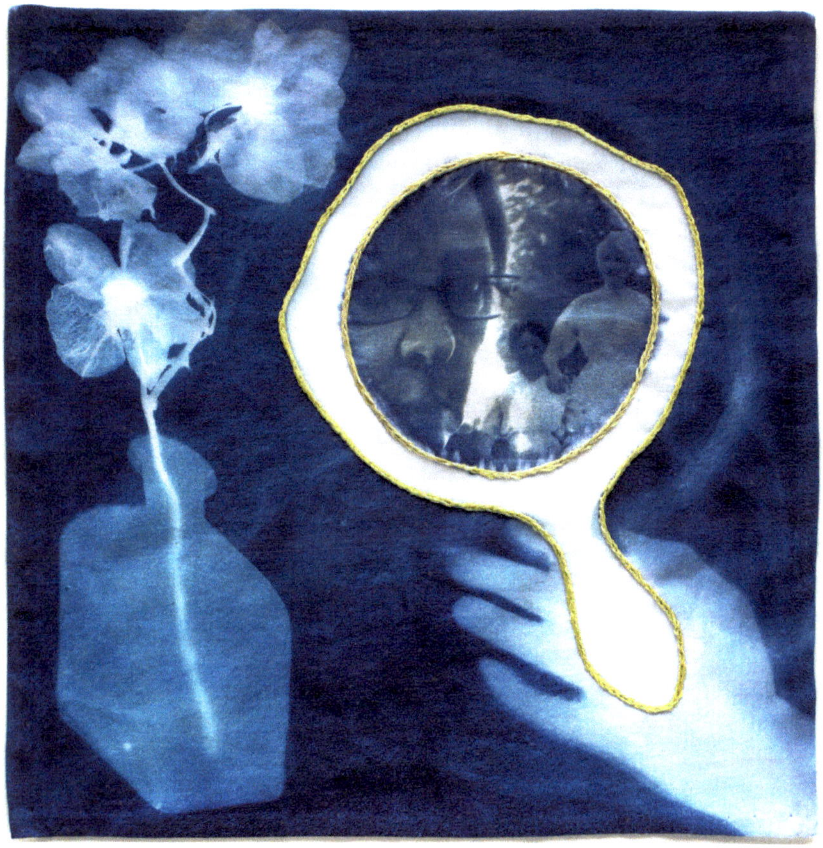

Figure 15.24. *Genetic Memory*, traditional cyanotype on 100% cotton with embroidery, photogram and photographic composite, 8″x 8″ © Kimberly Chiaris 2016

Process

I often use papers or fabric that have a translucent or delicate physical quality. Japanese gampi and kozo papers are a favorite. I also like to use 100% cotton drafting vellum. I sometimes add silver and gold leafing to the front and back of my cyanotypes and these thin papers let the metal shine through the translucent paper, especially when I add varnish or shellac to these papers. Fabrics like cotton, silk, and linen are great because I can sew on them or sew them to another object. I sometimes apply embroidery and tatting to enhance or emphasize areas of the photographic image on the fabric. I also use various watercolor papers as well as Hahnemühle Platinum Rag. I use found papers from old books and music scores, old personal letters as well as inherited family treasures—handkerchiefs, tablecloths, napkins, pillowcases, doilies, etc. I am constantly looking for new and unusual objects to use for my cyanotype work. I like to push the boundaries and combine different materials together; constructing two-dimensional work, books, and sculptural work. I often combine different kinds of papers with cloth, incorporating thread, found objects, wood, bone, animal parchment, tin and more. The possibilities are endless.

I make my negatives with files on my computer that are a combination of images that I capture with a digital camera, copies of old family photos and old illustrations. I make composites in Photoshop. I also use found objects and organic matter to contact directly as photograms.

Everyone has a preference for the tonal ranges of their images. I've found that if I make my digital negatives with more contrast, it brings out the highlights and shadows. When I work on cloth, this helps to define the image. It also helps to clarify images that have multiple layers of information.

I use the classic cyanotype 25%/10% kit from Bostick & Sullivan. I use an exposure box that my (creatively handy and helpful) husband and I built to expose my images. Inside and underneath the top of the box we placed LED light strips. The exposure time is around 10 minutes. I will occasionally use the sun to expose. The exposure for sun is usually 8–15 minutes in full sun between 11AM and 3PM in Colorado.

I develop my exposed prints in running water until the yellow is mostly gone, about a minute, then fill the tray with water and add about ¼ cup of white vinegar and soak it for about 30 seconds. I empty the tray and fill it up with fresh water and add about ½ teaspoon hydrogen peroxide and mix well before putting my image back in to soak another 30 seconds. All this is done with a splash of this and squirt of that. I don't really measure it out. Then I wash with fresh running water for approximately 30 minutes.

Sometimes I will tone the images with wine tannin and strong coffee. This will darken the image so I first bleach my image to make it lighter with washing soda (which is very strong and only needs a few grains in a tray of water and dissolved well). I mix the wine tannin and the coffee in a photo tray. I add about a cup of very strong coffee with about a tablespoon of powdered tannin. It can be diluted more to slow the toning but I have found that it takes a while to tone well. I also have taken a water soaked print and painted the coffee and tannin onto specific areas of a print or held part of the print in the toner to get varying degrees of tone in the print. Then I wash the print for about 20 minutes and dry.

I frame, mat and display my art based on what each piece needs to highlight its qualities. If my work is two dimensional, I frame it in a shadow box frame. The image is mounted on mat board and foam core to make it float in the space. I want to emphasize the precious value of each piece and, occasionally, the delicate translucency of the paper. Each frame is handmade specifically for each piece. If it is shown in a gallery, I usually paint the frame black or white.

The cyanotype process has not been a difficult process to learn. Pay attention to the fiber content of your material. One time I used some fabric that had a blend of cotton and synthetic fiber in it. The synthetic fibers didn't allow the cyanotype to absorb well and it came out very light. I make sure to use an all natural fiber material now.

I often work with thin fragile paper that can fall apart in water easily. To keep the paper from falling apart or tearing, I use a paper making frame that has a fiberglass window screen stapled to a wooden open frame. I take the screen and turn it upside down and lay the paper on the screen. Then I carefully lay the screened frame in a tray of water and agitate it very gently to keep all the fibers from disintegrating.

One book that I made (see p. 198) hangs vertically using watercolor paper for the top structure or spine of the book, kozo paper for the accordion fold pages that unfold downward from the watercolor paper cover. Hanging over the top of the book structure are two silk scarves flowing down along the side of the accordion fold pages. This forms a house shape with a peak at the top (the spine of the book) and the silk fabric makes up the sides of the house shape. Cyanotype images are on the watercolor, kozo and silk fabric. The kozo paper is toned in wine tannin and coffee.

At times I use fabric for my cyanotype work. One of my pieces (see previous page) is made from object photograms exposed in the sun along with a digital negative and embroidery added later. I placed a hand mirror down directly on top of the sensitized cotton fabric with my hand holding the handle. I laid some dried flowers next to the mirror. I then placed a glass bottle near the fabric but not directly on it letting the shadow fall over it. This made it look like the flowers were coming out of a translucent vase. I developed the fabric and dried

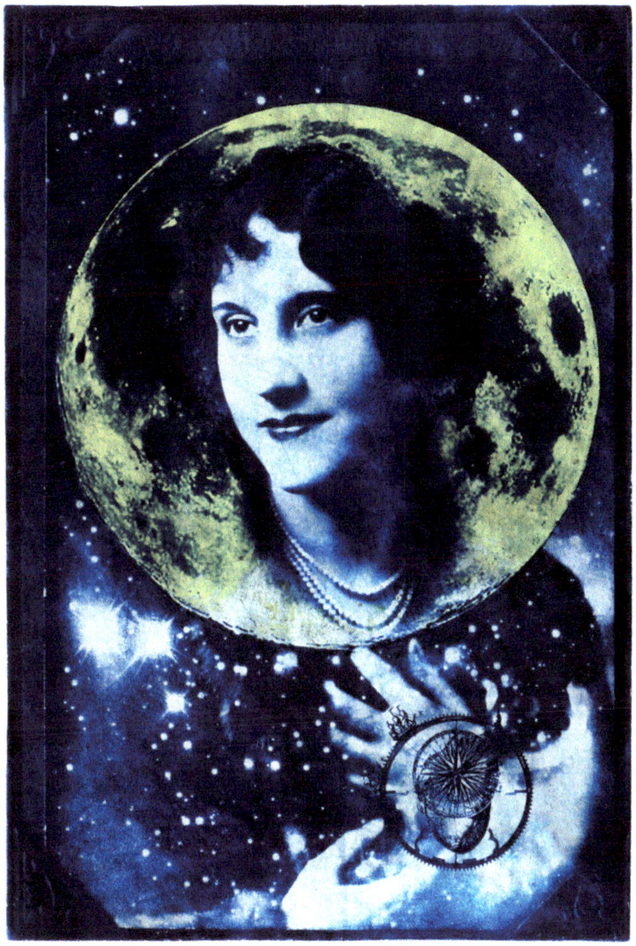

Figure 15.25. *Urania*, traditional cyanotype on Vidalon vellum with gold and silver leafing on verso and Krylon conservation spray shellac, 7.5" x 11" © Kimberly Chiaris 2018

it. Then I painted a circle of emulsion on the part where the hand mirror would be. After that circle of light-sensitive emulsion dried, I re-exposed that part by laying a Photoshop composited digital negative in that circle. When that was developed and dried, I embroidered around the outside of the hand mirror shape with yellow thread. The result looks like a person peering into a hand held mirror and viewing her ancestors in the distance.

Creative practice

An important part of my creative practice is to keep a journal. I write down dreams, notes about other artists that inspire me, quotes, sketches that help me to remember structural ideas, notes from tutorials, and conversations with friends. These seeds of ideas are the beginning of a long process that frequently culminates into a final work of art.

Artist statement

My work is about how trace memory, emotions, time and culture shape and shift our understanding of history. I blend contemporary ideas and images with personal and found historical artifacts to create newly imagined narratives that enhance the idea that recollection and interpretation are malleable. By combining contemporary digital processes with historical processes like cyanotype, I am creating new paradigms about how we remember, interpret and record our histories. I am making my own interpretation of historical artifacts for future generations.

Biography

Kimberly Chiaris' creative focus is mainly on themes about origin and identity and how trace memory, emotions, time, and culture shape and shift our understanding of history. Chiaris works with cyanotype and mixed media. She received her BFA in Photography from the Kansas City Art Institute. Her work has been shown in photography magazines and fine art galleries across the United States. She has taught numerous photo-based art classes and workshops at universities, colleges, school co-ops, museums, galleries, and nonprofit organizations as well as to individuals. She currently teaches cyanotype workshops through the Center for Fine Art Photography. To see more of her work, visit **kimberlychiaris.com**.

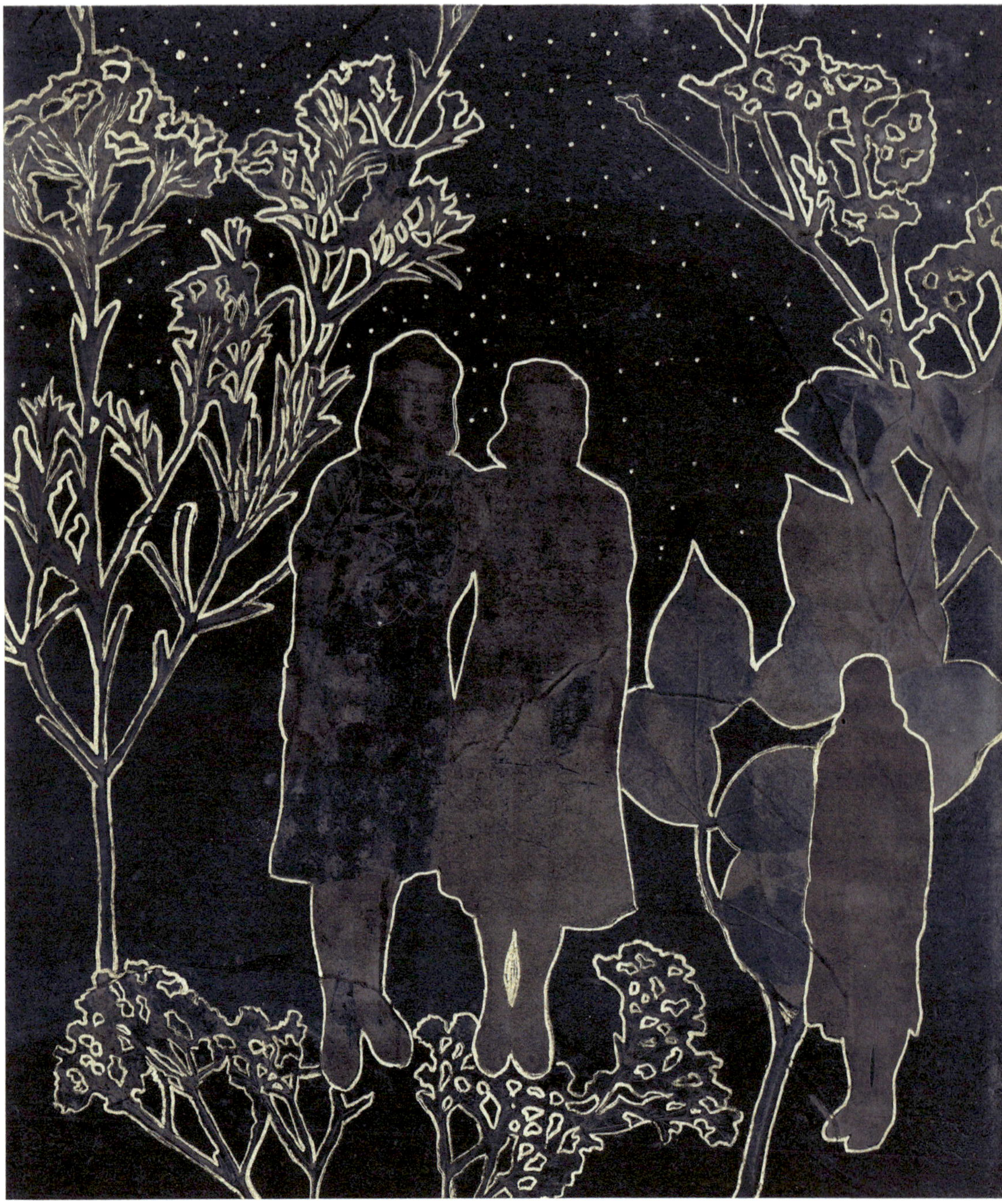

Figure 15.26. *Where Nothing is Broken or Lost*, cyanotype and gold leaf on Hosho paper, 11.25″ x 9″ © Monica Chulewicz 2018

Monica Chulewicz

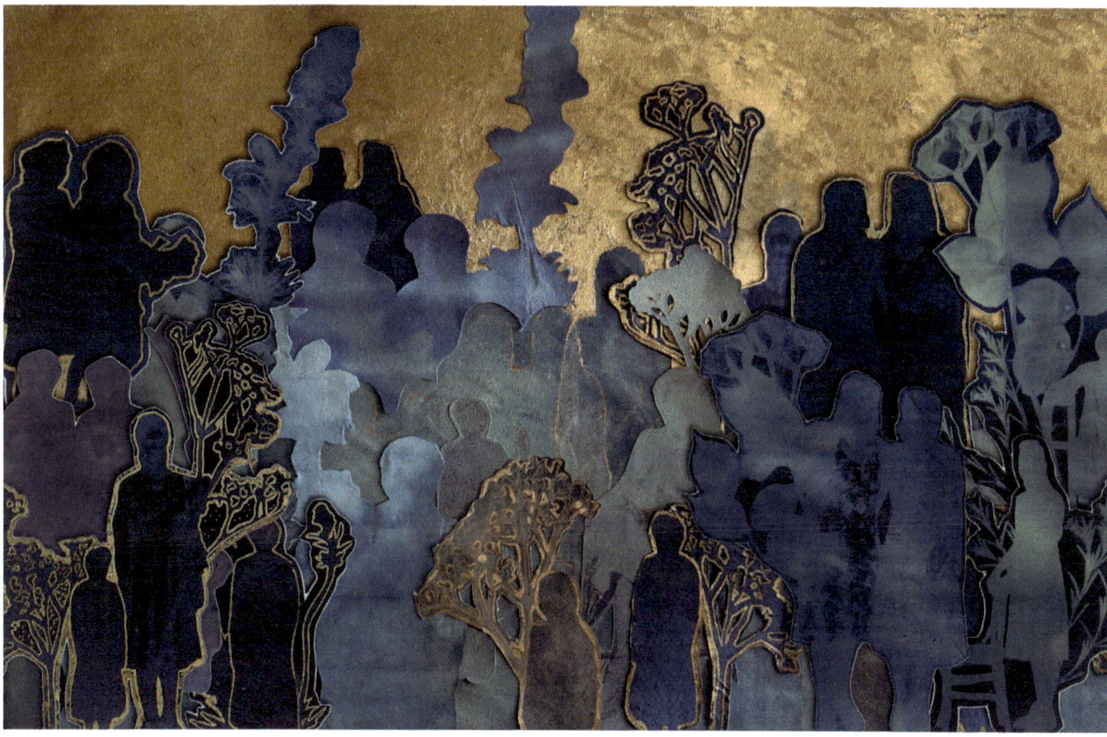

Figure 15.27. *Solace*, cyanotype on Hosho and Ingres papers, collage, gold leaf, 19˝ x 30˝ © Monica Chulewicz 2018

Process

I love the versatile nature of cyanotype. I work as a printmaker but I prefer to create unique 1/1 prints rather than a large edition. Using different exposure times or papers, as well as toning, is a huge benefit in this process. I aim to make each of my prints different from the one before it.

I've experimented with a few different paper types but my favorites, as of right now, are Hosho and Ingres. They are very different in the way they absorb and wash out the chemicals, the drying times, and the overall look of the finished print. Hosho really absorbs the chemicals significantly more, making the wash out and drying long. However, the longer the paper sits in the water, it loses strength and I have ripped some prints in the process. It seems to work best on smaller sizes due to this. Ingres, on the other hand, absorbs only a little, and washes out and dries fairly quickly. I enjoy experimenting with the different shades this paper comes in, as it affects the final print. The colors of both of these papers also vary. If exposed the same amount and using the same formula, Hosho papers print very deep blues and less contrast than Ingres.

My negatives are created by printing on transparencies in a laser printer. I work with digital photographs, as well as scanning flowers directly.

I use the classic formula. I have experimented with ratios; using double the amount of water, for example, leaves the finished product a lighter blue.

I use an exposure unit with UV bulbs and I typically expose 17 minutes, but it can vary slightly.

Depending on which paper I'm printing on, the wash time will be different, due to how much the paper absorbs the solution. I always develop with hydrogen peroxide, though. I occasionally tone with coffee but it depends how much I pour in and for how long. Each time is a little different.

My cyanotypes are matted on a white mat in a white, wooden frame.

When I was teaching myself cyanotype, I encountered problems in the process. One of the first mistakes I made starting out, I would stack sheets instead of spreading them across to dry (I have a small light-sensitive area). This led to some strange effects in the background, but not something I was fond of. After that batch of papers, I immediately stopped using this method and, instead, I spend a longer time coating papers so that I could make sure each piece is fully dry before continuing to the next. Now, I coat two large sheets of paper at a time, have two fans on them to speed up the drying process, and then repeat steps.

Not only was this a learning experience, but, I developed a series out of the process. What happened to the print that I first noticed this strange effect in? I almost tossed it, until a friend suggested I experiment to create something "different." This is where I began using gold leaf, covering the background in dots that look like distant stars, and the rest is history!

Creative practice

I use vintage photos of women, the silhouettes of them, and scans of live flowers in my cyanotypes on paper, each cut on its own as if it was to be collaged. Each of my paper cyanotypes are hand-editioned in gold leaf, which I hand draw using an 18k gold leaf pen. I have several of the same transparencies that I use over and over in my work, mixing and matching, as well as using different layouts of how I place them on my papers. A majority of my prints therefore end up as collages.

I have two modes of printing: a base print, and a collage print. The base prints have the background printed on in a specific layout. Meanwhile, the collage prints have my materials sporadically placed on the sheet of paper, placing them in such a way that I can fit as much as possible. While printing both of these, I don't have a plan on what my end product will be.

Once everything is dried is when the magic happens. I cut out all of the collage prints, I hand-edition all of my base prints, and I begin the process of laying out the collages. I let the process guide me.

One thing I've recently been experimenting with is with the exposure process, particularly using laser prints of the silhouettes, fully printed on one side. It seems that this reacts to the chemicals on the paper, leaving unique and different colors, such as rusty reds and purples. The colors are a lot more vibrant in the beginning, eventually fading slightly. It's still a work in progress with this experimentation but I'm enjoying the process as I develop it.

With each mishap during printing, instead of throwing out the print, I experiment with it. I certainly don't have anything left to lose in a cyanotype I'm already unhappy with.

Artist statement

Dark, moody, ethereal figures, intertwined with overgrown flowers and gold leaf, compose my *The Ache and the Light* series. I use Victorian era concepts of grief, mourning, and flower symbolism. The duality of flowers, how they are used as symbols of sympathy yet in celebration of life, is evident in my cyanotypes. In 2016, a dear friend in my close-knit disability community passed away. Amelia was a devoted Christian and found such solace in heaven. After her passing, I found myself constantly grieving, while simultaneously questioning life after death. With Amelia in mind, I hand-edition each cyanotype with gold leaf, a nod to the "golden streets" welcoming her, all while I depict my thoughts on our Great Beyond: our loved ones waiting for us, surrounded by the flowers we leave behind for them.

Biography

Monica Chulewicz is Polish-American artist working with prints on paper and fabric. While using vintage photos and materials, her themes of grief, mourning, the unknown, disability, and ableism stem from her experiences living with several chronic, rare, and progressive illnesses. She has received awards from the Kennedy Center, the Samuel I. Newhouse Foundation, and more. She has a BFA in Art, Art History, and Psychology and continues to live and work in New York. To see more of her work, visit **www.monicachulewicz.com**.

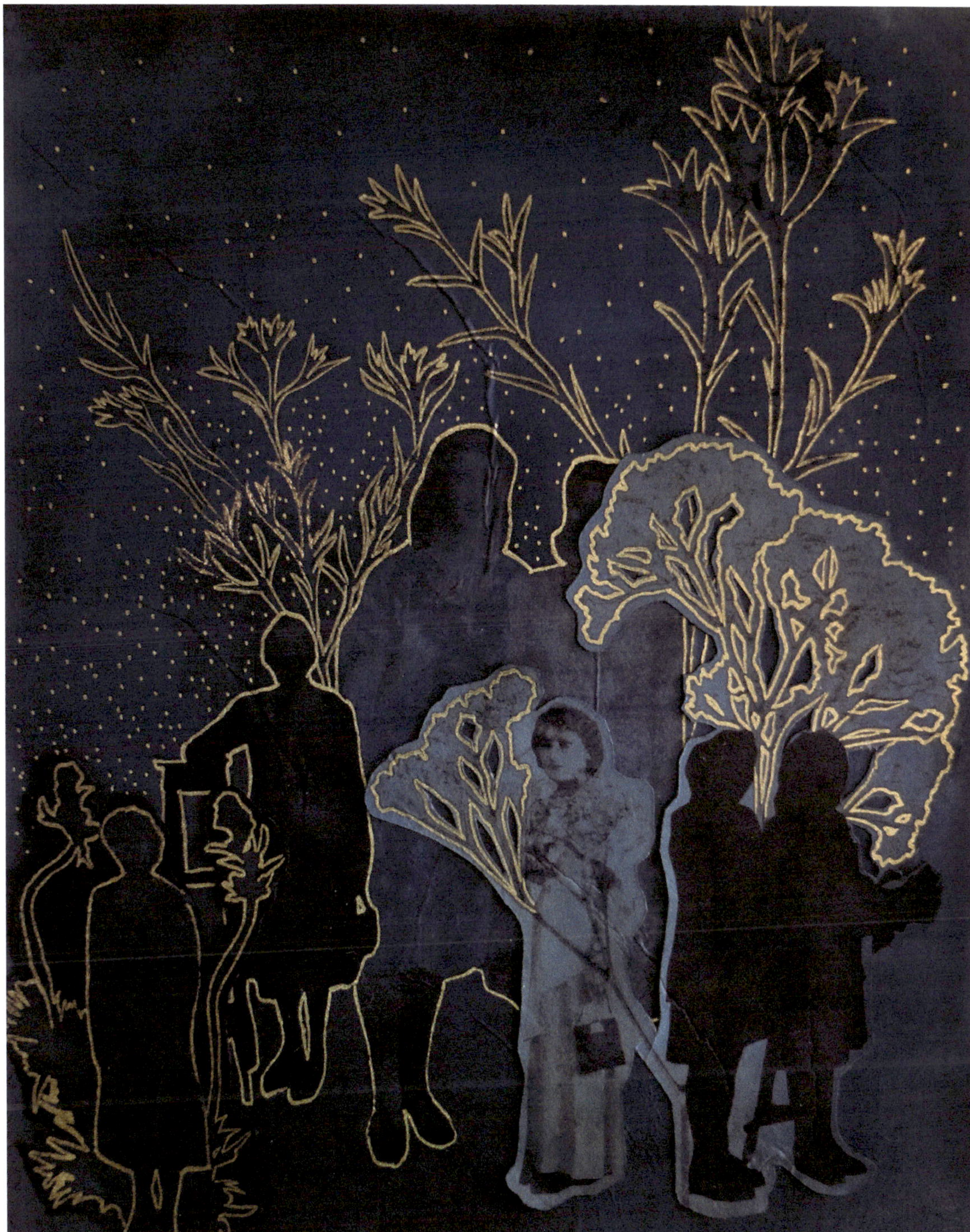

Figure 15.28. *Please Place Flowers Every Time You Visit (They Grow in Abundance Here)*, cyanotype and gold leaf on Hosho paper, 12″ x 9.5″ © Monica Chulewicz 2018

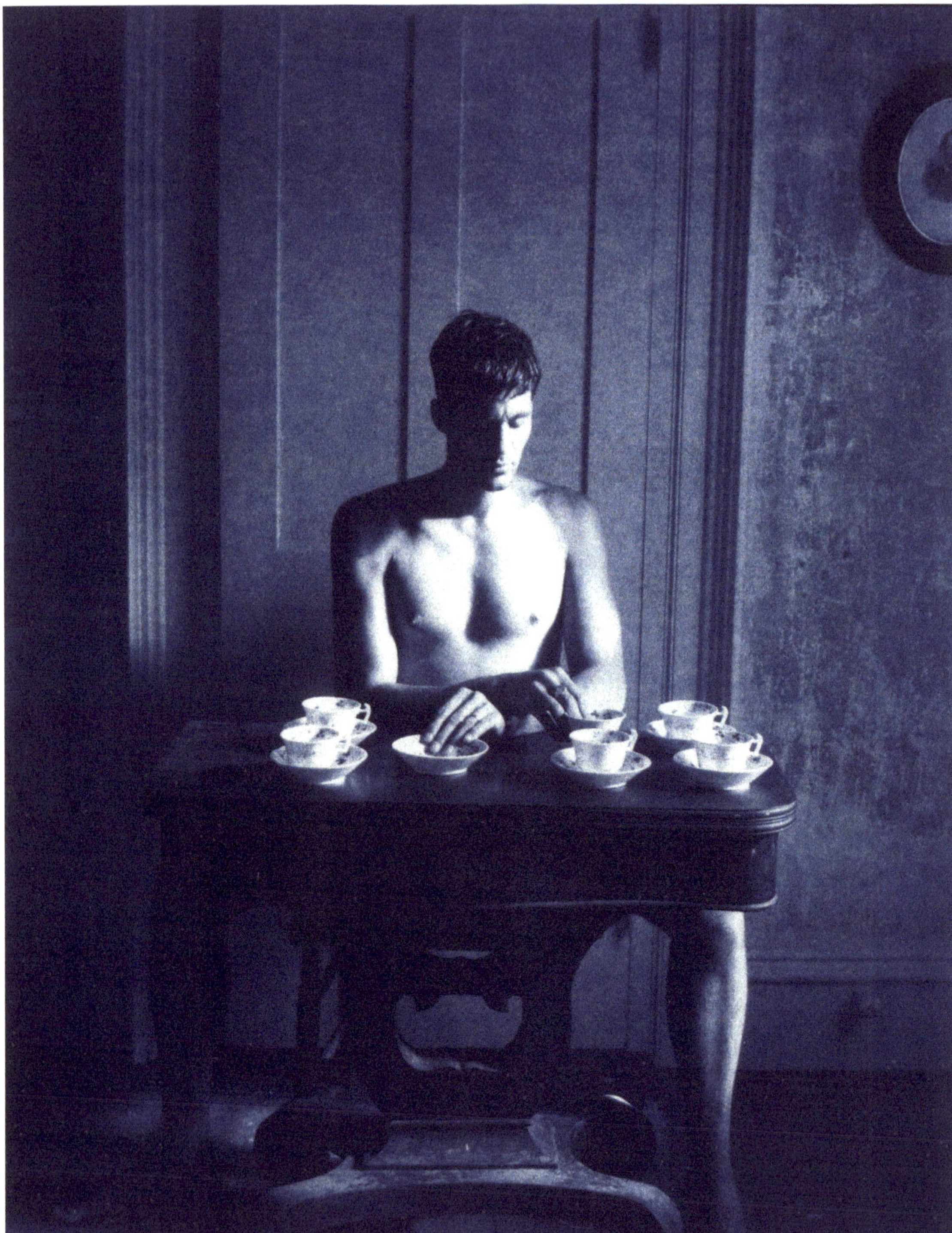

Figure 15.29. *Untitled, Self Portrait with Teacups, Stone Ridge NY* © John Dugdale 1995

John Dugdale

Creative practice

Twenty-six years ago, after working as a professional photographer for over a decade, I lost nearly all of my eyesight. Since then, like the first photographers who used water from a stream, light from the sun, and minerals from the earth to assemble their vision, I learned the process of creating a picture from the template of nature. To my surprise I found that the present could be preserved with the techniques and sensibilities of a more artisanal time. Like my heroes of photography, William Henry Fox Talbot, Sir John Herschel, and Julia Margaret Cameron, my focus turned to the joy of recording images of friends, family and personal belongings, and the delicate passage of day into evening. I now know that the eyes are merely instruments—the most splendid secrets of vision dwell in the heart and mind. This process has endeared me to a world of sense, quietude, and beauty that I was never able to fully journey through in my previous career as a commercial photographer. The result is a body of timeless photographs that I created in the spirit of the American Transcendentalists, Whitman, Dickinson, Thoreau, and Emerson.

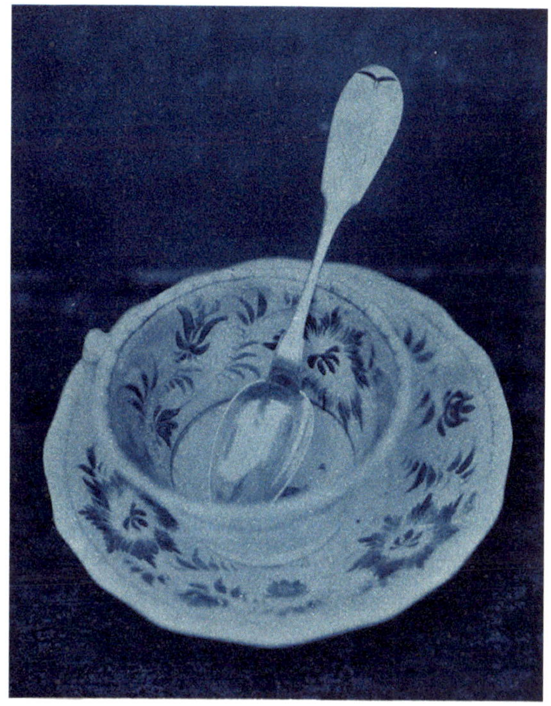

Figure 15.30. *Anne's Teaspoon, Stone Ridge NY* © John Dugdale 1996

Biography

John Dugdale is a remarkable and much-loved artist. Over the past twenty-six years, working in a state of near blindness, he has created photographs as if from his soul. Dugdale at age 58 has reached a notable level of artistic acclaim with his photographs in the Metropolitan Museum of Art, the Whitney Museum of Fine Art in New York City, among other noteworthy collections, and more than twenty solo exhibitions worldwide. Dugdale's theme is one of survival and the triumph of the creative spirit; his vision is of gracious beauty, emotionally rich and sensual. Dugdale was diagnosed HIV positive over twenty-six years ago and nearly struck blind, eventually becoming completely blind. Life forced him to see and photograph in a new and more personal way. His pictures are influenced by 19th century imagery, as seen in his dreamlike portraits or in his ordered still lifes. From image to presentation the artist's hand is apparent from start to finish. He began to use a 19th century large format camera producing classic, ethereal blue cyanotypes, a simple rustic process invented in 1842. He has also produced albumen and platinum prints. Born in Connecticut in 1960, Dugdale lives and works in his attic studio in Greenwich Village and his 19th century farmhouse located in Upstate New York. A successful commercial photographer, he graduated from the School of Visual Arts in 1984. His subjects are portraits and figure studies of his friends and family, flowers, still lifes of his collection of antique china and glass, and the rural landscape of his home. They are also intimate glimpses of private moments and personal spaces. There is a tranquil and timeless quality to his work. To see more of his work, visit **johndugdalestudio.com**.

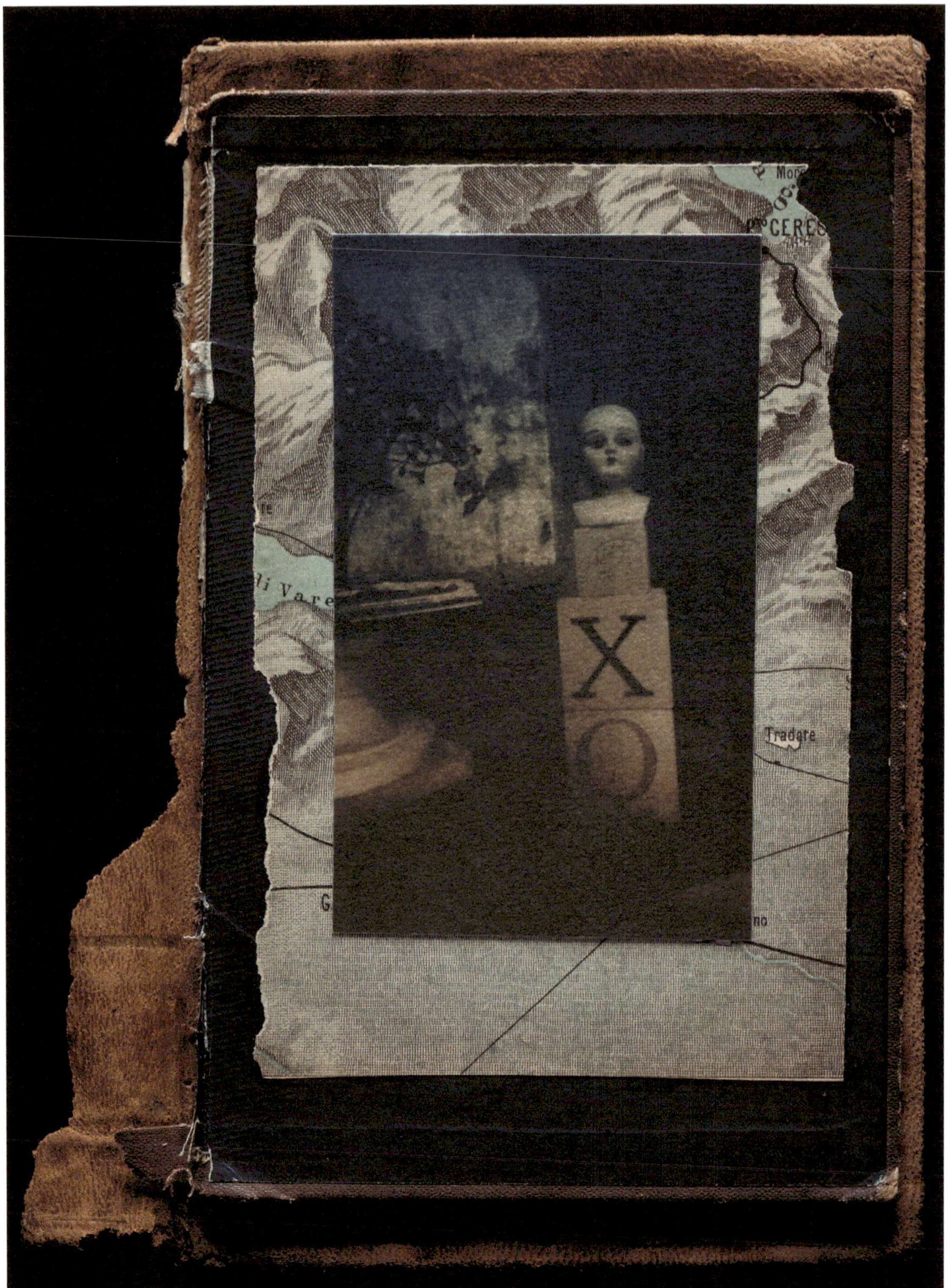

Figure 15.31. *Figure/Garden (constructed)*, pinhole cyanotype (toned), old map, old book boards, 8¼″ x 6⅛″ (irregular) © Jesseca Ferguson 2010

Jesseca Ferguson

Process

Cyanotype—often dismissed as child's play—can actually be one of the more challenging handmade photo techniques. As jazz musician Thelonious Monk once said, "Simple ain't easy!" In the handmade photography hierarchy (both in the 19th century and in today's renaissance of antiquarian processes), cyanotype somehow seems to rank below platinum/palladium, gold-toned salt prints, daguerreotypes, ambrotypes, etc. Perhaps it is cyanotype's underdog status that has appealed to me, but I also embrace the democratic nature of its straightforward, low-cost, easy-access, non-intimidating methodology. Truly anyone can make a cyanotype! But here is the challenge: Can you make a beautiful blue print? Or come up with an original interpretation of this 1842 process? Certainly the moody and sometimes melancholy tone of the blue image suits my purposes as well. The otherworldly aspect of blue can be haunting, as discussed by Rebecca Solnit in *A Field Guide to Getting Lost* or by Carol Mavor in *Blue Mythologies: Reflections on a Color*. And, of course, there is always that window of blue in Joseph Cornell's mysterious construction *Toward the Blue Peninsula: for Emily Dickinson* (c.1953).

Some of my favorite papers have disappeared or changed over the years. Now I try to find the right paper for a particular project, a practice which has led me to be very experimental and eclectic, especially when re-using 18th and 19th century book pages and endpapers, as in *A Small Book About the Moon* (2014). With *The Book that Printed Itself* (2017) I experimented with papers that would print well with cyanotype and would also work well as book pages. Some current favorites are handmade papers from St. Armand Mill in Montreal, Canada. Their natural linen drawing paper is a delight. Because these papers are handmade from linen and cotton rag, I must let go of notions of consistency. The papers vary from batch to batch, depending on what rags are available for making the pulp.

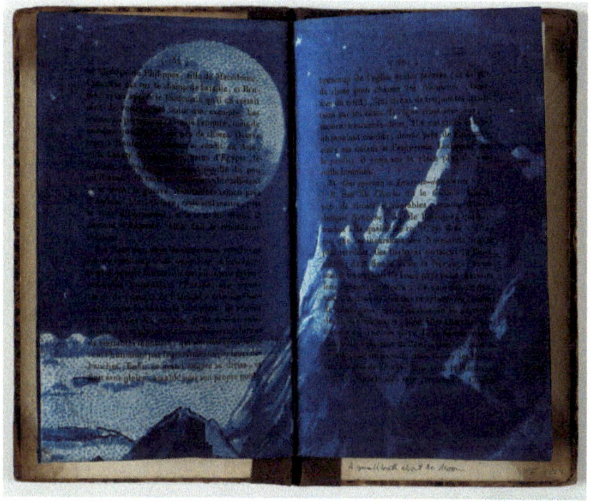

Figure 15.32. *A Small Book About the Moon*, cyanotype, 19th century book pages and boards, open 7″ x 8½″ (closed 7″ x 4¼″) © Jesseca Ferguson 2014

"The perfect print," and other ideals I once adhered to earlier, no longer seem relevant for what I am after lately.

For many years I worked with pinhole cameras to make large format negatives (4″ × 5″ up to 16″ × 20″) for contact printing. As beloved sheet films were discontinued (Polaroid Type 55, Agfa N31p, Kodak Ektapan, Efke, x-ray film), I began to scan found images and print them onto acetate. Poking around in old books or on the internet for images is almost as much fun as making pinhole images!

I use classic cyanotype and buy pre-measured dry pack kits from Bostick & Sullivan, Freestyle, Photographers Formulary or elsewhere—whatever works! I used to order bulk chemistry and mix my chemistry from scratch, following various formulas, but now keep it simple by adding distilled water to dry pack kits. I also like the portability of taking dry pack kits on the road and mixing the chemistry to use on location. Distilled water is key, just in case the local water is "off."

I use a wonderfully fast and efficient UV exposure unit from Edwards Engineered Products. I am now on my third UV unit. The first was made by a friend

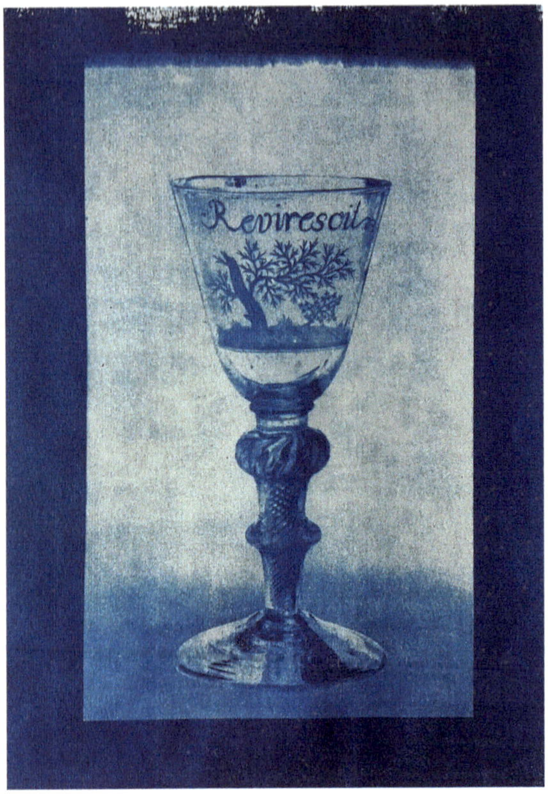

Figure 15.33. *Revirescit*, from *The Book that Printed Itself* series, cyanotype (from a waxed paper negative), 8¾″ x 5½″ © Jesseca Ferguson 2017

and eventually wore out. The next two were made by Jon Edwards and they are terrific. For simple acetates (positives and negatives), exposure times can be as short as one minute. However, for the waxed paper negatives used in *The Book that Printed Itself* series (ongoing from 2017), exposures can take 30–40 minutes, depending on the paper and the particular image.

I rinse the exposed print in running water until all unexposed cyan/yellow is gone, then immerse briefly in a hydrogen peroxide/water bath to oxidize/boost contrast. Then I rinse in running water for about five minutes using an old timey siphon washer.

Very occasionally I tone prints, using tannic acid. I get the premade kits from Bostick & Sullivan or bleach first with trisodium phosphate and then tone in really strong black tea. I am very experimental, testing by eye. I don't always like the loss of contrast that comes with toning, so don't do it often.

Creative practice

I love old abandoned books and have long been taking them apart to use their boards and pages in my collages, which incorporate my cyanotyped images. The handmade papers (often printed with letterpress) used in these books have vitality and strength, plus a beautiful surface quality, that I cannot easily find in modern papers. Happily, the St. Armand Mill in Montreal, Canada creates some papers that are reminiscent of these older papers and I delight in using them. The papers used in these 18th, 19th and early 20th century books are made of linen and cotton—ideal for cyanotype. I take apart the original books, print on their pages, then re-bind the cyanotyped pages into my own handmade books. There is a sense of the past and of the present in these works which I cannot get at in any other way.

Here is some practical advice discovered through working with antique book pages: These papers can be fragile when wet, so I sometimes cradle them on a sheet of Pellon (non-woven interfacing material) or plastic screening while processing them after exposure. I may drain them on an angled sheet of Plexiglas or the back of a photo tray until they are dry/strong enough to be handled. Coating old or otherwise delicate or fragile papers may require the addition of Tween 20 (10% solution), a surfactant/emulsifier which helps spread the cyanotype solution evenly on the surface of the paper.

Perhaps because I do not come from a photography background and never took a photography class, I bring a different attitude or sensibility to my work. I am interested in the unique object, rather than the editioned print, for example. I am intrigued by using mud, wood ashes or crushed flower petals to stain and pigment photo emulsions. I also enjoy collaborating with/being inspired by writers, such as Iranian writer Hossein Mortezaeian Abkenar. In my 2014–15 *Farsi Moon* series, for example, I printed Hossein's handwritten Farsi text in gum bichromate (pigmented with ashes, mud, clay). I then printed a 19th century scientist's image of the moon in cyanotype over the Farsi/gum bichromate image. This text came

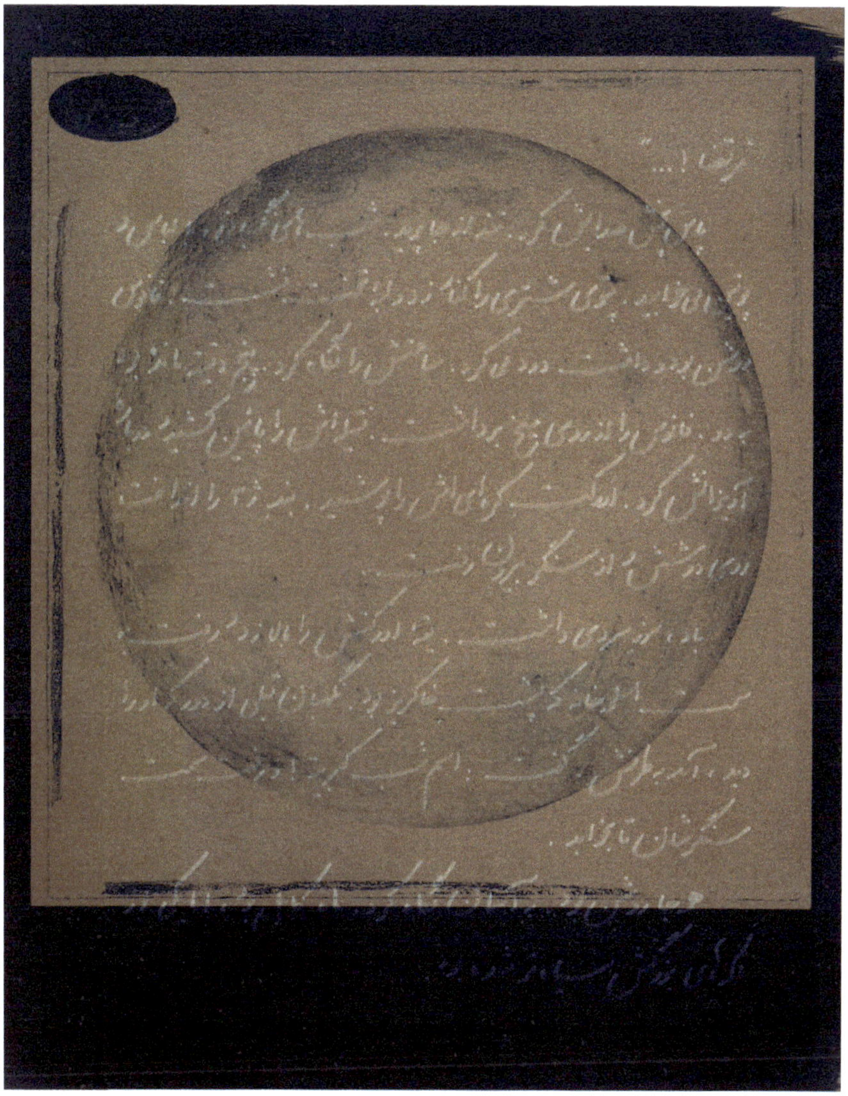

Figure 15.34. *The Scorpion, Chapter 6, Page 1 (Draper moon)*, cyanotype over gum bichromate (pigmented with wood ashes, watercolor), approximately 12″ x 10″ © Jesseca Ferguson 2014

from Abkenar's oneiric 2006 novel *The Scorpion*. In chapter six of this mysterious book, the moon is a character and a presence.

In 2017 I started *The Book that Printed Itself*, a series involving paper negatives made from found images (tipped into an old book). The images are of 16th and 17th century glass against a black background. I waxed them to render them translucent. These paper "negatives" are actually positives, but when printed in cyanotype, the resultant images appear positive. This was a roundabout way to produce negatives and harks back to the waxed paper negatives of the 19th century.

Biography

Jesseca Ferguson is interested in stories, dreams and memories, whether real, half-remembered or entirely imagined. Her images, books, and photo objects have been exhibited, collected, and published in the US and abroad. She teaches at School of the Museum of Fine Arts at Tufts in Boston, Massachusetts. To learn more about her work, visit www.museumofmemory.com.

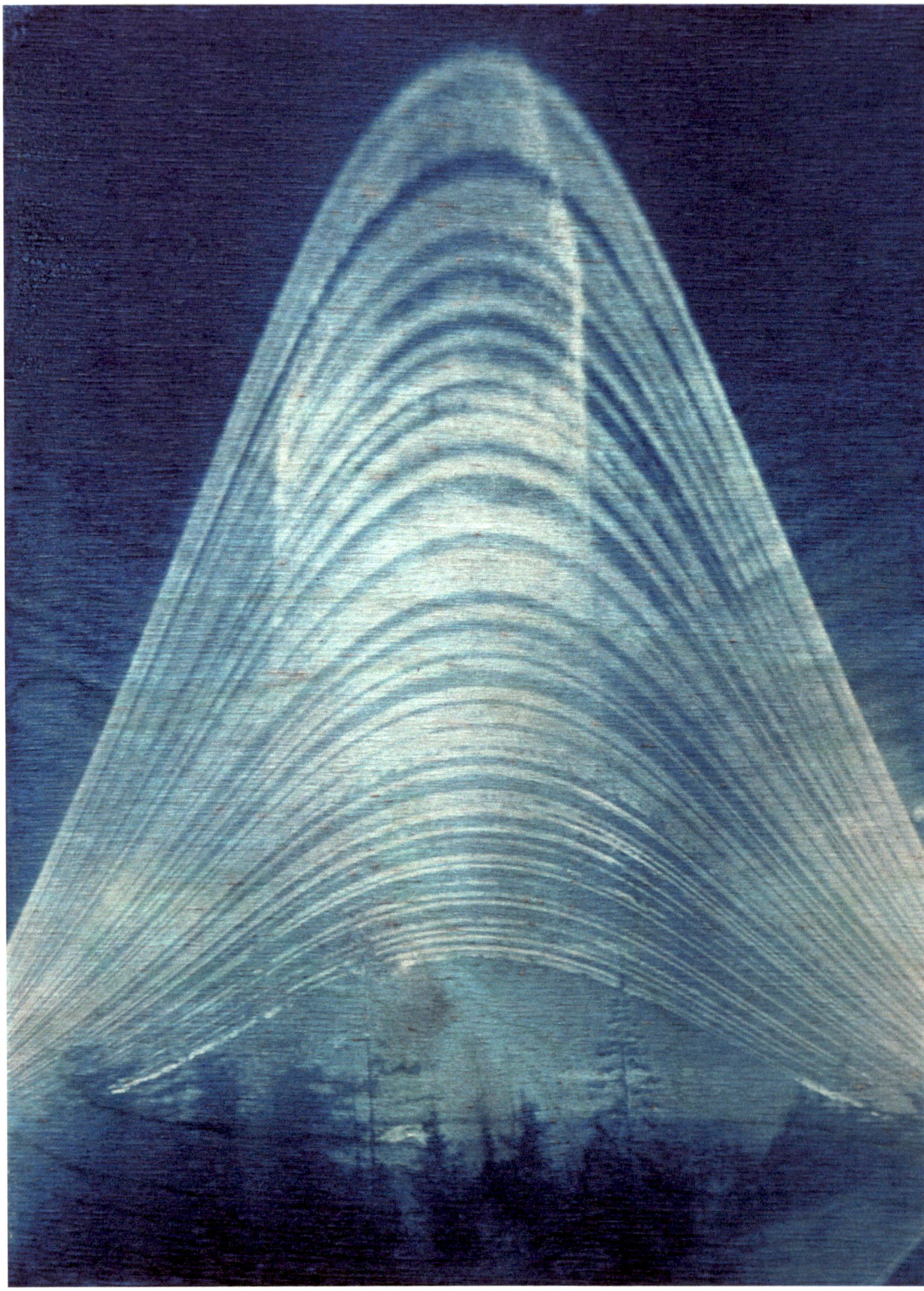

Figure 15.35. *Sun and Shore*, classic cyanotype on birch plywood developed in oxalic acid, 9″ x 12″ © Peter Friedrichsen 2018. We are not aware of the subtle daily changes of the sun's rays as the earth oscillates between the winter and summer seasons. With the use of a simple handmade pinhole camera and b&w photographic paper (scanned and made into a digital negative), we can capture these nuances as the sun makes its daily traverse; sun, cloud and seasons are all recorded. This is a solargraph recording of the sun and surroundings from December 20, 2008–June 21, 2009.

Peter Friedrichsen

Process

In my decision to choose an alternative printing process for my *Iceland* series, I wanted to emphasize the great abundance of water that exists both around and on the land. Low clouds frequently roll in from the surrounding Atlantic ocean, suggesting the cool blue tones inherent in the cyanotype process.

In choosing a surface for my prints, I was looking for something that would reflect the ruggedness of the landmass, something textured. Wood provides such a surface and leaves its own impression of grain and flow.

I print my negatives digitally on an Epson P800 printer. I occasionally use compensation curves but have not used them in the *Iceland* series. I use the classic formula containing equal amounts of 20% ferric ammonium citrate and 8% potassium ferricyanide w/v. In the *Iceland* series, I used the August sun with exposure times of 3–5 minutes (late August 11 AM–3PM) In other works, I have done exposures under UVBL for 20–25 minutes.

I develop in tap water containing 0.5% oxalic acid as it preserves the image intensity and helps with the clearing of highlights.

The wood on the surface of birch ply is shaved off of large birch logs in thin sheets. The effect is to create many horizontal fibers some of which break the surface. The ends of these fibers act as capillaries and will draw sensitizer deeply into the tips which results in significant staining. I overcome this problem by sizing the wood. I use 3 or 5 mm birch plywood as my image base since good quality ply is readily available and does not darken significantly with age. Before sensitizing, the wood is sized, otherwise highlights are stained quite heavily and reactions with wood tannins and other components can cause fogging.

Wood Preparation: I first lightly sand the surface with an abrasive hand sponge. All wood dust is then removed. I brush on one coat of full

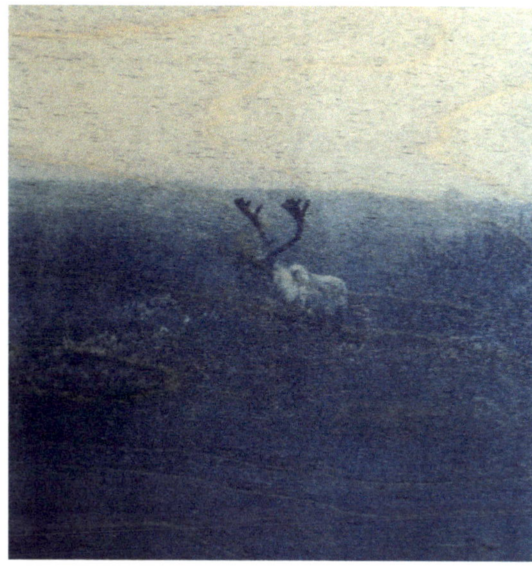

Figure 15.36. *Reindeer*, from the *Iceland* series, classic cyanotype on birch plywood developed in oxalic acid, 8″ x 8″ © Peter Friedrichsen 2016

strength Gamblin PVA size. The brushing and dipping of the brush into the size is continued until the wood surface stops forming dry spots resulting in a light sheen with the wood texture still partially visible. Over-sizing will cause adhesion problems later.

The sheet is laid flat until the sheen disappears after which it is hung vertically to dry for 24 hours. Once sufficiently dry, a very light sanding with a sanding sponge is applied once more to smooth lifted fibers.

I apply a second layer of a starch size. For the starch size, an approximate 7.5% solution of arrowroot or tapioca starch is made in a double boiler (7.5 g starch to 100 ml distilled water). The starch and water are mixed cold in a pot, then placed over a second pot of boiling water with continued stirring, resulting in a viscous mix. Once reaching a temperature close to boiling, it is covered and left on the double boiler for 5 minutes.

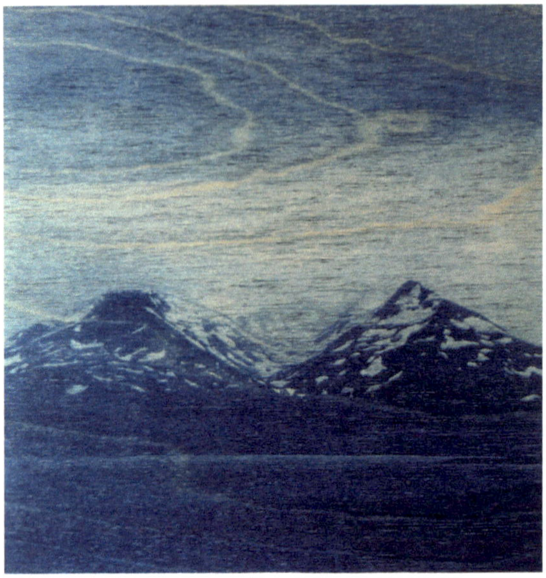

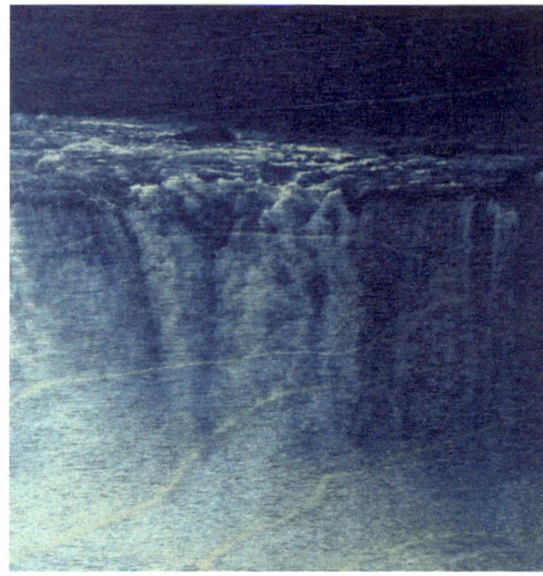

Figures 15.37 and 15.38. Left, *Mountains*; right, *Waterfall*, from the *Iceland* series, 8″ x 8″ each, classic 20/8 cyanotype on birch plywood developed in oxalic acid © Peter Friedrichsen 2016

The size solution is then placed in a blender and mixed for 3–4 minutes starting at a low speed and increasing the speed as it thins. The blending step thins the sizing making it brushable. The size will keep for about one week, after which it will start to separate.

Once cooled to room temperature the size has the consistency of heavy cream. I apply this by brush in the same manner as the Gamblin PVA size. A surfactant such as Tween 20 can be added to improve wetting of the surface. If the brush gets sticky on the surface, more size must be applied. When sufficiently brushed, the surface will look quite glossy while wet. The wood is left to dry horizontally for about an hour before it is hung and left to dry an additional 8 hours minimum. Once dry, the glossiness is reduced to a matte surface with a very slight sheen. The starch size acts as the receptive layer for the sensitizer and further inhibits staining.

I will frequently vary the sizing strength (4–9.5%), to manipulate the degree of grain staining. The sensitizer is now brushed onto the sized surface in a similar manner as the sizing. The starch sizing will start to absorb the sensitizer quickly and it will thicken it, causing resistance to brushing. This indicates that more sensitizer is required, so brushing and dipping is continued until the entire surface remains just brushable. The sensitizer can also be applied directly to the PVA sized surface, skipping the additional starch size; however, there will likely remain some open surface fibers which will draw in the sensitizer, resulting in more speckles after development.

After sensitizing, the wood surface is left to dry horizontally for about 15 minutes after which it can be hung to dry for 30 minutes, or a fan can be used to speed drying if the air-flow is clean. When sufficient time has elapsed, I blow dry the surface for several minutes with a hair dryer set on high. This removes residual moisture that would otherwise reduce the quality of the final print.

The wood surface is overlaid with a thin mylar sheet (0.05 mm) to protect the negative. The traditional formula used is somewhat hygroscopic and even more prone to sticking versus an application to paper since it sits more on the surface than actually within the fibers.

Thin ply will often warp to some degree due to the added moisture and drying. This warping makes close contact with the negative uncertain, resulting in a loss of sharpness in affected areas. The simplest

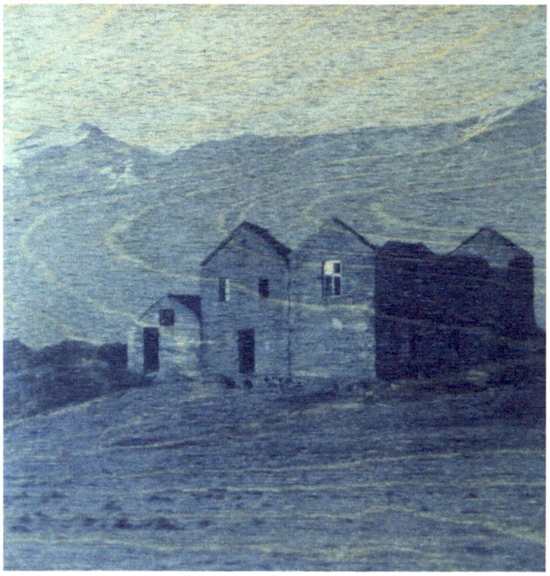

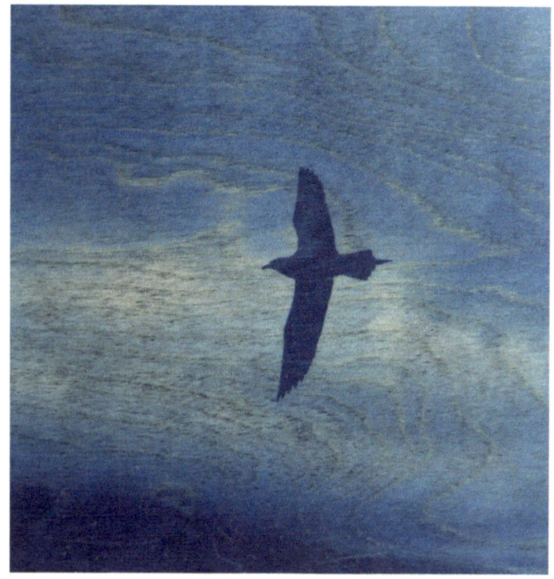

Figures 15.39 and 15.40. Left, *Farmhouse*; right, *Great Skua*, from the *Iceland* series, 8″ x 8″ each, classic 20/8 cyanotype on birch plywood developed in oxalic acid © Peter Friedrichsen 2016

remedy is to use a point UV light source which is approximated by direct sunlight, with exposure times of 3–5 minutes depending on conditions.

Once exposure is complete, the print is developed in a tray containing 0.5% oxalic acid in tap water for 3–5 minutes with gentle rocking. The wood is then rinsed in a clean tray with a flow of water for 5 minutes.

The surface when wet is very sensitive to abrasion, so it should never be touched until dry. To reduce warping of the final print, it can be clamped or nailed to a thick piece of plywood when damp, not wet, and left to dry for a day or two.

When working with wood, expect surprises due to the variability of the composition. To reduce the wasting of wood, the starch sizing can be applied to paper (leaving out the PVA sizing step) for practice runs, allowing for testing of the digital negative, exposure formulas, and development.

Creative practice

Alternative printing processes afford the artist a lot of freedom to experiment with formulas, techniques, and substrates but I always start into a new process conservatively, using traditional methods until I have practiced it sufficiently and reasonably understand the effects of different variables. Once I have gained sufficient skill, I consider manipulating the process to more effectively communicate my work as an artist.

Artist statement

Iceland's abundance of water, glacial ice, and frequent monochromatic skies informed my choice of the cyanotype process, lending its inherent cool blue tones to my captured imagery. As a country with vast fields of volcanic rock, the printing of my work onto wood also provides added emphasis to the ruggedness of this land.

Biography

Peter Friedrichsen is a photographer and printmaker based in Toronto, Canada. He uses a minimalist approach to isolate subject matter while defocusing the complexity that surrounds it. An endless experimenter, he has printed onto wood and metal to further amplify the subject and atmosphere of the captured moment. He has exhibited both locally and internationally. To see more of his work, visit **peterfriedrichsen.com**.

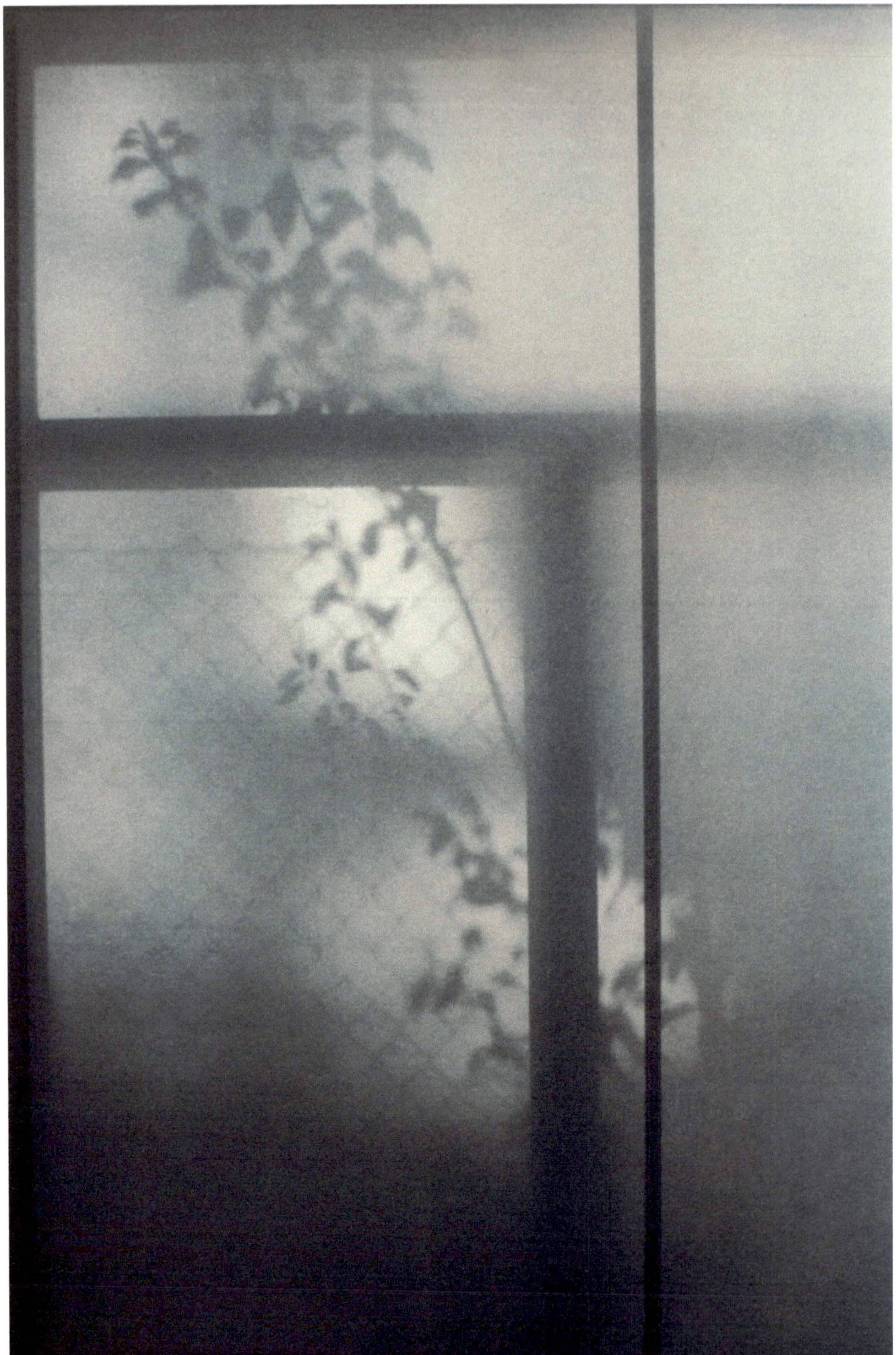

Figure 15.41. *Shadows #1*, cyanotype over platinum using duotone negatives, 10.5″ x 8″ © Annette Golaz 2017

Annette Golaz

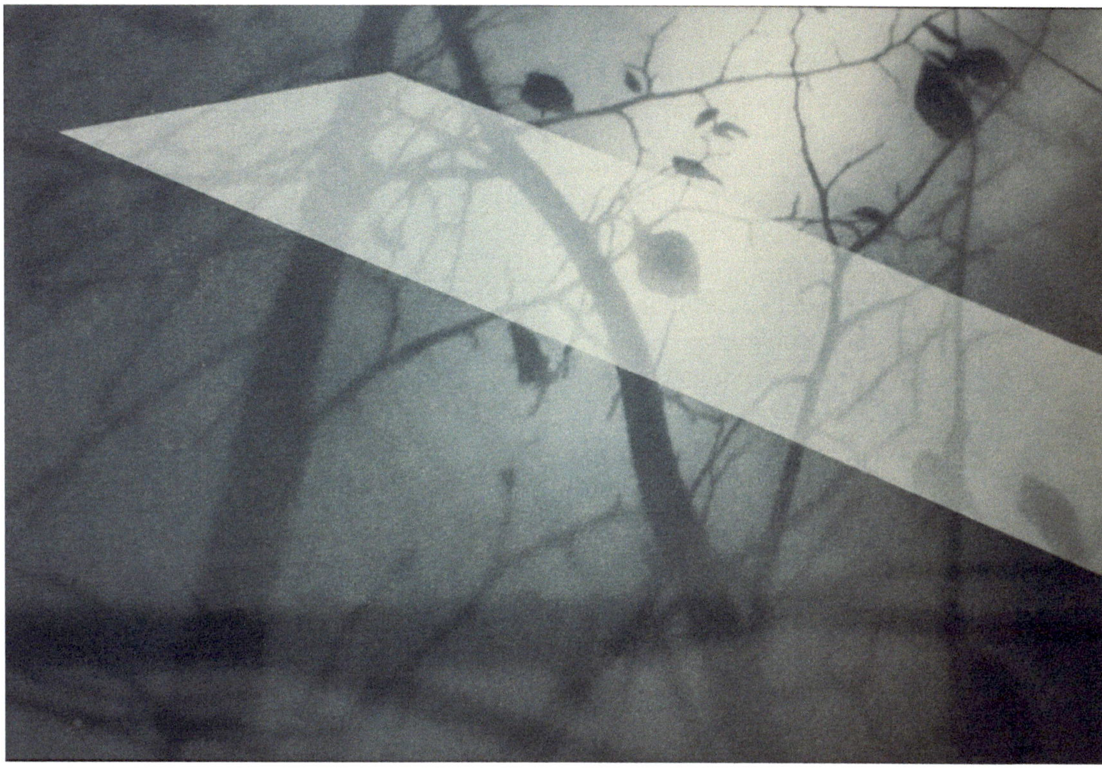

Figure 15.42. *Shadows #4*, cyanotype over platinum using duotone negatives, 10.5″ x 8″ © Annette Golaz 2017

Process

I have a preference for cool light and muted colors and all shades of blue—turquoise tending toward gray in particular. It was fairly obvious that the cyanotype would appeal to me. I fell in love with Sam Wang's duotone method. The interplay of warm blacks and cool blues when a cyanotype is printed over a platinum print touches my heart. To me it is so much more subtle and interesting than a pure black and white image.

My favorite paper is the Hahnemühle Platinum Rag. It is quite white, clears out fast and nicely, has a smooth surface and is quite sturdy. It provides a long tonal range, a nice Dmax, and beautiful blues.

I print my negatives on Pictorico Premium OHP with an Epson SureColor P6000. I had profiles made by a specialist for matte and glossy paper and I use the one for glossy paper to print the negatives. To prepare the negative I follow the instructions by Christina Z. Anderson from her book *Gum Printing* only I never sharpen the image. I use her curve for the classic formula but have developed a separate curve for the new formula.

I am very fond of the new formula created by Michael Ware (I buy the Formulary kit). It provides a much longer tonal range and prints with a finer grain than the classic formula if used on suitable paper. I use this formula for the first layer of my gum prints, too. Since it washes out quickly in water without an increase in contrast I have control over the intensity of blue in my base. To measure and mix the solutions A and B or the one solution with the one drop of citric acid I use syringes, reusing them until they don't move smoothly anymore.

I expose the new formula for 3 minutes in a vacuum UVBL exposure unit equipped with Philips 18 watt UV tubes which I bought on www.aliexpress.com for $330. The classic formula needs 5 minutes when printed on Hahnemühle Platinum Rag. When printing over a platinum print I expose roughly twice as long. I sometimes use Stonehenge paper which needs about 15 minutes' exposure time.

I develop for about 3 minutes in 2 liters of water mixed with 5 ml 50% citric acid, then rinse the print under gently running water for another 5 minutes. By then the Hahnemühle Platinum Rag has cleared. Then I leave the prints in a tray of cold water for another 10 to 15 minutes or for as long as it needs to get the desired intensity of the blue. Prints made with the new formula do not darken as much after 24 hours as does the classic formula. I check on the print regularly since left for too long in the water the blue fades too much or white spots can suddenly appear.

My favorite toning method is green tea bags. I bleach the print for a very short time in diluted household bleach so that some blue remains in the print and an interplay with the warm brown from the green tea and the cool blues from the cyanotype occurs. It looks particularly nice on portraits since it provides for a nice skin tone.

I like to use 3 mm mats and mount this on wooden panels coated with primer so that I can present them in shadow gap frames made for canvases. To mount the mats I use the Gudy 831 film for dry mounting by NESCH which is very reliable. For some prints I sandpaper the frames, coat them twice with iron paint and oxidize them. They then look very much like rusty iron frames. I use the Instant Rust Set by Modern Options. I never present my prints behind glass; the texture of the paper has to show.

New cyanotype does not work as well when printed over a platinum print. It washes off entirely. I use the classic formula instead. Most of the time I get a rather smooth result but at times the cyanotype prints unevenly. I still have not found out why that occurs.

I usually used a hake brush to coat the paper but then switched to a synthetic finer brush (that never loses hair). Suddenly I had something like snowflakes towards the borders of every print (see p. 155) and at first I did not understand why. It turned out that this brush did not absorb as much sensitizer. Perhaps too much coating can provoke it to crystallize. I coat, dry with a hairdryer and expose immediately.

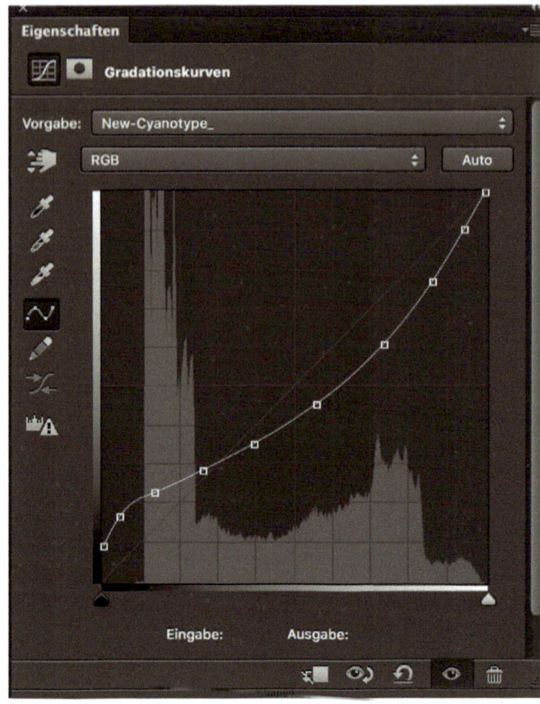

Figure 15.43. Golaz' new cyanotype curve

I print platinum for 90% of the normal exposure time. To determine exposure time I use the Stouffer step wedge. Somehow the platinum causes the paper to shrink permanently a tiny little bit. Therefore I print my cyanotype negative after the platinum print is ready so that I can print it 1 or 0.5 mm smaller in order to avoid any trouble with registration. I use the Na2 Platinum/Palladium Kit for Digital Negatives from Bostick & Sullivan. My developer is an ammonium citrate developer, which I warm up. The biggest challenge is black specks in the platinum print. I strongly believe that it has to do with the paper. Once I have a package with paper that does not produce specks I save it up for platinum prints.

The cyanotype is a wonderful first step into alternative photographic processes. With little and not very costly equipment one gets satisfying results and it spurs creativity and playfulness. There is a lot

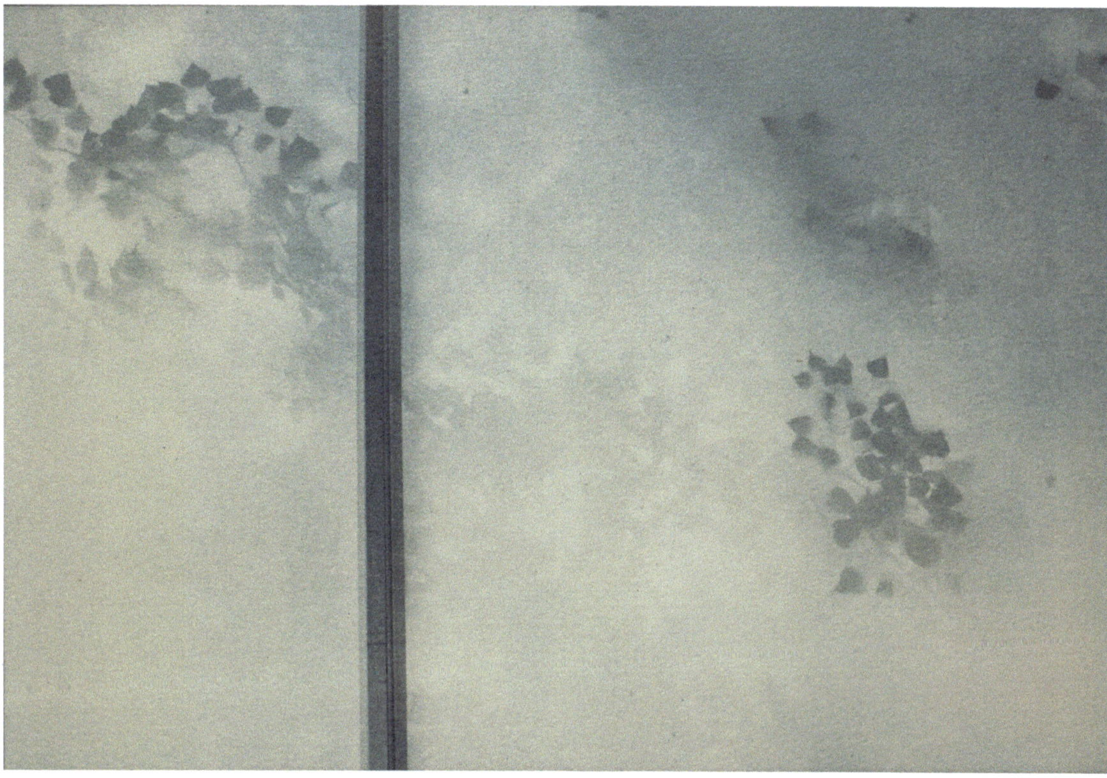

Figure 15.44. *Shadows #2*, cyanotype over platinum using duotone negatives, 10.5″ x 8″ © Annette Golaz 2017

of room to experiment with different substrates and toning methods. Chances are that the virus of alt processes will then cause a forever lasting infection.

Creative practice

In my photographs I like to depict contrasting elements and layers: organic forms and graphic lines for example. By printing a cyanotype over a platinum print I can add in a new layer of contrast between warm and cool colors to reinforce the statement. Having found alternative processes to reproduce my images a door has opened to a whole new world. The creative possibilities of interpretation and reproduction of photographs are boundless. I love the fact that after having taken a picture, only a small part of the work is done. I then choose a suitable process and substrate to print it and the perfect way to present it. It is very satisfying to be able to complete the whole process myself. Working in my studio is like meditation—no digital devices allowed with only soft music every now and then. Alternative processes are teaching me patience. Whenever I am in a hurry or want to print too many photographs at the same time something is bound to go wrong. The unpredictability of the processes makes the work a continuous challenge. My focus is to understand what is happening in the processes and to find new ways of improving my technical and artistic ability.

Biography

Swiss photographer Annette Golaz captures the poetry of everyday life in her work, with beauty often consisting of contrasts and conflicting layers. Golaz experiments with different techniques and outdated cameras. With a 20-year old digital device with 0.3 megapixels, for example, she challenges the belief that up to date equipment is the most important prerequisite for a successful photograph. To see more of her work, visit **www.agolaz.ch**.

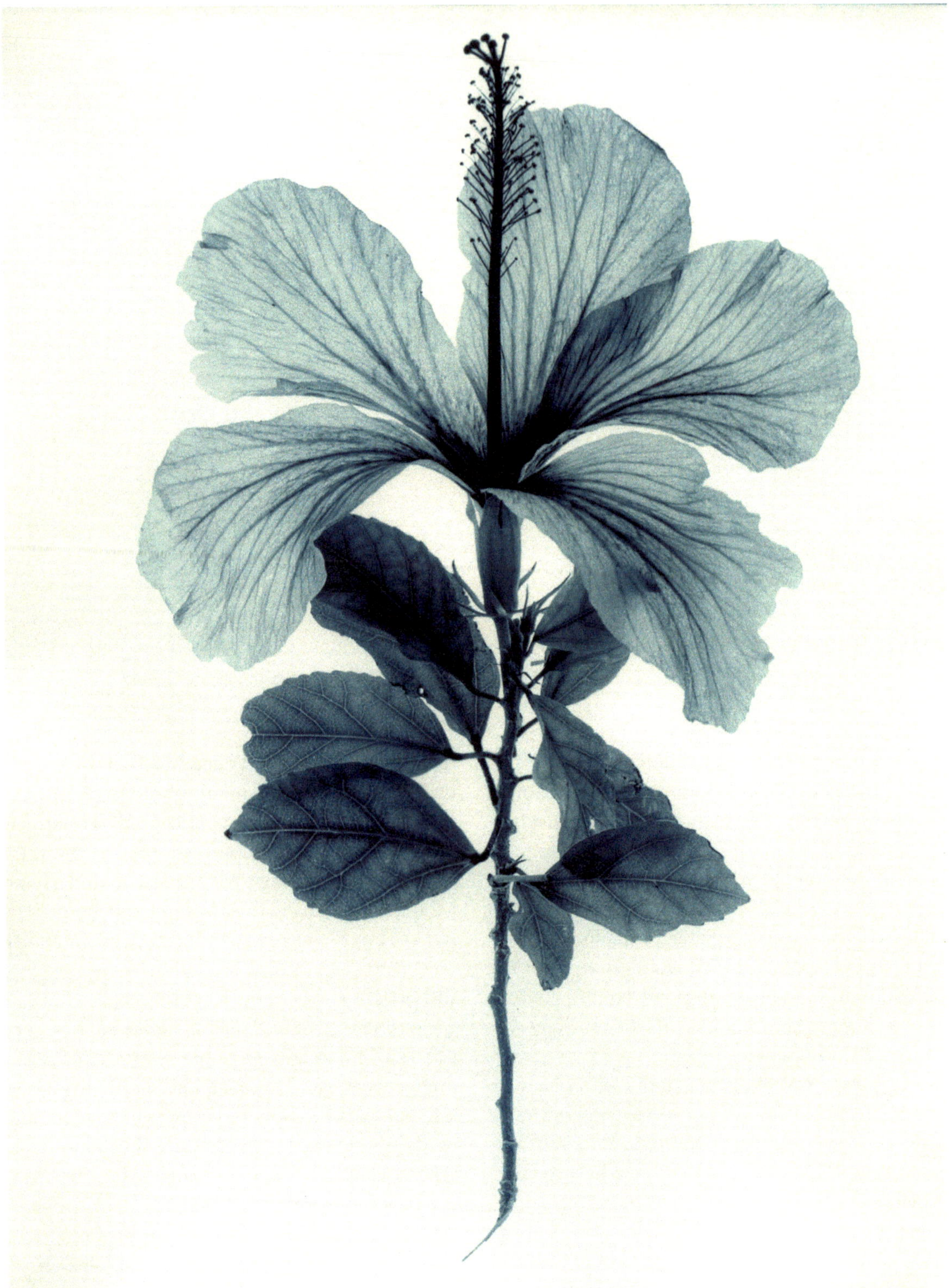

Figure 15.45. *Hibiscus Flower #1*, 10″ x 7.68″ © Jacek Gonsalves 2018

Jacek Gonsalves

Process

Cyanotype has always had an appeal to me because of its beautiful blues. It is quite a process to master if you are attempting to reproduce the same result several times. From the way one brushes the emulsion to how you wash the print, there are a multitude of intricacies one must remember and document for repeatable results. With my work I wanted to show that cyanotype isn't a process where the prints resulted in a contrasty image but one that has a multitude of tones. The way to do this is to spend a great deal of time calibrating the process; I do this using Roy Harrington's Quadtone RIP software.

I use Arches Aquarelle hot press 300 gsm watercolor paper, and print my negatives with an Epson 3800 printer on Folex transparency film. My cyanotype solution is the classic formula 25% FAC:10% PF. When making an 8″ × 10″ print I add 1.5 ml Part A and 1.5 ml Part B, with 3 drops of Tween 20. I built a UV unit under which exposures are 30 minutes. Once prints are exposed, they are put into a bath of water for 15 minutes face down and rocked gently in the tray. They are then put into a fresh batch of water for another 15 to 30 minute soak, and every 5 minutes the tray is gently rocked. I frame them with a white mat and white frame.

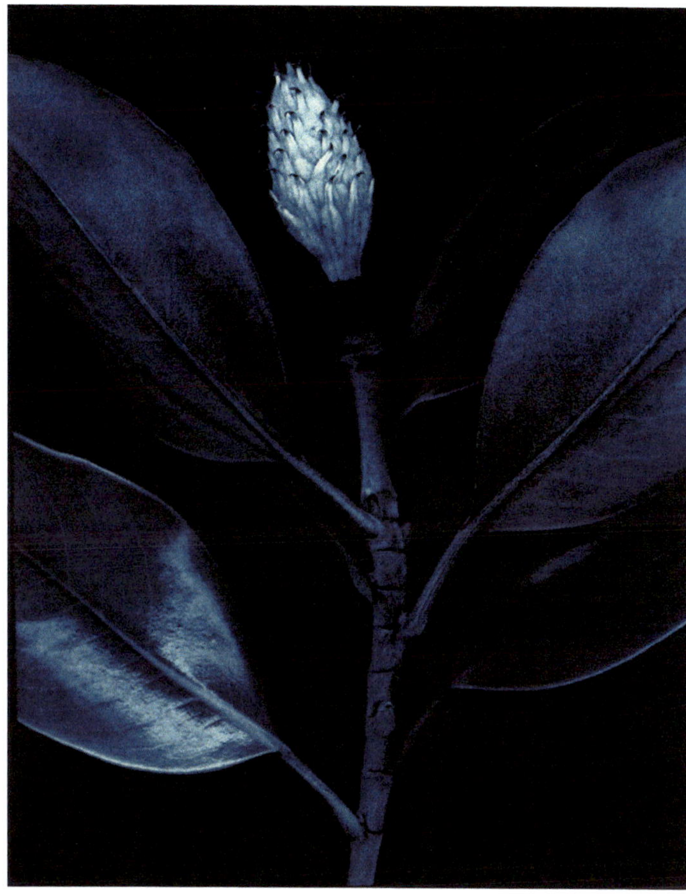

Figure 15.46. *Magnolia*, 10″ x 8.06″ © Jacek Gonsalves 2018

Artist statement

My work explores the subtleties of my suburban garden and the physicality of the plants I've cultivated. With influences as diverse as Katie Scott and Ernst Haeckel, I've studied these botanical elements with a slightly scientific lens. Each image merges the theoretical and scientific aspect of the cyanotype process with artistic expression.

Biography

Jacek Gonsalves (Perth, Western Australia) has practiced a number of creative processes including ceramics, photography, jewelry and drawing. After growing up between Kuwait, India and Australia, he has always been a keen observer of the world around him. His creative practice reflects the environment he inhabits and often illustrates this tapestry of cultural experience. He regularly participates in group exhibitions and his work has been acquired by a number of private collectors. To see more of his work, please visit **www.jacekgonsalves.com**.

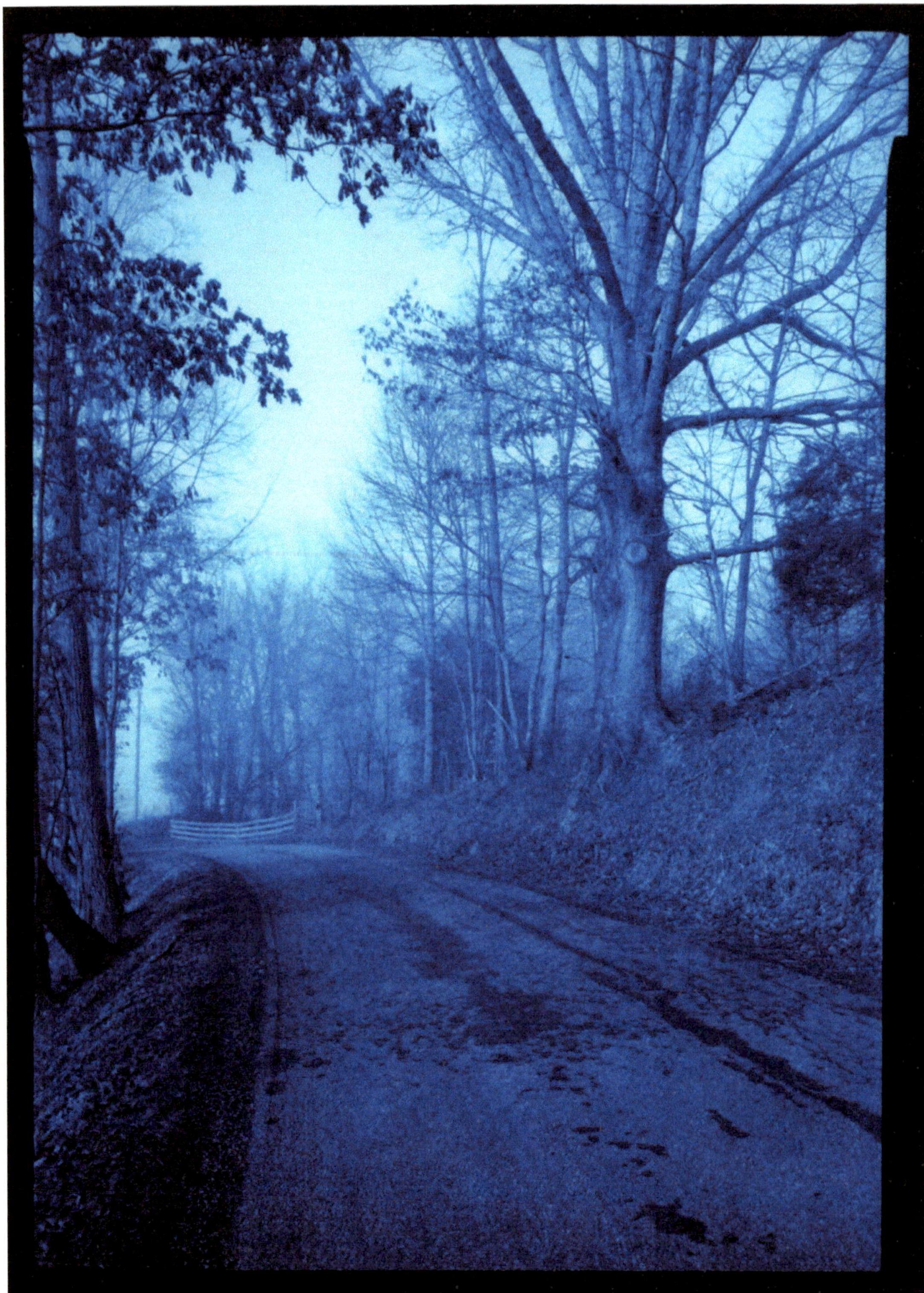

Figure 15.47. *White Oak and Fence*, new cyanotype printed from a 5″ x 7″ analog negative © John P. Jackson 2018

John P. Jackson

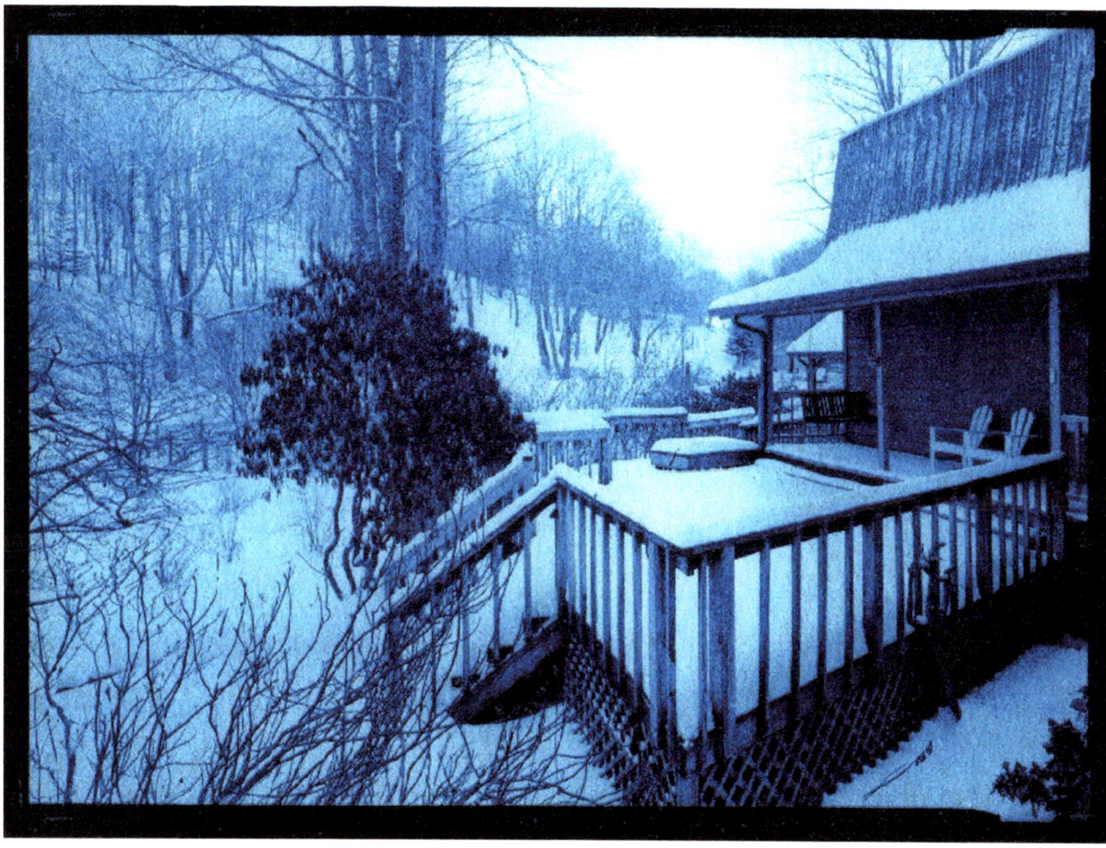

Figure 15.48. *Back Porch, Snow*, combination classic and new cyanotype printed from a 5″ x 7″ analog negative © John P. Jackson 2018

Process

The colors of the cyanotype process have always appealed to me. Producing a cyanotype print that is full tonality, as well as finely detailed, is a challenge that I enjoy.

I use Arches Platine, Bergger Cot 320, and Hahnemühle Platinum Rag papers. My negatives are made using panchromatic black and white film. The film formats I'm currently working with are 120 mm roll film, 5″ × 7″ sheet film and 7″ × 17″ sheet film. I make enlarged negatives in the darkroom using panchromatic film for both interpositive and enlarged negatives. Phil Davis's *Beyond the Zone System* and his other writings inform my entire picturemaking process. I use an incident light meter to determine exposure for film. A 4″ × 5″ Stouffer 21 step-wedge is used for testing both positive and negative materials. Paper (reflection) and film (transmission) densities are measured using an X-Rite 820 densitometer and the values are input into Davis's Winplotter software. The program generates characteristic curves and film development charts that are very useful in determining the exposure and development of the film negatives. BTZS is a "closed loop" system that works well for me.

I use new cyanotype with a drop of added 20% citric acid per ml of sensitizer. Also, I use a hybrid

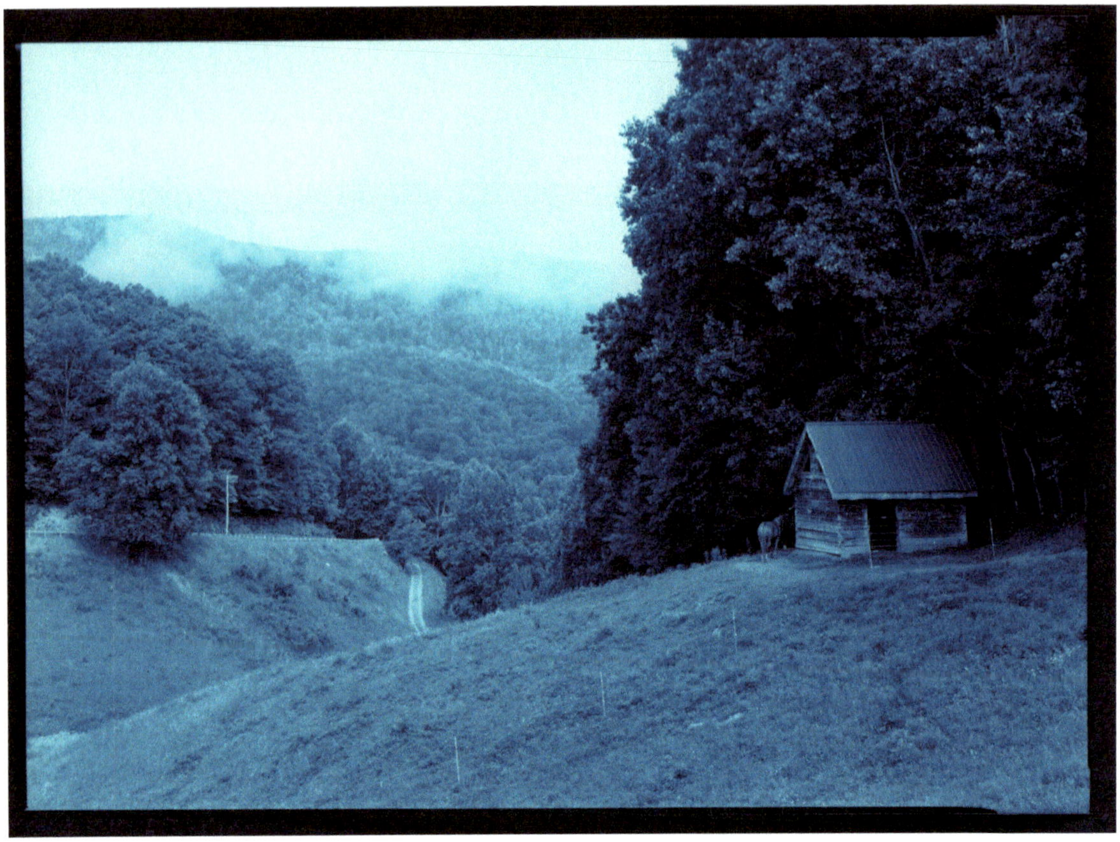

Figure 15.49. *July 4, 2017 Horse House*, combination classic and new cyanotype printed from a 5″ x 7″ analog negative © John P. Jackson 2018

formula that my photography mentor, Sam Wang, shared with me. It is new cyanotype + classic cyanotype (15%–25%FAC:10%PF) mixed 1:1 just prior to coating.

My exposure device is a 5000-watt commercial plate burner, a Teaneck Graphics floor model vacuum frame. Exposure times vary from less than 1 minute to about 15 minutes. After exposing, I develop for 1 minute in 1% HCL or 1% sulfamic acid while constantly agitating; then the print is moved into another tray with 1% citric acid for 4 minutes of gentle agitation. This development is followed by 20 minutes of washing. I hang the prints on a line in the dark room with clothespins.

I mat with 4-ply rag board for a back and a hinged over mat with an opening that is cut to show the edges of the image. For the 5″ × 7″ negatives I use an 11″ × 14″ back and over-mat with an opening of 6″ × 8″.

Problems with cyanotype I have encountered have been on occasion when using new cyanotype the print will show splotching. Careful pre-humidification of the paper and exposing the paper as soon as it is dry seems to help. Diluting the sensitizer is also effective. 2 parts sensitizer + 1 part water prevents splotching and only slightly diminishes Dmax.

For creative ideas, recently I have been testing cyanotype on cloth and Pictorico film using analog negatives to make the images.

Although cyanotype is considered easy, I find the process constantly challenging, rewarding and occasionally frustrating. The happy successes make the effort worthwhile.

Creative practice

My picture making is a personal need that involves constant interactive perception of the world around

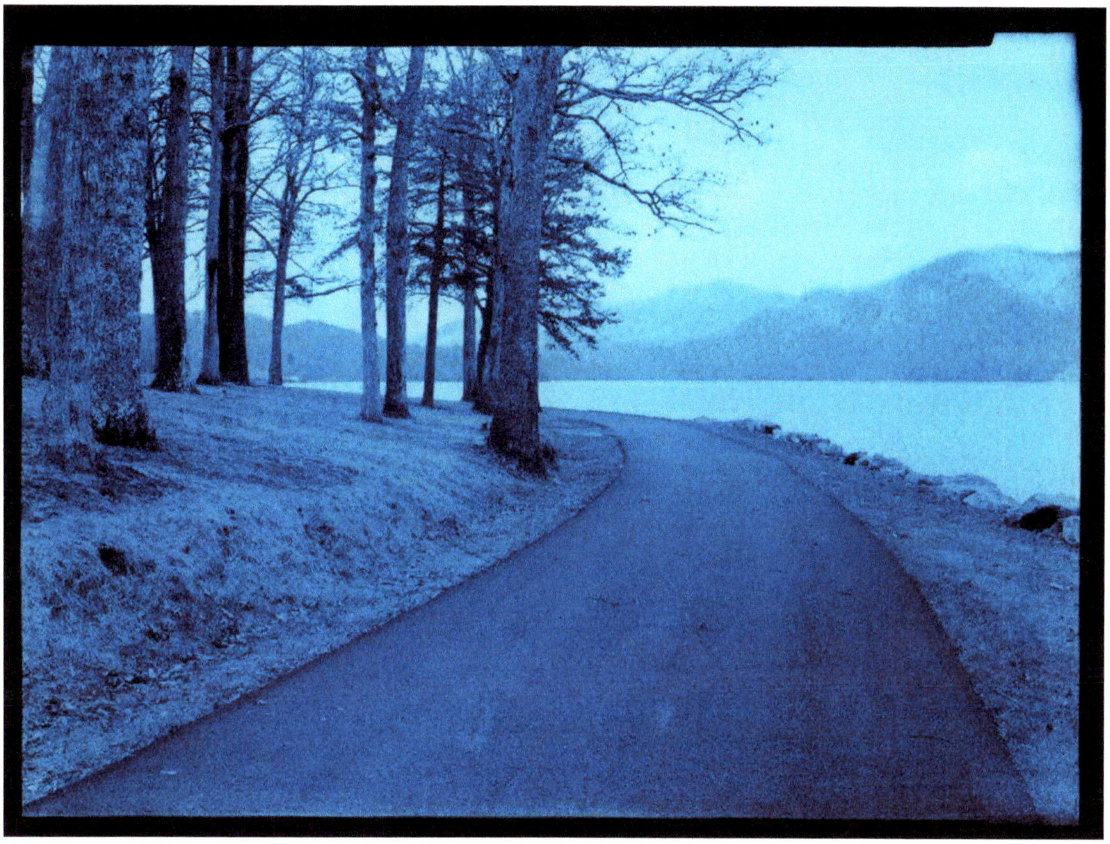

Figure 15.50. *Watauga Point*, combination classic and new cyanotype printed from a 5″ x 7″ analog negative © John P. Jackson 2018

me. It is like a "call and response" that challenges me and sometimes brings me joy. My wish to communicate with others, by sharing my pictures, motivates me to make prints. I practice "slow" photography because it suits my nature and the quality of analog photographic prints can be very beautiful.

Biography

John P. Jackson has been a passionate photographer for over forty years. In the mid-1970s he had the good fortune to study photography with Jerry Uelsmann and Todd Walker at the University of Florida, after which he graduated from the California Maritime Academy in 1980 with a BS degree in nautical industrial technology and an unlimited 3RD mate's license. From 1980 to 2013 he worked as a ship's deck officer on large cargo ships that sailed all over the world. He retired from the US Merchant Marine in 2013, last sailing as an unlimited ship's master (any tonnage any ocean) with his last command on a 980-foot Panamax container ship that was owned and operated by Maersk Lines Ltd. of Norfolk Virginia. In 2015, Jackson returned to photography. Over the last three years he has specialized in alternative photographic processes including cyanotype, platinum-palladium, gum dichromate, and carbon printing. Jackson photographs with analog film and prints all images by hand, including silver gelatin on manufactured photographic papers. To see more of his work, visit **www.ipernity.com/home/jpjackson**.

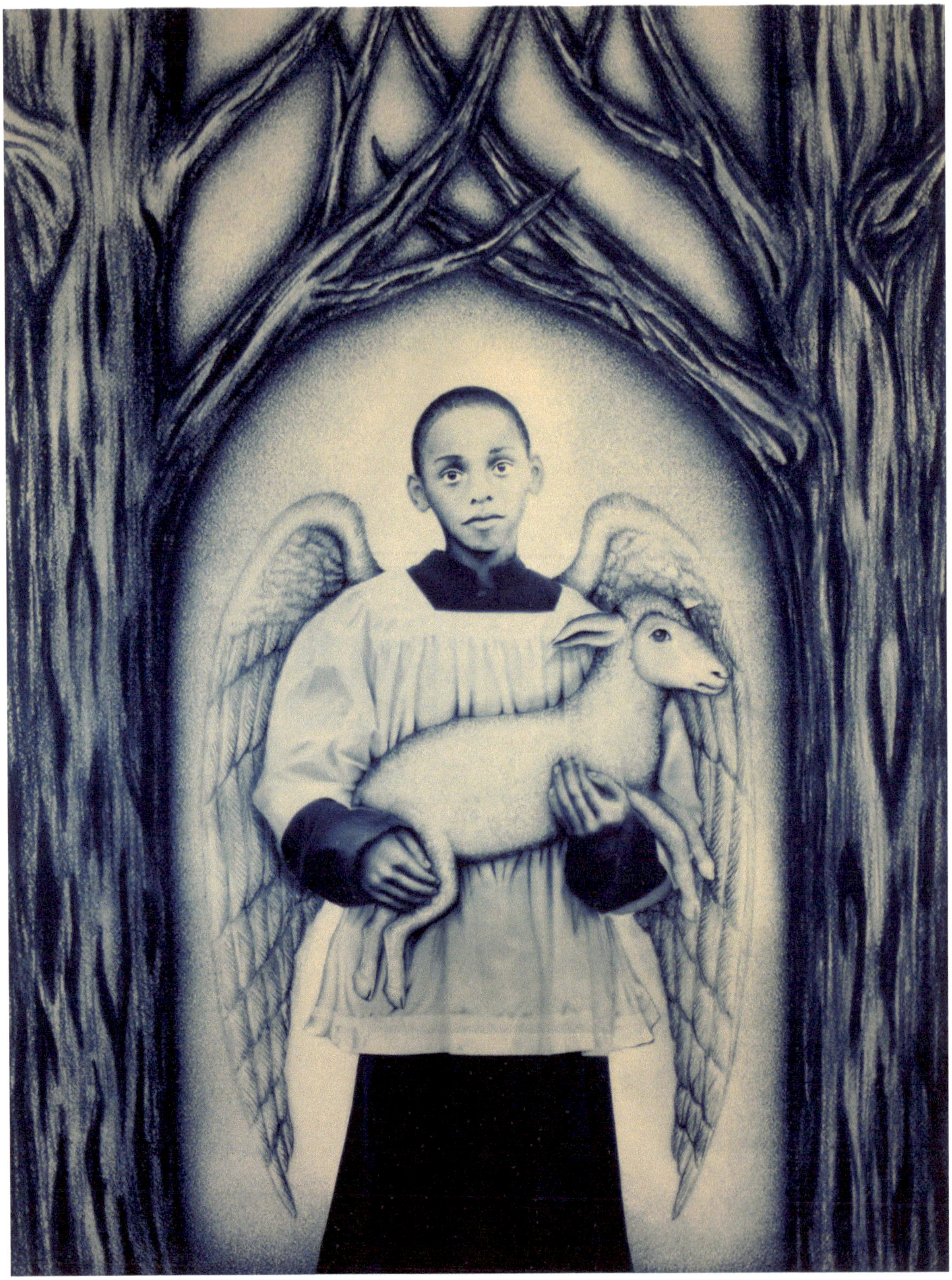

Figure 15.51. *What We Shall Become Has Not Yet Been Revealed*, cyanotype with hand drawing on vellum built into the negative, edition of 5, 52½″ x 40″ © Teresa James 2014

Teresa James

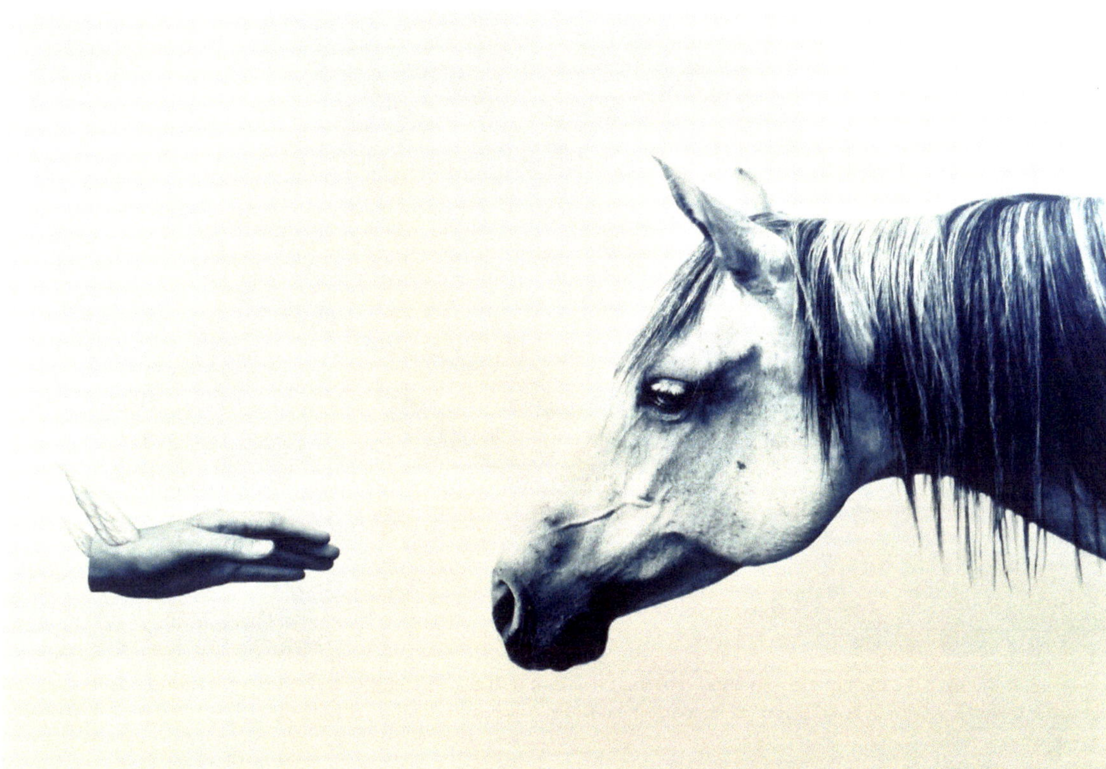

Figure 15.52. *Here We Breathe the Sweet Fragrance*, cyanotype and hand drawing with blue prismacolor, unique print, 27″ x 39½″ © Teresa James 2016

Process

The only paper I use is Arches hot press watercolor by the roll. It is very durable and has good light resistance. I have tried many other papers and have seen changes in the blues on final prints being left out into light even for a few days. I have never had that problem with Arches. Also for my large scale pieces it is important that it holds up when two people are carrying a 50″ × 127″ wet piece of paper to the drying station.

I use digital negatives and a custom curve and workflow that was made for me using the Precision Digital Negatives System with Curve Calculator III. For printing negatives I use an Epson Pro 9890 44″.

I have used both cyanotype formulas and found that for my work I can get really great results with the classic. It is so much easier to mix and having children and pets around my studio I feel much better about having less toxic chemicals in the shop. I do think though that the blues are very beautiful in the new solution and would recommend trying both to decide for yourself.

I have a custom made wooden exposure unit that is 48″ × 72″. It uses twenty-two 44-watt black light bulbs that are 48″ long. The light is directed downwards and is connected to an overhead base that is secured to the wall and ceiling. The table that sits underneath the bulbs is covered in black felt and is on wheels. It has a ¼″ piece of glass that sits

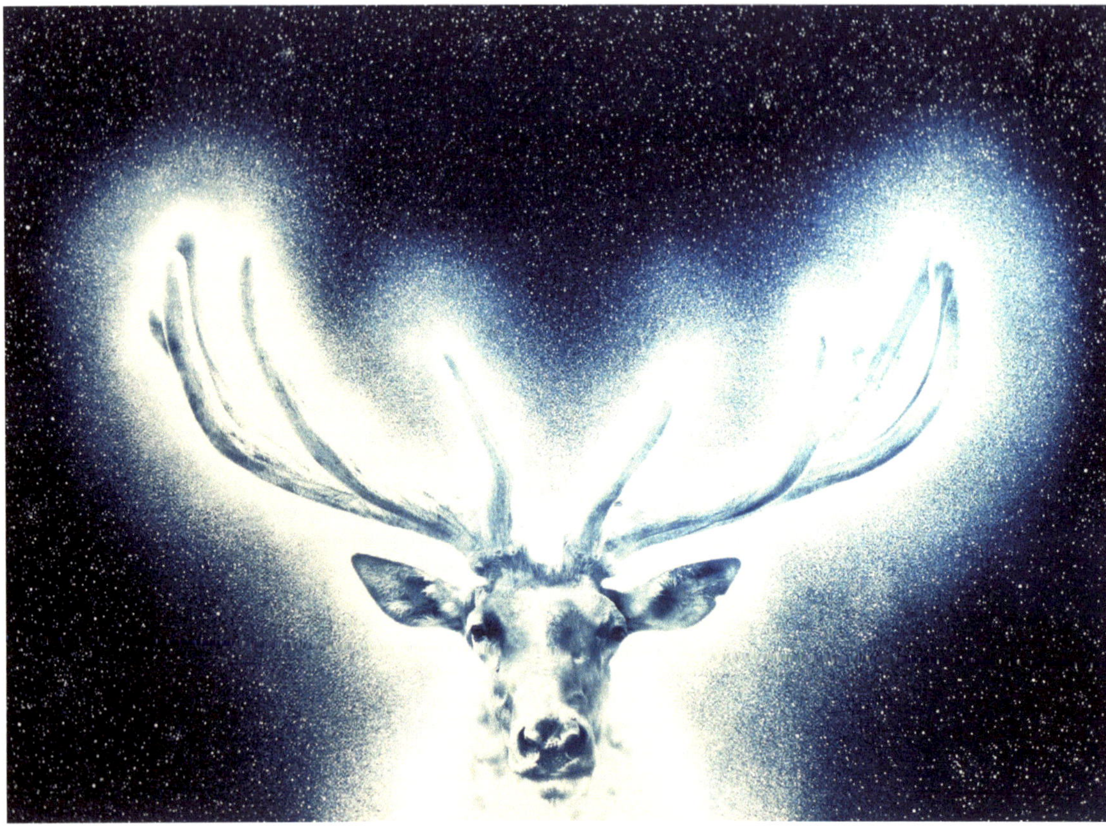

Figure 15.53. *Illumination*, cyanotype with hand drawing on vellum built into the negative, printed with a hand painted substrate overlay, edition of 5, 40″ x 56½″ © Teresa James 2015

on top and must be lifted with suction cup glass lifters. It does not have a vacuum seal but clamps are placed around the edges for good contact. Also the weight of the large piece of glass helps keep out any pockets of air. I expose the prints for 27 minutes.

I fill the sink with plain cold water and wash for approximately 10 minutes but I find that it is always best to use your eye. If the water is running clear and there are no pools or spots of yellow on the print it's ready to pull from the water. Some prints might need a longer or shorter wash depending on the imagery. My cyanotypes usually have a lot of white space so there is more chemical that comes off in the water and a longer wash might be necessary. Also, washing too long can take away some of the beautiful blue in small amounts. If you hold the paper up to the light and see faint water spots into the fibers, then it's time to pull it out of the rinse bath. These spots are more visible on the highlights of the image and when held up to the light but will disappear when the paper is dry.

For some of my larger prints I set up a makeshift pool in my front yard which consists of wooden sides hinged together, lined with a plastic tarp. I clamp the tarp to the wooden sides and fill it with the hose water. Washing exposed prints outdoors is before the sun comes up or just after dusk so I have to plan accordingly.

Before toning a cyanotype I let the print completely dry overnight after development. This gives the cyanotype time to dry down darker and helps you see more accurately what it will look like when toned. When toning with ammonia and coffee I first set up all my baths. Ammonia and

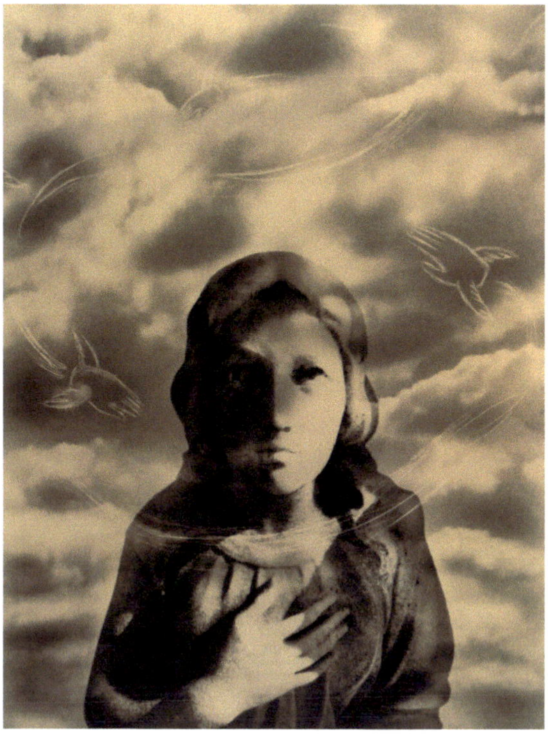

Figure 15.54. *Lend Me Your Heart*, cyanotype printed with hand-drawn acetate overlay, toned in ammonia and coffee, edition of 5, 34″ x 26″ © Teresa James 2017

water, coffee, and two water baths. On occasion I first paint with ammonia and water right onto the print with paint brushes, periodically blotting with paper towel and reapplying the solution. In the heavy blue areas it may take a few applications to see results depending on how strong your solution is. Lighter areas will bleach out more quickly so use a test print to get a feel for the technique and try different strengths of ammonia to water and record which one gives you the most control. When painting with ammonia the ratio I use is 20 ml of ammonia plus 140 ml of water. Next I submerge the print in a bath of ammonia and water for about a minute and then immediately into a clean water bath to stop the ammonia reaction. The print then goes into the coffee bath for 5–20 minutes; any longer, the paper isn't taking on any more tone. Lift the corner of the tray every few minutes to swish any coffee residue away and make sure there are no air bubbles. After the coffee bath I dunk the print into the second water bath. This takes off any excess residue and creates a nice even coffee tone. I blot the print by hand in-between blotter paper to prepare it for flattening. When I am toning multiple prints I keep only one print in the bath at a time. Your coffee bath will eventually take on some of the blue color and become murky so after two or three toned prints check to see if you need to make a new batch. When toning editions it is a good idea to keep an eye on this for consistency.

I have found that humidity control can play a big part on how your prints turn out. A little humidity is fine but too much can absorb into your paper and make the coating turn a dark green blue and uneven when it dries down, resulting in a dark and muddy blue final result. A nice chartreuse bright green coating when dry has produced the best outcome. Depending on what part of the world you live in I would invest in a dehumidifier. I always let my coated paper dry overnight and do not use paper more than a day old for best results.

I like to frame my cyanotypes as a projection float on a soft white mat with a white frame. Because of all the white space in my images it lets the eye be free from borders. Prints toned in coffee tend to look better with a warmer mat and darker frames. I always use UV-resistant Plexiglas for protection from direct light. It is not as heavy as glass and works great for shipping oversized pieces.

Creative practice

After experimenting with photogravure and etching on copper I was interested in creating even larger scaled pieces where I could still combine photography and my drawing. A photographer friend suggested I try cyanotype and that is when I started my research and began to expand my studio. I was already making prints with white ink on blue paper so the cyan color only made me drawn to the process even more. The blue and white assists my imagery of winged beings in portraying a heavenly or unseen world atmosphere.

I have had many obstacles come my way when making cyanotypes but most of them were the result of making larger scale prints. One problem in particular was how to coat such large pieces. I find that the hake brush works best. Try coating a 50″ × 127″ piece of paper with a 5″ brush, it will become overworked and take forever! After experimenting with puddle pushers and paint rollers I just couldn't get a coating I was happy with. The paint roller lays the chemical too heavy and coating rods always missed spots and made it very messy. I had an idea to create my own brush by combining multiple hake brushes together and handed this task off to my assistant at the time, Jonathan Sherman. He came up with a magnificent design which would allow me to coat enormous pieces of paper in just a few minutes. The brushes can also be changed out when they become too stained with chemical and no longer are acceptable for coating. Working extremely large has its own set of problems and after many years of trial and error I could be a good contact for those types of complications.

In Photoshop I use many layers to build into the negative because I'm typically using multiple photos for one image. I cut out any unwanted areas where I'm planning to draw on the print itself. Sometimes I scan in hand drawing on vellum to add to the image. I have also printed negatives with hand-drawn or painted transparency film and acetate overlays.

When my cyanotypes are developed and dried completely flat I draw on them with blue Prismacolors to blend the photography with hand drawing. My prints are unusually large for using a UV exposure unit and negatives printed on paper. I have come up with many ways to print oversized pieces using one negative but three different exposures for one final piece of art.

Artist statement

These cyanotypes are from a continuing three-part series that I have been working on for seven years. I have long explored spiritual and religious themes in making my art. Everyday people and animals, many of them my friends and their pets, are revealed as agents in the eternal battle of good and evil, their heroic spirits armored or winged. I want to lift the veil that obscures our double existence in the temporal and metaphysical worlds. Behind these works lie other invisible struggles: the painstaking labor to balance the generative forces of nature, human and machine, and the soul's resolve to manifest the optimism and grace in the face of life's trials. At their core, these are portraits of faith, bathed in a celestial blue.

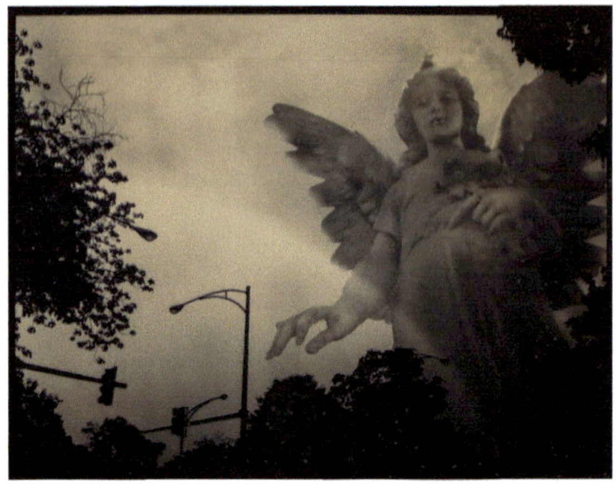

Figure 15.55. *Rainbow Angel*, cyanotype, hand painted with ammonia, toned in ammonia and coffee, edition of 5, 31″ x 39½″ © Teresa James 2016

Biography

Teresa James has been making prints for 30 years. She studied at the School of the Art Institute, Chicago, and became master printer for Chicago's Tony Fitzpatrick for thirteen years. James opened her atelier White Wings Press in 2005 where she specializes in cyanotypes and multi-color etchings on copper with photogravure. She had special equipment made to accommodate her grand scale cyanotypes, reaching as wide as 52″ and as long as 115″, stunning to see face to face. She combines numerous techniques and processes in her imagery, making hybrid prints that cause the eye to wonder where the photo ends and the drawing begins. To see more of her work, visit **www.teresajamesart.com**.

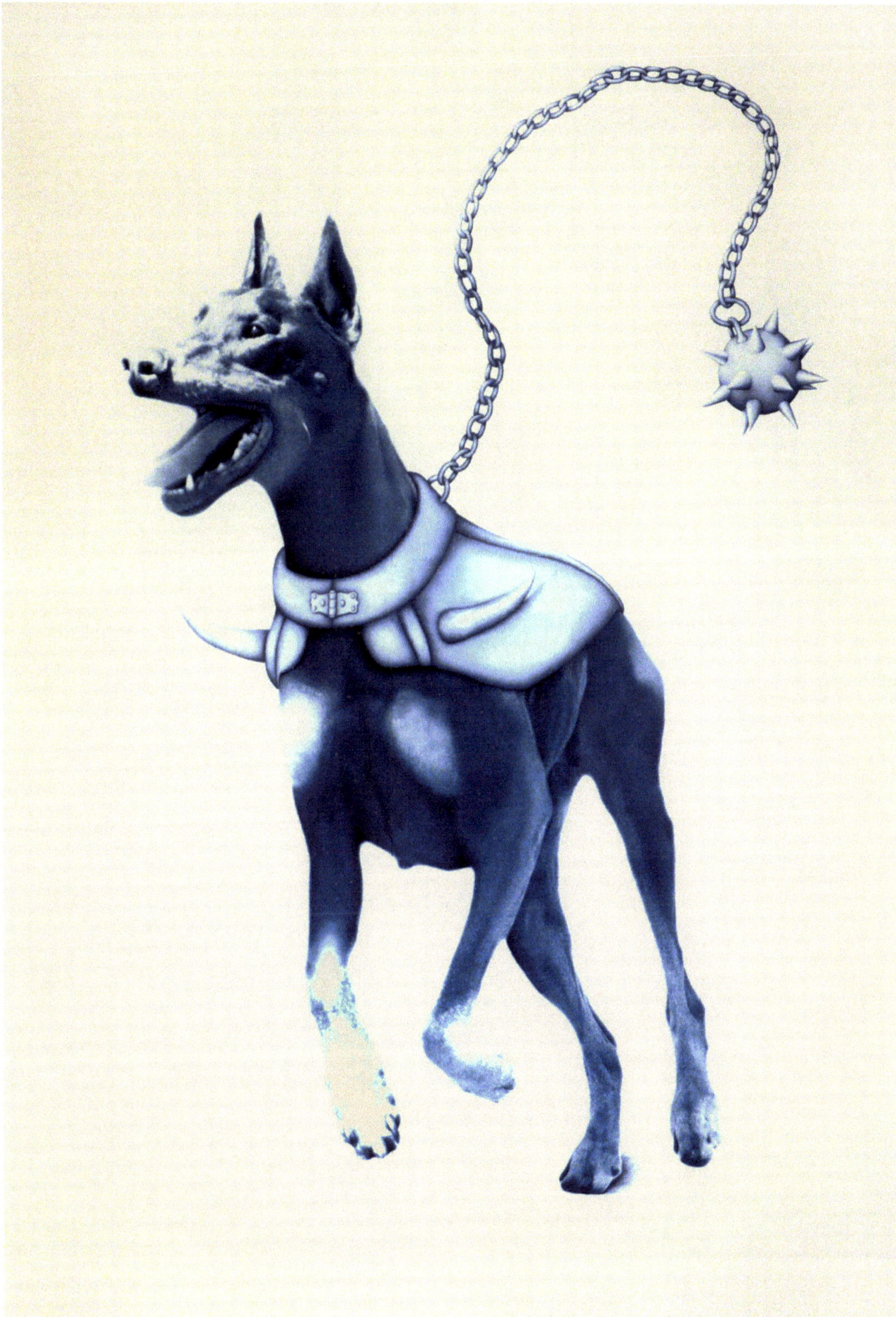

Figure 15.56. *Guinness*, cyanotype and hand drawing with blue Prismacolor, unique print, 58½″ x 41½″ © Teresa James 2014

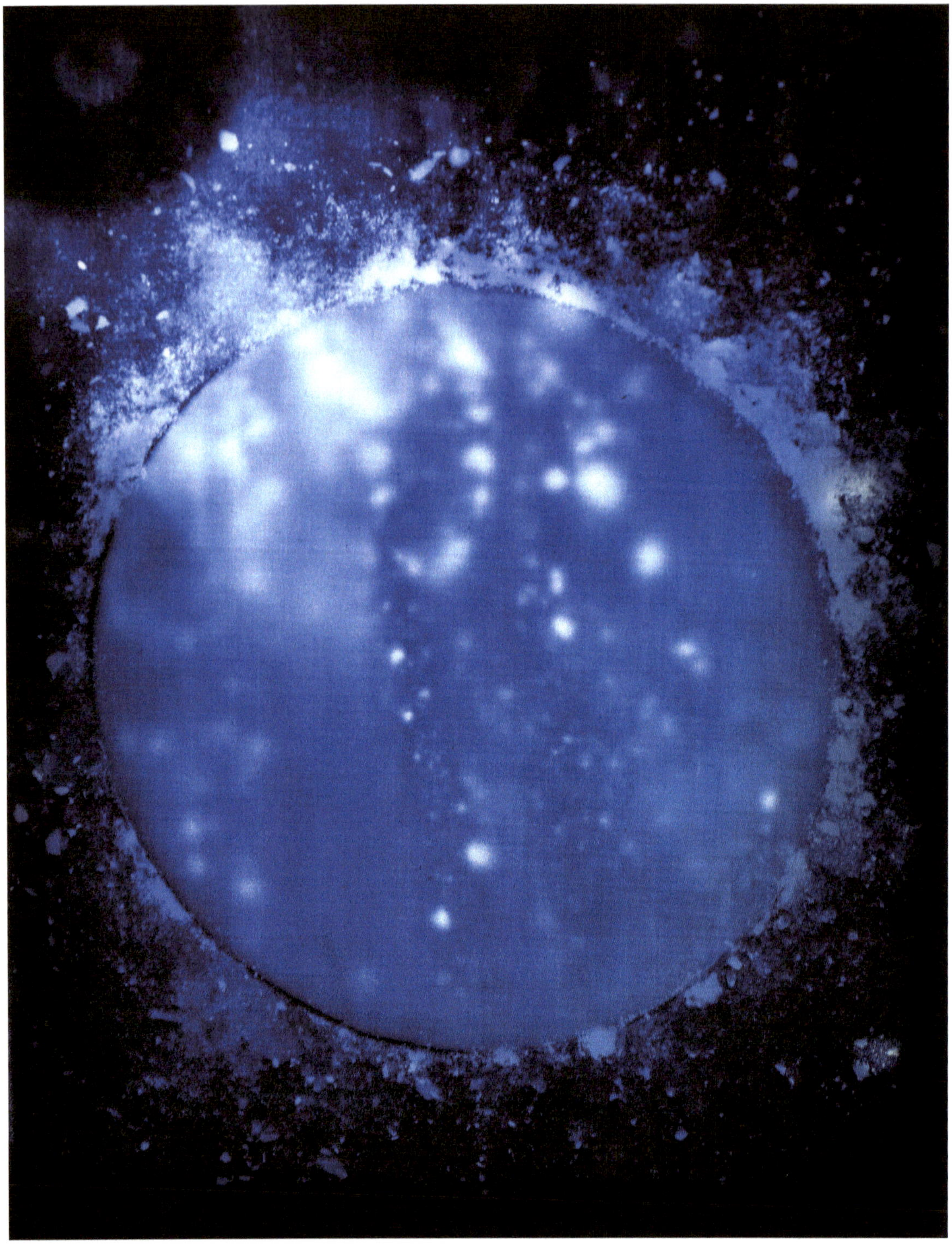

Figure 15.57. *Stardust and Ashes #100*, two *Stardust and Ashes* cyanotypes layered on top of each other and rephotographed on an LED light box, printed as a 32″ x 24″ archival inkjet print © Shannon Johnstone 2018

Shannon Johnstone

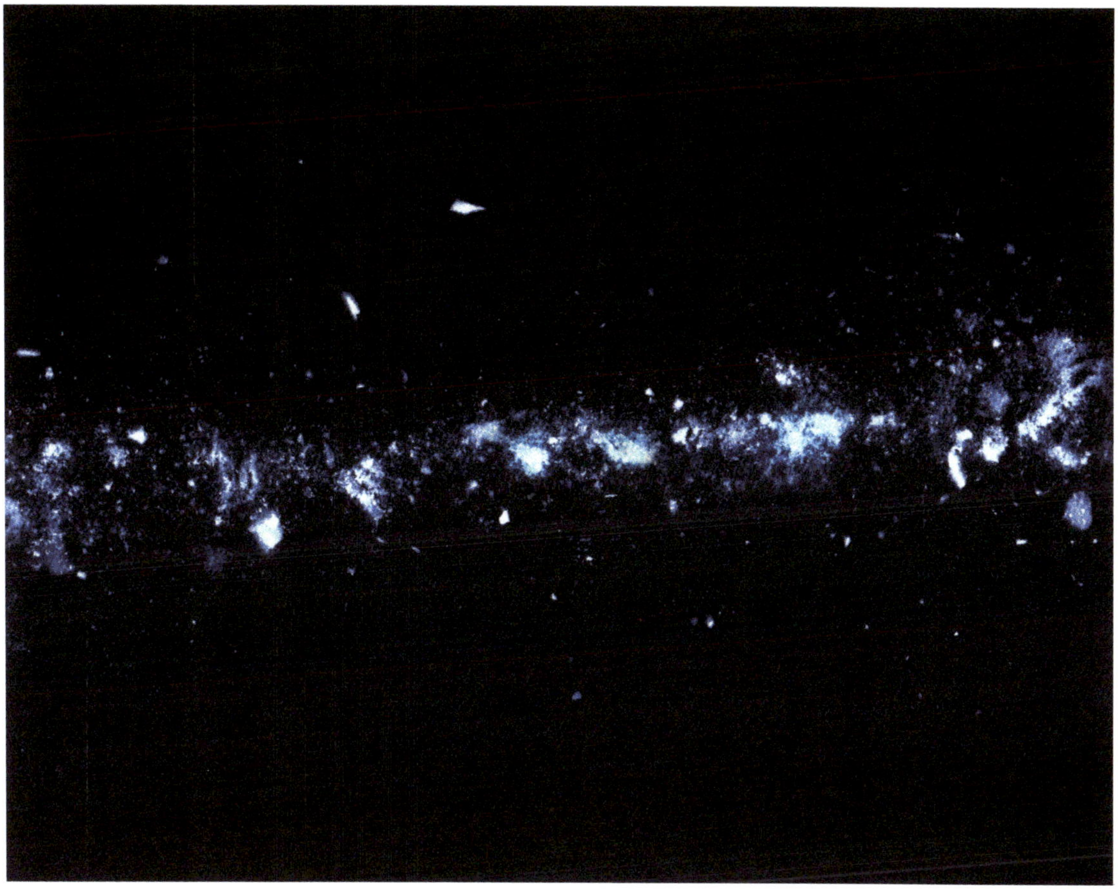

Figure 15.58. *Stardust and Ashes #82*, cyanotype on fabric, 8″ x 10″ © Shannon Johnstone 2018

Process

I use the ashes of euthanized shelter animals to create these images, not negatives, so each print is a monoprint. I sprinkle ashes onto the print and slowly add more during the exposure to vary the shades of blue throughout the image. Sometimes I blow on the ashes to move them around which creates varying shades of blue. I also use cardboard masks to dodge the light and create variations in tone.

As I am working, I am quite aware that I am working with deceased animals who were not lucky enough to find a home. It is really important to me to treat their ashes with respect and dignity. At the end of making a print, I return the ashes to the bag they are kept in. If there is any spillage or ashes that cannot be used, I plant them with the tree that holds the remains of my first beloved dog, Lula. The way I acquire ashes from the animal shelter is by asking.

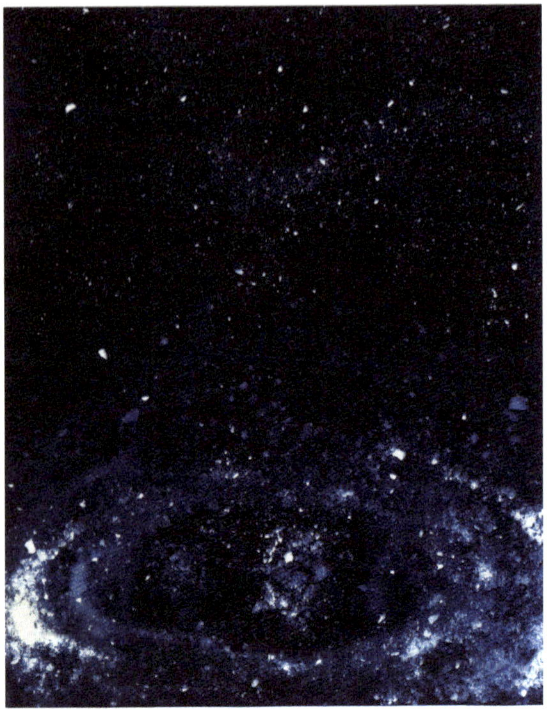

Figure 15.59. *Stardust and Ashes #96*, cyanotype on fabric, 8″ x 10″ © Shannon Johnstone 2018

I have been using Jacquard cyanotype pretreated fabric sheets. It is cotton sateen. Sateen absorbs dyes and the cyanotype coating better than regular cotton. I use the standard 8″ × 10″, as well as the mural sheets that are 5′ × 7′. Their products have worked extremely well in terms of ease of use and quality of the print. However, one thing to note about the Jacquard product is that the size of the sheets are approximate, and none are exact rectangles. This is perfect for me, but could be a frustration for someone who wants perfection.

Originally I made the decision to use fabric by mistake. I purchased a pack of the Jacquard product to do tests on. I bought it online and there was a sale, so of course I got the 30-sheet pack! As soon as I opened it I realized I had fabric instead of paper. At first I was disappointed, but I did a few tests and quickly realized the way the fabric moved and breathed matched what I was trying to say in the images. Also, as I hung them to dry on a clothesline, I discovered when light shines from behind them, they glow. I had a breakthrough moment while installing the work at an exhibition at The Front Gallery in New Orleans in 2017. The space was tight, so I decided to install one of the mural prints in the window. It was magical. By day, the cyanotype glowed and illuminated the gallery. At night, the cyanotype was lit from the lights in the gallery. Looking in from the outside, it looked as if a giant moon was glowing in the window. This is the direction I want to take this work now. I am leaning toward a more sculptural approach using light boxes, layering the cyanotypes, and hopefully more site-specific installations.

I expose by sun. Working in the sun is such an important part of my concept. I see the sun as adding life to these ashes, so I always work outside when making my cyanotypes. I create these images in my backyard, usually making the exposures between 12PM and 3PM. This is when the sunlight is directly over me and most intense. I tried to make these in twilight, but found it was hard to control the dodging and burning during those low intensity hours. I have also found that for where I live (North Carolina), the summer provides the most intense light. Sometimes I use the shade of a tree if I need to control my exposure over a longer period of time. This is especially important when working with mural prints since it takes more time to spread and work the ashes into different configurations.

Since I am sun exposing, the exposure times vary quite a bit with time of day, time of year, and if there are clouds. In general, the exposure times are about 45 minutes in direct sunlight. The shortest exposure has been 20 minutes, and the longest 4 hours. I judge the exposure by the shade of purple to copper-brown that the fabric turns. The more copper-brown the fabric turns, the darker the blue will ultimately become. If there is any moisture on the ground, that can also change the exposure as well. To combat this, I place the fabric in a darkroom tray while exposing. This has two functions; first it blocks any wind from blowing the cyanotype over (one hazard of working outdoors), and second, it lifts the print off the ground so that the humidity and moisture from the cement don't seep through. Since each of my cyanotypes are unique, I want some exposures to be much darker than others for variation. I imagine each like a different universe.

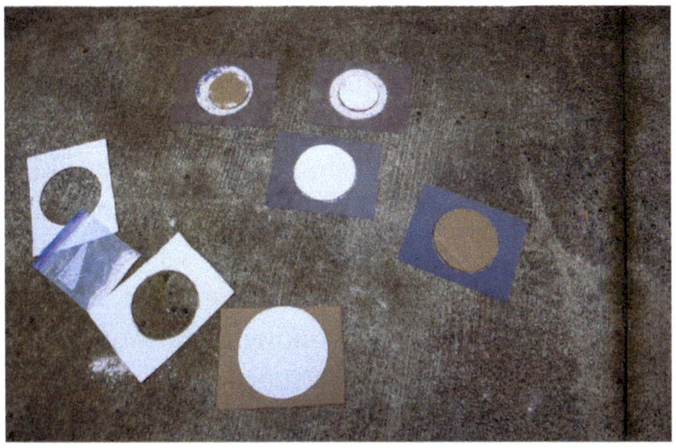

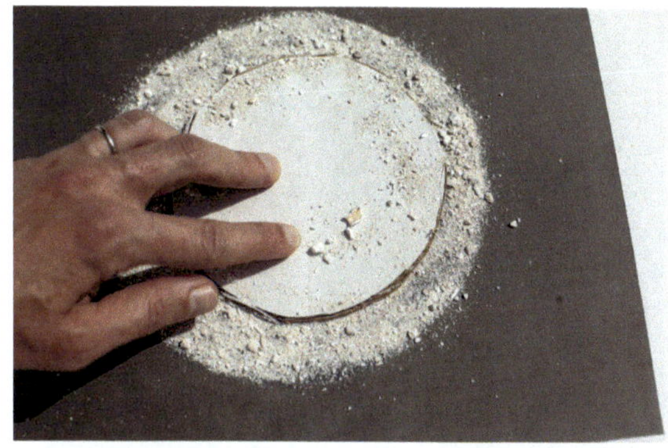

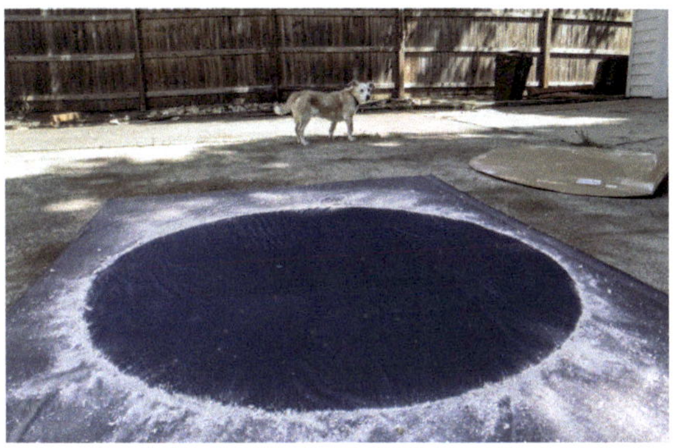

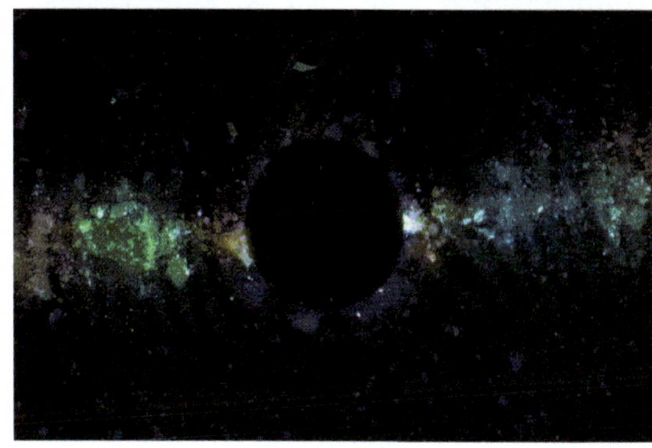

Figures 15.60–15.64. Top left and right, some of the masks Johnstone uses in the making of her prints. Middle left, creating *Stardust and Ashes #63* mural print with the supervision of Jorge. Middle right, *Stardust and Ashes #102*, two prints scanned and layered in Photoshop with different blending modes and layer effects used, printed 24″ x 32″ as an archival inkjet print. Bottom right, installation view at The Front Gallery, New Orleans, Louisiana, September 2017. Images © Shannon Johnstone 2018.

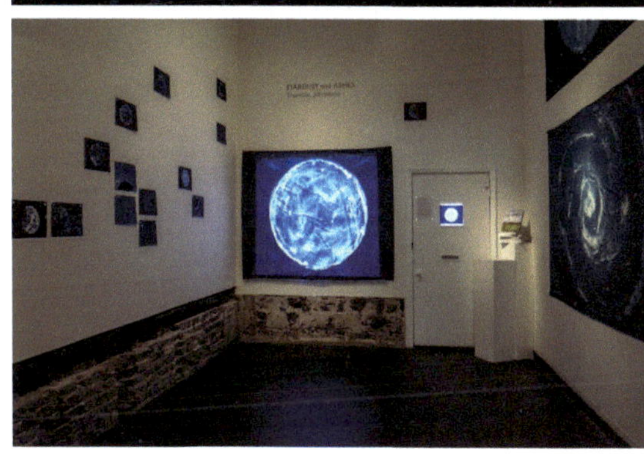

My advice for working with cyanotypes in sunlight is to be patient, and not expect precision. It is possible to achieve exacting results with cyanotypes, but not while working with it outdoors. There are too many variables that cannot be controlled. My experience has taught me that working outdoors with cyanotype is a much more fluid process than working with silver-based materials in a darkroom, but it is also much more forgiving.

A hazard of working outside is heat and sweat. North Carolina gets pretty hot and humid in the summer, so often my sweat drips onto the prints as I make them. The sweat starts the processing of the print, and leaves a strange blemish mark. When this first happened I was really frustrated, but then I realized this was another dimension to the project. Like the sun, I hoped to breathe life into these ashes which is why I use my hands and breath to manipulate them. My sweat, like my breath and the sun, is another way to incorporate life.

I develop the prints by washing them in our garage sink in an old Tupperware container with running water for about 20 minutes. The entire time the print is washing, I turn it over, swish it around, pull it out, rinse both sides, and then repeat this process over and over. Working with fabric makes this process easy since the sateen fabric is so durable.

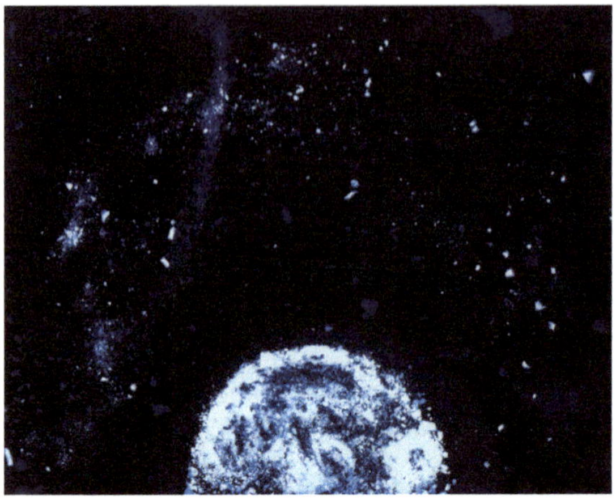

Figure 15.65. *Stardust and Ashes #78*, cyanotype on fabric, 8″ x 10″ © Shannon Johnstone 2018

The edges do fray a bit with rigorous washing, but I don't mind that. I wear Neoprene gloves when washing my cyanotype prints. Occasionally I will add a little hydrogen peroxide to deepen the blue in an image. The hydrogen peroxide speeds the oxidation process and immediately shows what the print will look like in a few days after the oxidation process is complete.

When I add hydrogen peroxide I follow these steps: first, I make sure the container is full of water, then I turn the water off, add the hydrogen peroxide and agitate it with my gloves still on so that it is thoroughly mixed, and then dunk the print and move it around in the water. I don't measure the hydrogen peroxide other than using the cap to add one or two capfuls to the water in the Tupperware container. Afterward, I always wash the print for a few more minutes and hang them to dry on a clothesline in my backyard.

So far my favorite way to display this work has been with magnets. This allows the viewer to see the images breathe and move and they are not constrained by a mat and frame. Before hanging the work, I iron and starch the prints. Then I place a flat tack or nail in the wall and adhere a small silver magnet. Then, I place the print on top of the first magnet and adhere a second magnet on top of the print. This method secures the work and does not leave a hole or a mark on the print.

I am really interested in making *Stardust and Ashes* more sculptural and focusing on site-specific installations. Some of the things I have been experimenting with is using the cyanotypes like a negative, and projecting light through them. I have also been scanning the cyanotypes, compositing them digitally, and then printing large archival inkjet prints. I am also thinking about sewing the fabric together in a cylinder shape and illuminating the prints from within so they hang like tiny lanterns. The fact that the work is on fabric gives these photographs a three-dimensional quality that is unlike anything I have done so far. I want to push that as far as I can.

Creative practice

I have a love for all animals, and special affinity for dogs. The tragedy of animal overpopulation is heartbreaking and touches me deeply. It is a topic that has been central to my photographic work for the past decade. We have four dogs and two cats and they are family members. After my father passed away unexpectedly, I wished to make a portrait of him and where he is now. I found an old solar kit my mom had given me and I used that along with some of his ashes to make his portrait in the sun. The result looked like a constellation, which was very fitting—something that glows at night and is always with you. I thought about what a gift it is to be able to mourn for someone. I couldn't help but think about all the pets who go into animal shelters and never come back out. The humans they belonged to don't mourn for them. They die, and it is as if their existence didn't matter. I longed to create something that might memorialize them. This is how *Stardust and Ashes* came about. The cyanotype is a perfect fit for this idea because in addition to the deep blues and archivalness of the cyanotype process, I am able to work with elements that reflect core concepts in the work such as sunlight and animal remains. In other words, I love the conceptual inverse of the cyanotype for this work—the daylight gives way to a night sky, and death produces lasting life.

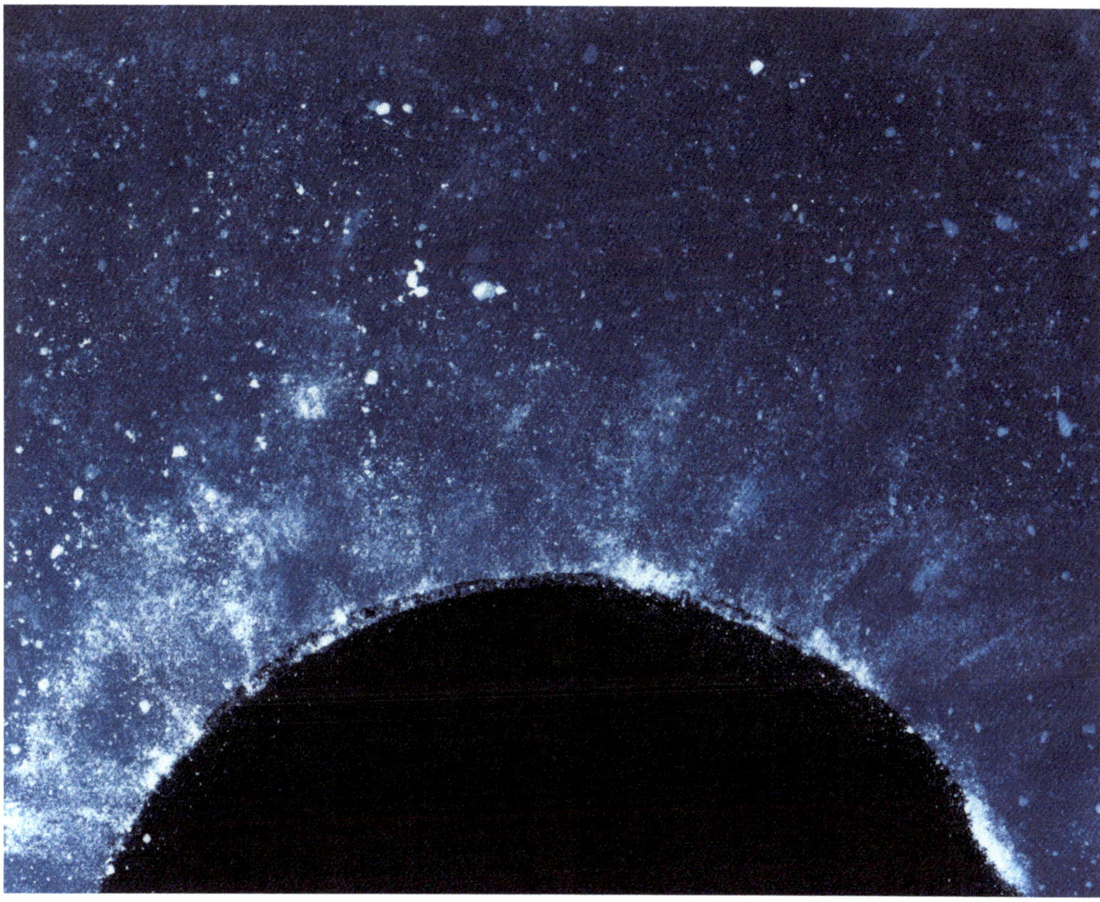

Figure 15.66. *Stardust and Ashes #55*, cyanotype on fabric, 8″ x 10″ © Shannon Johnstone 2017

Artist statement

The nitrogen in our DNA, the calcium in our teeth, the iron in our blood, the carbon in our apple pies were made in the interiors of collapsing stars. We are made of starstuff. —Carl Sagan, *Cosmos*

For the past decade I have been working with homeless pets and exploring ways to visualize the tragedy of animal overpopulation. I made these cyanotypes with the ashes of euthanized homeless animals from an animal shelter's crematorium. These animals died with nobody to mourn their passing, except maybe a few overwhelmed shelter workers. Using my own breath and fingers to manipulate the ashes, I work the ashes into celestial configurations while the sun exposes the cyanotype turning the negative space Prussian blue. With these images I mourn the passing of thousands of our forgotten companions, and remind us that we are all connected and headed for the same fate: reduced to dust and returned to the stars.

Biography

Shannon Johnstone's photographic work deals with themes that reclaim what has been discarded and make visible that which is hidden. Her project, *Landfill Dogs*, has been featured in national and international exhibitions and magazines, and was most notably on ABC world news with Diane Sawyer, and CNN.com. Johnstone is a tenured professor at Meredith College in Raleigh, North Carolina. To see more of her work, please visit **shannonjohnstone.com**.

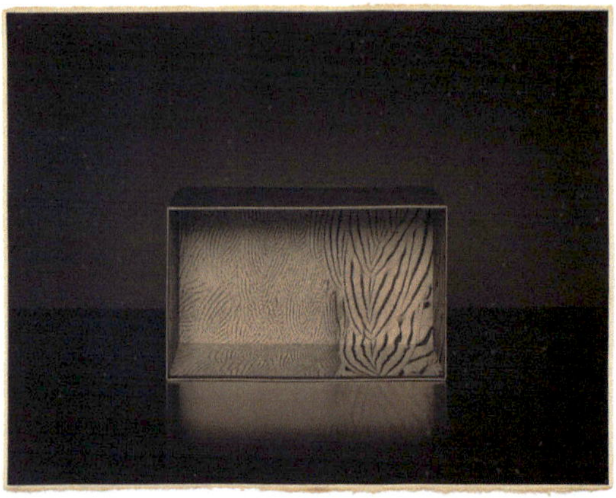

Figures 15.67–15.72. Top to bottom, left to right, *Alexander McQueen*, *Christian Louboutin*, *Mezlan*, *St. Laurent*, *Ferragamo*, *Prada*, from the *Le Scarpe* series, new cyanotype toned with wine tannin, printed on Arches Platine, 22″ x 30″ © Max Kellenberger 2018

Max Kellenberger

Process

All images for *Le Scarpe* are printed on 22″ × 30″ full sheet Arches Platine. For *Blues* I used 22″ × 30″ Arches Aquarelle 140 hot press bright white.

Images for *Le Scarpe* were captured on a digital camera, adjusted in Photoshop, and printed with an Epson 9900 onto Pictorico Ultra Premium OHP using Ron Reeder's digital negative method.

For *Le Scarpe* I use Mike Ware's new cyanotype formula from Photographer's Formulary, mixed and applied according to package directions. Prior to coating the paper, I pre-wet the paper with distilled water, allowing the water to soak into the paper for approximately two minutes before applying the cyanotype solution. Coating is done with a Sterling Studio Blender brush. For *Blues* I used the Photographer's Formulary classic cyanotype, mixed and applied according to package directions.

I use an Olec Olite printing lamp with a Douthitt vacuum frame, with a 15-minute exposure. I do a plain water wash for 20 minutes, adding a splash of hydrogen peroxide after the first 5 minutes of the wash.

Le Scarpe prints are first toned in powdered wine tannin (mixed ½ oz to a liter of water) for 1 hour, rinsed, and then a very dilute sodium carbonate bath for 1 minute.

I prefer to display these prints floated in a large frame to display the deckled edges and emphasize the physicality of the print. I find that the toned prints, especially, are very influenced by the color and intensity of light they are displayed in, and find they look their best in diffused daylight.

The biggest problem I encountered, especially with *Le Scarpe* and its large area of continuous tone, is getting the large sheets coated absolutely evenly with the cyanotype solution. I found that pre-wetting the paper very evenly with distilled water, letting it partially dry to a dull sheen, helped the cyanotype absorb evenly into the paper. Being cautious and methodical in applying the cyanotype and carefully going over the paper two or three times allowed the cleanest and most even application.

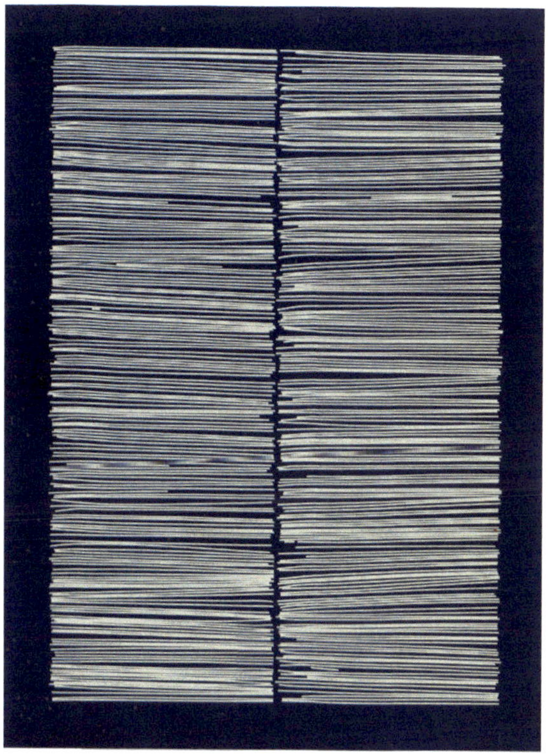

Figure 15.73. *Spaghetti*, from the *Blues* series, classic cyanotype on Arches Aquarelle paper, 22″ x 30″ © Max Kellenberger 2018

My main advice to anyone interested in any traditional printing processes is to take the time to experiment and figure out what works for you as a practitioner and your particular variables. What may work for one artist in one location may not work for everyone, so be flexible and curious as you create your own practice. And take lots of notes! Write on the back of the prints as you make them with the date and all of the information that went

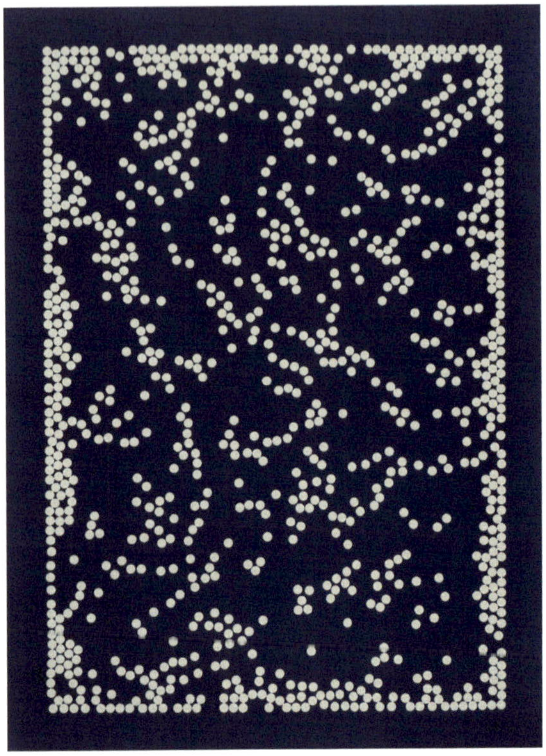
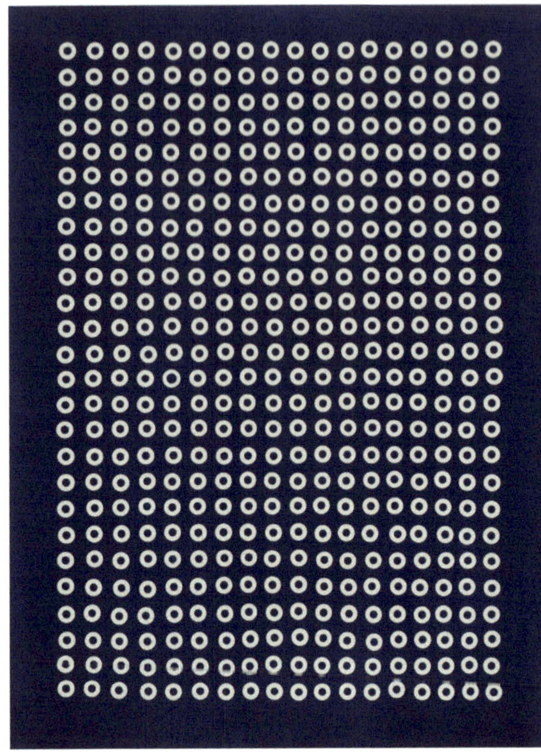

Figures 15.74 and 15.75. Left, *Aspirin*; right, *Washers*, from the *Blues* series, classic cyanotype on Arches Aquarelle paper, 22″ x 30″ © Max Kellenberger 2018

into making that print. My last piece of advice is to have patience. Every new process has a learning curve; there will be problems to solve, and moments of frustration. Just keep going.

Creative practice

Over the many decades I have been practicing photography, my largest realization is how to trust my instincts. There are thoughts and images which keep coming back to my mind which want to be expressed, and I've learned to listen to those thoughts. I don't always know the hidden meaning of these ideas until after they've been expressed in my work, but the images created from these moments are true to myself and my artistic sensibility. It's important to find ways to express what's inside, and make work that reflects my own personal vision.

The other side of that is also learning how to be objective about your own work. Not every idea ends up being successful and you must learn to reject what doesn't work. Some ideas have taken years of thinking, trying, reimagining to come to their final realization. And some have never made it. The important thing is to keep shooting, keep creating, keep experimenting. And edit, edit, edit.

Artist statement

For my series *Blues* I was inspired to create a series of systematic portraits of ordinary objects that came in boxes. Removing the objects from the boxes, be it pasta or washers or aspirin, and creating some sort of order from the chaos, I wanted a simple reproduction, a one-to-one copy of these ordinary objects—exactly what a blueprint was used for since its invention. Making this series in cyanotype seemed to be the perfect correlation of process and concept.

Le Scarpe was the result of many years of chasing the idea of capturing the emptiness found inside

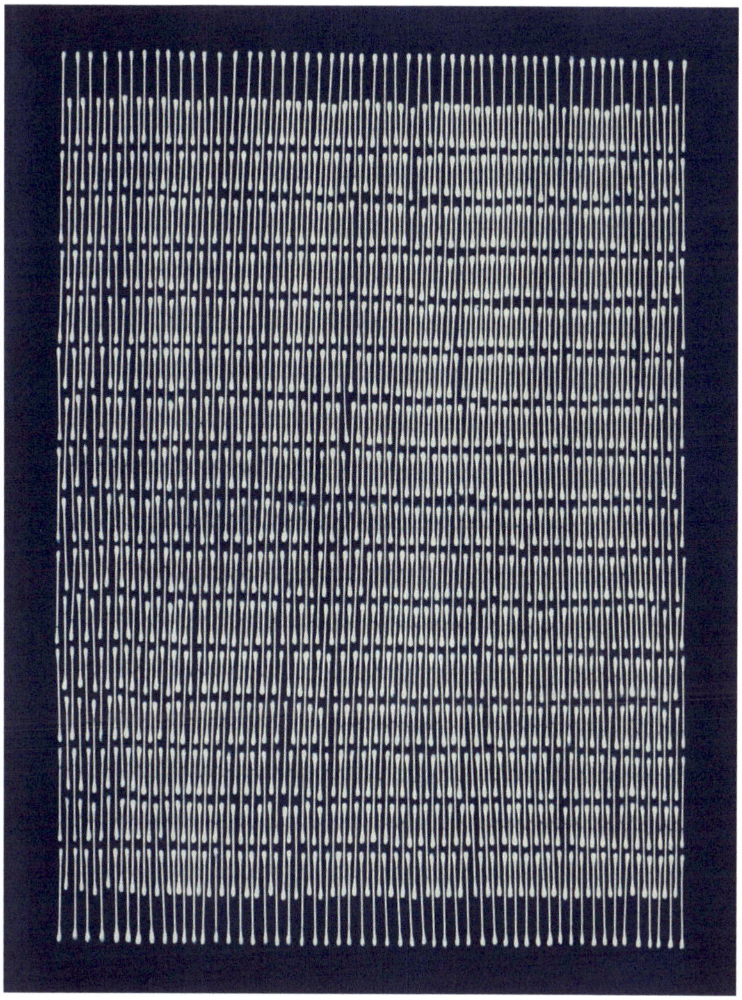

Figure 15.76. *Q-Tips*, from the *Blues* series, classic cyanotype on Arches Aquarelle paper, 22″ x 30″ © Max Kellenberger 2018

a shoebox. In the end I was drawn to the toned cyanotype process which captured the feeling of impermanence and transience of being that I saw in these empty spaces and added an element of physicality to the finished piece. The handmade imperfection and unrepeatabilty of the process brings a greater emotion to the print, and the split-tone warm/cool color echoes the sense of being and nothingness, life and death that I see inside these empty boxes.

While *Blues* is well within the classic cyanotype canon, I feel that *Le Scarpe*, being more conceptual in origin, is a more modern use of the process and takes advantage of the unique characteristics to add a heightened emotional element to the final prints.

Biography

A native of Switzerland, Max Kellenberger has been exhibiting work in the United States and Europe since 1972. His recent photography exhibitions include *A One Man Show* at Gallery 16 in San Francisco, *Blues* at Smith Anderson North in San Rafael, *Quietude* at Corden|Potts, *One: Unique Photographs* at Klompching in Brooklyn, NY, *Four Photographers* at TUTTIARTlucerne in Lucerne, Switzerland, and most recent *Le Scarpe* at Corden|Potts in San Francisco. He has been published internationally in photography magazines and books, most recently *The Polaroid Project: At the Intersection of Art and Technology*. Kellenberger resides in San Francisco, CA.

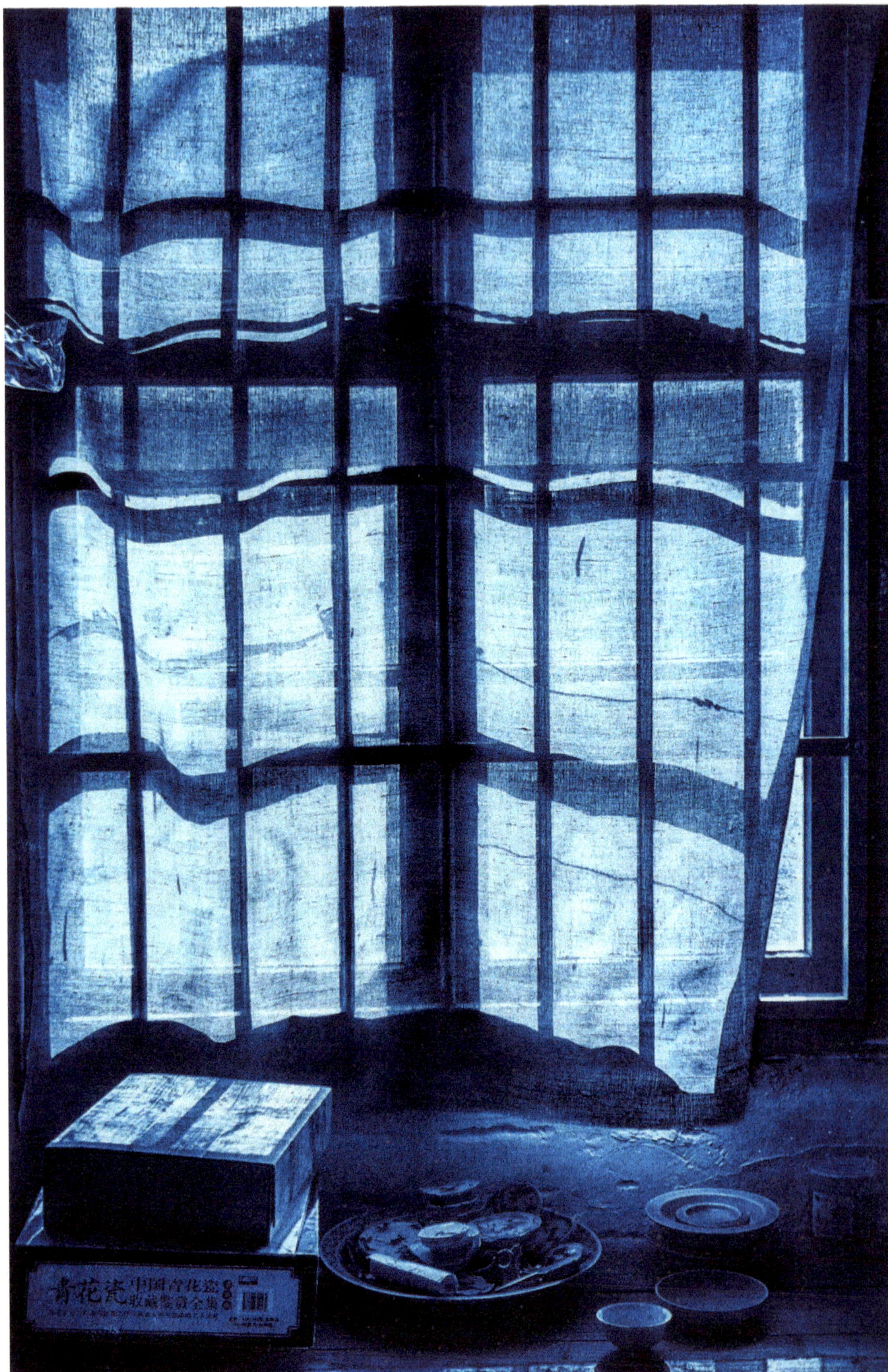

Figure 15.77. *Chengdao*, new cyanotype developed in 4% citric acid on Rising Stonehenge, 10″ x 14″ on 14″ x 20″ © Sandy King 2018

Sandy King

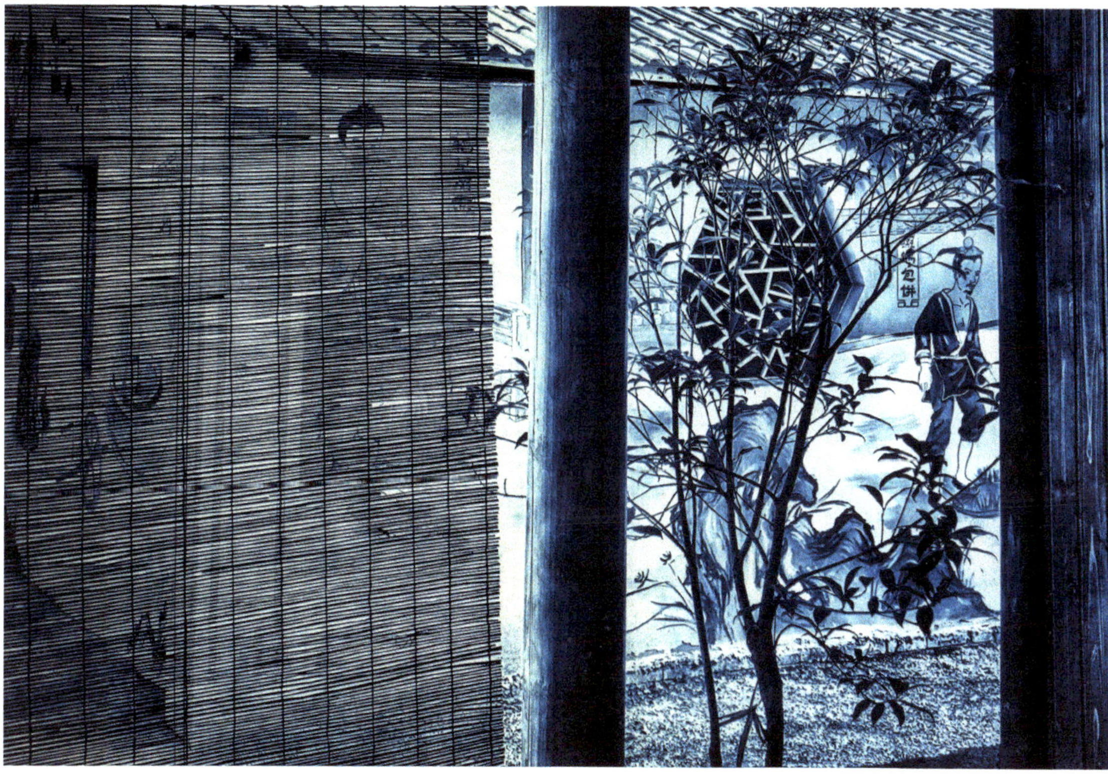

Figure 15.78. *Curtain*, new cyanotype developed in 4% citric acid on Rising Stonehenge, 10″ x 14″ on 14″ x 20″ © Sandy King 2018

Process

These prints were made with the new cyanotype process on either Masa, Stonehenge Rising, Arches Aquarelle or Canson Montval. My procedure with cyanotype is to coat the paper and expose as soon as it is dry to the touch. The sensitized paper was exposed with a bank of UVBL tubes, with exposure times of about 2.5 minutes printing with digital negatives. All of the prints were developed for 10 minutes, with constant agitation, in a bath of either 4% citric acid or 1% hydrochloric acid.

Creative practice

I am a fine art photographer from South Carolina interested primarily in the landscape, both its pristine state and as impacted by human intervention, as an abstraction of reality. My work is printed in monochrome because the elimination of color forces the viewer to concentrate on the essential elements of light, form and texture. Content is not unimportant in my work because I am attracted to beautiful and noble subjects, but ultimately the subject is less important than how it is depicted.

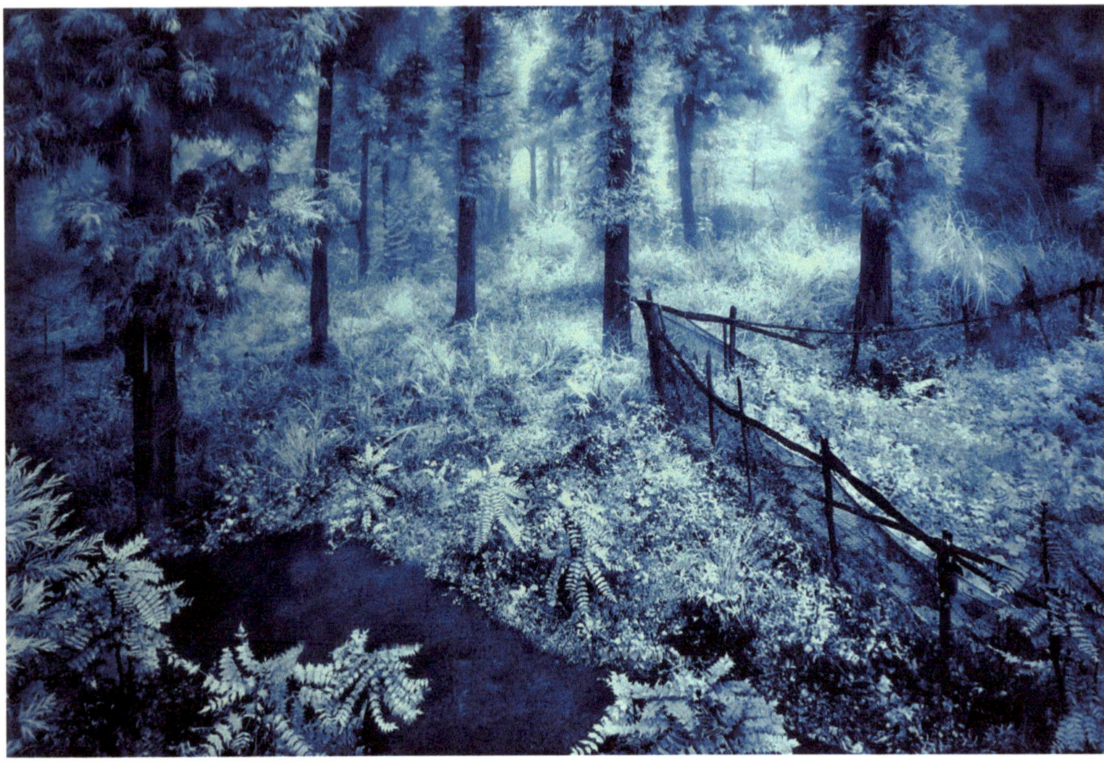

Figure 15.79. *Heavy Fog*, new cyanotype developed in 4% citric acid on Masa, 10″ x 14″ on 14″x 20″ © Sandy King 2018

Making photographs is an act of discovery, both in capturing the image in the field and in printmaking. In the very early days of photography viewers were fascinated by the detail of the photographic process. One such statement by an unknown Parisian reviewer was made describing the details in a daguerreotype, published in Alphonse Donné, "Le daguerreotype," *Le Voleur* (July 15, 1839): "The most delicate objects, the small pebbles under the water by the shore, and the different degrees of transparency they give to the water, everything is reproduced with an unbelievable exactitude. But our astonishment redoubles when we take a magnifying glass and discover, especially in the foliage of the trees, an immense amount of such fine detail that would be impossible for the naked eye to capture on its own."

In my own work there is a constant joy in the act of discovery, whether it be finding an extraordinary subject in the natural landscape, or in printmaking when a small detail in the scene is revealed that was not noticed when capturing the image.

I am captivated by the ability of the lens to gather light from objects in the world, and to expand this image to a detailed and precise rendering of nature. Photography always begins with something that is real, and this for me is one of its greatest strengths, and what sets it apart from other forms of the Visual Arts, like painting, which have an internal origin.

Printmaking is as important for me as seeing and discovery. Much of my work is done with 19th century handmade photographic processes, including carbon transfer and platinum and palladium. I am fascinated by these historical tools, and with analog photography, but I am also intrigued with the creative possibilities of digital work and print with digital negatives. The result is that my work is hybrid in nature, and attempts to combine the best features of analog and digital.

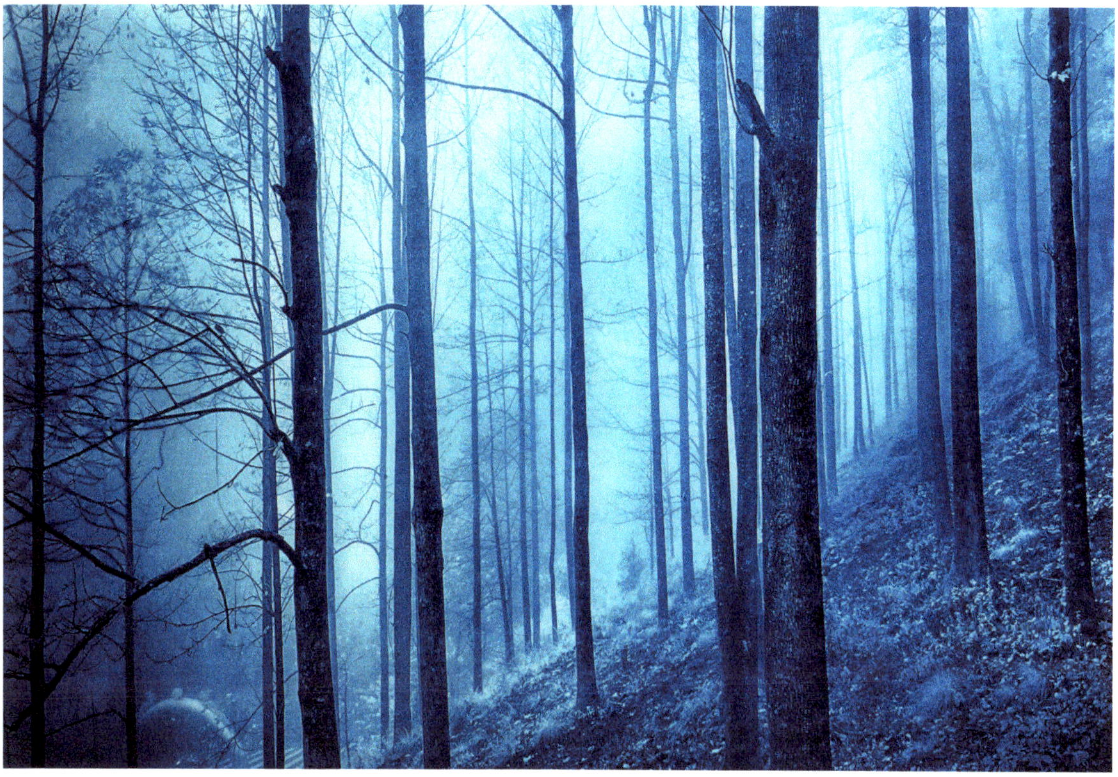

Figure 15.80. *Trees in Fog*, new cyanotype developed in 4% citric acid on Rising Stonehenge, 11″ x 14″ on 14″ x 17″ © Sandy King 2018

Biography

Sandy King taught at Clemson University from 1971 until he retired in 2006. He has printed with a variety of alternative printing processes since the early 1980s, including cyanotype, carbon transfer, kallitype, platinum/palladium, and Van Dyke. For over two decades, King has devoted much of his time to carbon printing and his personal work has been published in magazines such as *Photovision*, *Silvershotz*, and *View Camera*. He has conducted carbon-printing workshops in the US and in numerous countries abroad, including Canada, China, Mexico, Spain and Turkey. He authored numerous publications, including scholarly works on the history of photography as well as technical articles and books on handmade printing processes. His most recent publications include, *Handcrafted: The Art and Practice of the Handmade Print*, co-authored with Christina Z. Anderson, Zhong Jiaming, and Sam Wang, Zhejiang Photographic Publishing Company, Hangzhou, 2014, and *The Carbon Print Process*, coauthored with John Lockhart, 2017. To see more of his work, visit www.sandykingphotograhy.com.

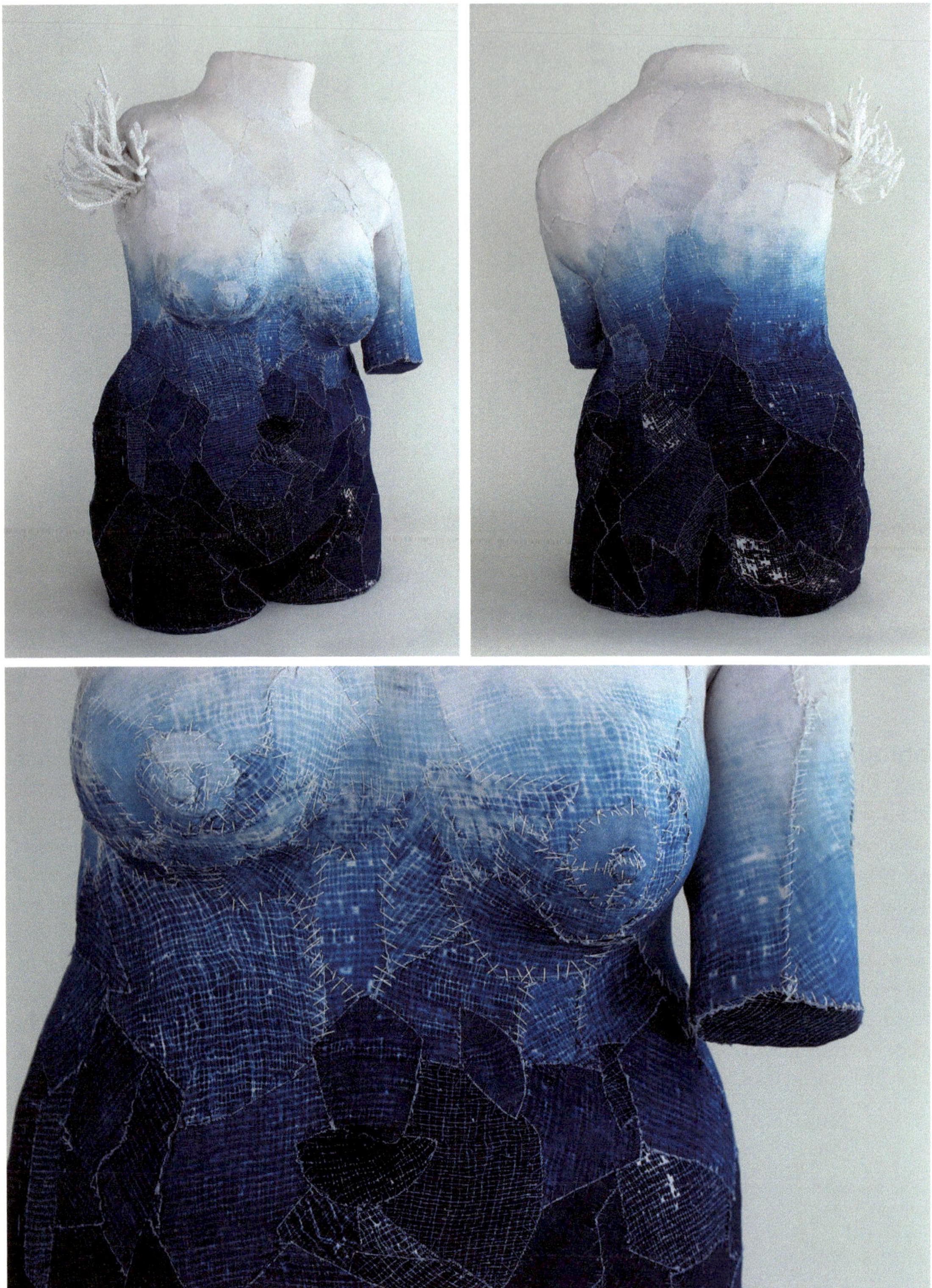

Figures 15.81–15.83. *Tidal Bather 1*, wood, plaster, mesh, felt, cyanotype prints on cotton, wire, cord, thread, 20″ x 27″ x 15″ © Tasha Lewis 2017

Tasha Lewis

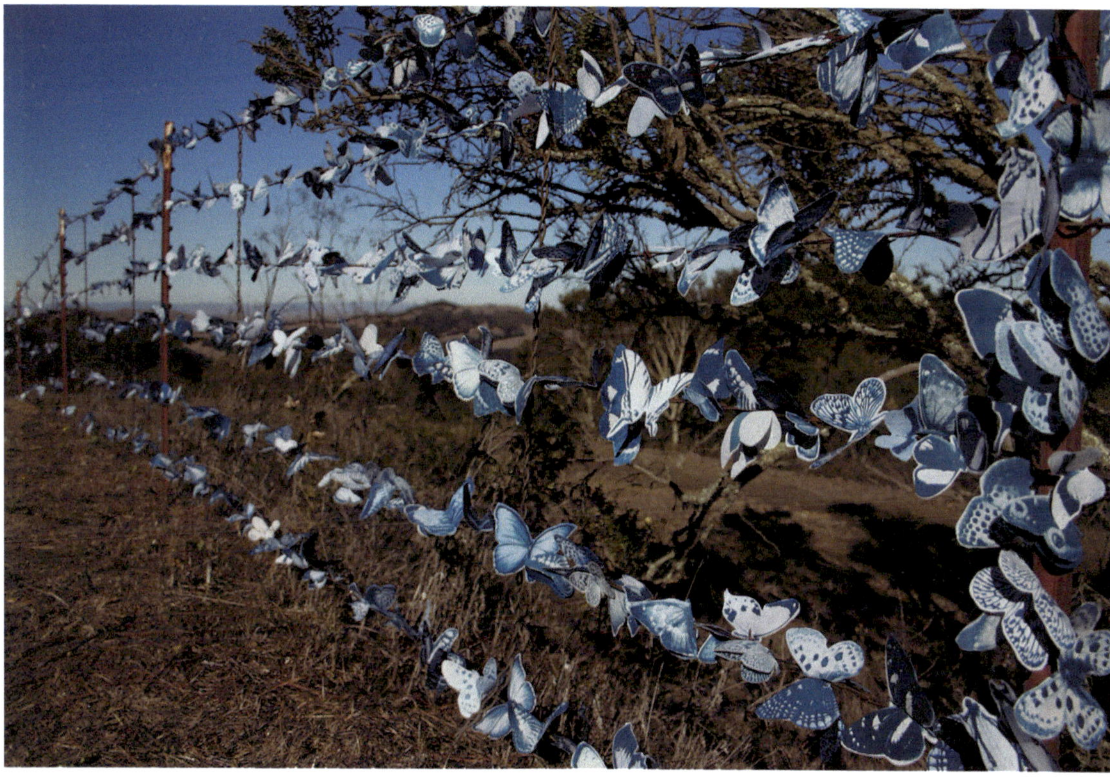

Figure 15.84. *The Swarm*, site specific installation, Berkeley, California, stiffened cyanotype prints on cotton, magnets © Tasha Lewis 2012

Process

I use cotton cloth from **Blueprintsonfabric.com**. I use a mixture of digital transparencies that I print at a copy shop from images manipulated in Photoshop, and handmade negatives which I paint on acetate in the studio. I have also done some work with simple photograms made with found specimens or vinyl stencils.

I use an ultraviolet light box in the studio, and the sun when I can. My exposure times are between 8 and 15 minutes, with the glass being about 8″ from the light source. I wash all my prints out in a bucket and add a cap of hydrogen peroxide before the final wash.

I have toned cyanotype using Bostick & Sullivan's cyanotype bleaching Solution A 14% potassium carbonate solution and cyanotype toning Solution B 40% tannic acid solution one after the other to create a purple color. I have also had success dyeing my fabric prints red, orange, and peach with RIT acid dyes. The whites of the cyanotype took the color of the dye as did the blues turning them purple to black.

One unique thing I do is in an earlier series I discovered that I could make double-sided objects (like my magnetic cyanotype butterflies) by creating a registered sandwich of two negatives with the cyanotype fabric in the middle. You print each side and then the final result will mimic the natural world (I have also made leaves this way).

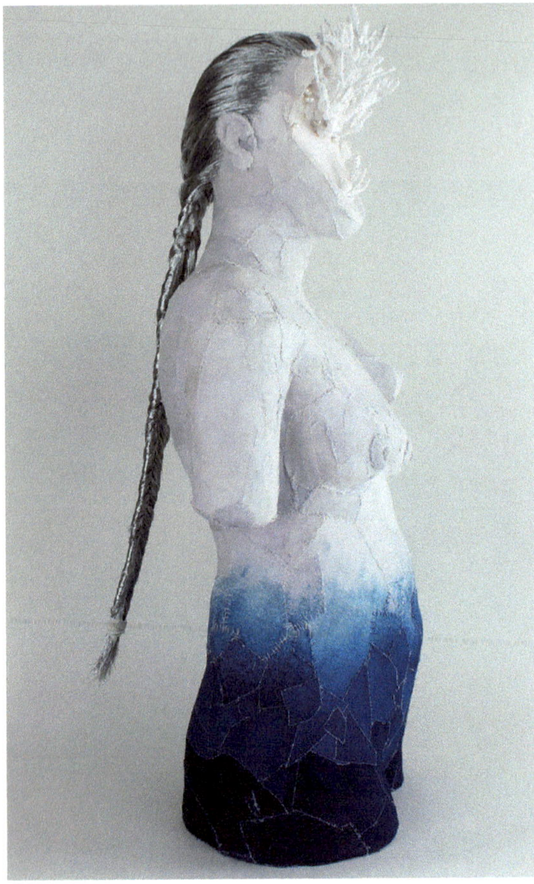

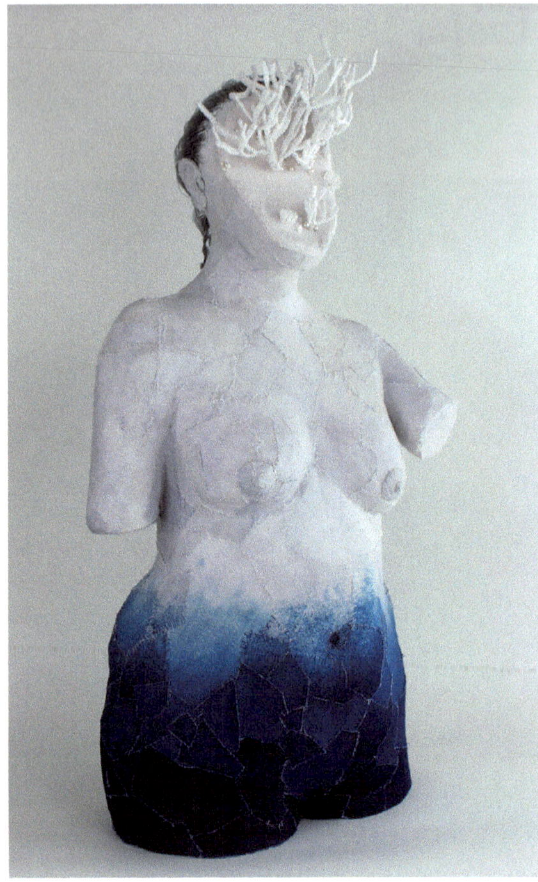

Figures 15.85 and 15.86. *Tidal Bather 2*, wood, plaster, mesh, felt, cyanotype prints on cotton, wire, cord, glass beads, thread, 18″ x 37″ x 15″ © Tasha Lewis 2017

Creative practice

I started out as a printmaker and a photographer and discovered alternative processes late in high school. I found cyanotype and kallitype to be fusions of printmaking and photography. I loved layering and augmenting images, hand-painting the emulsion, and making collaged negatives. In college I honed in on the potential for augmenting the dimension of a photograph by crafting forms that were then covered in cyanotype prints on fabric. While I later discovered other means of textile design, cyanotype was so familiar and so direct, it became an exceptional way to explore this new process.

In a more practical sense, I was drawn to cyanotype-coated cotton because you could print huge sheets of it with giant negatives without worrying about having a large enough tray to develop paper, or about improperly painting the emulsion. Accordingly, all my energy could be channeled into making exciting negatives, and manipulating the final cyanotype photograph after it was developed.

While there have been many artists before me to explore the sculptural potential of cyanotype on cotton, I believe I am the first to integrate them into a taut skinlike layer around a solid form.

I approach cyanotype as one step or element in my process and attempt to emphasize its strength in conveying complex surface textures while not defining the work with the structural limitations of fabric.

While I also use other means of surface decoration for my textiles, cyanotype will always be a favorite process for its immediacy, its beautiful color, and its historical roots.

Artist statement

I am drawn to the cyanotype process both because of its history and because of its flexibility. In the beginning, my inspiration was drawn from an investigation into the cultural/scientific/historical context in which the cyanotype process was born. Popularized by scientists, and botanists in particular, the cyanotype is intrinsically tied into the scientific recording boom of the late 19th and early 20th centuries. These are the times of the curiosity cabinet, the prints of Anna Atkins and a rush of explorers/scientists to colonial lands only to bring back specimens from foreign ecosystems. The cyanotype is a process of documenting. The resultant image—which is the basic fodder for all of my work—is a kind of scientific stand-in for the actual object in question. It is the trace of the original. In this way, like cyanotype's use for building blue prints in more recent centuries, my work is formed as the re-presentation of something real; it is somehow not quite the object itself.

As I have developed this technique, my work has become most interested in this tension between the qualities of the surface—where everything is hand sewn with a needle and thread—and the illusory power of the photographic base—where the image is created digitally and then printed in the sun.

I have used my magnetic cyanotype butterflies to organize a global collaborative street art project during which 200 participants from 45 different countries created, photographed, and then de-installed ephemeral installations of blue swarms on metal objects around their homes.

Biography

Tasha Lewis is a mixed-media sculptor living and working in Jersey City, New Jersey. She has shown across the country and has self-published two books: *Illustrating James Joyce's Ulysses in Eight Weeks*, and *Swarm the World, a Global Collaborative Street Art Project*. For more details on her process please visit her website **www.TashaLewis.info** and follow her on Instagram **@TashaLewisArtist**.

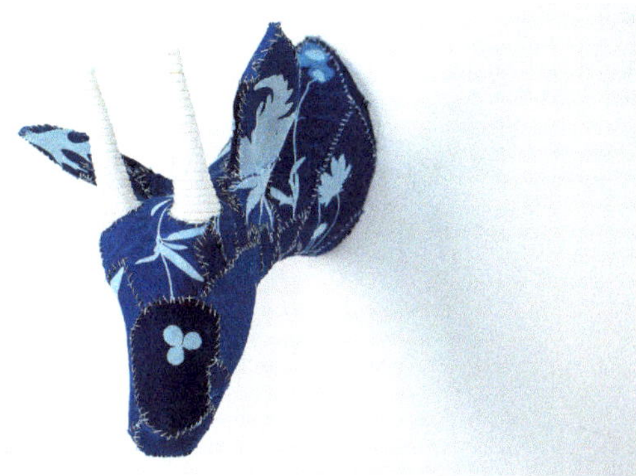

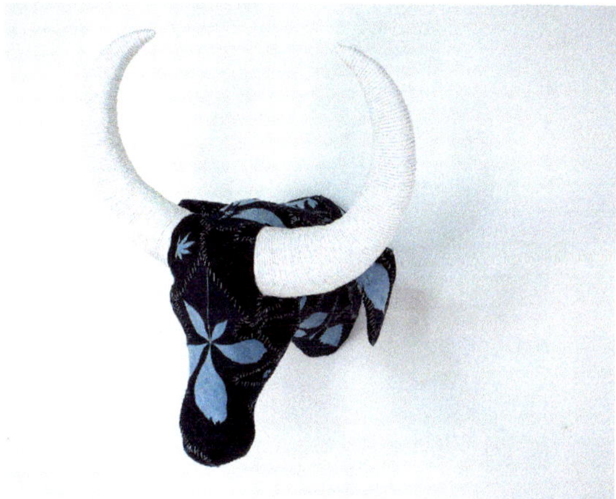

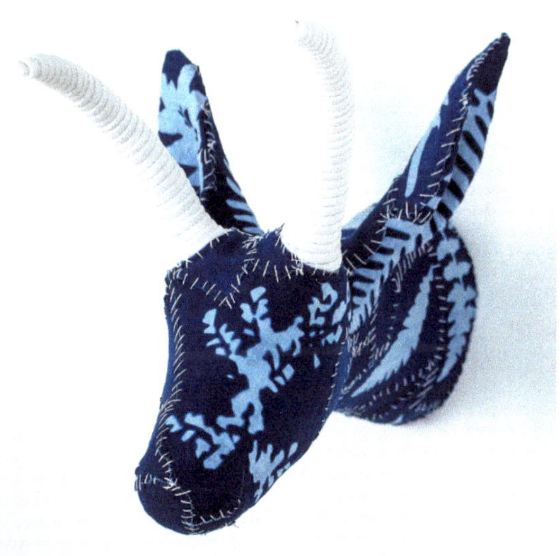

Figures 15.87–15.89. From top to bottom, *Botanical Beast 2* (10″ x 9″ x 12″), *Botanical Beast 3* (27″ x 24″ x 18″), *Botanical Beast 1* (10″ x 8″ x 12″), wood, plaster gauze, cyanotype prints on cotton, wire, cord, thread © Tasha Lewis 2017

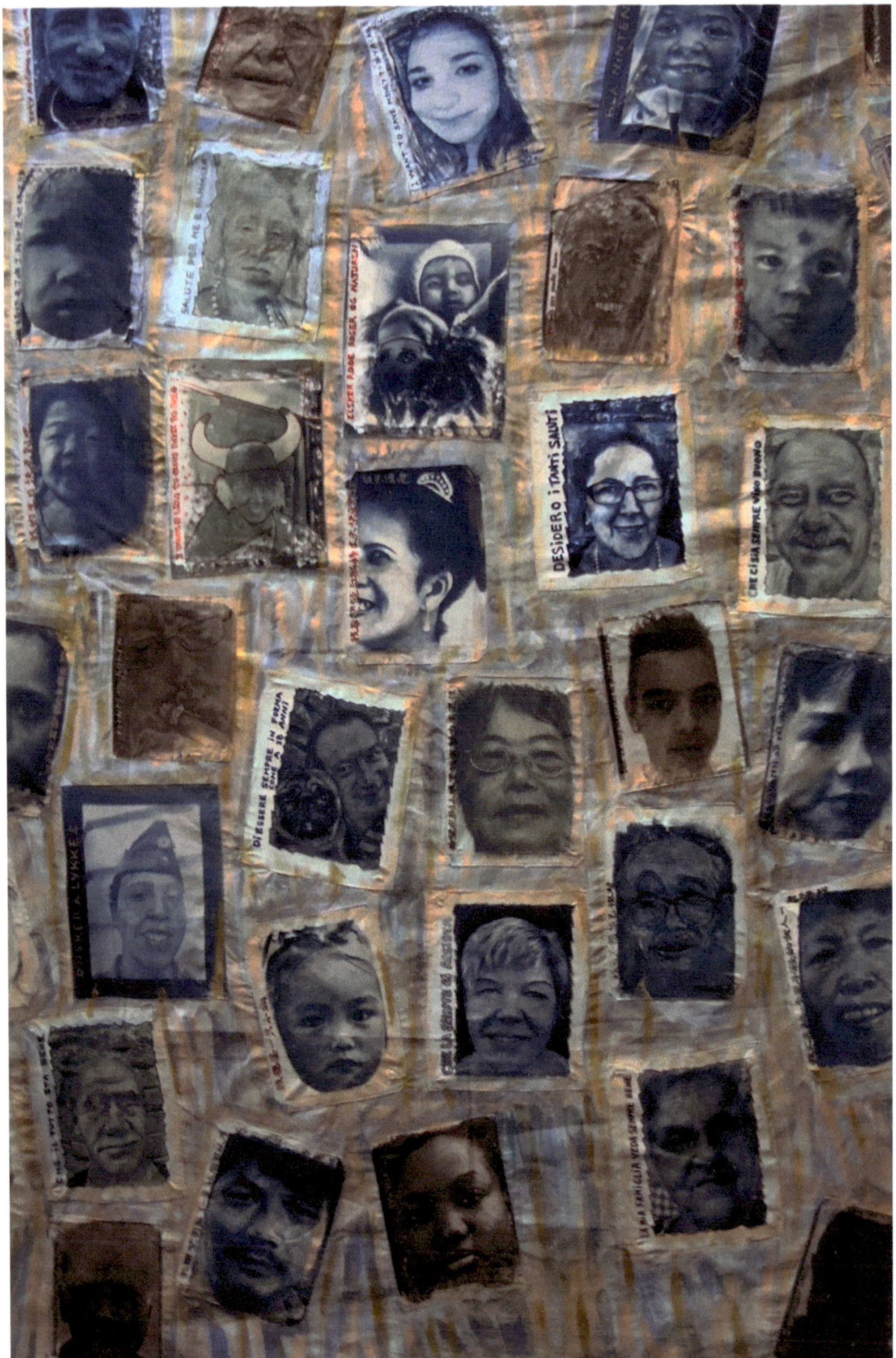

Figure 15.90. *Moskva* detail © Galina Manikova 2015

Galina Manikova

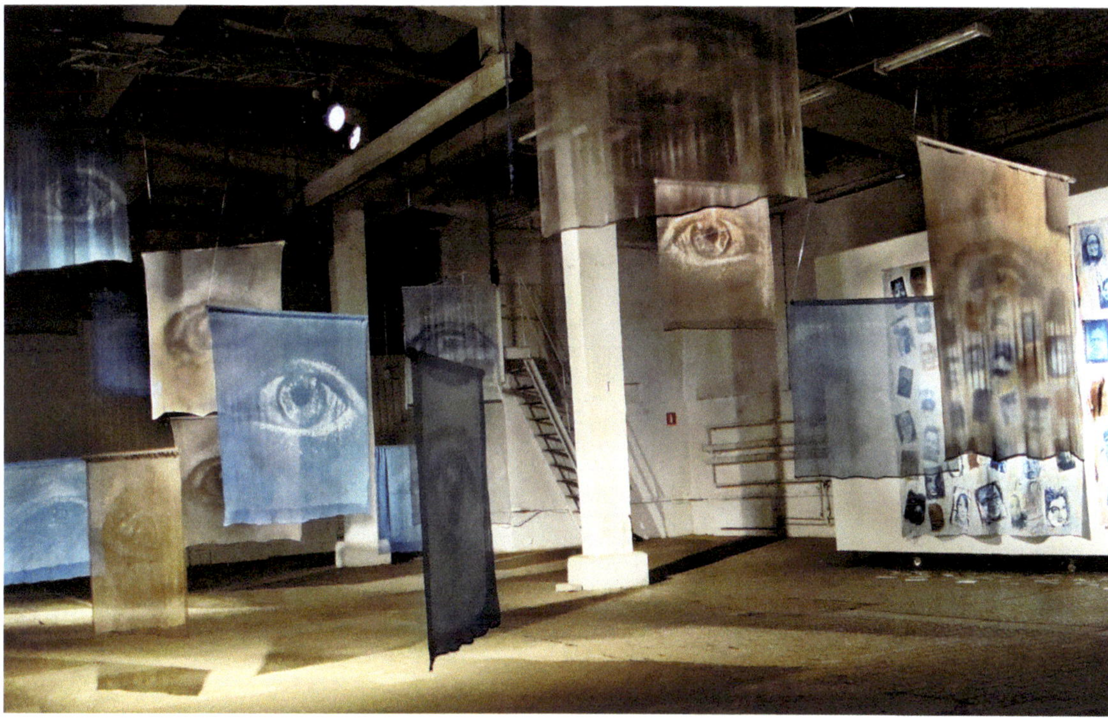

Figure 15.91. *Moskva* © Galina Manikova 2015

Process

Cyanotype is cheap, easy and versatile. It can be toned to other colors. It is graphic and can have different expressions. I usually do not work on paper, but glass, ceramics, textiles and wood. I make negatives in camera, in the darkroom and digitally. I use my own cyanotype formula, which is a variation of the classic one. I adjust the amount of FAC according to the material used, sometimes doubling the amount if it is to be fired on glass or ceramics. I use all different types of light sources, from sun light to UVBL and also a vacuum unit for some projects. I develop in just water, and tone with all sorts of recipes, from soap to chemicals.

Creative practice

I was born on a Japanese island of Sakhalin, grew up in Moscow, was educated at an art academy in Jerusalem and have lived in Norway since 1986. Sakhalin, taken over by Russians after the war, was exile, but better than inside the barbed wire. My parents were military doctors, both of them. It was impossible for them to raise children on Sakhalin, so I grew up with my grandparents in Moscow, in a big apartment behind the Bolshoi Theatre, where my grandmother worked as a stage designer. In a dramatic rupture with my father, who had survived Stalin purges and returned to Moscow, I escaped to Israel at nineteen, where I lived for thirteen years. In Israel, at Bezalel Art Academy, I specialized in photographic methods on ceramics, which is my

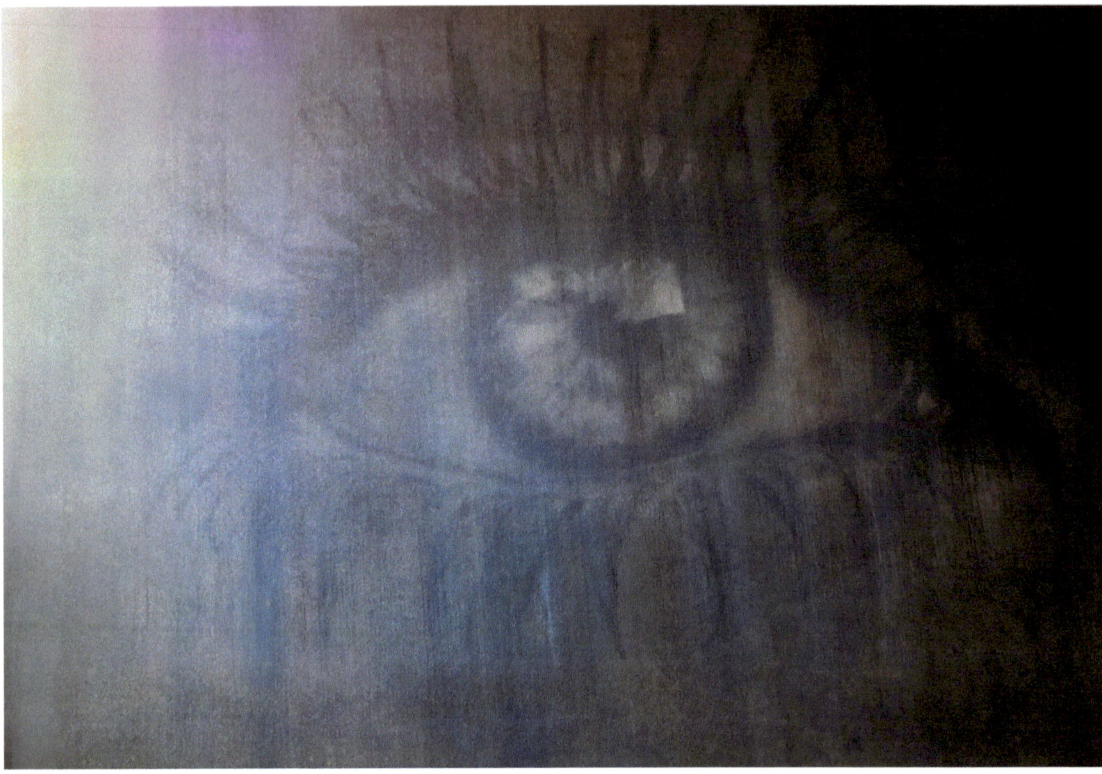

Figure 15.92. *Moskva* © Galina Manikova 2015

field of expertise. Over the years, I have had a series of exhibitions and teaching projects in major cities such as Berlin, Copenhagen, Paris, San Francisco and Oslo. They show the diversity of my work, using different materials and techniques.

My culture has been influenced by many traditions, and my national feelings are confused. Globalization and migrations of nationalities in Europe have created a new group of people, who belong to the same race as me, the race of hybrids, of mixed masters. Perhaps that makes me a true cosmopolite. I have worked a lot on the subject of identity, and lately about age, getting older, and being a woman.

Artist statement

These series were created 2015–2016 and have resulted in two large exhibitions, *Eyesee*, an installation in Moscow in August–September 2015, and *Oenene Som Ser* in Norway in September–October 2016. I started working with images of eyes in 2013 during my residency at the Art Peace Hotel in Shanghai. Maybe I got my inspiration from the many different faces and eye forms of the Chinese people from different ethnic groups. There were so many unfamiliar combinations of faces and eye forms and colors! Maybe it was my own approaching blindness and helplessness in regard to moving around and seeing the world. I have been taking thousands and thousands of images.

When I came back home to Norway, the process of developing a concept for an exhibition or a project began. I wanted to transform eye forms into something else and decided to do so by enlarging the images and transferring them onto semitransparent materials. At such a transition the quality of the images would make all the difference. I started with building a super big format macro camera. Creating a special lens has been quite a big challenge. I solved it by converting an old Russian aerial spy lens and turning it the wrong way, an achievement I am still proud of. The next step was to capture enough different people in order to have a sufficient variety of forms and colors. Each

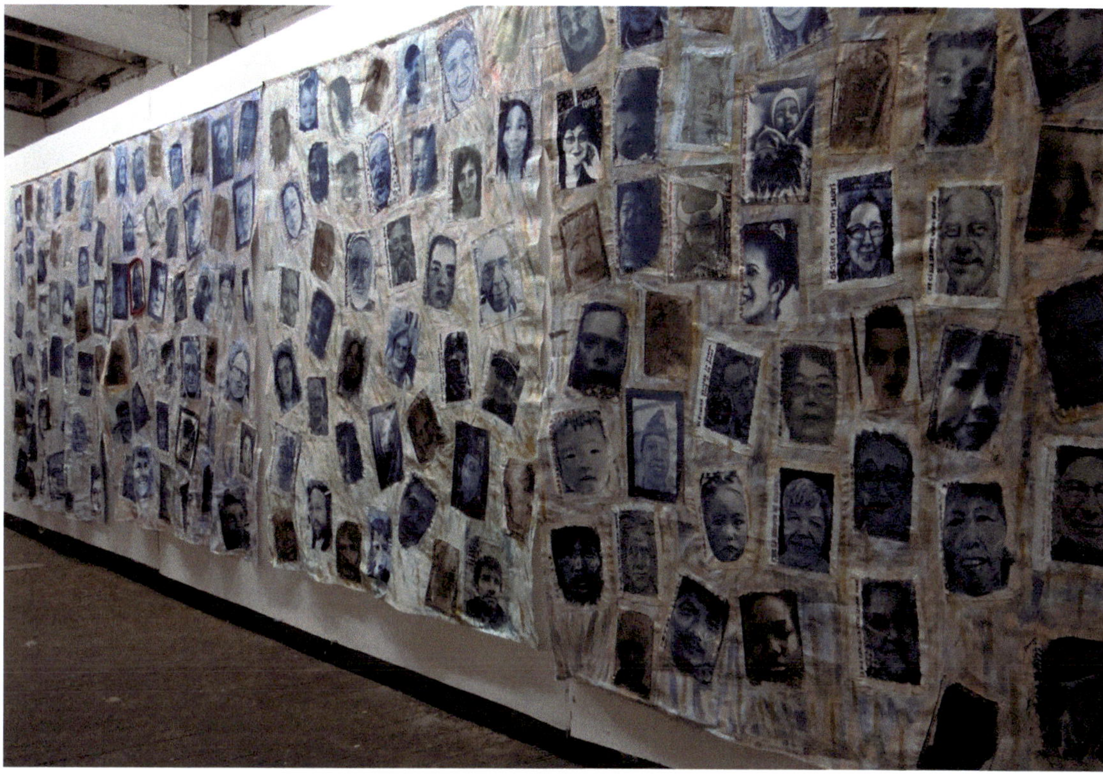

Figure 15.93. *Moskva* © Galina Manikova 2015

person has been photographed with three different cameras, from large format to video. I could only take one big picture per day as it had to be carried directly into the darkroom. In-camera direct negatives on film were processed and manipulated in the darkroom as soon as they were shot. Every one of the 30 or so images has been tray developed and toned. This took time. I only ruined one of these very expensive film plates. My negatives are 150 × 150 cm.

The silk was soaked in two different emulsions, dried and contact printed with the large negatives in the sun. I prepared a wooden plate, covered it with a sponge mattress and a glass plate. After the handmade emulsions dried, silk had to be placed on the frame, stretched with pins and covered with the glass plate, all in the dark, and then the whole sandwich was carried out into the sun, exposed, carried in and disassembled. The silk had to be washed, dried and ironed later.

I have used silk as it was easier to transport. The images breathed and moved in the big room. They were visible from both sides. The qualities of detail and texture were exquisite at different distances and light conditions. The pictures were turning around and responding to the movements of the visitors in the room.

I received amazing response from the public to this project. Several articles resulted, and the exhibition was shown on national TV and discussed on the radio. People kept coming back to see it again and again. There was even one visitor who surreptitiously spent the night there in order to feel it and to think.

For the next exhibition in Sandefjord, Norway, I made special forms out of plastic tubes and bent metal pipes to stretch the silk. Each picture had to be sewn first, stretched on the frame and covered with silk ropes around the frames in order to make a nice finish. I also used metal thread formed as an eye drawing and hung overhead lenses in the middle.

The theme of the exhibition is communication, observation and connection between people. I aimed to reconnect my own background and share

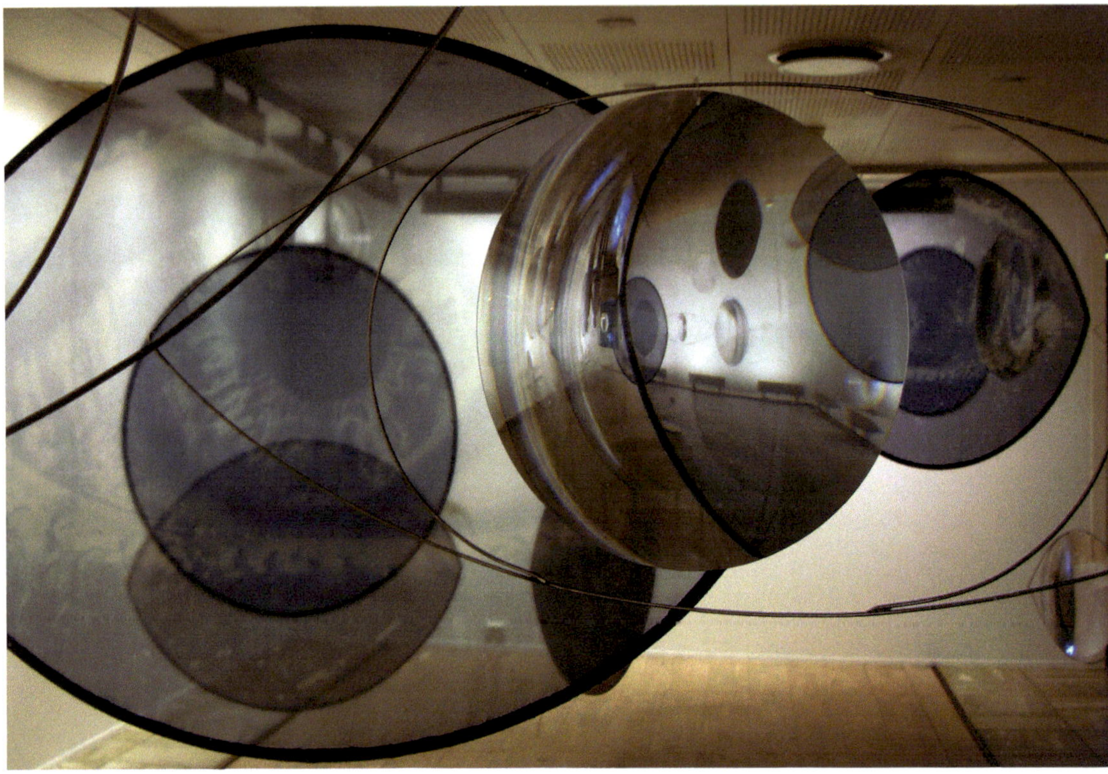

Figure 15.94. *Sandefjord* © Galina Manikova 2015

my reactions and outsider views with the others. This project was about myself, about my life in a foreign landscape, about being exposed and vulnerable, about watching and being watched. Eyes symbolize experience and everyday struggles during over 40 years spent outside of Moscow, often shaded by segregation and isolation.

The project includes a range of photographs of different eyes, printed on the transparent silk and presented as abstract three-dimensional forms hanging from the ceiling so that they create a fundamentally new landscape. The motives and signs in this space are overlapping each other and are projected on the live figures moving around in this new landscape—the public. The public is offered an opportunity to leave their comments and wishes on the paper notes, which are placed in the special holes in the silk, and then scanned and put on the interactive so-called "Wailing wall" created specifically for this purpose. See **www.wailingwall.no**. In this way the public is invited to participate in creating the exhibition as a part of a web project which will be its final destination. Dreams and desires of individuals are woven into one fabric that affects and involves the audience, transforming the entire space of the exhibition into a single interactive arena, a kind of "visual world," a dreamspace full of illusions, impressions and possibilities for the visitors to get engaged and involved into the art process.

Biography

Galina Manikova likes to call herself a citizen of the world; she was born on the Japanese island of Sakhalin in 1953, lived in Moscow until 1973, graduated from the art academy Bezalel in Jerusalem in 1978, and has lived and worked in Norway since 1986. Manikova has had 50 different personal and group exhibitions during her nearly 40-year career as an artist and a photographer. Her specialization is in historic photographic processes and the transfer of photographic images to different materials like glass, metal, textile and ceramics, as well as video and computer art. To see more of her work, visit **www.galina.no**.

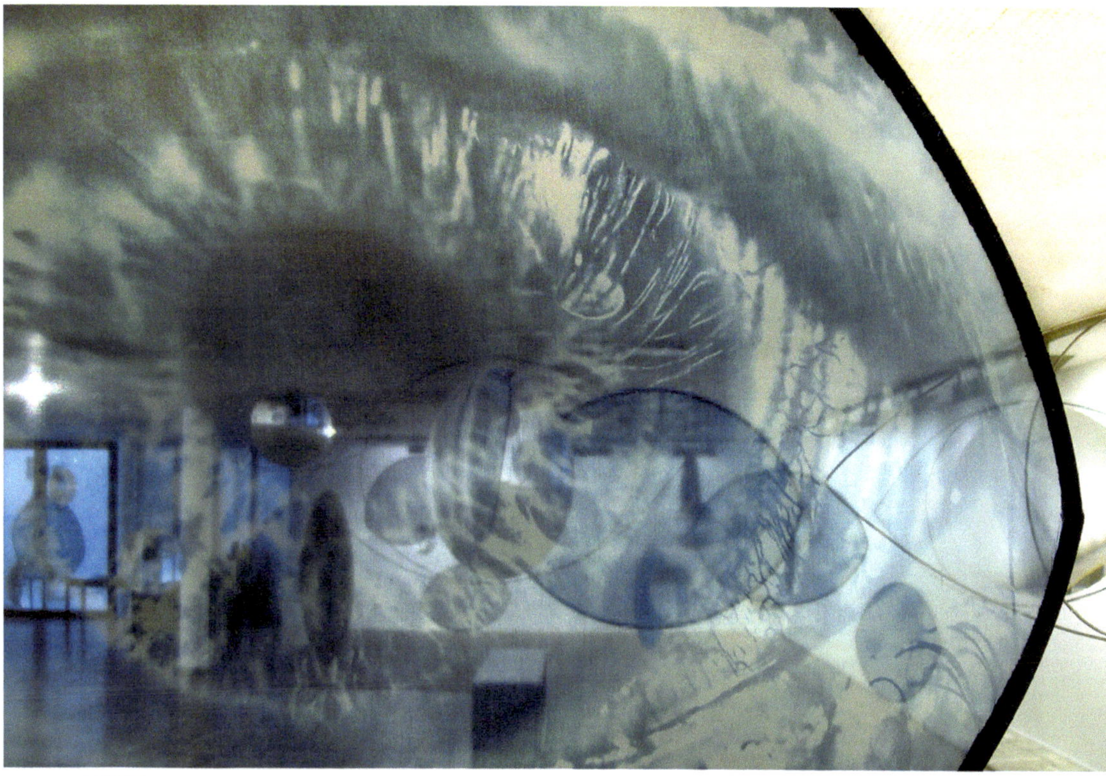

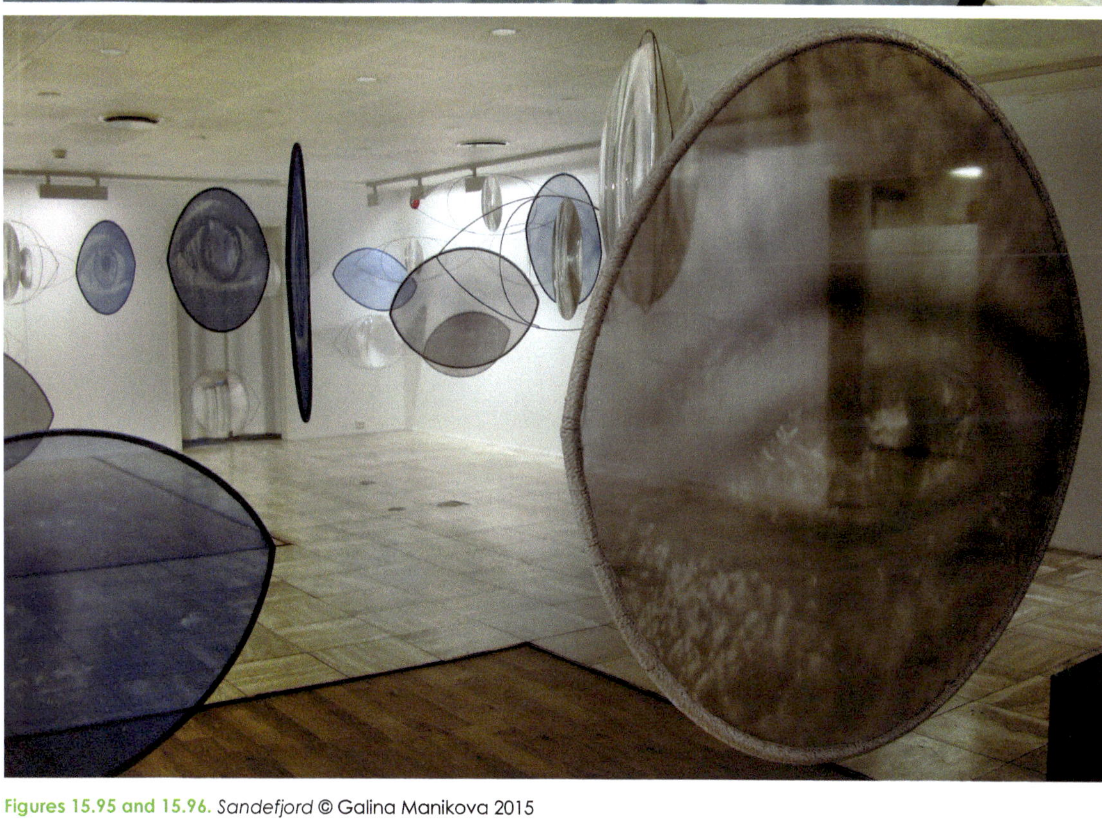

Figures 15.95 and 15.96. *Sandefjord* © Galina Manikova 2015

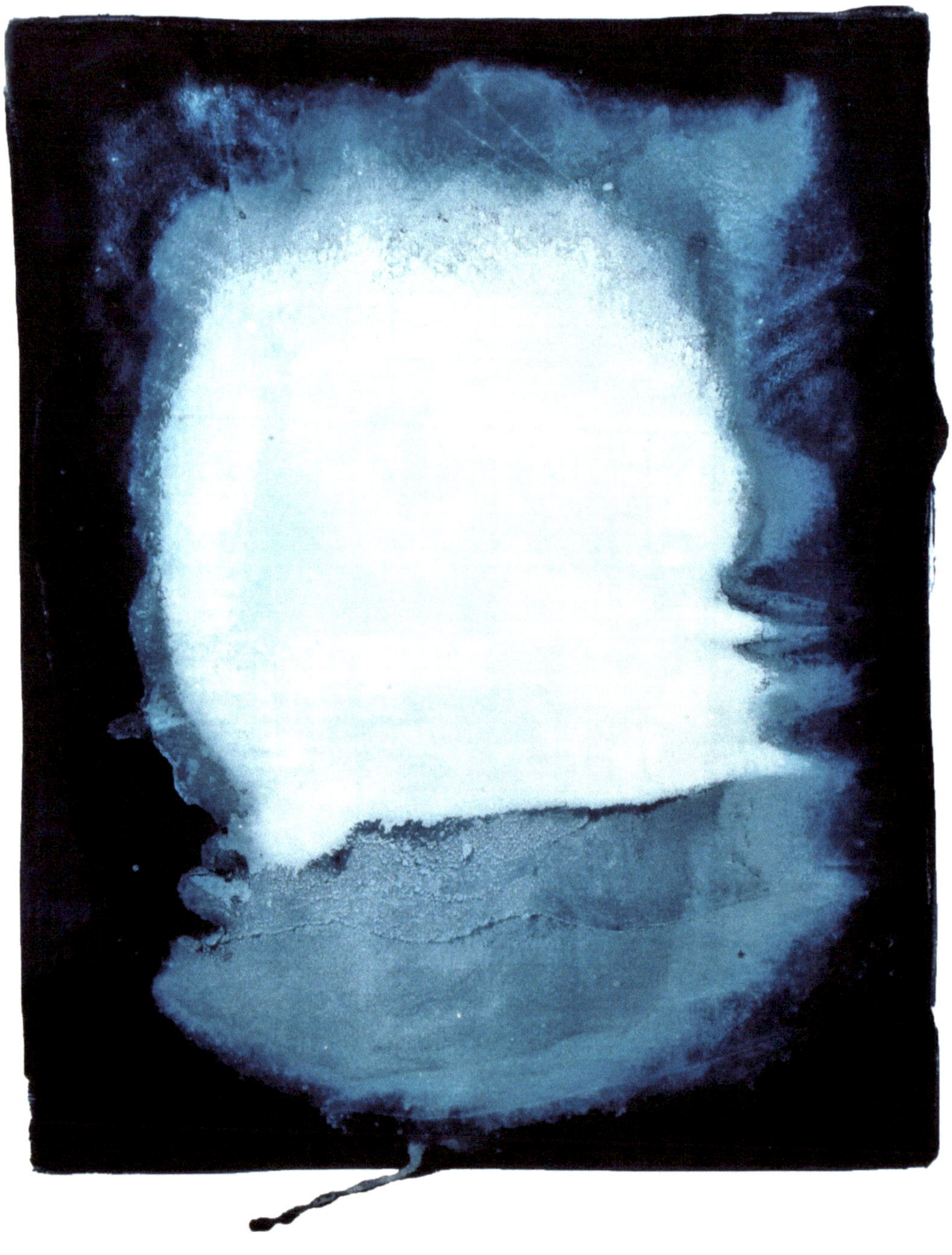

Figure 15.97. *Untitled*, from the *Prussian Blue* series, 56 x 67 cm © Constanza Isaza Martínez 2010

Constanza Isaza Martínez

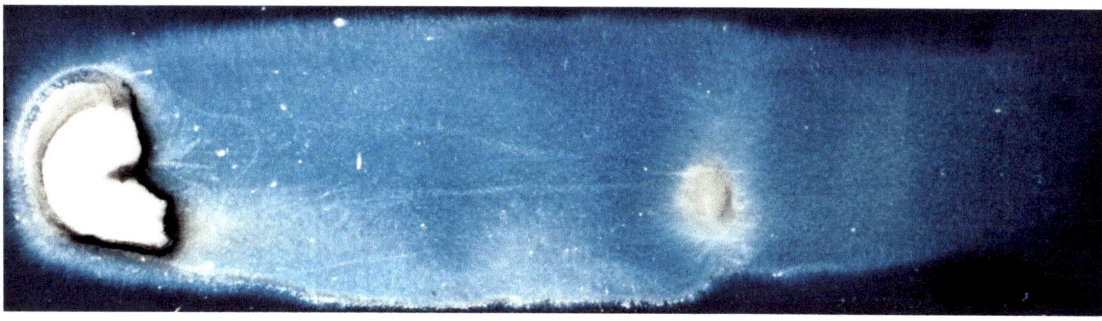

Figure 15.98. *Untitled*, from the *Prussian Blue* series, 28 x 76 cm © Constanza Isaza Martínez 2010

Process

I began working on this ongoing series in 2009, when I started researching historical photographic printing processes. Previously, I did a lot of work with silver gelatin printing, but cyanotype was the first process I encountered where I could make my own photographic prints from scratch, from weighing the chemicals to finishing the prints. The beauty and versatility of the process really appeals to me, as well as its slightly rough-around-the-edges feel and the seemingly endless scope for experimentation. In my day-to-day work, I use processes like photopolymer gravure and platinum printing, which demand a high level of precision and skill, as well as being very expensive. I find it creatively liberating to work with cyanotype, let go of the perfectionism I normally aim for, and allow myself free rein to experiment.

I have tested hundreds of different papers for cyanotype, so it's really difficult to pick a favorite. For more abstract work like the *Prussian Blue* series, I like to use Rives BFK and Fabriano Artistico, among others. For prints from digital negatives, I like Simili Japon, Hahnemühle Sumi-e, Zerkall, and Arches Aquarelle. I also love working with Japanese papers like Awagami Hosho, which prints a really beautiful deep blue.

I make my negatives using a Photoshop adjustment curve made with software created by Peter Moseley.

I start with the classic cyanotype formula for all my cyanotypes. I use equal parts of a 25% w/v solution of ferric ammonium citrate, and a 10% w/v solution of potassium ferricyanide. In the *Prussian Blue* series, this varies quite a bit, as I often end up using many different dilutions on one print.

I use the sun for my *Prussian Blue* prints, and a Xeros vacuum exposure unit when I'm working with negatives. Exposure times in the sun vary from 5 minutes to several hours or even days. My Xeros unit is super fast because I changed the glass to a piece of low-iron glass, so I can expose a cyanotype in less than 2 minutes!

I normally just rinse in tap water but I occasionally rinse first in a very weak solution of citric or acetic acid—approximately 1%. I always use a hydrogen peroxide bath before my final wash.

I rarely tone my cyanotypes as I work with a wider range of other processes, so if I want a color other than blue, I'll make a Vandyke, kallitype or platinum print. However, I have used oak tannin, tannic acid, black tea and green tea, and I like green tea the best as it doesn't stain the paper too much.

For me, part of the attraction of working with cyanotype or other handprinted processes is the physicality of the final print, so I like to emphasize this when I display prints. I love printing on paper with deckled edges, and float-mounting the print in a box frame so you can see the paper edges.

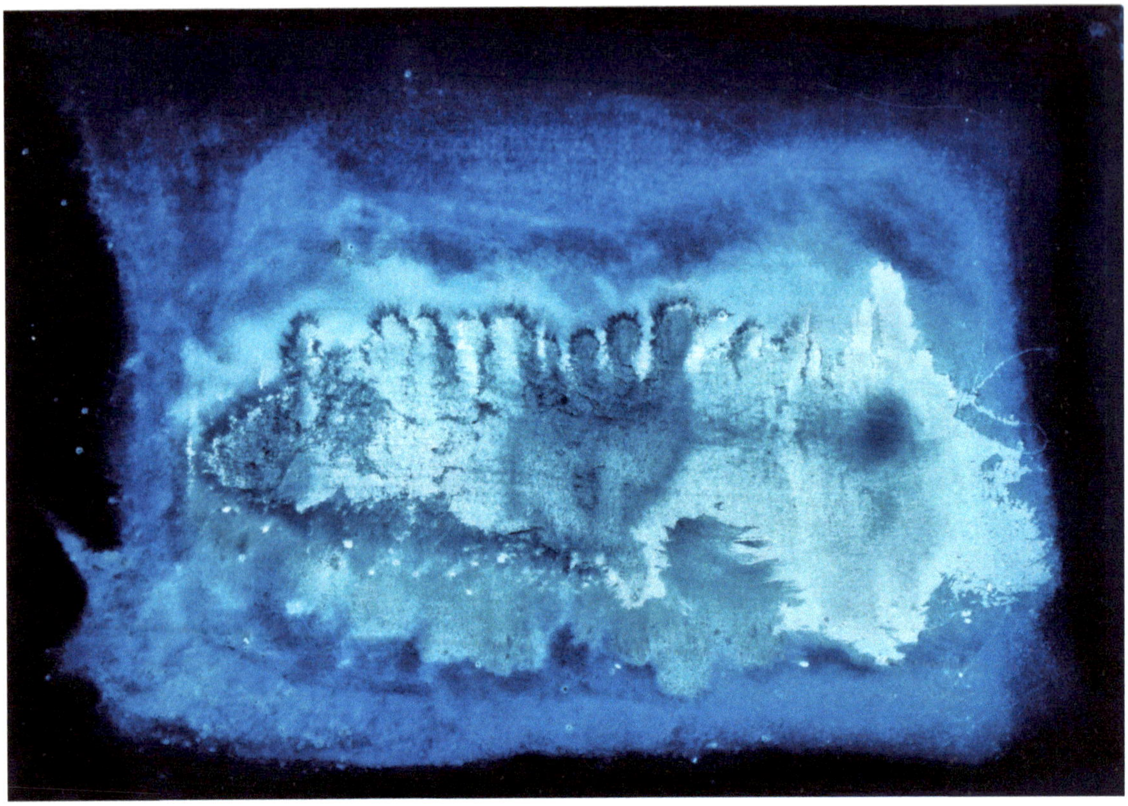

Figure 15.99. *Untitled*, from the *Prussian Blue* series, 56 x 76 cm © Constanza Isaza Martínez 2010

The biggest problem with cyanotypes, for me, is their inherent graininess, especially in the high midtones. Paper choice seems to make a huge difference, as well as an initial rinse in citric acid.

Creative practice

In the *Prussian Blue* series, I'm turning the cyanotype process in on itself to explore the nature and essence of the process itself. I don't use negatives of objects, but work with the chemistry itself in many layers and different dilutions on the paper. I have also made some very large cyanotypes on fabric with Andrés Pantoja—the largest one made to date was around 8 × 15 m, which we made in collaboration with Melanie King. It's really challenging to work at that scale, but it's worth it for the impact the final piece delivers!

My cyanotype tip would be not to take it too seriously. When you work with a process like platinum printing, the cost of the materials and the necessity of being precise and very clean in order to avoid contamination can really limit your creativity. Cyanotype, on the other hand, is so inexpensive, easy, safe, and fun, that you can allow yourself to experiment without worrying about the cost or risks involved. This is why cyanotype is one of my all-time favorite processes.

Artist statement

The *Prussian Blue* series explores the expressive potential of the photographic medium through its most basic elements: light, chemicals, and substrate. By reducing the medium to these three components, the restrictions that photographic representation traditionally carries are eliminated. Without a camera, a negative, or the obligation to represent the material world through a lens, the prints produced explore the essence of the medium and allow the surface of the photograph, usually overlooked, to become the subject of the work.

The history of the invention of photography is populated by natural scientists, chemists, and

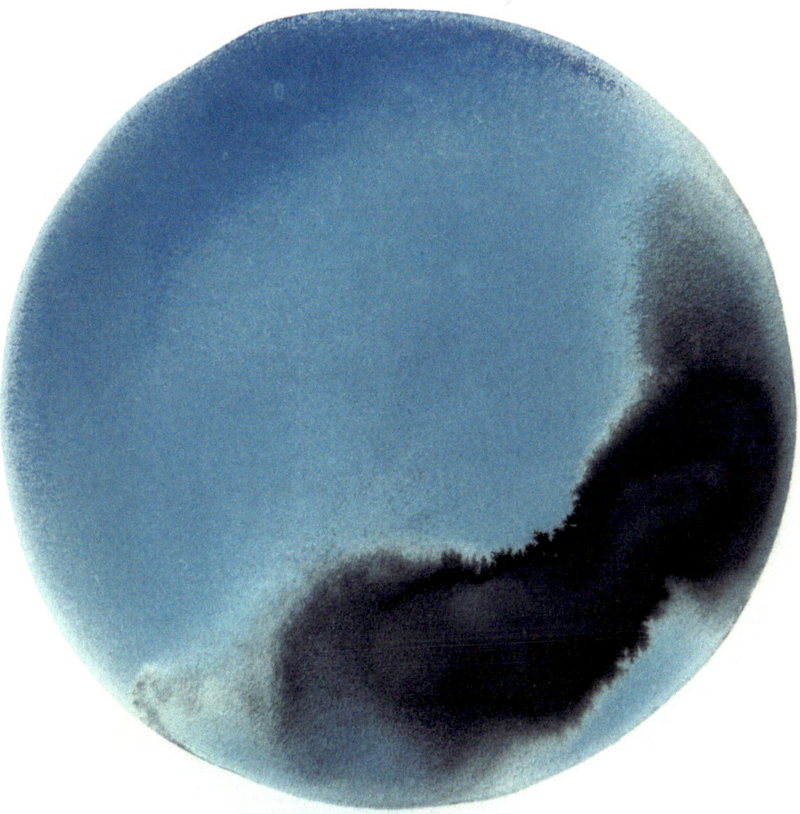

Figure 15.100. *Untitled*, from the *Prussian Blue* series, 26 x 26 cm © Constanza Isaza Martínez 2015

astronomers, whose scientific research required the precision of photographic representation, and therefore served as the catalyst for the invention of photography. Sir John Herschel—the inventor of the cyanotype process—was a true polymath who conducted research in chemistry, astronomy, and optics, among other fields.

The *Prussian Blue* series takes the cyanotype process as a starting point and uses the chemistry and paper alone to make images without negatives or camera. The images remain photographic due to the process and chemicals used, but differ from more visually familiar forms of photography because they are made without cameras or negatives; using the chemistry itself as an ephemeral "negative" that is destroyed during processing. The images—which draw attention to the surface and nature of the photographic image—allude to the process of investigation, experimentation, and analysis which is a fundamental aspect of both science and photography.

Biography

Constanza Isaza Martínez is a visual artist based in London, working primarily with photography. She studied Photographic Arts at the University of Westminster, and in 2012, she received her MA in History of Art from the Courtauld Institute of Art. Constanza's practice is characterized by an interest in the materiality of the print, the artistic possibilities of historical photographic processes, and the process of making as central to the meaning of the work. Constanza lives and works in London, where she co-owns Lux Darkroom, a space dedicated to teaching and researching historical photographic processes. She regularly exhibits her work both in the UK and abroad. To see more of her work, visit **www.constanzaisaza.com** and **www.luxdarkroom.co.uk**.

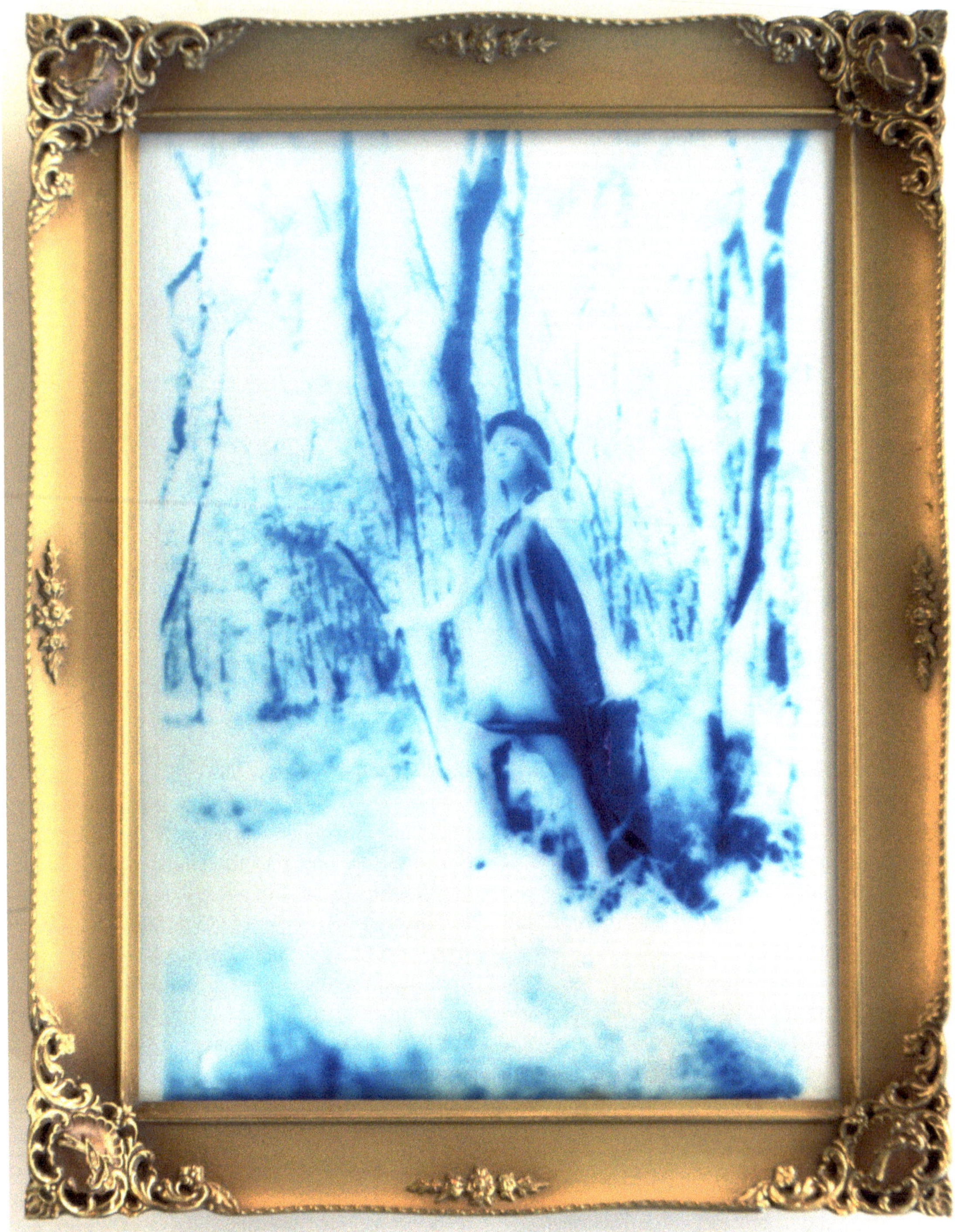

Figure 15.101. *Wald*, cyanotype on curved glass, 22.5 x 28 cm © Sehera Nawaz 2017

Sehera Nawaz

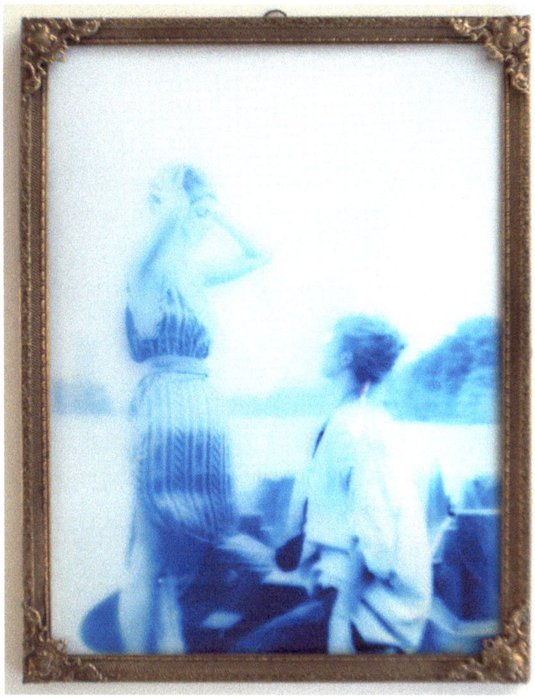

Figure 15.102. *Verehrung*, cyanotype on curved glass, 18.5 x 24.5 cm © Sehera Nawaz 2017

Process

While creating some glass cyanotypes I was looking for nice frames at the flea market and came across antique ones that have fitted curved glass. It occurred to me that this small distance between the glass and the white background can be used to create a shadow in sharp light. When the viewer changes perspective it creates an effect of uncertainty, as a ghostly image of the print can be seen on the white paper in the background. To get these images onto the glass I had to create an enlarger to project the negatives; contact printing was not an option with the curved glass.

The two-color series is an ongoing effort to strive towards more colorful cyanotypes. I use black tea to dye the first layer. The stone paper (cradle-to-cradle paper made from stone) is very important in this process as it doesn't change size in water.

Cyanotype is the underestimated process that makes more sense now than ever: it is environmentally friendly but it seems to me that it was never given the same research attention as silver halide or even other iron III based photographic processes. When I came across it, I was looking for a process that didn't need a special developer or fixer, but rather things that were readily available. I played with other natural photographic techniques like anthotype but after reading Mike Ware's *Cyanomicon* I was convinced cyanotype was everything I wanted. At first it seemed so easy to use but digging deeper there are still so many untried things and unanswered questions: it is so versatile and can still be pushed to new limits.

Starting out I used all kinds of papers, but soon was drawn towards more experimental surfaces like glass and wood. Glass is fascinating, also because you do not have the structure of the paper intervening with the print. Although the pigment is the same the blue appears slightly different and the highlights are much more visible. I was looking for a paper that could have the same effect and came across stone paper. It almost feels like plastic, but is made out of stone and a natural binder. It is a paper that is produced with the cradle-to-cradle principle in mind, which is for me a perfect fit for cyanotype. As the pigments of cyanotype cling to natural materials I am using the stone paper that has 80% stone.

A problem that occurred especially when using stone paper, but also with other kinds of paper is that when enough emulsion has sunk into the support, the excess emulsion just floats on top of the paper and therefore, when exposed, cannot hold on to any organic structure and is washed away during the process of development. I started using dry

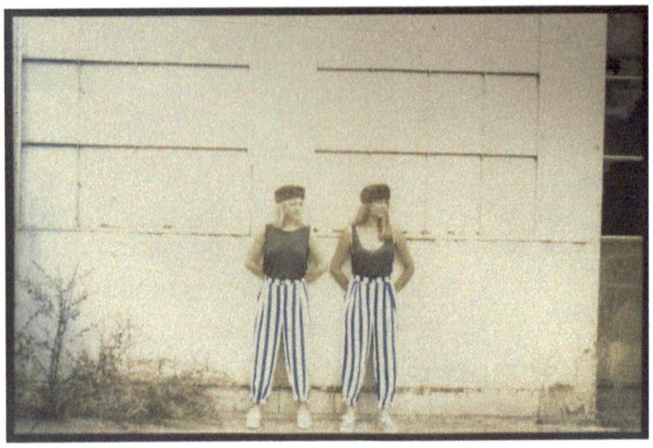

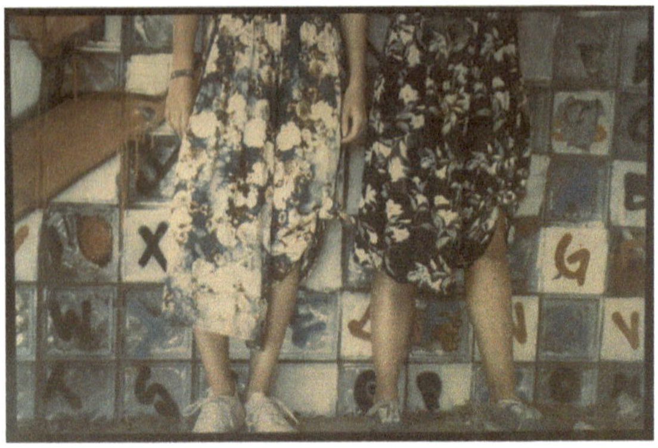

Figures 15.103 and 15.104. Left, *Two #2*, two-colored cyanotype on stone paper, 41 x 29.5 cm; right, *Two #1*, two-colored cyanotype on stone paper, 41 x 29.5 cm © Sehera Nawaz 2018

pieces of cloth to remove the excess emulsion and got much more detailed prints due to this practice.

Stone paper behaves quite differently compared to normal paper. The coating has to be done very evenly, otherwise one can see differences on the final print. It is sensitive to touch after exposure, but the images are clear and the details are beautiful.

When using paper I still use inkjet printed negatives that I convert from my analog scans. I don't like the process but it is still the fastest way to try something out. Some of my prints are just directly from medium format black and white film, but when shooting in color I need to scan them. I am striving for a completely analog process chain and for curved glass it was possible to achieve it by disassembling a 35 mm enlarger that I rebuilt to be used with a 395 nm LED. The most important thing was to change the lens of the enlarger to an old camera lens that does not have a UV-filter/coating on it. It has room for improvement, as 395 nm is not a very intense UV radiation, but it works and one can project onto curved surfaces with it.

My other exposure device is a self-built light box with six UV fluorescent tubes at a 5 cm distance. With normal paper and inkjet printed negatives it requires 2–3 minutes to expose. Stone paper only requires 1–2 minutes.

The cyanotype formula I use for paper is classical: 25 g FAC + 100 ml water and 10 g PF + 100 ml water separated for storage and mixed 1:1 before applying it to the surface. If they start to mold I sometimes strain them through a coffee-filter. For development I just use running water until the paper is clear, but usually I spray the print with 3% hydrogen peroxide after development to immediately see the results.

For toning I use instant coffee, black or green tea. I have also experimented with red wine. Black tea is used to get a reddish tone in the highlights and coffee to get a deeper black in the shadows. If you leave the print in coffee you can get a very cold black tone. Sometimes toning brings back highlights that were previously not visible, as the tannins stick to any Prussian-blue they can find and intensify and/or darken even the slightest contrast in the highlights. Frequently I also use washing soda to get brownish tones.

For my two colored prints I split a scanned color negative into cyan/magenta/yellow and usually use the cyan and magenta channels to print, but depending on the image it is sometimes better to use cyan and yellow. After exposing the magenta negative I use black tea to color it. Trying many different types of black tea I found a type of Ceylon black tea to be the best, but not every Ceylon tea brings out a strong red tone. After toning, I use washing soda to bleach away any trace of blue. Washing out the soda is crucial, as the blue layer would react to any leftover bleach. The second blue layer is where the magic happens. It is quite tricky

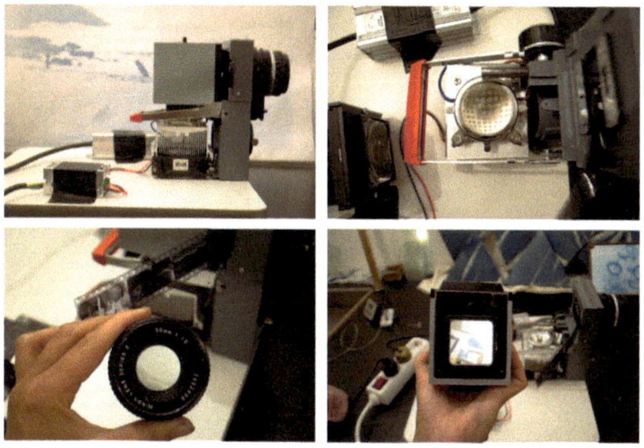

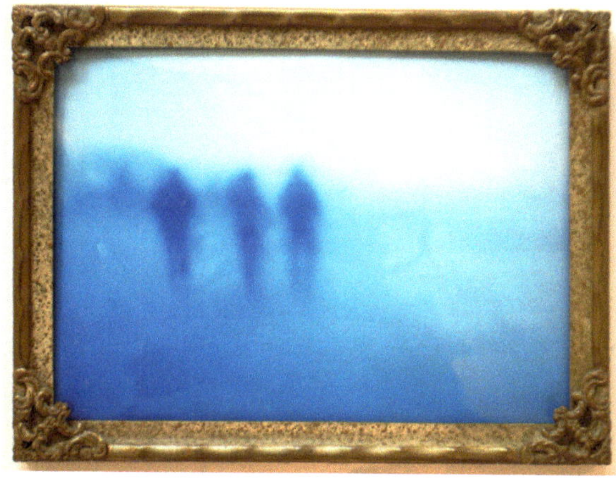

Figures 15.105–15.109. Left, Nawaz' enlarger setup; right, *Feld*, cyanotype on curved glass, 19 x 14.5 cm © Sehera Nawaz 2017

to place the negative exactly on top of the first print, but it helps that the stone paper does not change its size during development in water. I usually use a little less time to expose the cyan channel, but it totally depends on the image and the negative. Sometimes I use a little washing soda to slightly lighten the blue.

For glass cyanotypes I use a formula found in *Cyanotype: Historical and alternative photography* by Peter Mrhar who suggests: 1 part alcohol, 1 part 20% gelatine and 1 part mixed emulsion. Let your gelatine never go above 40°c and let it sit to get rid of bubbles. Bubbles are really a thing to look out for and dust is your mortal enemy. Using distilled water helps. I use this formula also to print on ceramic or even wood, because if the wood is too dense, the water-based emulsion will soak into the surface very unevenly.

In Berlin, there are quite a few second hand stores that sell old wood or even antique brass frames. I love using them for my cyanotypes, especially when I can reuse the old glass to print on.

Creative practice

I always feel inspired by light that reacts and light that travels in strange ways, caustics, reflections, refractions. When using new techniques it is crucial to me to understand them as much as I can and be very curious about cause and effect. I research and try to find any new ways of thinking about it. In the end though, it is important to let go of all these expectations, as the best prints happened, when I was not pushing the limits, but knowing about them and letting the process be whatever it can or wants to be. This feels important especially to cyanotype, as the best images sometimes are accidental and the result might not be easy to explain.

My studio is located in Berlin, Neukölln. In the cellar of our small shop I created my little cyanotype heaven, where I regularly host intimate cyanotype printing workshops to get the word out about this fantastic photographic printing technique.

Biography

Sehera Nawaz is an analog and alternative photography artist, specialized in cyanotype. Having in-depth knowledge about the behavior of light through studying and publishing in the field of film rendering/computer science, since 2015 she has used her knowledge to research and explore old photographic printing techniques. Nawaz' studio is at Mahlower Eins, Berlin Neukölln. To see more of her work, visit **http://solarlamp.de**.

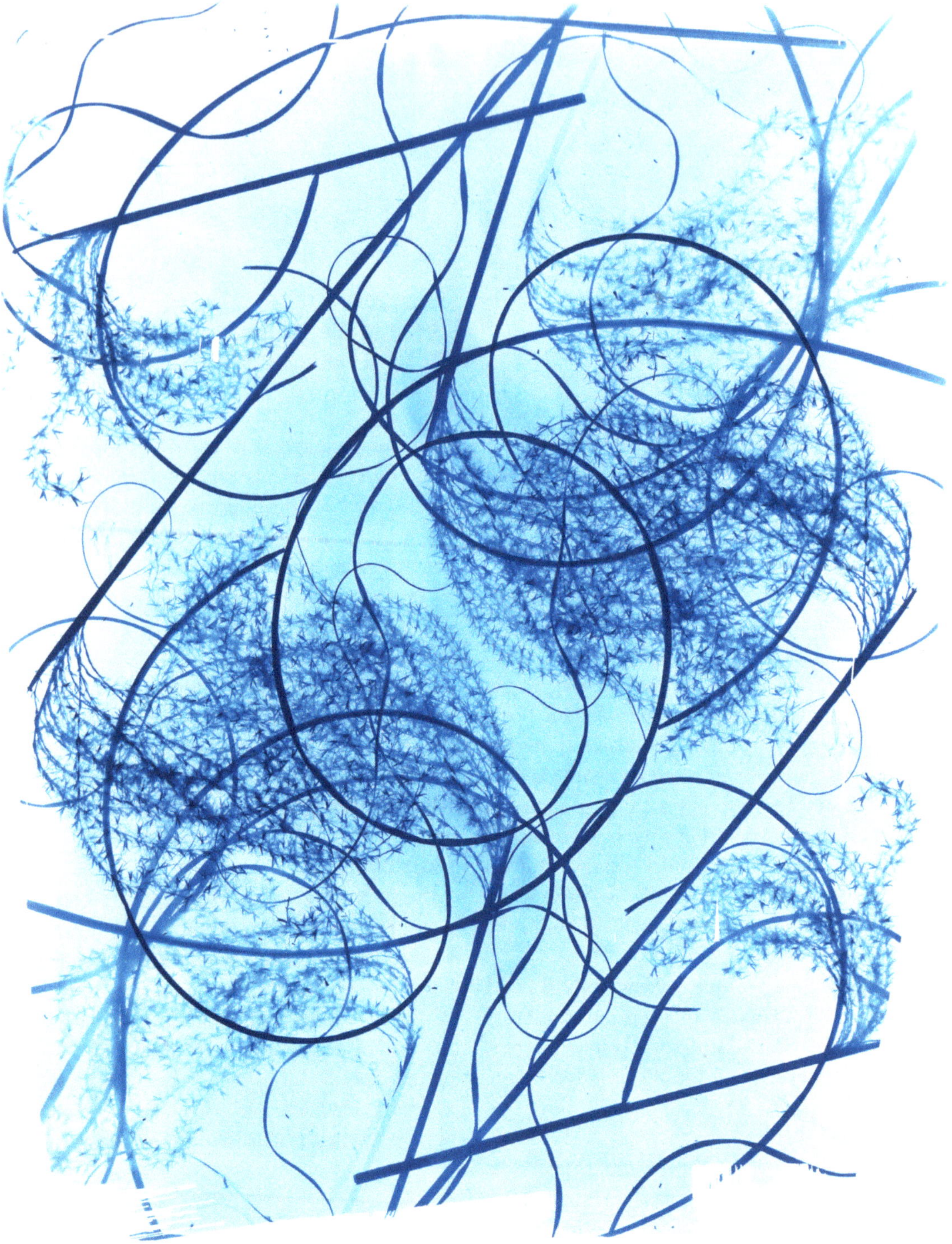

Figure 15.110. *Nature/Chaos*, 11˝ x 14˝ © Zeva Oelbaum 2005

Zeva Oelbaum

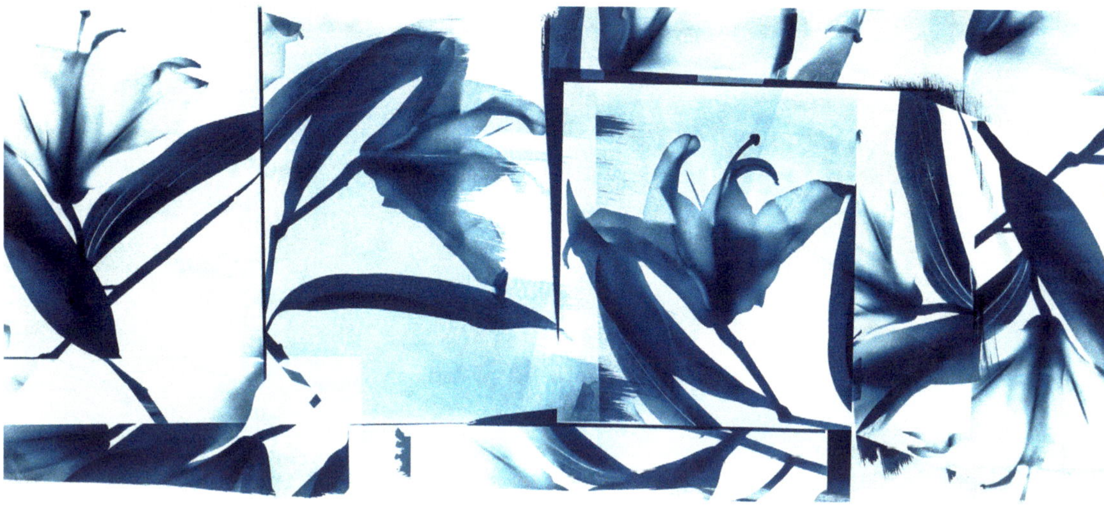

Figure 15.111. *Lily Composition*, 26.5″ x 11.5″ © Zeva Oelbaum 2005

Process

I use the classic cyanotype formula and worked primarily with Crane's watercolor paper until it was discontinued. Since then, I've experimented with lots of different papers, but don't have a favorite.

I create large film negatives, using an enlarger as a light source. My exposure times vary from 8 seconds to 30 seconds. I layer the objects directly onto the film, and control the negative density through the exposure. That allows me to control the intensity of the blue on the final cyanotype.

I use a UV lightbox to expose the final print. My lightbox exposures vary widely, from a few seconds to as many as 5 minutes, depending on the density of my films.

I have experimented with toning cyanotypes using black tea for a vintage look, but I prefer my work to have a clean, modern feeling.

I prefer for my cyanotypes to be floating in the frame or matted with an 8-ply simple white mat.

While working, I became fascinated with natural fractals, or a never-ending simple pattern that repeats itself at different scales to create infinitely complex patterns. As a result, I started working with multiple exposures. I would coat the paper with the cyanotype chemistry, expose the paper and process normally, then repeat the process several times on the same sheet of paper.

Nature is full of fractals and they are infinitely complex. Similar patterns recur at progressively smaller scales and seemingly chaotic patterns are actually highly ordered and controlled. A snowflake is an example of this. The term "fractal" was first used by mathematician Benoit Mandelbrot in 1975.

Creative practice

I love cultures. With an undergraduate degree in anthropology, I have always been intrigued by the convergence of culture, history and visual media—both photography and film. In addition to my work as a photographer, I am also a documentary filmmaker and my specialty is working with archival footage and early cinema. Using primary source material, found objects and historic

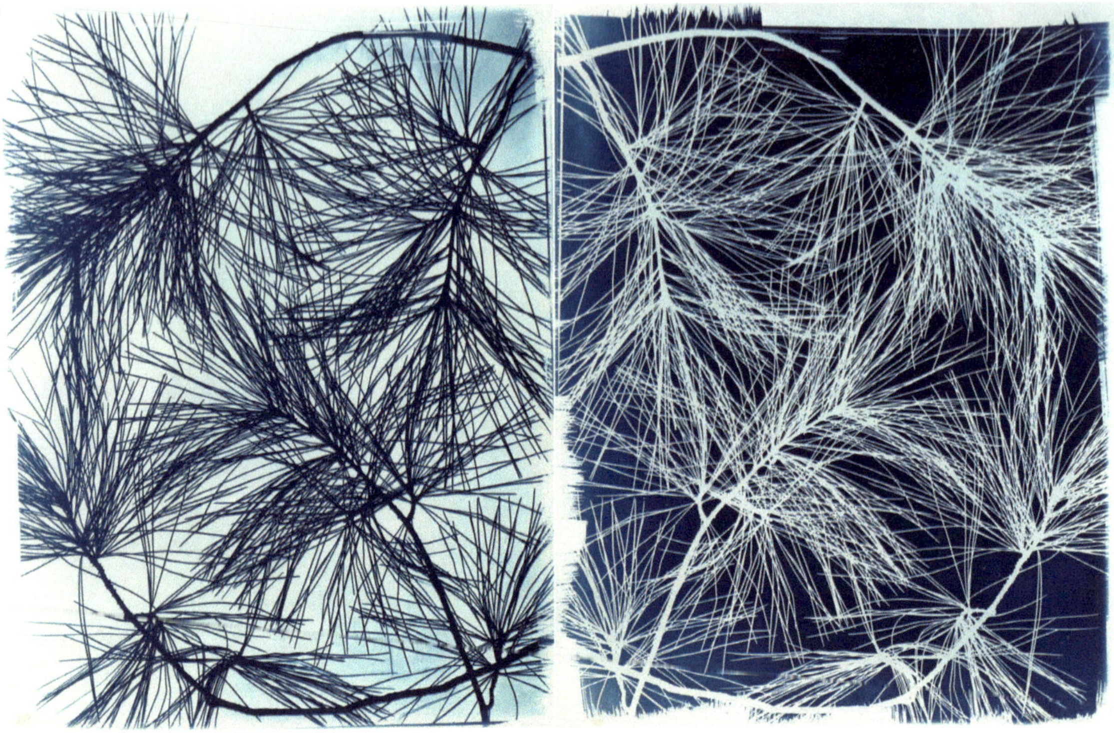

Figure 15.112. *Pine Diptych, 22˝ x 11˝* © Zeva Oelbaum 2001

processes, I am drawn to stories of women who are underappreciated or little known, despite stunning accomplishments.

Much of my work has centered on the late 19th century. I am fascinated by the Victorian and Edwardian urge to travel to every corner of the globe and collect specimens of the natural and ancient world. It was within this era of collection and categorization that the processes for photography and film were discovered and found their *raison d'être*. Individuals such as Dr. Albert Kahn of Paris sent out an army of photographers and filmmakers around the world in the early 1900s to "bring back the world." Kahn's "Archives of the Planet" hold nearly 80,000 autochromes and hundreds of hours of footage. Botanists traveled to every corner of the world collecting samples, as did naturalists and hunters, such as Theodore Roosevelt. In fact, the subject of my recent documentary film, "Letters from Baghdad," was a pioneering British explorer, archaeologist and diplomat who helped draw the borders of Iraq after World War I and established the Iraq Museum. Gertrude Bell (1868–1928) was also an extraordinary photographer and used a panorama camera in the early 1900s to document ancient sites throughout the Middle East. Eerily, many of those sites have now been destroyed or damaged by ISIS.

One of Bell's colleagues was the German archaeologist Ernst Herzfeld, who used the cyanotype process to proof his photographic negatives of ancient sites throughout Iraq, Syria and Iran. I love the fact that cyanotypes were inexpensive and versatile and could be produced anywhere with ease, since they only needed to be rinsed in pure water.

My interest in the cyanotype process grew when I learned that it holds an important place in the history of photography. I was very surprised to learn that British Botanist Anna Atkins (1799–1871) published the first book illustrated entirely with photography in 1843. She hardly has enjoyed the

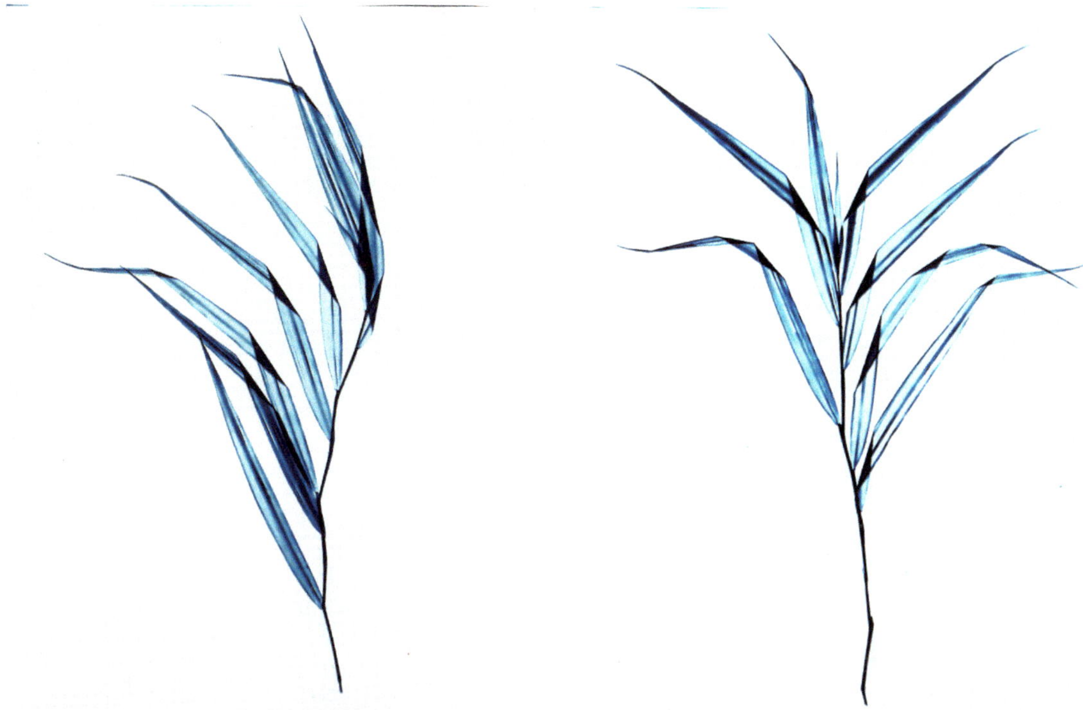

Figures 15.113 and 15.114. Left, *Cyanotype Grasses No. 20*; right, *Cyanotype Grasses No. 19*, 11″ x 14″ © Zeva Oelbaum 2004

same name recognition as Sir John Herschel or Sir William Henry Fox Talbot. Her cyanotypes inspired me to create a body of work on the natural world.

Biography

Zeva Oelbaum is the author and photographer of *Blue Prints: The Natural World in Cyanotype Photographs* and *Flowers in Shadow: A Photographer Rediscovers a Victorian Botanical Journal*, both published by Rizzoli. She is the recipient of two NEH grants. Her work has been reviewed in such publications as *ArtNews* and *The New Yorker*. Her recent documentary film "Letters from Baghdad" will broadcast on PBS in December, 2018. Her photographs are in numerous international public collections including the Bibliothèque Nationale de France, Museum of Fine Arts, Houston, and the Brooklyn Museum. To see more of Oelbaum's work, visit **zevaoelbaum.com**, **LettersFromBaghdad.com**, and **RuthGrubertheemovie.com**.

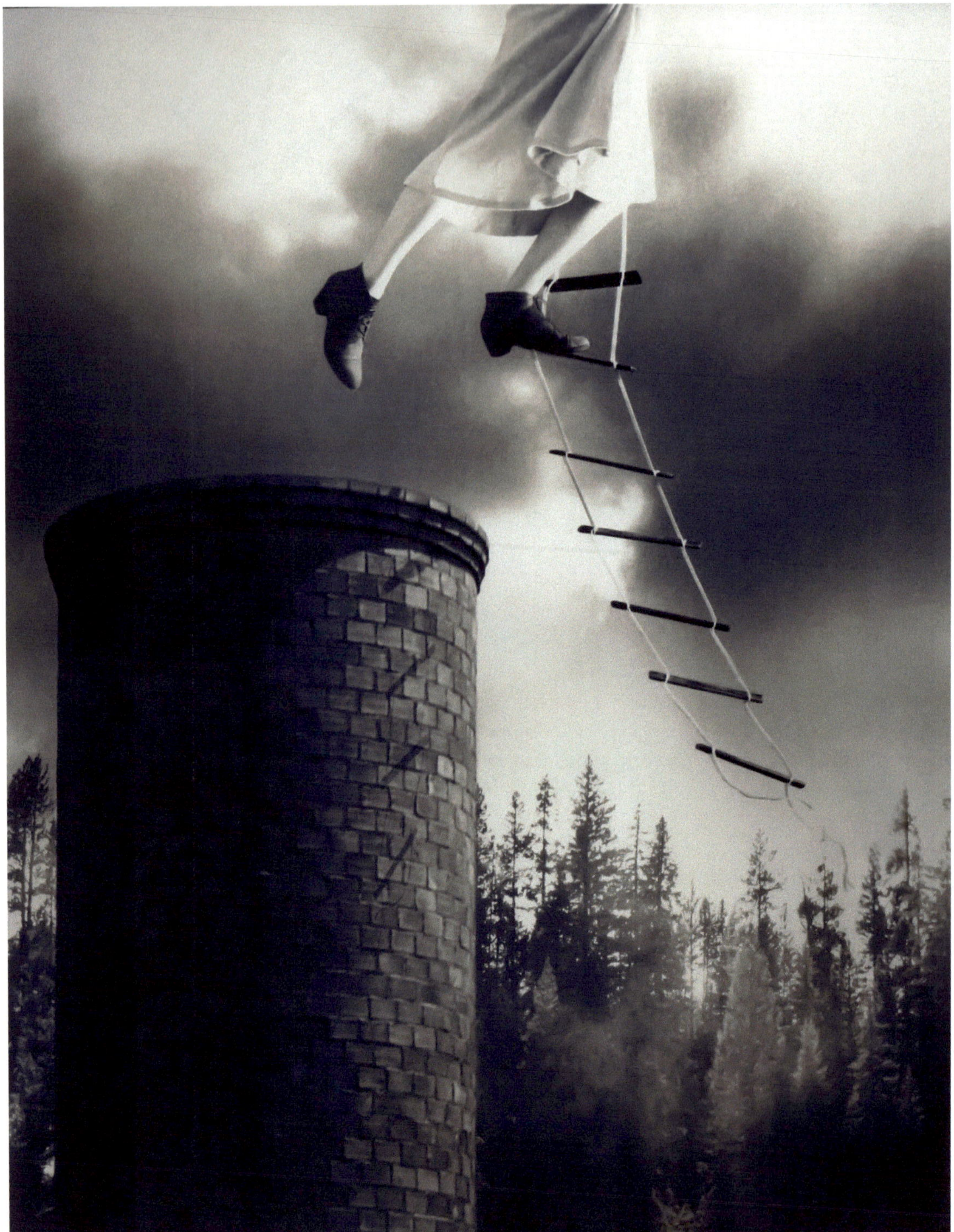

Figure 15.115. *Flight*, cyanotype toned with tea and wine tannin and a bleaching agent (usually highly diluted silver gelatin paper developer or sodium carbonate), 11″ x 14″ © Emma Powell 2013

Emma Powell

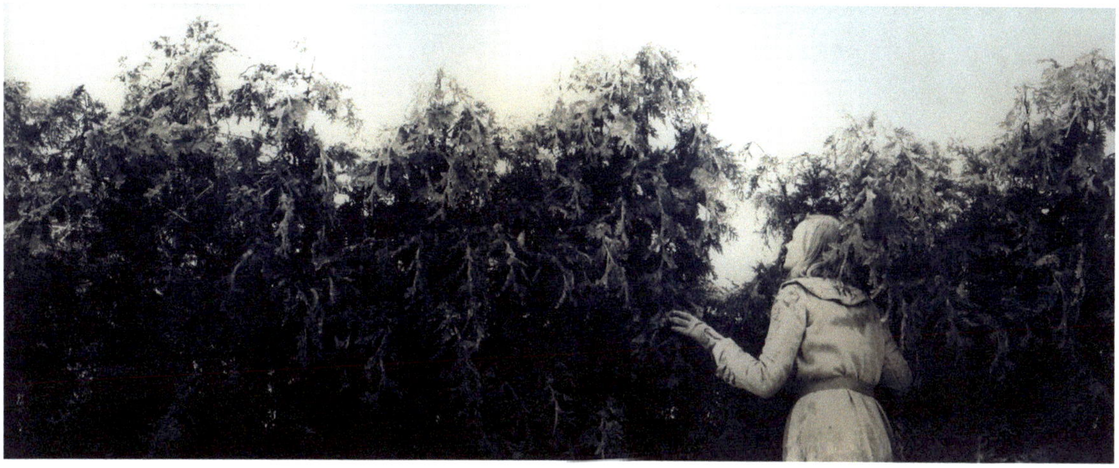

Figure 15.116. *On the Other Side*, cyanotype toned with tea and wine tannin and a bleaching agent (usually highly diluted silver gelatin paper developer or sodium carbonate), 8″ x 20″ © Emma Powell 2014

Process

For this project I was experimenting with creating fictional spaces. I was working on emphasizing metaphor over conveying reality. The blue color is unnatural and enabled me to build a dream world in which to set my scenes. At the time I was also moving from job to job and I did not have regular access to a traditional darkroom. I was able to set up a way to use this process wherever I was.

Though it may not be the most high quality paper, Strathmore Bristol is very sturdy and predictable; it puts up with a lot of abuse from multiple washings and toning.

I use the classic formula. I mix it all together at the same time and use it up quickly. I mix a minimum of 24 hours before printing.

I have used many different devices. Most of this series was printed with an old metal halide UV plate maker designed for printmaking processes. I have recently started using an integrated light meter from **www.lightmeasure.com** to get accumulated light readings to compare light sources and expose outside.

After the cyanotype is dry, I place the print on a large piece of Plexiglas and pour the toning chemicals or water over the surface. I have found the bleach is difficult to wash out when the print is submerged and this causes problems when it dries. This also gives me control and allows me to more selectively tone parts of the print, and also dodge and burn. My process steps are: 1. water about 5 minutes to re-wet the print; 2. tea/wine tannin bath 2–5 minutes (submerged or poured over the print); 3. bleach with diluted silver gelatin paper developer (100 ml to 2000 ml water) or sodium carbonate, poured on the surface of the print, not submerged as a bath—sometimes I will dilute this even more to slow the process and give myself more control; 4. pour on tea and let it sit for a few minutes until the tone returns; 5. water wash for 10–20 minutes, making sure not to wash too long or the print will wash away. The print will dry darker.

I have tried repeating the bleaching and toning steps multiple times. This changes the color in extreme ways. The wine tannin produces almost purple colors. This is harder to control, and I find I lose the highlights quickly. For my workflow I only

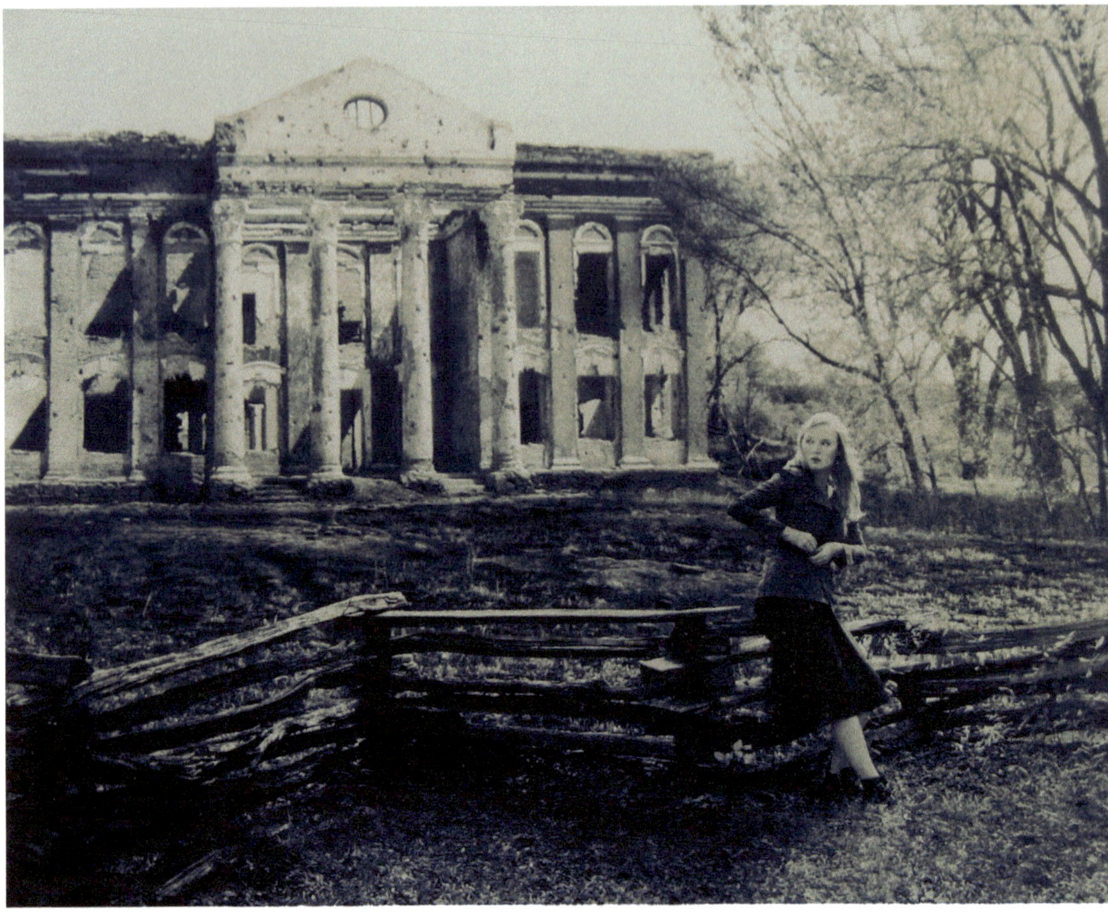

Figure 15.117. *Shelter*, cyanotype toned with tea and wine tannin and a bleaching agent (usually highly diluted silver gelatin paper developer or sodium carbonate), 11˝ x 14˝ © Emma Powell 2013

bleach once. Also, when toning I don't completely change the color. I keep some of the blue in the low-key values and mainly tone the highlights, creating a split tone effect. Cyanotypes can be easy and is a great process for beginners, but printing consistently toned cyanotypes can be very difficult.

Creative practice

In my artwork I often use alternative chemical photographic processes, chosen intentionally to evoke another place or time that is removed from the every day, adding layers of narrative meaning through historical references or material qualities. These chemical techniques give me the freedom to explore experiential methods of image-making, and allow for chance and happy accidents. I describe working with alternative processes to be a form of collaboration, because there are often surprises that could not have been anticipated. I also incorporate digital manipulation in my work in order to expand the narrative possibilities of my imagery. Digital collage allows me to make scenes that would be impossible to capture in reality. I then use the chemical process to produce the final piece. This distracts from the digital manipulation in the negative. The physical print and historic style allows the viewer to more easily suspend their disbelief and accept a fantasy perspective. This is most likely because the resulting image looks more like an illustration than reality. I am also drawn to historic processes because they enable me to play with space between fantasy and history.

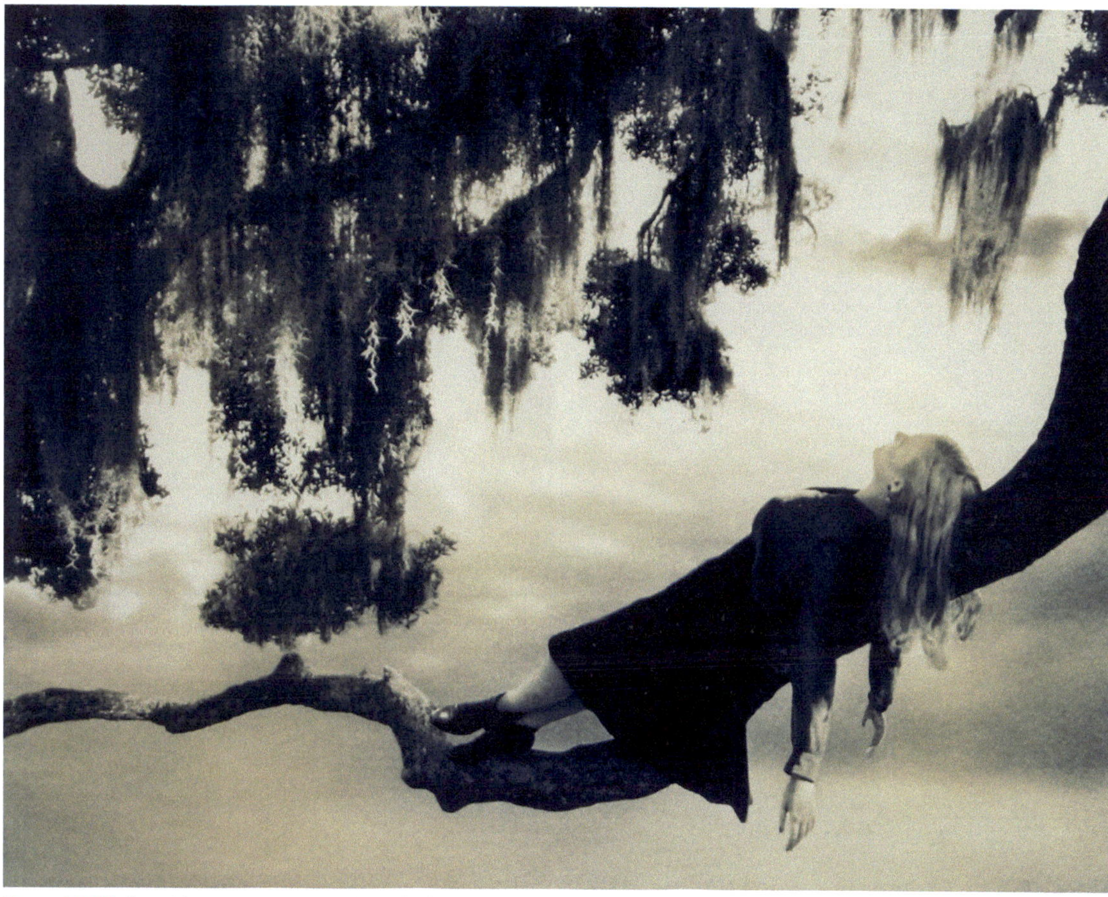

Figure 15.118. *Spanish Moss*, cyanotype toned with tea and wine tannin and a bleaching agent (usually highly diluted silver gelatin paper developer or sodium carbonate), 11″ x 14″ © Emma Powell 2012

Artist statement

From my earliest days, I have had a difficult relationship with sleep. As a child, I avoided it at all costs, especially at night. To get me back to bed, my father used to tell me stories. They were not traditional children's bedtime stories but invented tales that began on our quiet street and journeyed down open drains to a dream-world of caverns, forests, and oceans full of unexpected animals and dangers. The story would always find its way back to the real world and end where it had begun, hopefully, but doubtfully with me that much closer to sleep. *In Search of Sleep* recreates this shadowy realm and allows me to explore my real-life questions, from personal dramas to romantic doubts. The cyanotype process, with its distinctive blue tones, visually traverses the distance between waking and sleeping. These images are toned with tea and wine to both dull the blues and add warmth. Tea, wine, cyanide (potassium ferricyanide)—all three of these substances relate to different levels of consciousness that often mirror the mental states evoked by my photographs. *In Search of Sleep* creates a visual lullaby that allows me to safely explore what I love, what I fear, what I remember, and what I imagine.

Biography

Emma Powell is an Assistant Professor of Art at Colorado College in Colorado Springs. Powell received her MFA in photography from Rochester Institute of Technology. Her work often examines photography's history while incorporating historic processes and other devices within the imagery. Powell's photographs have been exhibited widely, both nationally and internationally. To see more of her work, visit www.emmapowell.photo.

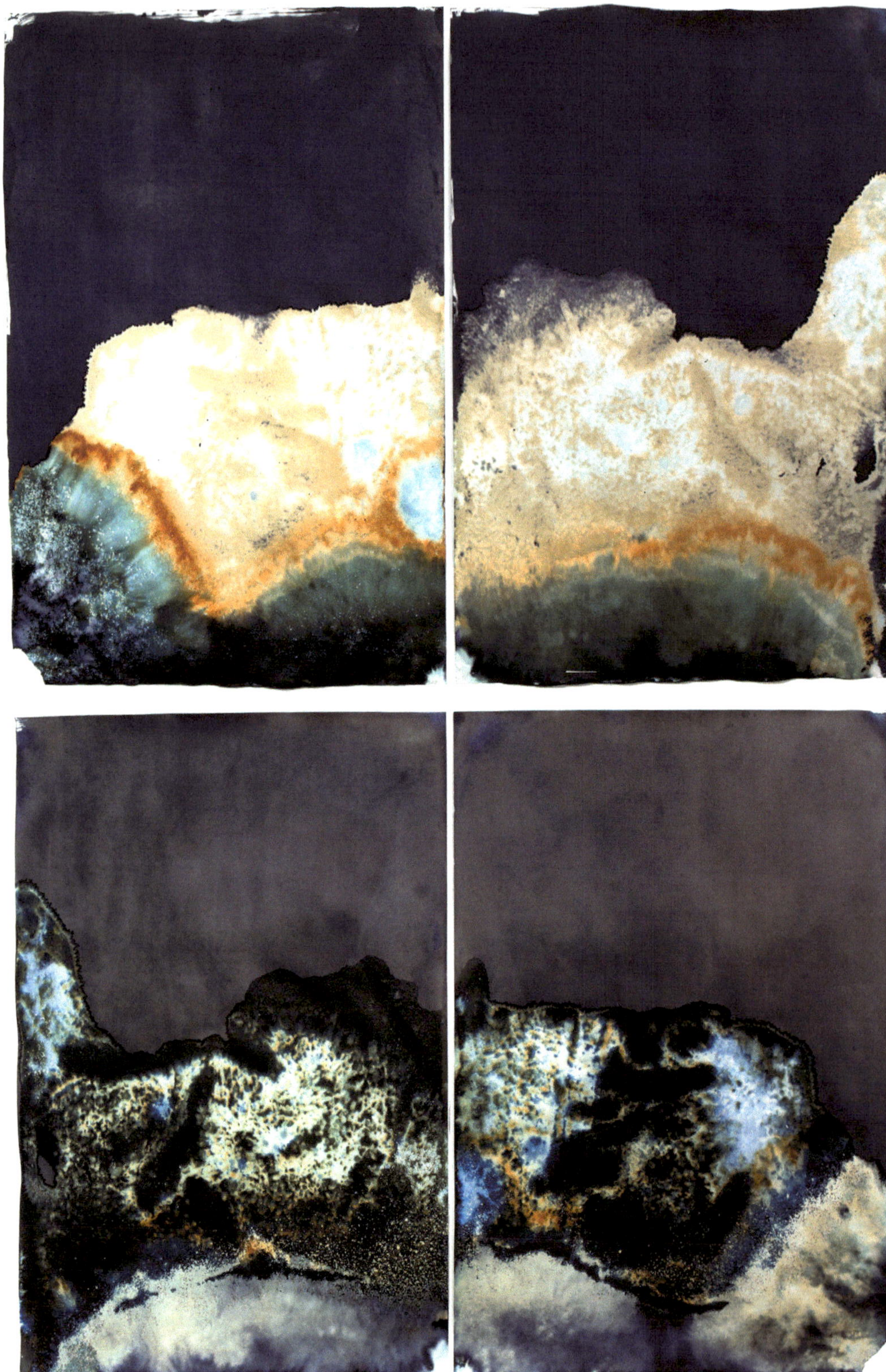

Figure 15.119. *Littoral Drift #18 (Recto/Verso, Diptych, Rodeo Beach, Sausalito, CA 07.21.13, Three Waves, Buried and Flooded)*, unique cyanotype, 30″ x 50″ © Meghann Riepenhoff 2013

Meghann Riepenhoff

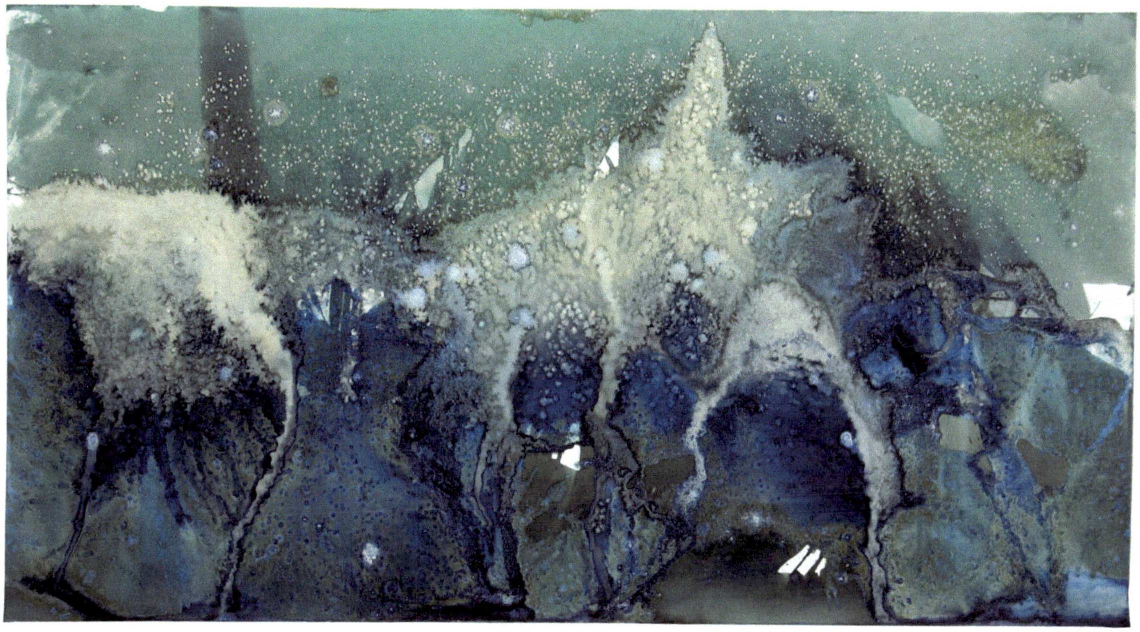

Figure 15.120. *Littoral Drift Nearshore #502 (Bainbridge Island, WA 04.01.16, Two Waves and Salt, Scattered and Poured)*, unique cyanotype, 24˝ x 48˝ © Meghann Riepenhoff 2016

Process

I use Strathmore Bristol for its strength and durability. I like hot press papers, with a very smooth surface that helps to punctuate the detail in my work.

I make cameraless cyanotypes, and use Bostick & Sullivan's classic cyanotype formula. My exposures are from the sun, not UV boxes, so my exposure times vary greatly depending on the weather. Typically an average exposure is around four hours; I'm intentionally drastically overexposing.

My pieces are developed to the degree that water from the landscape comes in contact with them. As my work exposes and dries, it records subtle topographies from the landscape. Other than what naturally occurs in the environment (such as pH, sedimentary impact, etc.), I do not tone the work.

Some of my work is framed—hinged to recessed mat boards and floated in a deep frame that allows for the three-dimensional nature of the work. I also make two-sided pieces that I suspend from the ceiling with custom steel clamps and rare earth magnets. Other work is large-scale, and is tabbed directly to the wall of an exhibition space.

I imagine my issues are extremely esoteric, but perhaps humorous, to share. My list of problems includes dogs and kids running over my prints, waves pulling prints out (only to have generous surfers help recover them), and pieces becoming frozen into the landscape.

Creative practice

The first female photographer, Anna Atkins, was also a pioneer in botanical/scientific illustration. She used specimens collected from the shorelines, and made cyanotype photograms with them. Her

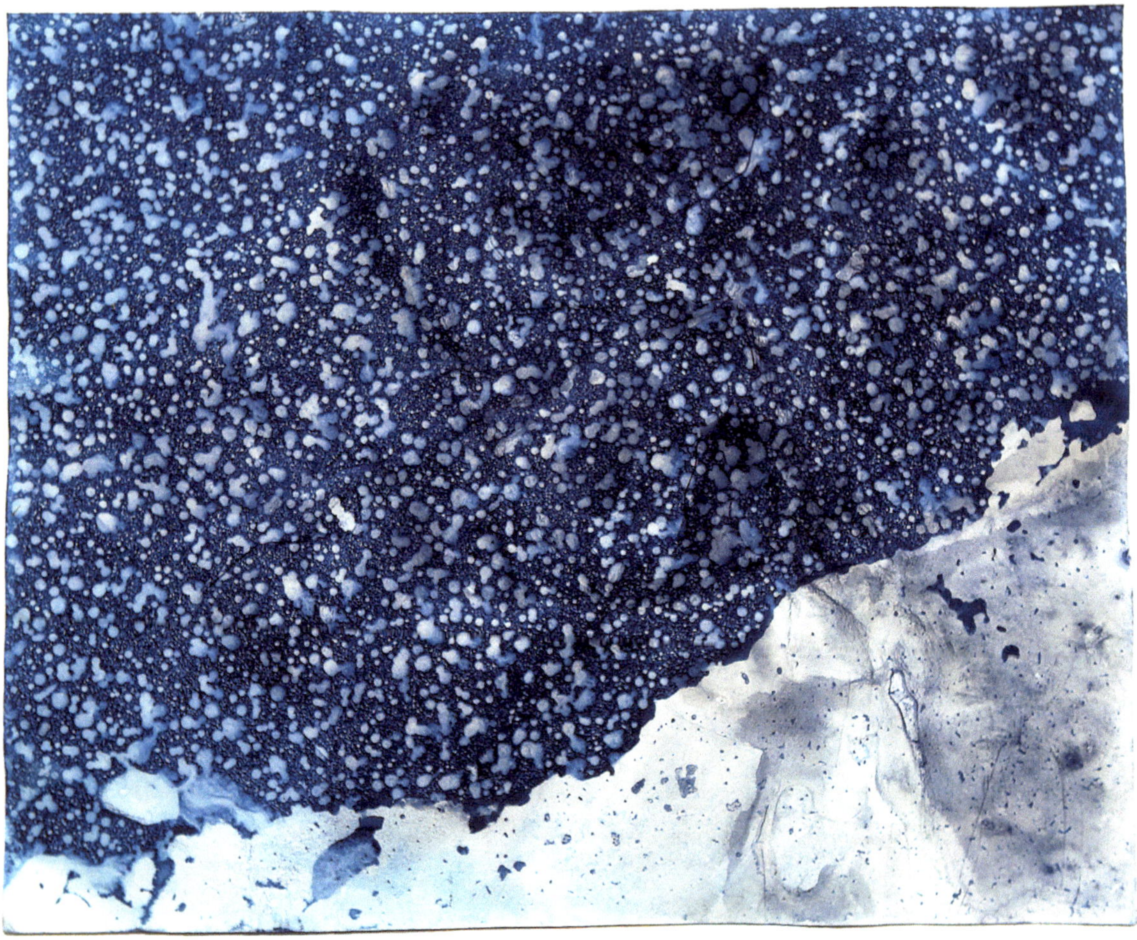

Figure 15.121. *Littoral Drift Nearshore #409 (Kohler Center, Sheboygan, WI 01.07.16, Mixed Precipitation and Border Ice)*, unique cyanotype, 19″ x 24″ © Meghann Riepenhoff 2016

minimal, elegant compositions offered a new way of seeing, and a new way of understanding the world. Previously scientific illustrations were hand-drawn approximations, until Atkins' *British Algae* introduced a new way to record the environment around us.

I'm influenced by Atkins' work, which illustrates the earliest relationship between photography and shoreline. In my work, I revisit the shoreline with cyanotype, but make a new kind of image of the landscape, where elements in the environment are both subject and process.

In keeping with nodding to Atkins' early work, I make prints in and of the environment around us. I work directly in waves, surface water, waterfalls, precipitation—primarily where aquatic and terrestrial landscapes meet.

Artist statement

This work stems from my fascination with the nature of our relationships to the landscape, the sublime, time, and impermanence. Entitled *Littoral Drift*, a geologic term describing the action of wind-driven waves transporting sand and gravel, the series consists of camera-less cyanotypes made in collaboration with the landscape and the ocean, at the edges of both. The elements that I employ in the process—waves, surface water, wind, and sediment—leave physical inscriptions through direct contact with photographic materials.

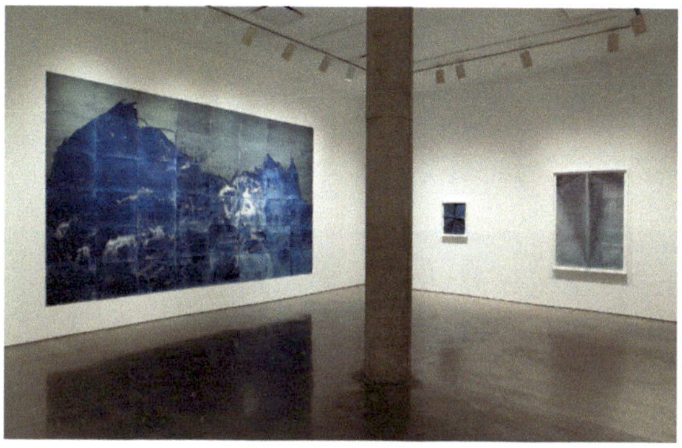

Figure 15.122. *Littoral Drift* (solo exhibition) at Yossi Milo Gallery, 2016

A sister series to *Littoral Drift*, *Ecotone*[1] also engages dynamic photographic materials in the landscape, but collaborates with precipitation rather than ocean waves or surface water. Rain, snow, ice, fog, etc. chemically activate the photographic materials, while they expose via the residual sunlight that exists even in the heaviest storm.

Biography

Meghann Riepenhoff is a 2018–2019 Guggenheim Fellow. Her series *Littoral Drift* and *Ecotone* have been exhibited at the Denver Art Museum, The New York Public Library, the Museum of Contemporary Photography, and Yossi Milo Gallery. Publications include *The New York Times*, *ArtForum* and the *Wall Street Journal*. Collections include the High Museum of Art, Albright-Knox Gallery, and the San Francisco Museum of Modern Art. To see more of her work, visit www.meghannriepenhoff.com.

1. "The area where two or more distinct habitats adjoin is called an ecotone. Because it is a border zone where multiple sets of resources and opportunities become available, an ecotone tends to support greater biological diversity than either of the systems it mediates between. Delineating an ecotone, however, can be problematic. Like habitats and ecosystems, ecotones are not self-defining, as, for example, individual species are. They are human constructs, which derive their shape and character from the qualities their observers find most salient...In scientific terms one might say that the world is composed of gradients with relative discontinuities. Put simply: things change nearly everywhere, and so nearly every place is the edge of something and shares the qualities of an ecotone." — William deBuys from *Home Ground: Language for an American Landscape* (San Antonio: Trinity University Press, 2010).

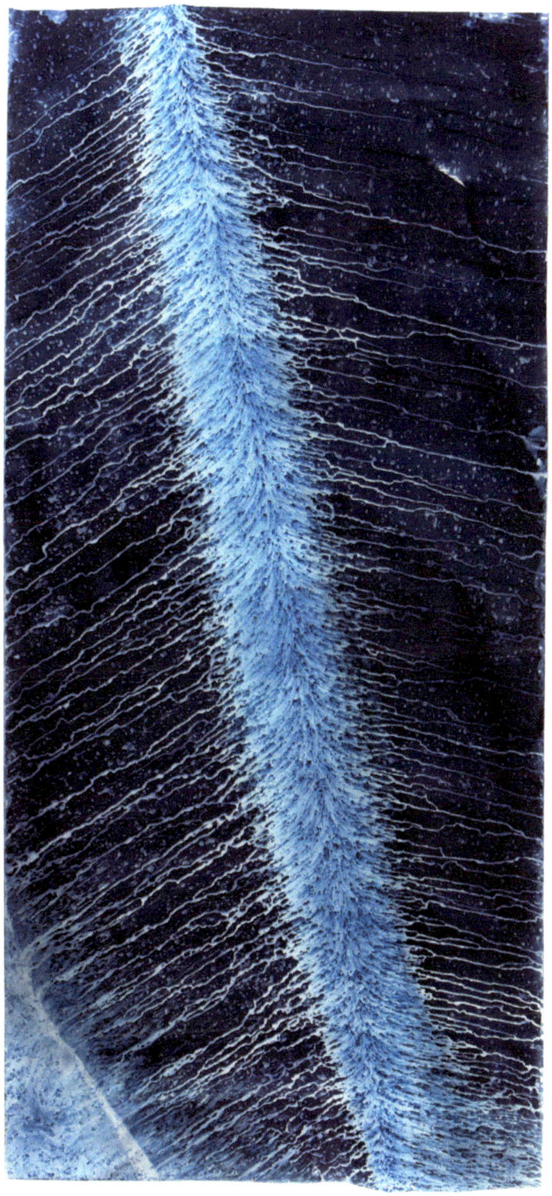

Figure 15.123. *Ecotone #377 (Bainbridge Island, WA 03.23.18, Draped on Windfall Branch, Showers and Hail)*, unique cyanotype, 42″ x 90″ x 1″ © Meghann Riepenhoff 2018

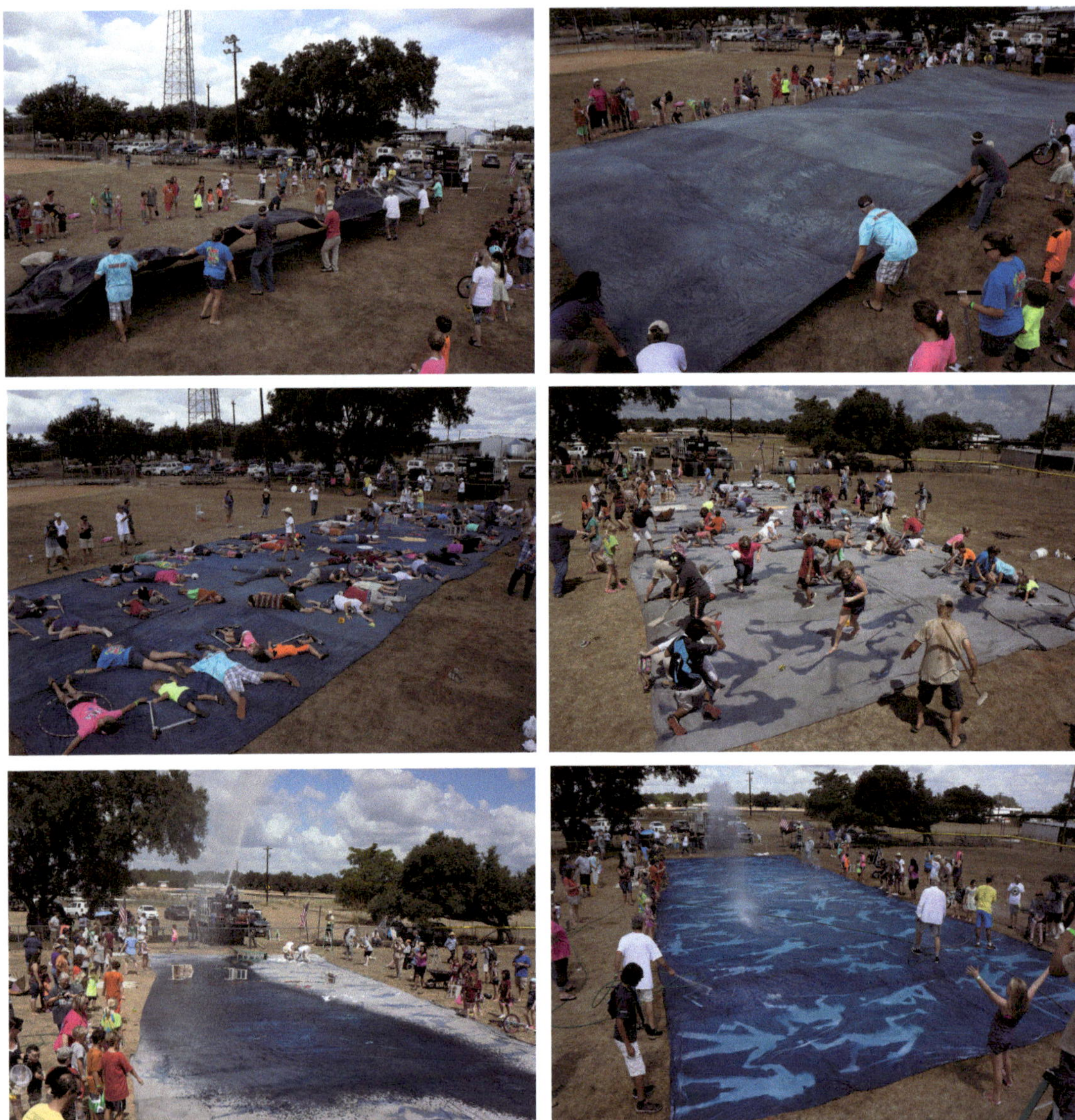

Figures 15.124–15.129. Making of *Big Blue* in the softball field, Johnson City, Texas, September 19, 2015. Photographs © Vicky Stromee 2015

Judy Sherrod July 6, 1956—July 28, 2017

Founder of Shootapalooza (shootapalooza.com), a "murmuration of artists," mostly women, who come together to make work, fun, and memories

"SHOOTAPALOOZA IS ABOUT PEOPLE, not photography. It's a collective, creative, mindset. It's an evolving, dynamic entity that is constantly re-inventing itself. It is amorphous. It is flubber.

"The first gathering was held in Port Aransas, Texas. It was an experiment. I didn't know if anyone would show up. Sixteen people did. From seven states and the District of Columbia. My job was to set the stage and get out of the way.

"shootapalooza truly created itself. It's about learning together and teaching each other. We had Photo Labs. Not workshops, but photo labs—short, hands-on, learning sessions. It's about trying new things and not being scared of screwing up.

"The genesis: I was in Pass Christian, Mississippi in December, 2012, working with S. Gayle Stevens. We discussed inviting our Mississippi and Louisiana friends to the coast for a weekend of photography and general rambunction. Our host, Barney Adams, said, 'Oh yeah, a real shootapalooza!' We never got the rambunction going, but that evening I looked up shootapalooza.com, confident it was already registered. Lo and behold, it was available so I grabbed it. A year later I asked the members of my Facebook group if they thought a photography festival on the Texas coast in February was a viable idea. The answers were sufficiently positive, so I put the thing together.

"shootapalooza isn't really about photography. It's about creativity. It's about expression. It's about connecting the dots until you burst out of the box. I mentioned earlier that shootapalooza is an evolving entity, constantly re-inventing itself. It's partially due to the inventiveness of its artists. And then there's this weird part that seems to come floating down out of the wherever." — Judy Sherrod[1]

1. Quoted from "Kevin Tully Interviews Judy Sherrod about Shootapalooza" in September 2014 *SouthXSoutheast* photomagazine (http://www.buildsxsemagazine.com/2014/09/kevin-tully-interviews-judy-sherrod-shootapalooza/).

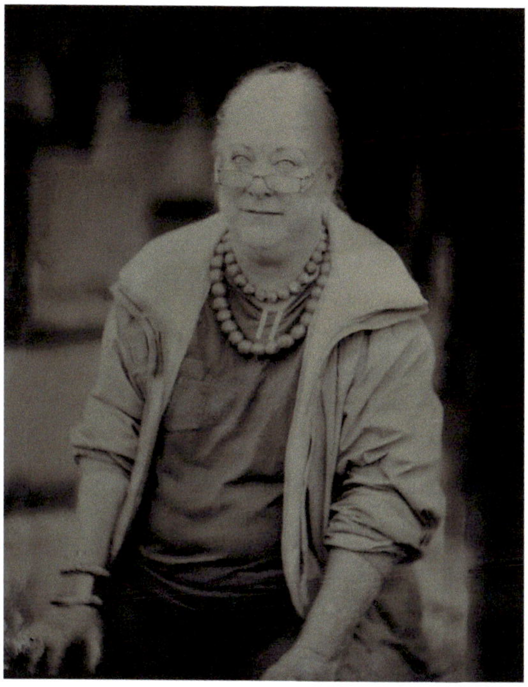

Figure 15.130. Judy Sherrod © S. Gayle Stevens 2012

Founder of World Cyanotype Day (worldcyanotypeday.com) September 19, 2015, and the largest cyanotype print in the world on that day. *Big Blue* was 2700 square feet (*Big Blue* has since been surpassed by a 2977.72 square foot cyanotype created by Stefanos Tsakiris in Thessaloniki, Greece, on September 18, 2017)

"JUDY RESEARCHED WORLD WHATEVER DAYS and found out there wasn't a World Cyanotype Day so Texas again—there had to be a World Cyanotype Day with a cyanotype print that had to be the largest, so she talked to Linda at the blueprint fabric place and asked about getting a huge piece to break the record. Amanda and Kevin at A Smith Gallery helped secure the help of the Hill Country Science Mill in Johnson City and Amanda, Kevin, DorRae Stevens, and Linda Haddock sewed the fabric together at A Smith Gallery." — S. Gayle Stevens

"We rolled out the treated fabric, all 2700 square feet of it, pulled off the protective black plastic sheeting, grabbed the edges of the greenish cotton fabric and nineteen howling men, women and children pulled like hell. It billowed out like one of those huge flags favored by marching bands. The assembled crowd of eager participants cheered and whistled. I hollered, 'Get on it!' Grandmothers, mothers, fathers, teenagers and small children rushed to take their places on the baking, collective art project; sprawling helter-skelter from one end to the other—it looked like a massive accident of joy. There were little children with baby dolls and shovels. Women with parasols and hammers. A recently Medicare qualified gallerista in galoshes, holding her father's walking cane. The Mayor with an owl. A disembodied hand floating above a chair. Five high school volleyball players, hands linked, looking like vicarious sky divers. A small pair of crutches belonging to a retired polio patient, her first set as a child. A strapping young boy holding sunflowers. A man with a large plywood disk, looking like a dancing pumpkin head. A young man appearing to march and play his trombone, his wheelchair folded, trailing behind. His mother lying behind him. Her right arm stretched out toward him, appearing to either be letting him go or encouraging him on. I will never forget the look of proud, wide-eyed anticipation in the eyes of the father of the wheelchair bound young man as he placed the trombone on his son's lap and wheeled him out of the shade into that bright, hot smiling moment."— Kevin Tully, A Smith Gallery, Johnson City, Texas

Founder of World Cyanotype Day Facebook group

"If you've ever been to Texas, you know a little something about tumbleweeds. You can never really tell where they're coming from. Or where they're going. They just appear, already in motion, propelled by some unseen force and on their way to somewhere.

"And if you've ever been to Texas, you might also know a little something about dust devils. They're towering, swirling energy, unpredictable and likely to disappear before you get a good look at them.

"So if you know a little something about Texas and tumbleweeds and dust devils, then you might be able to grasp a little something about Judy. She's a force to be reckoned with, that's for sure.

"Judy started the World Cyanotype Day Facebook group and made me an administrator. Little did I know how much I would benefit from that. She was always doing stuff like that—setting others up for success and then slipping quietly into the background and on to another project."— Db Waltrip

And friend...

"I came to know Judy October 2014 at the Center for Photographic Arts in Fort Collins, Colorado. The infancy of Shootapalooza was attending an opening of *Alternative Processes*. I walked over to four women huddled together mid-gallery. I knew their faces but had never met them: Judy Sherrod, Carol Dass, Becky Ramotowski and Heather Oelklaus. Judy looked out over her spectacles and I said, 'So, what's the huddle? A grand scheme to take over the world?' We all started to giggle one at a time, until the huddle opened up into a madcap of laughter. Judy brought me along to their budding Shootapalooza event of brief instructive slideshows, Big Burgers and beer.

"Judy and I developed a very personal, private relationship, confiding in each other on deeper matters and slung away at wild dreams of huge, bigger than life photography projects. I told Mo Jude, 'You'll not be able to avoid my camera!' She really disliked having her picture taken, but I intruded like paparazzi. I documented HerSelf at Shootapalooza events and visits. She kept many a portrait set aside. She called me Wild Irish Rogue with loud introductions.

"Judy became driven to cyanotype after the first Shootapalooza conference in Port Aransas. We saw advancements in Europe. 'Jude, we can't let the Brits get away with that largesse and no friendly competition!' She: 'Well just how big do you want it?!' 'Aye, 30 by 90 feet,' said I. 'We're talkin' Guinness!' said she.

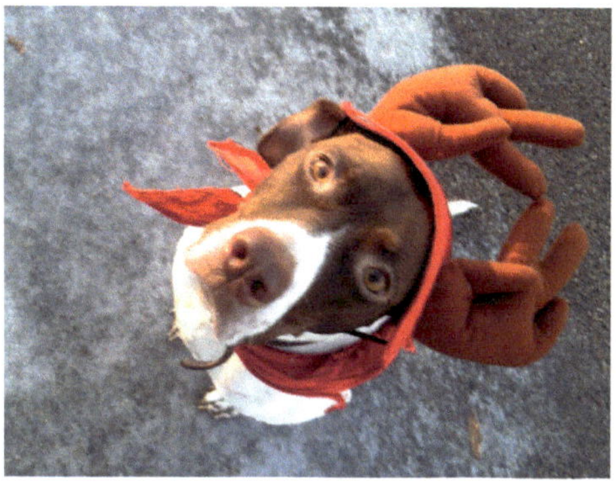

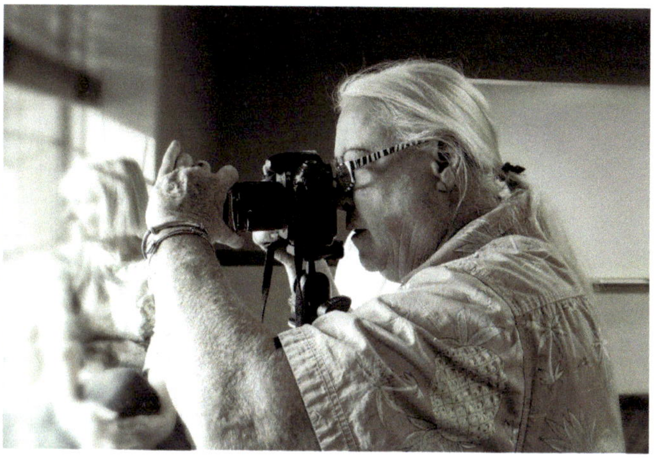

Figures 15.131 and 15.132. Left, Judy's beloved dog "B" © S. Gayle Stevens. Right, Judy Sherrod at the first Shootapalooza, Port Aransas, Texas © Mariana Bartolomeo 2014

"Becky, Jude and I made drawings. We all—Amanda, Kevin, DorRae, Linda—took on the *World's Largest: Big Blue* in Johnson City. In the middle of the night, out in the ballfield, Kevin, Amanda and I rolled up 'the leviathan' in black plastic to ready it for the next day. Hands on hips, Jude was ecstatic at the vision coming to life.

"Mo Jude visited MySelf and Stuart in Tucson many a time. The last was to leave *Big Blue* with us just prior to Shootapalooza, Arizona 2017. She wanted me to re-sensitize and re-develop *Big Blue* to gain a darker blue, lost to exposure on the Texas iron-impacted soil. 'You're the only one I know who can fix it!' said she.

"...Ahh, Jude... why didn't you call the doc when I told you? Insufferable, you are. Bein' a year your senior...YourSelf my little sister; Listen to me! True, it is! It's a dribblin' darkness that near drowns out the memory of your hollerin' cackle.

"Well, here's to our best aged house Bushmills! 1 or 2 fingers? I say. 'Two,' says she.

"I squash a whole lime down into the bottom of her glass, flick in a bit o' black pepper, stir a slow stir... and with no perceivable wince it's down the hatch and off to sleep in our guest room with Miss B. Good night, Miss B. I hope you are spoiled as ever! And, Mo Jude, I see you in the evening star. And, I'll see you there in the future. Fáilte, tanaí am mo chara."—Mariana Bartolomeo

"JUDY AND I MET AT PHOTONOLA 2010. She had a portfolio case made out of the top and bottom of a pinhole camera she was working on. I was intrigued. I also met Anne Berry, Ann Marie George and Lori Vrba. The five of us became When Pigs Fly. I frequent Pass Christian, Mississippi and Judy was only 5–6 hours away in Wichita Falls, Texas. She came a-visitin' and asked if we could make a tintype with her new 16″ × 20″ pinhole camera. I said I don't have a silver bath for 16″ × 20″ but I have an 11″ × 14″. So we shot an 11″ × 14″ plate of Helen Davis' house where I have a darkroom.

"Judy, coming from Texas, said, 'What is the largest wet plate pinhole tintype anyone has made?'

"'As far as I know no one has made a mammoth plate pinhole tintype,' I replied.

"And so *Nocturnes* was born. For two years we trudged across the sand with the 20″ × 20″ pinhole camera Judy had built for our quest. We celebrated our plates with toasts of Prosecco and our 'private' rating system of the finished plates—which consisted of how many 'fucks' we uttered as we stared at our new creations. 49 plates, and a friendship born.

"Can't believe you are gone Friend." — S. Gayle Stevens

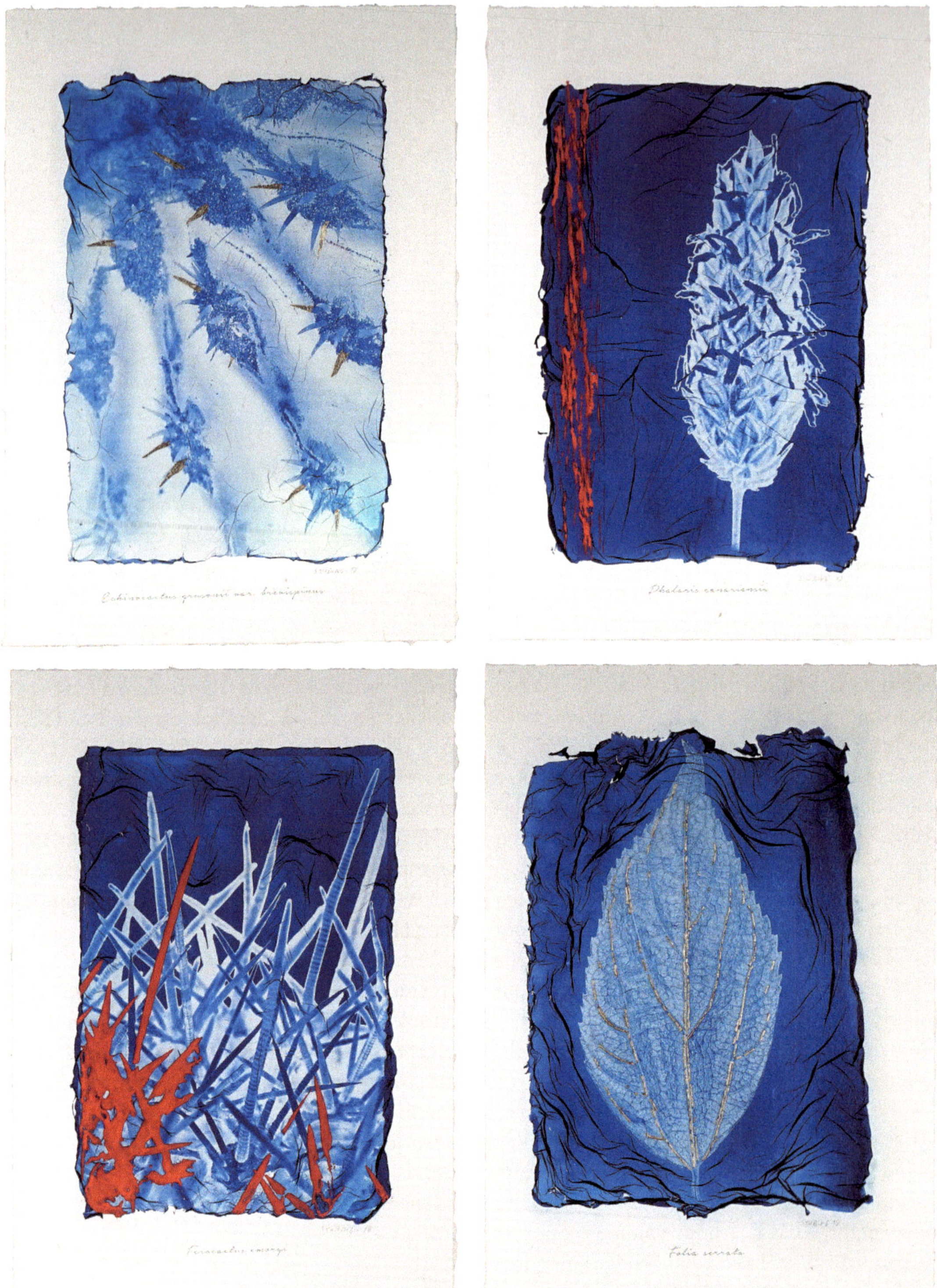

Figures 15.133–15.136. Clockwise from top left, *Echinocactus Brevispinus*, *Phalaris Canariensis*, *Folia Serrata*, *Ferocactus Emoryi*, cyanotype on gelatin, emulsion lift on paper, gold leaf or acrylic paint, and dry ink jet transfer, 70 x 50 cm © Gerardo Stübing 2018

Gerardo Stübing

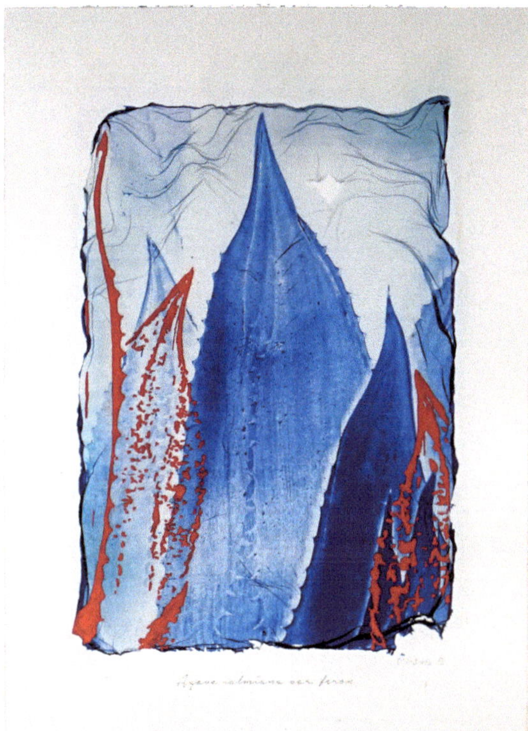

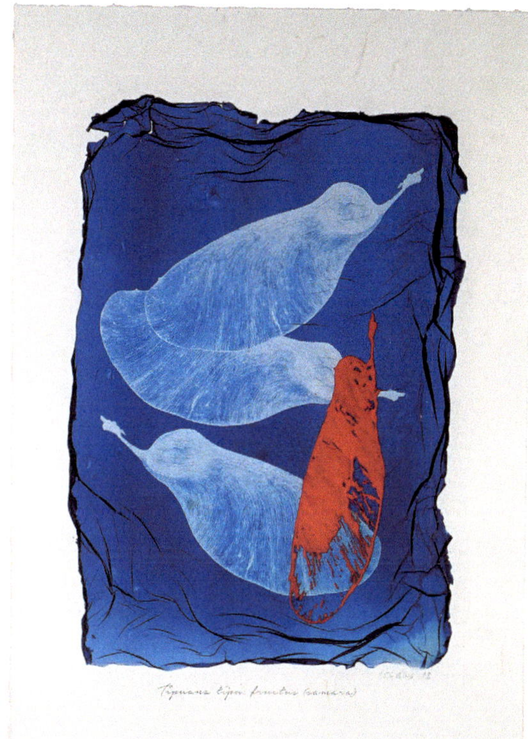

Figures 15.137 and 15.138. Left, *Agave Salmiana*; right, *Tipuana Tipu*, cyanotype on gelatin, emulsion lift on paper, acrylic paint and dry inkjet transfer, 70 x 50 cm © Gerardo Stübing 2018

Process

These prints are on handmade Japanese mulberry paper. I use the 20/8 classic cyanotype formula with added oxalic acid, expose for 2 minutes, 16 seconds under UVBL. I expose the cyanotype on a gelatin layer on glass, develop in running water for 20 minutes, and then transfer the emulsion lift to paper.

Biography

Gerardo Stübing (b.Valencia 1957) is a Valencian artist of German origin who has had a long career and recognition in the field of Botany, a science he has taught as Professor of Botany for 30 years at the University of Valencia. Stübing is currently pursuing his doctorate in Fine Arts at the Polytechnic University of Valencia. His first contributions to the field of art were photographic illustrations in various books about plants and ecosystems which he has also authored. His interest is pictorial works that initially show a clear geometric tendency that evolves progressively towards biomorphology. Stübing's work can be defined as an ambiguous abstraction inspired by plants and natural landscapes, for which he uses non-conventional materials. Stübling has exhibited in over 50 competitions, in which he has won several awards and prizes, and nine solo shows. His work has been collected by the University of Valencia, International University of Andalusia, Jiloca Studies Center, Bancaja de Segorbe Foundation, British Institute of Seville and the Ministry of Agriculture. To see more of his work visit **www.gerardostubing.com**.

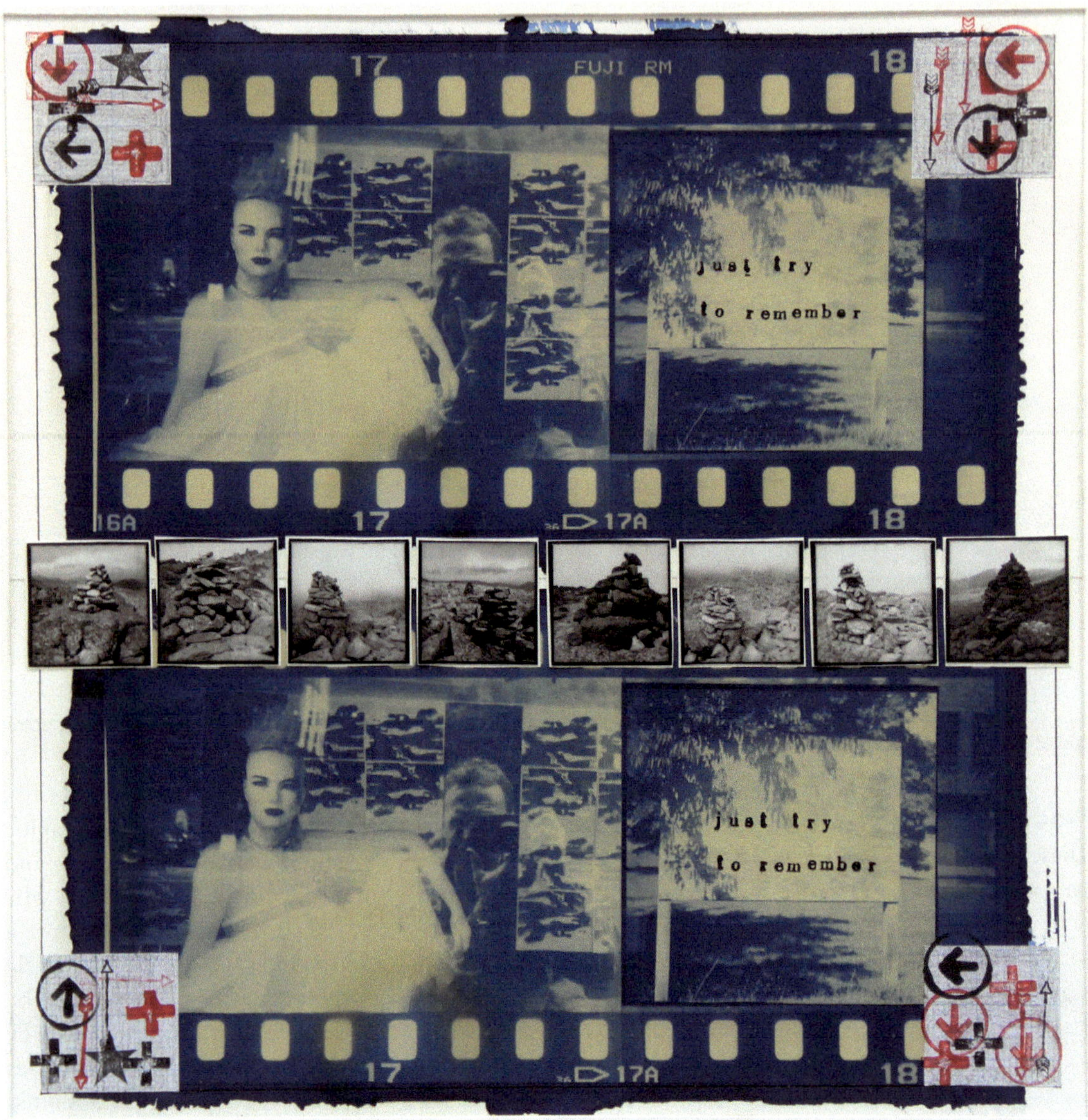

Figure 15.139. *Just Try To Remember, 2018*, cyanotype photographs, silver gelatin photographs, graph paper, ink, and graphite on paper, 20″ x 20″ © David Underwood 2018

David Underwood

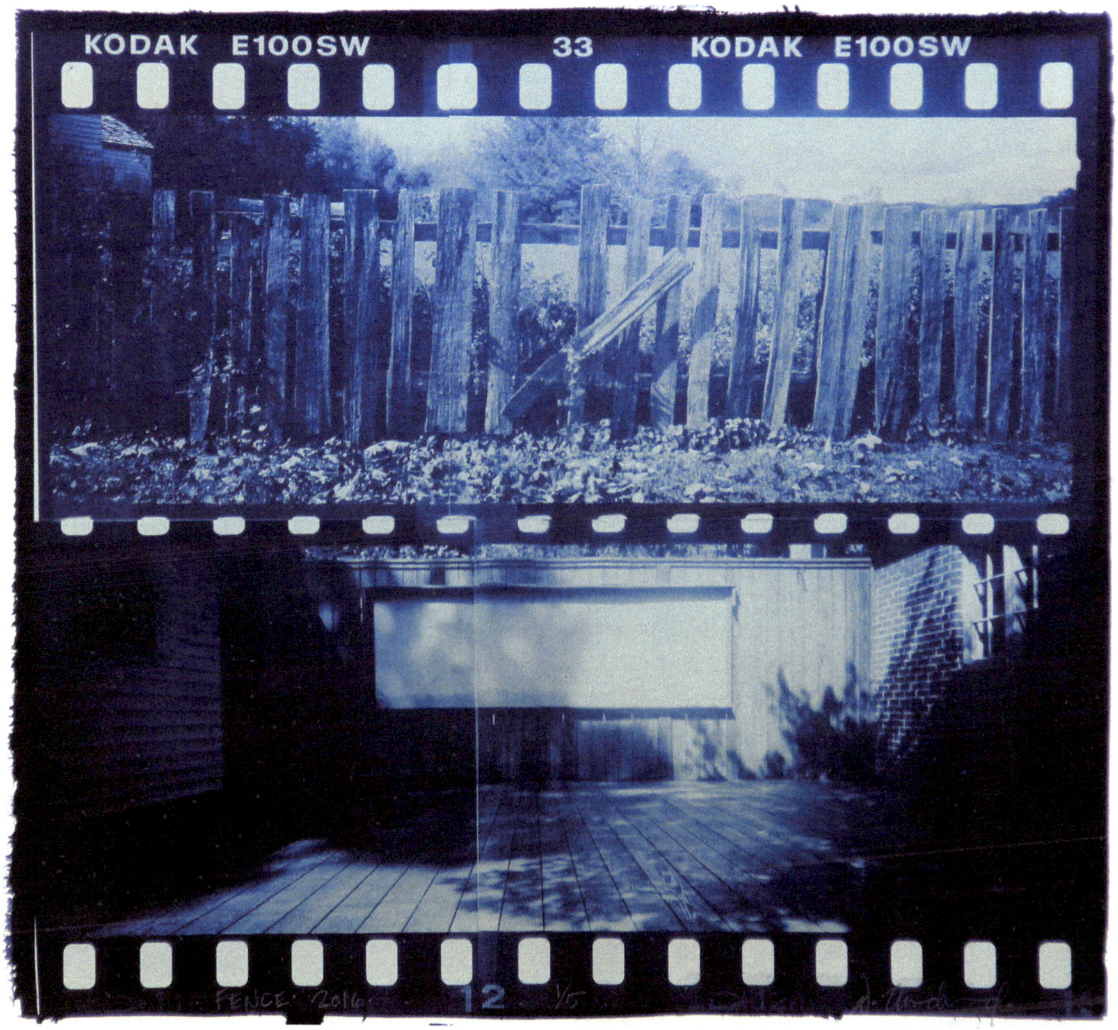

Figure 15.140. *Fence*, original images shot on Ektachrome film in a Hasselblad X-pan 35 mm panoramic camera, then enlarged on 4 pieces of analog film to make the negative, new cyanotype, printed in an edition of 5, 16.5″ x 18.5″ © David Underwood 2016

Process

I usually print cyanotypes on Arches cold-press watercolor paper, which is readily available in various sizes. I most often start with color transparency film photographs, which I then enlarge in my analog darkroom onto Arista Ortho Litho 8.5″ × 11″ or 10″ × 12″ sheet film. I then tape several of these B&W negatives together with transparency tape to make large composite negatives. I allow the tape and the obvious joints between the different images to show on the resulting cyanotype, which helps articulate the fact that I am combining images, and therefore the artwork becomes somewhat about photography itself and the process of photography.

I use Photographers' Formulary New Cyanotype, mixed to 150 ml working solution (halfway between the manufacturer's recommendations of heavy concentration and maximum dilution of 200 ml). I expose my cyanotypes in a contact frame, outside in the sun. My exposure times are totally by intuition, by watching the clouds, the sun, and the cyanotype emulsion material change color. My exposure times vary by the season and the weather, usually 3–15 minutes.

I develop by standing the print against a large sheet of Plexiglas in my darkroom sink and manually spraying it with tap water until there is no more blue washing off the paper, and then I submerge the print in a running water wash for 20–30 minutes. I don't use any chemicals whatsoever in the processing or washing.

I don't usually tone my cyanotypes, but some of my students have had success by lightly toning with concentrated tea or coffee.

If I am displaying a composite cyanotype by itself (not combined with other mixed-media), I will usually give it a standard archival white mat and frame it under Plexiglas with a traditional metal or wood frame. However, I often combine cyanotypes with other types of photographs (silver gelatin prints or color digital prints) and mixed-media such as painting, drawing, collage, and/or text.

My most common problem is that the highlight areas of a cyanotype might have a slight brown or yellow tint. This seems to usually be caused by inadequate washing.

Creative practice

I print my cyanotypes in short editions. I prepare my large negative, coat 8 to 10 sheets of good paper with cyanotype emulsion, and spend most of a sunny day printing all the coated paper in one long session (it helps to have good and predictable weather). After processing and washing all the copies, I then cull out a few of the bad prints (due to poor guesses of sun exposure or due to variances in emulsion thickness) and end up with a good edition of perhaps 5 to 7 copies.

I learned early on to get comfortable with failures. It is OK to spend a day attempting to make good photographic prints (of any kind), only to be dissatisfied at the end of the day and throw away the day's work. The key is to get back to work the next day, with the previous day's lessons learned, and move forward to keep creating.

Artist statement

For more than 30 years, I have been making composite photographs and mixed-media artworks which include multiple original photographs, with the intention of expanding the vocabulary of photography beyond the limitations of single frame imagery. Using digital, analog, and alternative photographic techniques, sometimes combined with collage, drawing, and/or painting, I make unique assemblages of original imagery that are sometimes photographed all of the same subject, but most often photographed many miles and perhaps many years apart. Combining different photographs into new compositions allows for expansion of content, implied meaning, and expressive aesthetics.

Biography

Over a 30+ year career of making photographic and mixed-media artworks, David Underwood has held 30 solo exhibitions and had work included in more than 125 group exhibitions. His work is included in the permanent collections of 16 museums, including the California Museum of Photography and the Smithsonian American Art Museum. A native of Columbus, Ohio, Underwood is Professor of Art at Carson-Newman University, where he has been teaching Art and Photography for 28 years. He is the editor and publisher of the 2014 book, *Rich Community: An Anthology of Appalachian Photographers*. To see more of his work visit **www.underwoodartworks.com**.

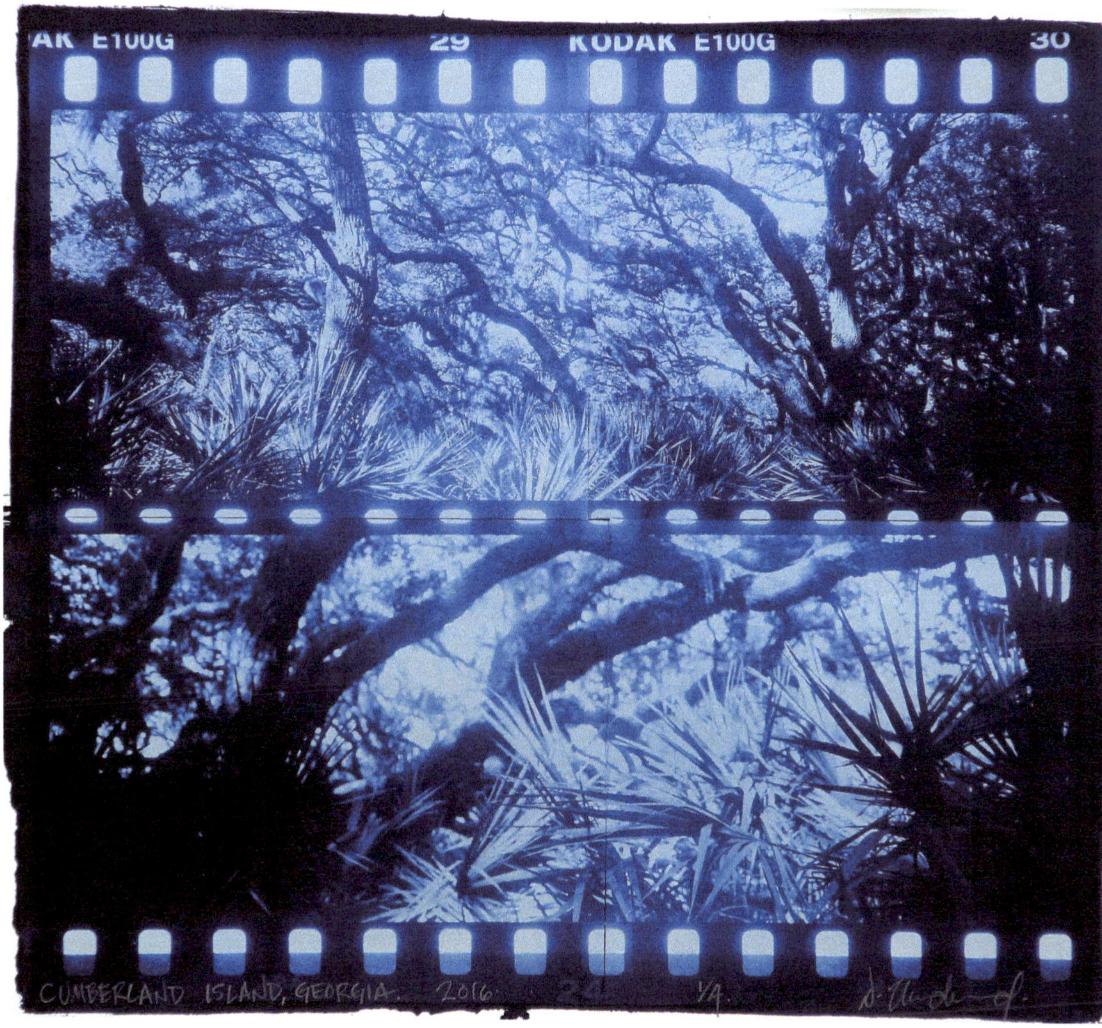

Figure 15.141. *Cumberland Island, Georgia*, original images shot on Ektachrome film in a Hasselblad X-pan 35mm panoramic camera, then enlarged on 4 pieces of analog film to make the negative, printed in an edition of 4 with new cyanotype, 16.5″ x 18.5″ © David Underwood 2016

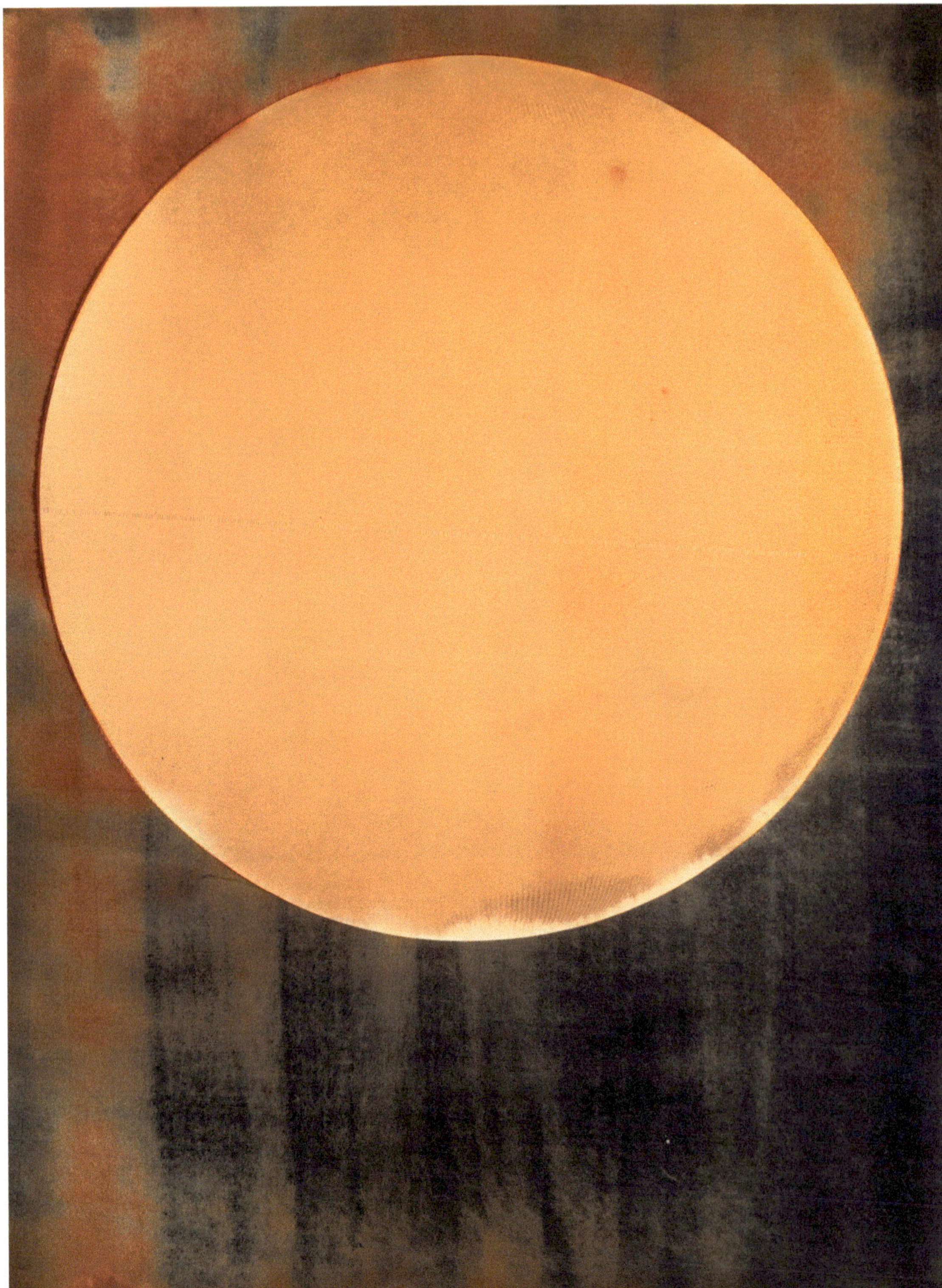

Figure 15.142. *Portal 112*, Rives BFK first coated with cyanotype and exposed with image A on the inside of the circle for 40 minutes, washed and dried and then coated with Van Dyke brown and exposed on the outside of the circle for 8 minutes, washed and dried, allowing the inside of the circle to slightly expose while processing, 30″ x 22″ © Harrison D. Walker 2018

Harrison D. Walker

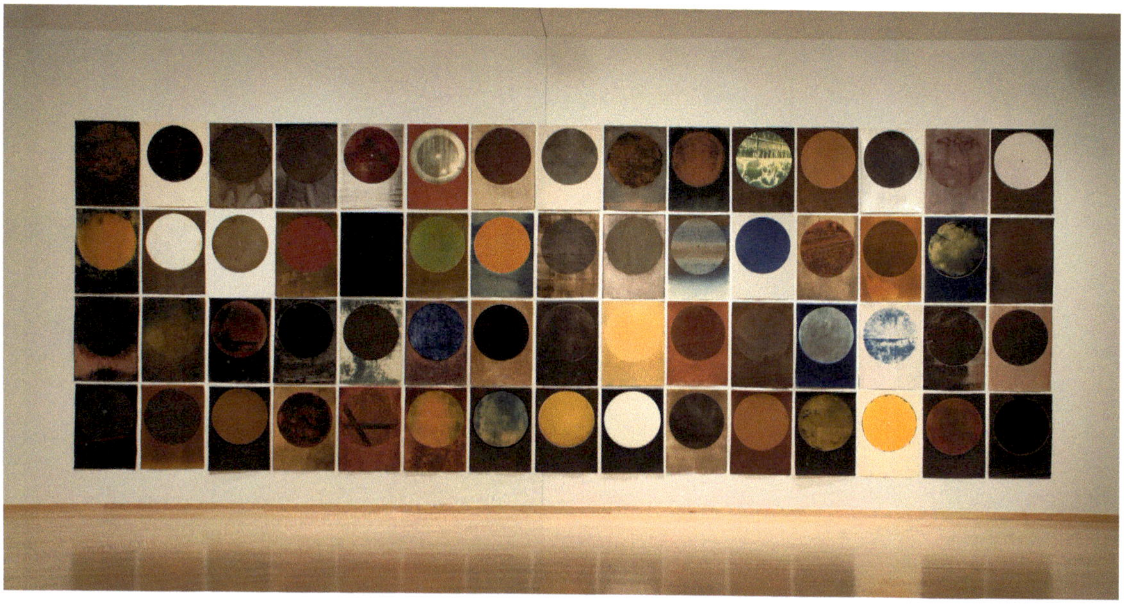

Figure 15.143. *Portals*, Installation View, found images and objects, acrylic paint, charcoal, conté crayon, cyanotype, etching ink, film developer, graphite, gum bichromate, liquid light emulsion, lithographic ink, paper, pencil, pigments, sodium carbonate, silver nitrate, tannic acid, Van Dyke brown, and wax, 30″ x 22″ each © Harrison D. Walker 2016–2018

Process

Cyanotype was one of the first photographic processes, other than silver gelatin, that really sucked me into the handmade photographic print. I use cyanotype because it is relatively simple to mix the chemistry, expose, and process and has a great latitude in terms of combination printing with other materials as well as toning. I am fascinated by the variability of outcomes—within the range of obtainable blues—to the way the cyanotype emulsion interacts with others.

I typically use Rives BFK for most of the *Portals* because it is really durable and can withstand multiple cycles of processing; however, for more representational images I like Arches Aquarelle and Fabriano Artistico because they have a finer grain yielding a generally sharper image.

I either start with a scanned silver gelatin negative or digital image capture printed on Pictorico. I typically print through the Quadtone RIP software on an Epson 3880 or 9900.

I use 62.5 g FAC to 10 oz tap water + 22.5 g PF to 10 oz tap water. I most often use the sun; ideally I would expose for about 40–60 minutes in open shade on a sunny day. I will sometimes set in direct sunlight for a few minutes if the exposure is slow, but living in Georgia, direct sunlight can cause humidity issues when exposing. I prefer a slower, less intense, exposure as opposed to blasting it in direct hot sun.

I place the print in still water for a few minutes initially face up, then flipped face down. After 2–3 minutes I put it in a gently running water bath until the water runs clear and the highlights are white, approximately 10–15 minutes. I don't typically use additives at any stage of processing. I hang smaller prints on a clothes line to dry and use screens to dry larger prints, making sure the screen is at a 45° angle to avoid pooling of residual water.

For a standard cyanotype I do not tone. I prefer the deep blues from a properly exposed print. I do manipulate some prints with sodium carbonate,

Figure 15.144. *Portal 113*, Rives BFK coated with cyanotype and exposed on the outside of the circle for 3 hours, washed and dried, soaked in a 10% sodium carbonate solution, washed and dried, 30″ x 22″ © Harrison D. Walker 2018

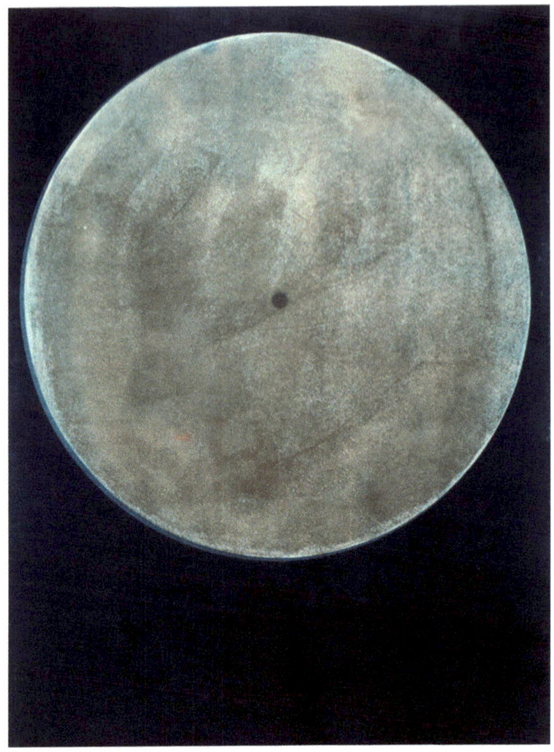

Figure 15.145. *Portal 011*, Strathmore watercolor paper coated with cyanotype and exposed on the outside of the circle for 12 minutes and the inside of the circle for 3 minutes, washed and dried, coated with liquid light emulsion and exposed for 1½ minutes on the inside of the circle, processed in Sprint developer, rinsed, and fixed in Sprint fixer, washed and dried, coated with cyanotype and exposed on the outside of the circle for 15 minutes, washed and dried, 30″ x 22″ © Harrison D. Walker 2016

film and/or paper developer diluted at varying ratios, tannic acid, and tea/coffee. With this, I am not so structured. It is more of the cooking method, a pinch of this or that, doing most of it by eye.

All of the *Portals* are hung with linen pockets, attached on the back of the print at the top of each corner. I place a magnet inside the pocket which attaches to a nail in the wall. This allows the print to float freely. For smaller prints I love a white mat and discreet white wooden frame.

Living in Georgia can cause problems with humidity. To overcome this I dry the print for an hour or so in the dark usually with a box fan in the room. After it feels dry to the touch I almost always hit it with a cool hairdryer for a minute or two, on both the front and the back. I rarely coat paper and allow it to languish for multiple days before exposing. I will also try to avoid exposing in direct sun, which can cause condensation on the printing frame if exposed in hot weather.

Creative practice

I use cyanotype more like a printmaker or a painter might, focusing less on representational imagery and more on the materials itself. Within *Portals*, I am thinking about these emulsions as washes, colors, surfaces, and textures and am interested in the ability for these photographic materials to elicit various emotions and perceptions. The series has a secondary component called the *Portal Instruction*

Manual that tracks each material and step used to create each printed object. I am interested in the variable outcomes that can come from factors such as climate, geographical location, the varying hand from artist to artist, temperature, etc.

Artist statement

My work is driven by interests in the intersections of space exploration and war strategies, often visually referencing the celestial, the post-apocalyptic, and the otherworldly. I employ the visual alchemy of printmaking, drawing, and photographic materials to create forms that evoke an experiential and emotional viewing—understood through the meditative act of looking.

Portals is a multi-component project exploring color, chemistry and psychological perception. *Portals* is made of three sections: *Portals*, a series of print variations of a repeated circular form intended to explore how the viewer perceives variations in texture, surface, color, image, and time; *Footnotes*, a living collection of prints made from archives, both personal and public, exploring the other worldly, space technologies and the evolution of the V-2 rocket, which provides context and influences to *Portals*; and *Manual*: a catalog of *Portals*, including a photographic reference and the materials and procedures to create each printed object. Conceptually, *Portals* engages with two primary concerns: the evolution of the V-2 rocket, and photography itself. I was born in Huntsville, Alabama, the "Rocket City," where after WWII the US further developed Germany's V-2 rocket under the guidance of its creator, Wernher von Braun. This technology was instrumental to the first lunar landing, and continues to play a major role in the development of the US Missile Defense Program and NASA today. Though not explicitly referenced in the imagery of the *Portals*, the V-2 rocket serves as a touchstone throughout the project, a connection made visible within *Footnotes*. Most of the work is not lens-based, yet much of it is inherently photographic. The actions of light (and pressure) are fundamental in both making and understanding this work. The use of light-sensitive emulsions in tandem with other materials evokes a depth of tactility and surface that is often absent in contemporary lens-based arts. Light, and its interactions with these materials, is at the heart of each *Portal*.

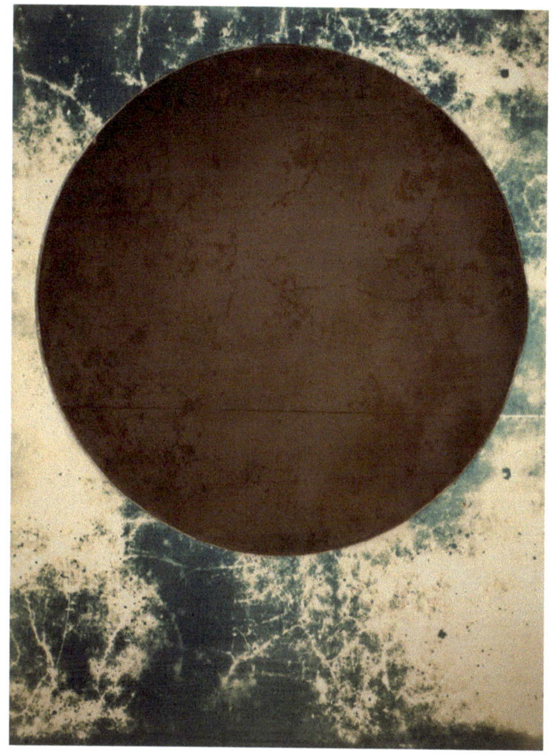

Figure 15.146. *Portal 057*, Rives BFK coated with cyanotype and expose 6 minutes on the inside of the circle and 18 minutes on the outside of the circle with image C covering the entire print, washed and dried, inside of the circle coated with Van Dyke brown and exposed for 5 minutes, fixed in a 3% sodium thiosulfate solution for 2 minutes. washed and dried, soaked in Sprint film developer diluted 1:4 for 3 minutes. washed and dried, 30˝ x 22˝ © Harrison D. Walker 2016

Biography

Harrison D. Walker (b. Huntsville, Alabama) is an artist living in Athens, Georgia, working as a printmaker and freelance project consultant. He received his MFA at Tyler School of Art in Philadelphia, Pennsylvania. Walker is represented by Corey Daniels Gallery and Candela Books + Gallery. To see more of his work visit **www.harrisondwalker.com**.

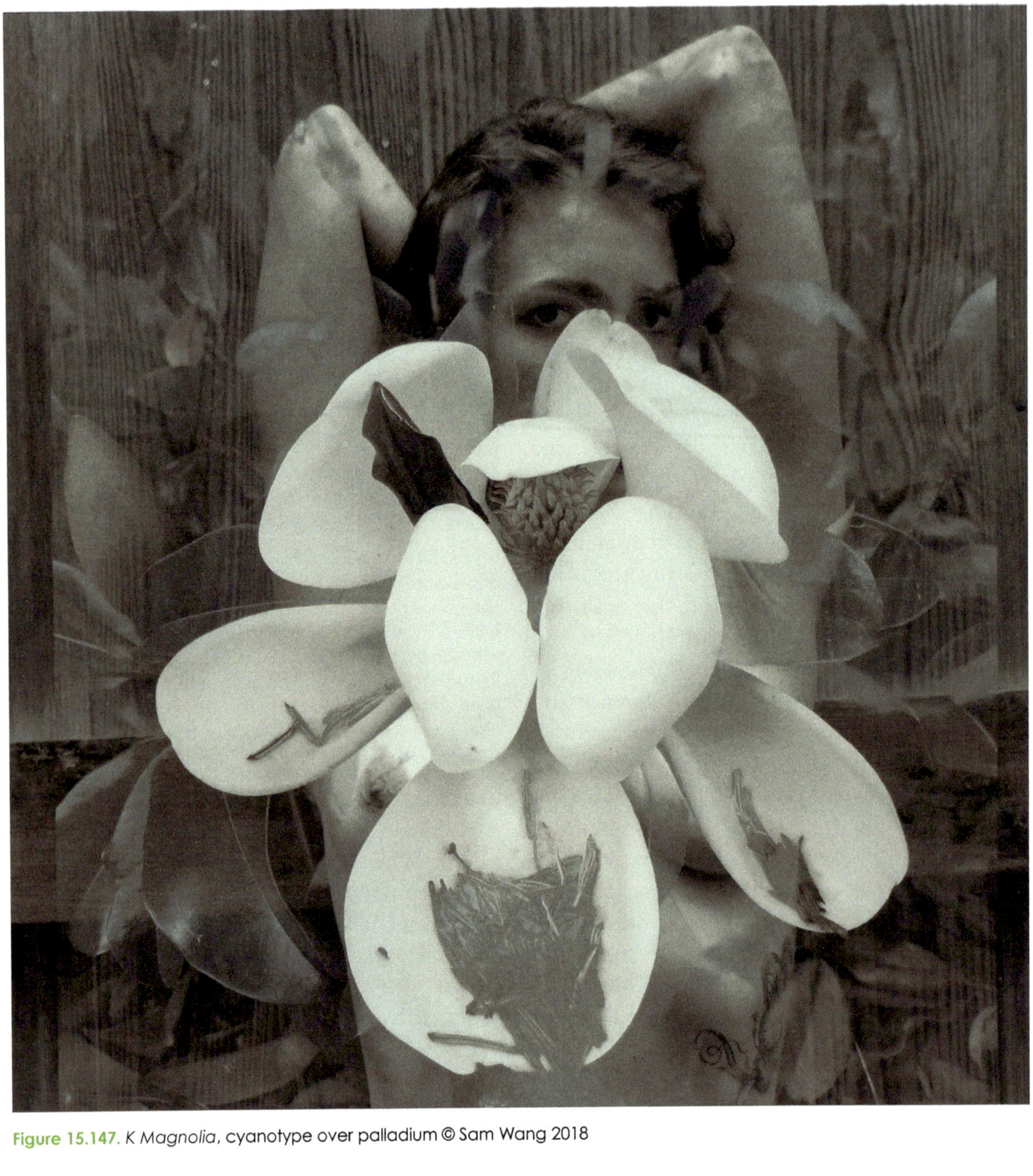

Figure 15.147. *K Magnolia*, cyanotype over palladium © Sam Wang 2018

Sam Wang

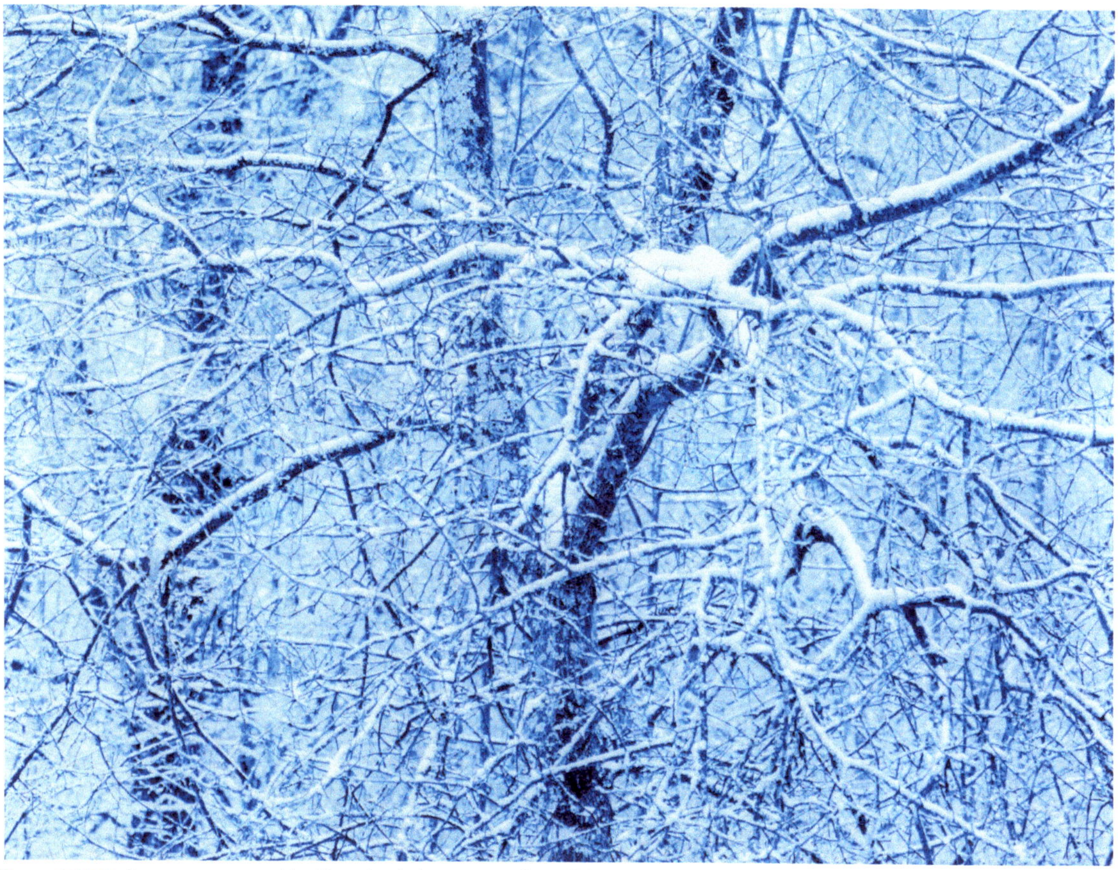

Figure 15.148. *Snowscape*, combination classic/new cyanotype © Sam Wang 2018

Process

The image can come from anywhere: I've used cell phone images, scanned film negatives, scanner generated files, and of course mostly images from the digital camera. I look for images that might fit the process. Then in Adobe Photoshop I do the usual preparations of conversion to black and white, adjusting tonality and sharpness, etc. Then I save it as a Gamma 2.2 negative which I print on Fixxons transparency material.

The paper I use is pH neutral, so the only preparation is to keep it in a humidified environment. Thanks to Mark Nelson for his advice many years ago, I have learned to always humidify the paper that I intend to coat, whether in platinum/palladium or cyanotype or whatever else. The paper is placed on a screen over a tray of water, and covered by another tray over it upside-down.

The amount of chemical solution required is calculated to be about 1 drop per 2 square inches. So for an 8″ × 10″ area it is 20 drops classic cyanotype plus 20 drops new cyanotype, and 2 drops of 40% citric acid to prevent fog. This is mixed just before use and coated on the humidified paper with a

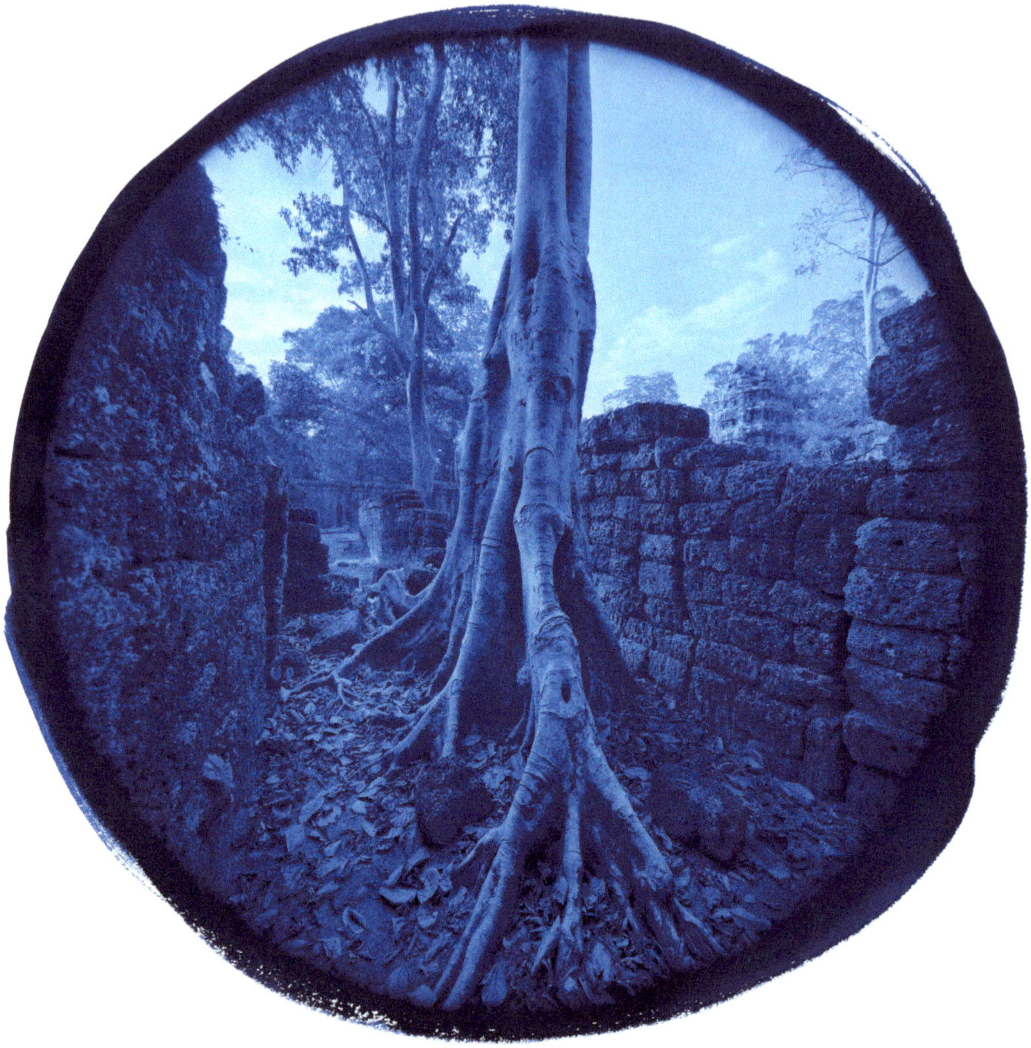

Figure 15.149. *Tree at Angkor Thom*, combination classic/new cyanotype, 20˝ x 20˝ © Sam Wang 2018

magic brush. The coated paper is laid flat on some old newspaper for a few minutes, until it's no longer glossy, then hung on a line to dry in the dark. I make sure it is thoroughly dry before exposure.

Creative practice

Handcrafted processes were always a very important part of my work, partly because of my background in the other art fields such as painting, drawing, and digital graphics, but mostly because they slow down the creative process and allow time for incubation of ideas and demand step-by-step discoveries of ways to solve problems. It began from the very early years of my career, in the late 1960s with photo-silkscreen and pinhole and cyanotype, to casein and gum in the 1980s. In terms of imagery, each process seemed to demand different ones from the day-to-day stash of images that accumulate. I usually prefer to deal with technical matters on the conscious level and let my subconscious deal with what they might mean.

Biography

Sam Wang (b. China 1939) received his MFA in photography from the University of Iowa in 1966 and taught photography and art with the computer at Clemson University in South Carolina

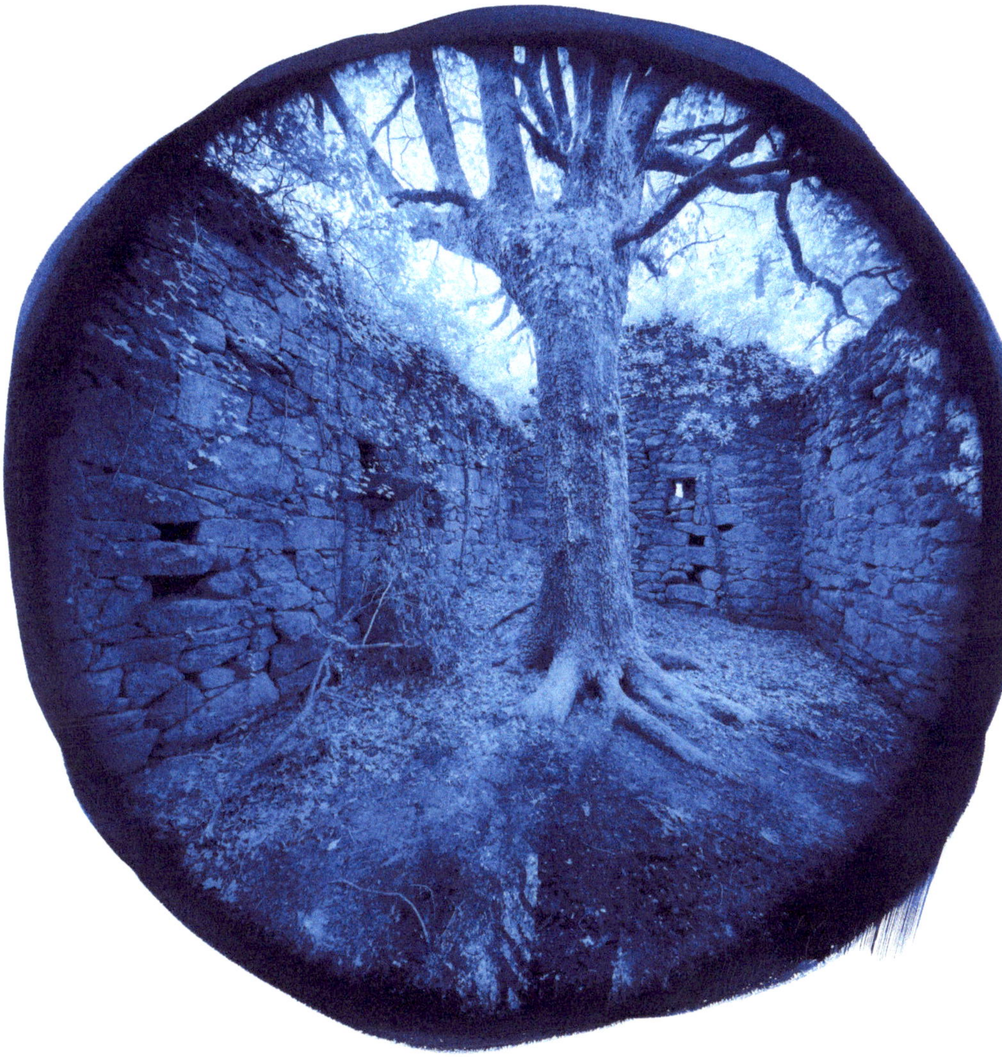

Figure 15.150. *Tree in Spanish Ruins*, combination classic/new cyanotype, 20″ x 20″ © Sam Wang 2018

until his retirement as Alumni Distinguished Professor in 2006. He was the Honored Educator at the 2007 SPESE Conference, and received the Elizabeth O'Neil Verner Award from the South Carolina governor and the South Carolina Arts Commission in 2012 for his contribution to the arts in South Carolina. Wang's most recognizable work includes round images that utilize the entire image field of wide-angle lenses. He began with designing and constructing a 5″ × 7″ camera in 1978, grafting a medium format wide angle lens to a 5″ × 7″ back, and continued with numerous different sizes and configurations over the years, including the recent entries into digital and infrared. A monograph of his work, *Sam Wang—Four Decades of Photographic Explorations*, was published by Jiangsu Arts Publishing in China in 2010. He has also coauthored a book on general photography entitled *Photography and Expression*, and on alternative printing processes entitled *Handcrafted: The Art and Practice of the Handmade Print*, both published in China in 2014. Wang's works have been exhibited and collected internationally. To see more of his work, visit www.samwang.us.

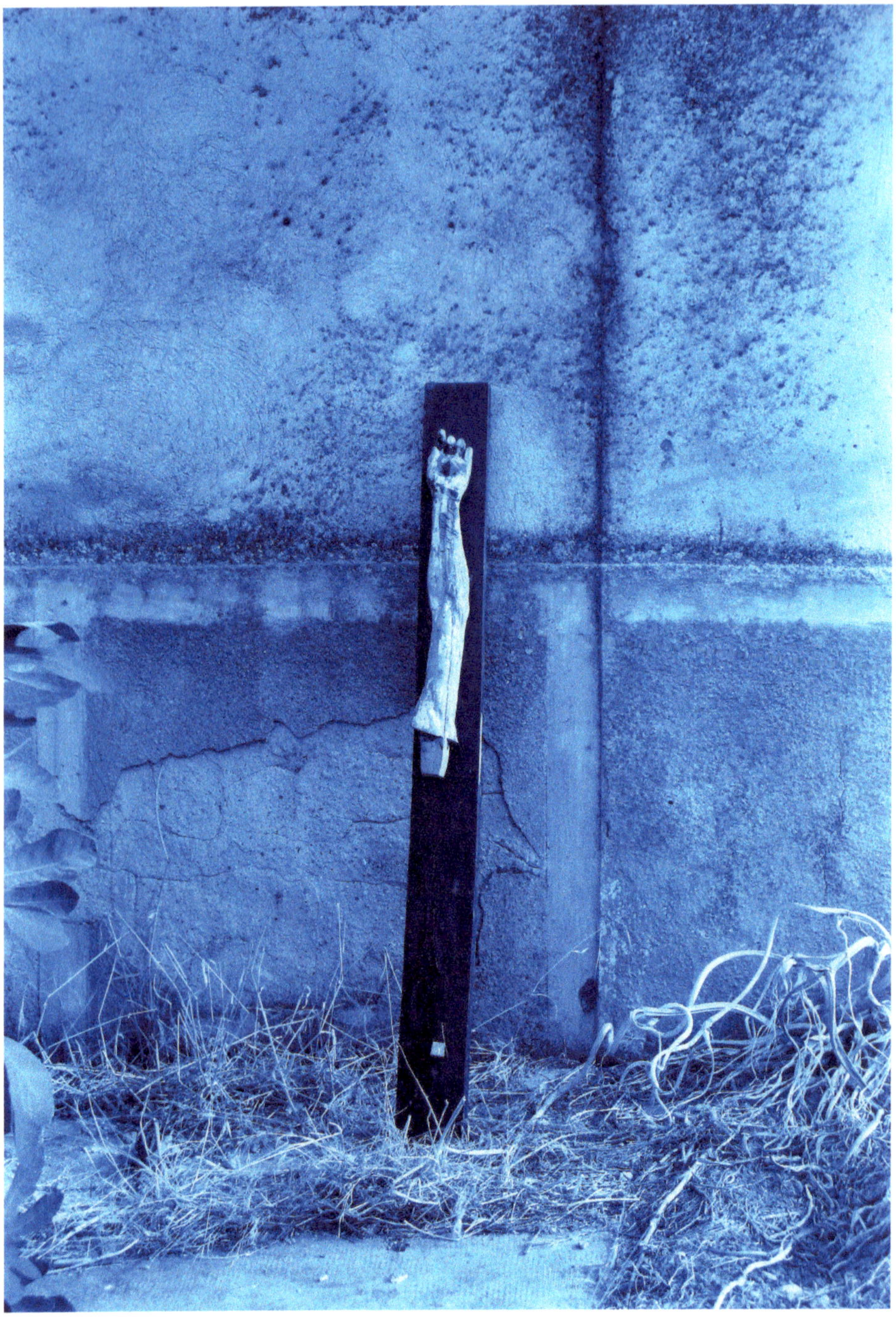

Figure 15.151. *Fragment from a Crucifix, Tudia, Sicily,* 284 x 408 cm © Mike Ware 2009. "Throughout history, the beautiful Italian island of Sicily has been a crossroads of civilizations. There are contradictions here: this fertile but troubled land can be shattered by earthquakes and volcanoes; damaged buildings may be left derelict, and even whole towns deserted. The warm-hearted people too are split by differences —social, political, and religious. The powerful Sicilian sun lights up my visual metaphors that form contrasts with the dark and sometimes violent undercurrents of past history. The fragment from a desecrated, broken crucifix lay on the ground, until placed respectfully against the wall of the nearby derelict church, when it suddenly took on a new, paradoxical significance—by making the communist salute. The wooden block cemented into the stained wall, in grazing light, is reminiscent of a place of execution. The abandoned effigy with the 'exploding finger' has another disturbing property—it is articulated."

Mike Ware

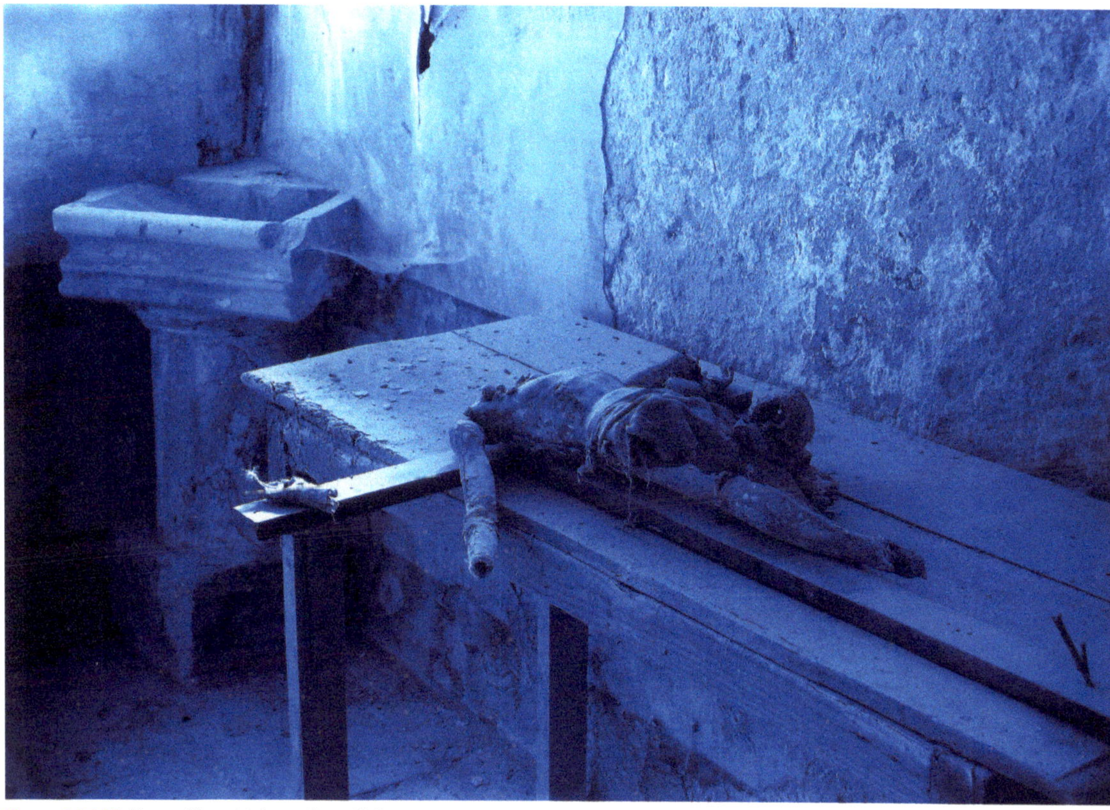

Figure 15.152. *Deposition, Noto Antica, Sicily*, 410 x 285 cm © Mike Ware 2007

Process

Generally, I work in the landscape and have a liking for small islands, so one is never far from the sea, and the blue of cyanotype has often seemed appropriate to such subject matter—ever since the days of Anna Atkins. However I also enjoy using the blue medium in a surreal sense, because—apart from sea and sky—it is such a rare color in nature.

My favorite papers are Buxton and Herschel, once handmade by Ruscombe Mill, but no longer available. I also use Arches Platine [Note: page references to *Cyanomicon II* will be included here for easy reference; the book is available as a free download at Ware's website, and is a must-read if you haven't already. For paper, see *Cyanomicon II* §7.4.4 p. 214]. My negatives are printed on Permajet Digital Transfer Film from digital files using an Epson Photo R2880. Digital files come from a digital camera or scans (Epson 4990) of 6 × 9 cm rollfilm negatives in my earlier silver-gelatin archive [see *Cyanomicon II* §6.6 p. 188 for digital procedures].

I use new cyanotype; I invented it [see *Cyanomicon II* §7.4 p. 209].

For smaller prints up to 10″ × 12″ I use a Philips Cleo facial tanning unit; for larger prints, a NuArc exposure unit [see *Cyanomicon II* §6.4.4 p. 183].

It is important to me that, when matted, the images still show clean edges, and a clear margin, so I always mask when printing. I do not "show the brushmarks" [see *Cyanomicon II* §6.4.3 p. 181 for reasons].

I develop for the first 10 minutes in dilute mineral acid—preferably nitric, 0.2 to 1% v/v, and then a 20-minute water wash [see *Cyanomicon II* §7.4.8 p. 216]. The only toner I use—for some of my work—is lead acetate [see *Cyanomicon II* §8 p. 233, §8.6.1 p. 237].

As for cyanotype issues, I have had no problems on Buxton or Herschel papers because my erstwhile hand-papermaker observed a practice of "scrupulous cleanliness." I'm not sure that this is always the case in larger paper mills, and I have seen problems of unaccountable "blotchiness" in cyanotypes printed on machine-made commercial papers. I believe this may be due to contamination by micro-organisms—"biofouling" either in the mill or stored paper. This is a problem that needs further urgent research.

My new cyanotype process has an exposure scale in the region of 2.1–2.4, i.e. 7–8 stops, depending on the wet processing. This is the same as platinum-palladium, argyrotype, new chrysotype, or salt printing, and can therefore use the same negatives—as an inexpensive proofing medium for these costly processes. In contrast, the traditional cyanotype has a much shorter exposure scale and can only be used with negatives prepared with a density range suited for grade 2–3 silver-gelatin or gum bichromate printing (*ca.* 0.9–1.2 or 3–4 stops).

Creative practice

It is my belief—in the absence of any other—that our ability to observe the world is the only intelligible purpose that we can attribute to it. A chemist like me is just some molecule's way of getting round to thinking about itself, even though it did take four billion years. Observation feeds the roots of both the twin trees of science and art in the human landscape. My aim is to enlist the science of photochemistry to enhance the art of photographic expression. Science is a way of looking at the world from the point of view of nobody; whereas Art is always a way of looking at the world from the point of view of a unique somebody, and this is true even of the seemingly 'objective' medium of photography. In working out the blind struggle of evolution, our role as intelligent observers is uniquely important (at least until we find out, for sure, if there are any other intelligent observers elsewhere). Photography provides an archival means of preserving those observations that hold significance for us, and of registering the world's impact on our human condition. And I do not refer only to the "heavily meaningful" kind of significance: the human condition also embraces significant humor, triviality and fun as well as the whole gamut of resonances springing from the simile, metaphor and allegory of our poetical endowment, which science cannot admit. Photography enhances our ability to see, and it induces us to make observations that might otherwise prove elusive. Paul Nash wrote of "...the peculiar power of the camera to discover formal beauty which ordinarily is hidden from the human eye." It is the delight in this act of disclosure that unites me with my fellow photographic observers of so many different backgrounds, tastes, beliefs and personalities.

Biography

Mike Ware is a graduate and D.Phil. of the University of Oxford, who taught and researched in chemistry at the University of Manchester, UK, until 1992. He then committed himself to the independent study of the science, history, art, and conservation of alternative photographic processes, especially those based on the photochemistry of iron, known as siderotypes. His research into methods of printing in noble metals, such as the platinotype, palladiotype, chrysotype, argyrotype, and the new cyanotype, has been published in over 60 articles in both the scientific and popular literature, and his studies of early photographic processes appeared in the academic periodical, *History of Photography*. He has supervised postgraduate research in photograph

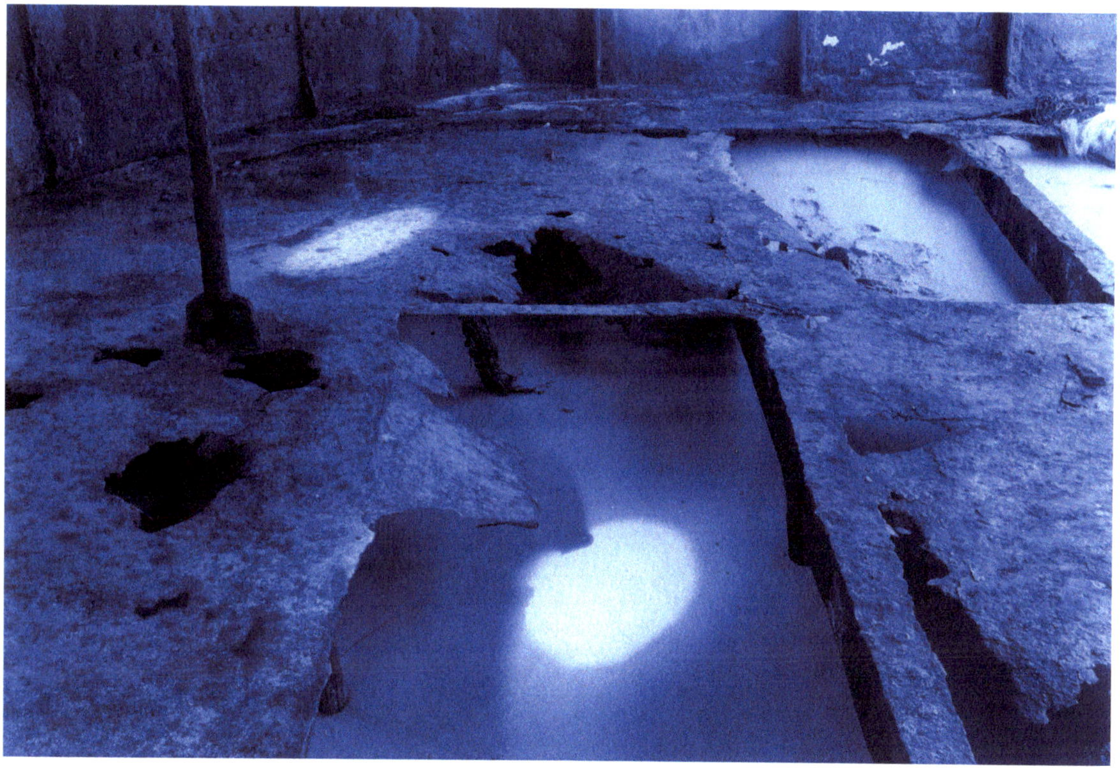

Figure 15.153. *Blockship Interior, South Ronaldsay*, 238 x 158 mm from an original 6 x 9 cm negative scanned and printed digitally with an Epson Photo R2880 © Mike Ware 2005. "The Orkney Islands lie off the north-western tip of Scotland where the Atlantic Ocean meets the North Sea. In two World Wars, the Isles protected Scapa Flow, the home anchorage for the battle fleet of the British Royal Navy. Derelict hulks were sunk in the narrow sea passages between the Islands as blockships to inhibit the access of enemy submarines, but they did not prevent the catastrophe of October 1939, at the outbreak of World War II. The German submarine U-47 penetrated Scapa Flow and torpedoed the Navy's capital ship HMS Royal Oak, at anchor, causing her to sink with the loss of 833 men. The dark violet-blue of lead-toned cyanotype has an elegiac quality for rendering the rusting interiors of these tragically ineffective blockships."

conservation and consulted for National Museums and Galleries in the UK and USA. In 1990 he was awarded the Hood Medal of the Royal Photographic Society. The history and conservation of the first photographs on paper is the subject of his books: *Mechanisms of Image Deterioration in Early Photographs* (1994), and *Cyanotype: the History, Science and Art of Photographic Printing in Prussian Blue* (1999); both published by the Science Museum of London. His later monographs, *Gold in Photography* and *The Chrysotype Manual*, were published in 2006. The American Institute for Conservation of Historic and Artistic Works awarded him its Special Recognition for Allied Professionals in 2016. By way of a counterbalance to scholarly activity, Ware has exhibited and lectured on his personal photographic work and conducted printing workshops in the UK, Europe, the USA, and Australia; in 1995 he appeared on BBC television in the Open University series *The Chemistry of Creativity*. Ware regards photography as a natural meeting ground for science and art, and his *raison d'être* is devoted to bridging the gap between those Two Cultures by enlisting chemical science to enhance the art of photographic expression. To see more of his work visit **http://www.mikeware.co.uk**.

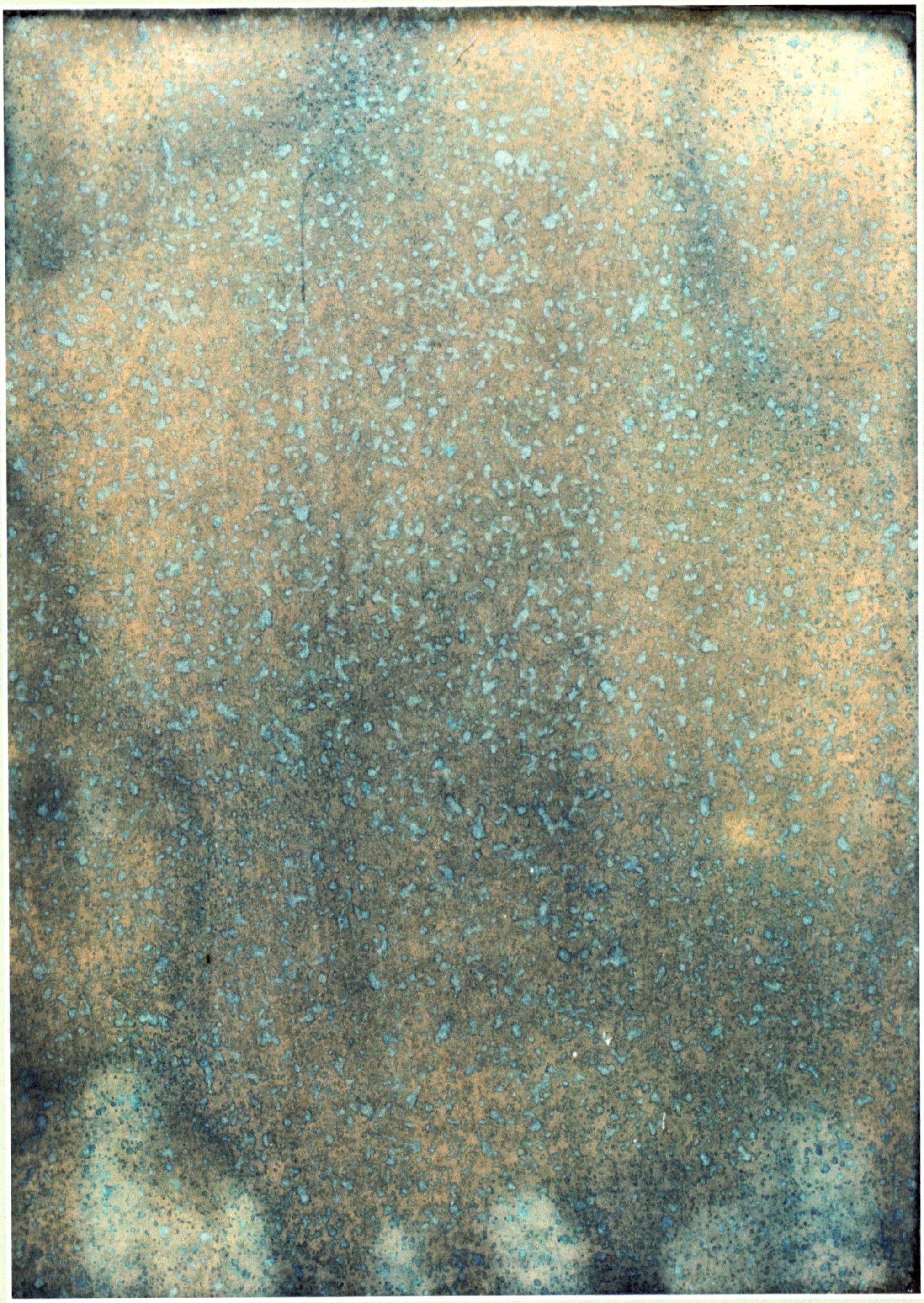

Figure B.1. *Rain I*, 11″ x 15″ © Patricia Howard 2018. "The *Rain* series was made by setting coated paper outdoors in the rain for 24 hours. The drops formed a pattern and also developed the image. These images were made during a time when it had rained every day for two weeks and I was growing increasingly frustrated with both the weather and lack of sunlight for making photographs." Patricia Howard is a photographer and educator based in Alexandria, Virginia. She is currently a studio arts instructor with the Smithsonian Associates and was previously an Assistant Professor of Art at Juniata College in Pennsylvania. Howard's work often focuses on the concept of home and she utilizes a variety of processes both traditional and digital. To see more of her work, visit **howardpatricia.com**.

Bibliography

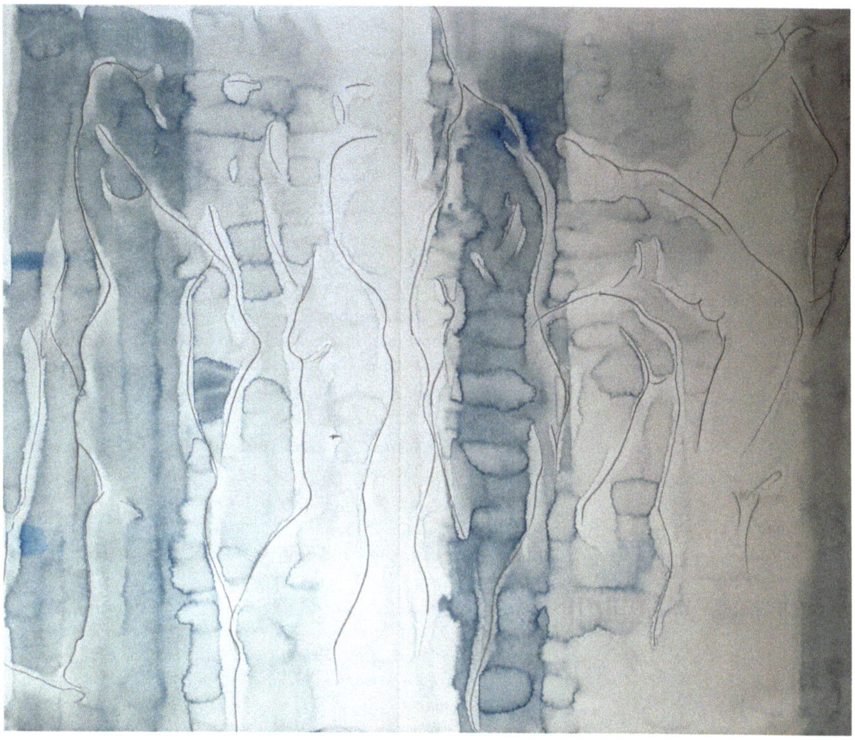

Figure B.2. *Water Woman*, pencil and cyanotype, 15″ X 18″ © Sarah Van Keuren 2018

"Blue Prints on Cloth" in *Popular Photography*, Vol. II. Boston, Massachusetts: F. R. Fraprie, April 1914, p. 279.

"Blue Prints on Linen" in *The Photographic Times*, Vol. XXIX. New York: The Photographic Times Publishing Association, June 1897, p. 302.

"Ferro-Prussiate Paper" in *The Photographic News, The Journal for Amateur Photographers*, Vol. XLI. London: Photographic News, January 15 1897, p. 48.

"Lantern Slides in Prussian Blue" in *The Photographic Times*, Vol. XXIX. New York: The Photographic Times Publishing Association, June 1897, p. 301.

"Notes on Blue Prints" in *Popular Photography*, Vol. II. Boston, Massachusetts: F. R. Fraprie, January 1914, p. 135.

"Pellet's Process" in the appendix "The Encyclopaedic Dictionary of Photography" in *The Photographic Times*, Vol. XXIX. New York: The Photographic Times Publishing Association, 1897, p. 339.

"Printing Process" in *The Photographic News, The Journal for Amateur Photographers*, Vol. XLI. London: Photographic News, April 2 1897, p. 222.

"Restoring Old Blue Print Paper" in *Popular Photography*, Vol. II. Boston, Massachusetts: F. R. Fraprie, December 1914, p. 120.

The American Annual of Photography and Photographic Times Almanac for 1890. New York: Scovill & Adams Company, 1890.

"The Cyanotype Copying Process" in *The Photographic Times*, Vol. VII. New York: Scovill Manufacturing Company, August 1877, pp. 175–176.

The Modern Way in Picture Making: Published as an Aid to the Amateur Photographer. Rochester: Eastman Kodak Co., 1905.

"The Workroom-Cyanotype" in *The Photographic Journal of America*, Vol. LV. Philadelphia: Edward L. Wilson Company, Inc., February 1918, pp. 92–93.

"Titles on Blue Prints" in *Popular Photography*, Vol. II. Boston, Massachusetts: F. R. Fraprie, December 1913, p. 104.

"Rae." "Printing in Salts of Iron" in *The Photographic Times*, Vol. XXIX. New York: The Photographic Times Publishing Association, February 1897, p. 102.

Andés, Louis Edgar. *The Treatment of Paper for Special Purposes*. London: Scott, Greenwood & Son, 1907.

Antonini, Marco, Sergio Minniti, Francisco Gomez, Gabriele Lungarella, and Luca Bendandi. *Experimental Photography: A Handbook of Techniques*. New York: Thames & Hudson, Inc., 2015.

Arnold, Brian. *Alternate Processes in Photography, Technique, History, and Creative Potential*. New York: Oxford University Press, 2017.

Arnow, Jan. *Handbook of Alternative Photographic Processes*. New York: Van Nostrand Reinhold Co., 1982.

Barnier, John. *Coming Into Focus, a Step-by-Step Guide to Alternative Photographic Printing Processes*. San Francisco: Chronicle Books, 2000.

Batchen, Geoffrey. *Emanations: The Art of the Cameraless Photograph*. New York: DelMonico Books, 2016.

Bingham, Robert J. *Photogenic Manipulation Part 1*, 7th ed. London: George Knight and Sons, 1850.

Blacklow, Laura. *New Dimensions in Photo Processes*, 5th ed. Boston: Focal Press, 2018.

Bolle, Oscar. "Prints in Turnbull's Blue" in *The British Journal of Photography*, Vol. XLIV. London: Henry Greenwood & Col, July 16 1897, pp. 459–460 and July 30 1897, p. 487 (translated from the *Photographisches Wochenblatt*), and the same article in *The Photographic Times*, Vol. XXIX. New York: The Photographic Times Publishing Association, December 1897, pp. 579–580.

Brothers, Alfred. *Photography: Its History, Processes, Apparatus, and Materials*, 2nd ed. London: Charles Griffin and Company, 1899.

Brown, George E. *Ferric and Heliographic Processes: A Handbook for Photographers, Draughtsmen, and Sun Printers*. New York: Tennant & Ward, 1900 [First edition 1900, reprinted 1905, this edition n.d.).

Brown, Ruth. *Cyanotypes on Fabric. A blueprint on how to produce… blueprints!* East Yorkshire: SC Publications, 2016.

Burns, Nancy Kathryn and Kristina Wilson. *Cyanotypes: Photography's Blue Period*. Worcester Massachusetts: Worcester Art Museum, 2016.

Coe, Brian and Mark Haworth-Booth. *A Guide to Early Photographic Processes*. London: Victoria & Albert Museum, 1983.

Crawford, William. *Keepers of Light*. New York: Morgan and Morgan, 1979.

Croucher, J. H. *Plain Directions for Obtaining Photographic Pictures, etc.* Philadelphia: Henry Carey Baird, 1855.

Duchochois, P. C. *Photographic Reproduction Processes*. New York: The Scovill & Adams Company, 1891.

Eder, Josef Maria. *Die Lichtpausverfahren, die Platinotypie und Verschiedene Copirverfahren Ohne Silbersalze (Cyanotypie, tintenbilder, Einstaubverfahren, Urancopien, Antrhakotypie, Negrographie, etc.)*. Halle: Verlag von Wilhelm Knapp, 1899, "Lichtpausen mittels Cyanotypie," pp. 238–247.

Emerson, P. H. *Naturalistic Photography for Students of the Art*. London: Dawbarn & Ward, Limited, 1899.

Enfield, Jill. *Jill Enfield's Guide to Photographic Alternative Processes, Popular Historical and Contemporary Techniques*. Burlington, Massachusetts: Focal Press, 2014.

Enfield, Jill. *Photo-Imaging, A Complete Guide to Alternative Processes*. New York: Amphoto Books, 2002.

Eshbaugh, Mark L. *Alternative Photography Processes: A Worker's Guide*. Massachusetts: RMR Press, 2006.

Fabbri, Malin and Gary. *Blueprint to Cyanotypes*. Alternativephotography.com, 2006.

Farber, Richard. *Historic Photographic Processes, A Guide to Creating Handmade Photographic Images*. New York: Allworth Press, 1998.

Fisher, George Thomas. *Photogenic Manipulation: containing plain instructions in the theory and practice of the arts of photography, calotype, cyanotype, ferrotype, chrysotype, anthotype, daguerreotype, thermography*. London: G. Knight, 1843.

Flynn, Deborah. *Imaging with Light Sensitive Materials, Cyanotype, Van Dyke Brown, Gum Bichromate, and Casein Printing*. New York: VSW Press, 1976 (no page numbers).

Frederick, Peter. *Creative Sunprinting. Early Photographic Printing Processes Rediscovered*. London: Focal Press, 1980.

G. E. B. "Imitation Platinotypes" in *The Photographic News, The Journal for Amateur Photographers*, Vol. XLI. London: Photographic News, March 26 1897, p. 206.

Gassan, Arnold. *Handbook of Contemporary Photography*, 3rd ed. Athens, Ohio: Handbook Co., 1974.

Gernsheim, Helmut. *The History of Photography from the Camera Obscura to the Beginning of the Modern Era*. London: Thames and Hudson, 1969.

Hamblin, R. A. "Hints to Beginners on Choosing a Printing Process" in *The Photographic News, The Journal for Amateur Photographers*, Vol. XLI. London: Photographic News, July 16 1897, p. 456.

Harrison, Jerome. "The Chemistry of Silver Printing, Printing with Salts of Iron—Cyanotype and Kallitype" in *Photographic Times*, Vol. XXI, September 4 1891, pp. 440–442.

Hawkes, Lieut. H. P. *Photography in a Nutshell, by "The Kernel."* London: Iliffe, Sons & Sturmey Ltd, 1898.

Herschel, John F. W. "On the Action of the Rays of the Solar System on Vegetable Colours, and on Some New Photographic Processes" in *Philosophical Transactions of the Royal Society of London* (132), June 16 1842, pp. 181–214. http://www.jstor.org/stable/pdf/108152.pdf?refreqid=search%3Aa9dea82d3bbaeba38520acecdc79f69b

Hewitt, Barbara. *Blueprints on Fabric, Innovative uses for cyanotype*. Loveland, Colorado: Interweave Press, Inc., 1995.

Himes, Charles F. "Blue-Print Solution Will Keep" in *The Photographic Times*, Vol. XXIX. New York: The Photographic Times Publishing Association, May 1897, pp. 223–224.

Hirsch, Robert. *Transformational Imagemaking, Handmade Photography Since 1960*. Burlington, Massachusetts: Focal Press, 2014.

Hirsch, Robert. *Photographic Possibilities*, 3rd ed. Boston: Focal Press, 2009.

Holmes, G. Lewis. "'Blue' Printing" in *The Photographic Times*, Vol. XXI. New York: Scovill Manufacturing Company, February 18 1891, pp. 81–82.

Hoppé, E. O., C. S. Coombes, F. Low, J. Littlejohns, W. F. Slater, E. A. and G. R. Reeve, and H. P. Maskell and others. *Photography*. New York: Doubleday, Page, & Company, 1911.

House, Suda. *Artistic Photographic Processes*. New York: American Photographic Book Publishing, 1981.

Howell-Koehler, Nancy. *Photo Art Processes*. Worcester, Massachusetts: Davis Publishing Inc., 1980.

Hunt, Robert. *A Manual of Photography*. London and Glasgow: Richard Griffin and Co., 1857.

Hunt, Robert. *Photography: A Treatise*. London: John Joseph Griffin and Co., 1851.

Hunt, Robert. *Researches on Light, An Examination of All the Phenomena Connected with the Chemical and Molecular Changes Produced by The Influence of the Solar Rays, Embracing All the Known Photographic Processes, and New Discoveries in the Art*. London: Longman, Brown, Green, and Longmans, 1844.

Hunt, Robert. "Chromo-Cyanotype, a New Photographic Process" in *The London, Edinburgh, and Dublin Philosophical Magazine and Journal of Science*, Vol. XXIV, January–June 1844, pp. 435–439.

James, Christopher. *The Book of Alternative Photographic Processes*. New York: Delmar Cengage, 2016.

John Michael Kohler Arts Center. *The Alternative Image*. Sheboygan, Wisconsin: John Michael Kohler Arts Center, 1983.

Kennel, Sarah. *In the Darkroom, An Illustrated Guide to Photographic Processes Before the Digital Age*. New York: Thames and Hudson, 2010.

Kosar, Jaromir. *Light Sensitive Systems: Chemistry and Application of Nonsilver Halide Photographic Processes*. New York: John Wiley and Sons, 1965.

Langford, Michael. *The Darkroom Handbook*. New York: Alfred A. Knopf, 1992.

Lietze, Ernst. *Modern Heliographic Processes*. Rochester: Visual Studies Workshop Press, 1974, a reprint of the 1888 original.

Lizzard, Abe. "Positives in Blue (Cyanotypes)" in *Anthony's Photographic Bulletin*, Vol. XXIII. New York: E. & H. T. Anthony & Co., June 11 1892, pp. 342–345.

MacLean, Hector. *Popular Photographic Printing Processes*. London: L. Upcott Gill, 1898.

Martin, Ralph. 'Toning Blue Prints" in *Popular Photography*, Vol. II. Boston, Massachusetts: F. R. Fraprie, May 1914, pp. 312–315.

Mavor, Carol. *Blue Mythologies, Reflections on a Colour*. London: Reaktion Books Ltd., 2013.

McDonald, John and Melba Smith Cole. *How to Make Old-Time Photos*. Blue Ridge Summit: Tab books, 1981.

Mrhar, Peter. *Cyanotype, Historical and Alternative Photography*. [Location and publisher not identified], 2013.

Museum of Fine Arts, Boston. *Blue: Cobalt to Cerulean in Art and Culture*. San Francisco: Chronicle Books, 2015.

Neblette, C. B. *Photography, Its Principles and Practice*, 4th ed. New York: Van Nostrand Co., Inc, 1946.

Neblette, C. B. *Photography, Its Principles and Practice*, 2nd ed. London: Chapman & Hall, Ltd, 1931.

Nelson, Mark. *Precision Digital Negatives for Silver and Other Alternative Processes*. Elgin: Little Joe Press, 2004.

Nettles, Bea. *Breaking the Rules: a Photo Media Cookbook*, 3rd ed. Urbana, Illinois: Inky Press Productions, 1992.

Newman, Thelma R. *Innovative Printmaking, The Making of Two- and Three-Dimensional Prints and Multiples*. New York: Crown Publishers, 1977.

Oelbaum, Zeva. *Blue Prints. The Natural World in Cyanotype Photographs*. New York: Rizzoli, 2002.

Persinger, Tom. *Photography Beyond Technique*. Massachusetts: Focal Press, 2014.

Petzold, Paul. *Effects and Experiments in Photography*. New York: Focal Press, 1973.

Pitt, Marion. *Practical Guide to Photography*. London: Marion and Co., 1880.

R. W. K. G. "Ferro-Prussiate Dodges" *in The Photographic News, The Journal for Amateur Photographers*, Vol. XLI. London: Photographic News, November 26 1897, p. 778.

Reilly, James. *Care and Identification of 19th Century Photographic Prints*. Rochester: Eastman Kodak, 1986.

Rexer, Lyle. *Photography's Antiquarian Avant-Garde*. New York: Harry N. Abrams, 2002.

Ribemont, Francis, Patrick Daum and Phillip Prodger, eds. *Impressionist Camera: Pictorial Photography in Europe, 1888–1918*. New York: Merrell, 2006.

Roche, T. C. *How to Make Photographs: A Manual for Amateurs*. New York: E. & H. T. Anthony & Co., 1886.

Sanderson, Andrew. *Handcoloring and Alternative Darkroom Processes*. Switzerland: Rotovision, 2002.

Schaaf, Larry J. *Out of the Shadows. Herschel, Talbot, & the Invention of Photography*. New Haven: Yale University Press, 1992.

Schaaf, Larry J. *Sun Gardens: Victorian Photograms by Anna Atkins*. New York: Aperture, 1985.

Schaefer, John P. *The Ansel Adams Guide, Basic Techniques of Photography, Book 2*. Boston: Little Brown and Company, 1998.

Seigel, Judy ed. *The World Journal of Post-Factory Photography*. New York: Post Factory Press, April 1998–April 2004.

Skeel, Adelaide. "Blues" in *The American Annual of Photography and Photographic Times Almanac for 1889*. New York: Scovill Manufacturing Company, 1889, pp. 48–50.

Skeel, Adelaide. "Something More About the Blues" in *The Photographic Times*, Vol. XXI. New York: Scovill Manufacturing Company, January 30, 1891, pp. 54–55.

Society for Photographic Education. *Some Various Processes as Demonstrated at the Process Fair at The National Conference of the Society for Photographic Education, Asilomar, California*, 1978.

Stulik, Dusan C., and Art Kaplan. *The Atlas of Analytical Signatures of Photographic Processes: Cyanotype*. Los Angeles: The Getty Conservation Institute, 2013.

Sutton, Thomas. *A Dictionary of Photography*. London: Sampson Low, Son and Co., 1858.

Tennant, John. A., ed. "The 'Blue Print' and its Variations" in *The Photo-Miniature*, Vol. 1, No. 10, January 1900, pp. 481–514.

Towler, John. *The Silver Sunbeam: A Practical and Theoretical Textbook on Sun Drawing and Photographic Printing, Comprehending All the Wet and Dry Processes at Present Known, with Collodion, Albumen, Gelatin, Wax, Resin, and Silver*. New York: Joseph H. Ladd, Publisher, 1864.

Valenta, Eduard. "The Use of Various Iron Salts in Printing Processes" in *The British Journal of Photography*, Vol. XLIV. London: Henry Greenwood & Col, February 26 1897, pp. 134–135; same article in *The Photographic News, The Journal for Amateur Photographers*, Vol. XLI. London: Photographic News, March 5 1897, p. 152, and in *The Photographic Times*, Vol. XXIX. New York: The Photographic Times Publishing Association, May 1897, pp. 237–238.

Van Keuren, Sarah. *A Non Silver Manual. Cyanotype, Brownprint, Palladium, and Gum Bichromate with instructions for making light-resists including pinhole photography*, 3rd ed. Woodlyn, Pennsylvania: Munro Printing, 2005.

Vogel, Dr. E. *Practical Pocket-Book of Photography. A Short Guide to the Practie of All the Usual Photographic Processes for Professionals and Amateurs*. London: Swan Sonnenschein & Co., Ltd, 1896 and New York: Macmillan & Col, Ltd, 1896.

Wall, E. J. *Photographic Facts and Formulas*. Boston: American Photographic Publishing Co., 1924.

Wall, E. J. "The Iron Salts I" in *American Photography*, Vol. XVI, November 1922, pp. 677–688; "The Iron Salts II" in *American Photography*, Vol. XVI, December 1922, pp. 766–774.

Ware, Mike. *Cyanomicon II, History, Science and Art of Cyanotype: photographic printing in Prussian blue*. Buxton: self-published, 2016. http://www.mikeware.co.uk/mikeware/downloads.html

Ware, Mike. *New Cyanotype, Siderotype Workshop Notes*. Buxton: self-published, 2009. http://www.mikeware.co.uk/mikeware/downloads.html

Webb, Randall, and Martin Reed. *Alternative Photographic Processes, A Working Guide for Imagemakers*. New York: Silver Pixel Press, 2000.

Wehrenberg, Charles. *Mississippi Blue, Henry P. Bosse and his Views on the Mississippi River Between Minneapolis and St. Louis, 1883–1891*. Santa Fe, New Mexico: Twin Palms Publishers, 2002.

Welu, James A. *Frederick Coulson, Blueprints of a Golden Age*. Worcester, Massachusetts: Worcester Art Museum, 2016.

Wensfield, Hugh. "Cyanotype Plates" in *Camera Craft, A Photographic Monthly*, Vol. XXVII. San Francisco, California: Camera Craft Publishing Co., August 1920, pp. 258–261.

Whitaker, Channing. "Apparatus for Printing by the Blue Process" in *The Photographic News*, Vol. XXVII. London: Piper and Carter, December 7 1883, p. 359.

Wike, B. H. "Choice of Papers" in *Popular Photography*, Vol. II. Boston, Massachusetts: F. R. Fraprie, July 1914, pp. 415–416.

Wilks, Brady. *Alternative Photographic Processes, Crafting Handmade Images*. New York: Focal Press, 2015.

Wood, John. *The Photographic Arts*. Iowa City: University of Iowa Press, 1997.

Woodbury, Walter E. *Photographic Amusements*, 9th ed. Boston: American Photographic Publishing Col, 1922.

Worobiec, Tony and Ray Spence. *Beyond Monochrome, A Fine Art Printing Workshop*. London: Surrey Fountain Press, 1999.

Worthington, J. C. and J. C. Millen. *The Photographic Primer, A Manual of Practice*. Riverton: The Riverton Press, 1897.

Index